Dalia Kriksciuniene · Anu Bajaj · Ajith Abraham
Editors

Bio-Inspired Computing

Proceedings of the 14th International Conference on Innovations in Bio-Inspired Computing and Applications (IBICA 2023) held during December 14–15, 2023
Volume 1: Smart Healthcare

Springer

Editors
Dalia Kriksciuniene
Vilnius University
Kaunas, Lithuania

Ajith Abraham
School of Artificial Intelligence
Sai University
Chennai, India

Anu Bajaj
Department of Computer Science
and Engineering
Thapar Institute of Engineering
and Technology
Patiala, Punjab, India

ISSN 2367-3370 ISSN 2367-3389 (electronic)
Lecture Notes in Networks and Systems
ISBN 978-3-031-78936-6 ISBN 978-3-031-78937-3 (eBook)
https://doi.org/10.1007/978-3-031-78937-3

This Springer imprint is published by the registered company Springer Nature Switzerland AG
The registered company address is: Gewerbestrasse 11, 6330 Cham, Switzerland

IBICA 2023 Preface – Volume 1

Welcome to the 14th International Conference on Innovations in Bio-Inspired Computing and Applications (IBICA 2023) and 13th World Congress on Information and Communication Technologies (WICT 2023). IBICA-WICT 2023 was held in 5 different cities namely Olten, Switzerland; Porto, Portugal; Kaunas, Lithuania; Greater Noida, India; Kochi, India and in online mode. IBICA-WICT 2023 is hosted and sponsored by the Machine Intelligence Research Labs (MIR Labs), USA and supported by Bennett University, Greater Noida, India.

14th International Conference on Innovations in Bio-Inspired Computing and Applications was organized to provide a platform for world research leaders and practitioners, to discuss the "full spectrum" of current theoretical developments, emerging technologies, and innovative applications of Bio-inspired Computing. Bio-inspired Computing is currently one of the most exciting research areas, and it is continuously demonstrating exceptional strength in solving complex real-life problems. The main driving force of the conference is to further explore the intriguing potential of Bio-inspired Computing.

13th World Congress on Information and Communication Technologies (WICT 2023) was organized to provide an opportunity for the researchers from academia and industry to meet and discuss the latest solutions, scientific results and methods in the usage and applications of ICT in the real world. Last few decades have experienced technologies that are evolving so rapidly, altering the constraints of space and time, and reshaping the way we communicate, learn and think. Rapid advances in information technologies and other digital systems are reshaping our ecosystem. Innovations in ICT allow us to transmit information quickly and widely, propelling the growth of new urban communities, linking distant places and diverse areas of endeavor in productive new ways, which a decade ago was unimaginable. Thus, the theme of this World Congress is "Innovating ICT For Social Revolutions".

IBICA-WICT 2023 received 656 papers from 36 countries and based on the review process we have accepted 263 papers. Each paper was reviewed by at least five or more reviewers, and based on the outcome of the review papers were accepted for inclusion in the conference proceedings. Proceedings were divided into Five Volumes. This Volume is focused on Smart *Health Care.* Artificial Intelligence has the potential to transform healthcare by improving patient outcomes, increasing access to care, and optimizing healthcare delivery processes while also addressing challenges related to data privacy, algorithm bias, and ethical considerations to ensure that AI is deployed responsibly and ethically in healthcare settings. This Volume comprises of 53 articles focused on Smart Healthcare deals with a wide range of healthcare topics including early disease detection/diagnosis, personalized treatment, predictive diagnostics, telemedicine, medical imaging, drug discovery, healthcare operations, etc.

First, we would like to thank all the authors for submitting their papers to the conference, for their presentations and discussions during the conference. Our thanks go to program committee members and reviewers, who carried out the most difficult work by carefully evaluating the submitted papers. Our special thanks to the following plenary speakers, for their exciting plenary talks:

- Christine Zarges, Aberystwyth University, UK
- Theresa Schmiedel, University of Applied Sciences and Arts Northwestern Switzerland, Basel, Switzerland
- Ke Feng, Singapore-ETH Centre, The National University of Singapore, Singapore
- Stefka Fidanova, Bulgarian Academy of Sciences, Bulgaria
- Diego Oliva, Universidad de Guadalajara, Mexico
- Sebastian Ventura, University of Cordoba, Spain
- João Pedrosa, INESCTEC, Portugal
- Nuno Bettencourt, Instituto Superior de Engenharia do Porto, Portugal
- Milan Tuba, Singidunum University, Serbia
- Dalia Kriksciuniene, Vilnius University, Lithuania
- Kusum Deep, Indian Institute of Technology Roorkee, India

We express our sincere thanks to the organizing committee chairs for helping us to formulate a rich technical program. Enjoy reading the articles!

Virgilijus Sakalauskas
Thinakaran Perumal
Ajith Abraham
Preethi Thekkath
Dalia Kriksciuniene
Yunchuan Sun
Ryotaro Kamimura
Sreela Sreedhar
Anu Bajaj

IBICA–WICT 2023 Organizing Committees

General Chairs

Virgilijus Sakalauskas	Vilnius University, Lithuania
Thinakaran Perumal	Universiti Putra Malaysia, Malaysia
Ajith Abraham	Sai University, India
Preethi Thekkath	Toc H Institute of Science & Technology, India

Program Chairs

Dalia Kriksciuniene	Vilnius University, Lithuania
Yunchuan Sun	Beijing Normal University, China
Ryotaro Kamimura	Tokai University, Japan
Sreela Sreedhar	Toc H Institute of Science & Technology, India

Web Services Chair

Kun Ma	University of Jinan, China

Publication Chair

Anu Bajaj	Thapar Institute of Engineering and Technology, India

Publicity Chair

Pooja Manghirmalani Mishra	Machine Intelligence Research Labs, India

Local Organizing Committee

Jesna Anver	Toc H Institute of Science & Technology, India
Leda Kamal	Toc H Institute of Science & Technology, India
Elsaba Jacob	Toc H Institute of Science & Technology, India

Mithu Mary George	Toc H Institute of Science & Technology, India
Mima Manual	Toc H Institute of Science & Technology, India
Ashly Jospeh	Toc H Institute of Science & Technology, India
Sajith P. J.	Toc H Institute of Science & Technology, India
Leema Mathew	Toc H Institute of Science & Technology, India
Baazil P. Thampy	Toc H Institute of Science & Technology, India
Sharath P. Raju	Toc H Institute of Science & Technology, India
Philo Sumi	Toc H Institute of Science & Technology, India
Jacob Jose Chemmankuzha	Toc H Institute of Science & Technology, India

International Program Committee

Abdelkrim Haqiq	Hassan 1st University, Morocco
Alberto Bugarín-Diz	Universidade de Santiago de Compostela, Spain
Alberto Guillén	University of Granada, Spain
Alberto Fernandez Hilario	University of Granada, Spain
Alexey Kornaev	Innopolis University, Russia
Alfonso Guarino	University of Foggia, Italy
Alfredo Cuzzocrea	University of Calabria, France
Alicia Troncoso	Pablo de Olavide University, Spain
Ambili P. S.	REVA University, India
Amelia Zafra	University of Cordoba, Spain
Amit Kumar Tyagi	Vellore Institute of Technology, India
Amutha S.	Vellore Institute of Technology, India
Ana de Almeida	Instituto Universitário de Lisboa, Portugal
Ana Madureira	Polytechnique of Porto, Portugal
Anazida Zainal	Universiti Teknologi Malaysia, Malaysia
Andre C. P. L. F. de Carvalho	University of Sao Paulo, Brazil
André S. Santos	ISEP, Portugal
Andrea Soltoggio	Loughborough University, UK
Andries Engelbrecht	University of Stellenbosch, South Africa
Anitha Kumari K.	PSG College of Technology, India
Antonello Florio	Politecnico di Bari, Italy
Antonio Arauzo-Azofra	University of Cordoba, Spain
Antonio Salmeron	Universidad de Almeria, Spain
Anurag Rana	Shoolini University, India
Arijit Ghosal	St. Thomas' College of Engineering and Technology, India
Arthi B.	SRM Institute of Science and Technology, India
Aryan Chaudhary	Bio Tech Sphere Research, India
Aureli Soria-Frisch	Starlab Barcelona, Spain

Azah Kamilah binti Muda	UTeM, Malaysia
Bac Le	University of Science, Vietnam
Bay Vo	Ho Chi Minh City University of Technology, Vietnam
Bernard De Baets	Ghent University, Belgium
Bernardete Ribeiro	CISUC, University of Coimbra, Portugal
Bhawna Goyal	Chandigarh University, India
Carlos Cotta	Universidad de Málaga, Spain
Carlos Pereira	Universidade de Santiago de Compostela, Portugal
Carlos Serro	ISCTE - Instituto Universitário de Lisboa, Portugal
Catarina I. Reis	Polytechnic of Leiria, Portugal
Cesar Hervas-Martinez	University of Cordoba, Spain
Chang-Yong Lee	Kongju National University, Korea
Chao-Chun Chen	National Cheng Kung University, Taiwan
Chun-Hao Chen	National Taipei University of Technology, Taiwan
Corrado Mencar	University of Bari Aldo Moro, Italy
David A. Pelta	Universidad de Granada, Spain
Debdatta Kandar	North-Eastern Hill University, India
Deepika Koundal	UPES, India
Deepti Chaudhary	Kurukshetra University, India
Devi Priya R.	KPR Institute of Engineering and Technology, India
Rajashree Shettar	RV College of Engineering, India
Amit Kumar Mishra	Jain University, India
Iqbal Sarker	Edith Cowan University, Australia
Saira Varghese	Toc H Institute of Science & Technology, India
Durai Raj Vincent P. M.	VIT Vellore, India
Durgesh Srivastava	Chitkara University, India
Eiji Uchino	Yamaguchi University, Japan
Elakkiya R.	BITS Pilani Dubai Campus, UAE
Elizabeth F. G. Goldbarg	Universidade Federal do Rio Grande do Norte, Brazil
Enrique De la Cal	University of Oviedo, Spain
Enrique Herrera-Viedma	University of Granada, Spain
Fabio Zambetta	RMIT University, Australia
Fernando Batista	INESC-ID and ISCTE-IUL, Portugal
Fernando Gomide	University of Campinas, Brazil
Francisco Herrera	Universidad de Granada, Spain
Gabriella Casalino	University of Bari, Italy
Gahangir Hossain	University of North Texas, USA

Gaurav Gupta	Shoolini University, India
George Papakostas	International Hellenic University, Greece
Gianluca Zaza	University of Bari, Italy
Giovanna Castellano	University of Bari, Italy
Gregorio Sainz-Palmero	Universidad de Valladolid, Spain
Guenther Raidl	Vienna University of Technology, Austria
Hajime Nobuhara	University of Tsukuba, Japan
Heba Afify	Biomedical Engineering, Egypt
Hsueh-Chan Lu	National Cheng Kung University, Taiwan
Hudson G. Medeiros	Federal University of Rio Grande do Norte, Brazil
Irwin King	The Chinese University of Hong Kong, Hongkong
Isaac Chairez	Tecnológico de Monterrey, Mexico
Isabel Jesus	Institute of Engineering of Porto, Portugal
Ivo Pereira	University Fernando Pessoa, Portugal
Ja-Hwung Su	National University of Kaohsiung, Taiwan
Janos Botzheim, E?tv?s	Loránd University Faculty of Informatics, Hungary
Janusz Kacprzyk	Polish Academy of Sciences, Poland
Jayaraman Valadi	FLAME University, India
Jerry Chun-Wei Lin	Western Norway University of Applied Sciences, Norway
Jesus S. Aguilar-Ruiz	Universidad Pablo de Olavide, Spain
Jesus Alcala-Fdez	University of Granada, Spain
Jimmy Ming-Tai Wu	Shandong University of Science and Technology, China
Jolanta Mizera-Pietraszko	Military University of Land Forces, Poland
Jose A. Gamez	Universidad de Castilla-La Mancha, Spain
Jose Jerez	Universidad de Malaga, Spain
Jose A. Lozano	UPV/EHU, Spain
Jose Maria-Luna	University of Cordoba, Spain
Jose Puerta	Universidad de Castilla-La Mancha, Spain
José Everardo B. Maia	UECE, Brazil
Juan Vidal	University of Santiago de Compostela, Spain
Ju-Chin Chen	National Kaohsiung University of Science and Technology, Taiwan
Julia Flores	University of Castilla - La Mancha (UCLM), Spain
K. Reddy Madhavi	Sree Vidyanikethan Engineering College, India
Kawuu W. Lin	National Kaohsiung University of Science and Technology, Taiwan
Kingsley Okoye	Tecnologico de Monterrey, Mexico
Kolla Bhanu Prakash	Koneru Lakshmaiah Education Foundation, India

Lalitha K.	Kongu Engineering College, India
Leandro M. Almeida	Universidade Federal de Pernambuco, Brazil
Leandro Dos Santos Coelho	Pontifical Catholic University of Parana, Brazil
Leonilde Varela	University of Minho, Portugal
Loan Nguyen	International University, Vietnam
Luca Di Gaspero	DPIA, University of Udine, Italy
Luciano Sánchez	Oviedo University, Spain
Luis Magdalena	Universidad Politecnica de Madrid, Spain
Luís B. Elvas	Instituto Universitário de Lisboa, Portugal
Luiz Zárate	Pontifícia Universidade Católica de Minas Gerais, Brazil
Manikandan V. M.	SRM University AP, India
Mansi Sharma	Amrita Vishwa Vidyapeetham, India
Manuel Fernandez Delgado	University of Santiago de Compostela, Spain
Manuel Lama	University of Santiago de Compostela, Spain
Manuel Mucientes	University of Santiago de Compostela, Spain
Maria Jose Del Jesus	Universidad de Jaén, Spain
Mário Antunes	Instituto Politécnico de Leiria, Portugal
Marley M. B. R. Vellasco	PUC Rio de Janeiro, Brazil
Matteo Gaeta	Università degli Studi di Salerno, Italy
Mauricio Ayala-Rincon	Universidade de Brasilia, Brazil
Mauro Birattari	Université Libre de Bruxelles, Belgium
Meera Ramadas	MIR Labs, USA
Michael Blumenstein	University of Technology Sydney, Australia
Michael N. Vrahatis	University of Patras, Greece
Michele Mastroianni	Università degli Studi di Salerno, Italy
Mika Sato-Ilic	University of Tsukuba, Japan
Nikola Kasabov	Auckland University of Technology, New Zealand
Olgierd Unold	Wroclaw University of Science and Technology, Poland
Oscar Castillo	Tijuana Institute of Technology, Mexico
Oscar Corcho	Universidad Politécnica de Madrid, Spain
Pascal Bouvry	University of Luxembourg, Luxembourg
Pascal Van Hentenryck	University of Michigan, USA
Patricia Melin	Tijuana Institute of Technology, Mexico
Paulo Carrasco	Alagrve University, Portugal
Paulo Moura Oliveira	UTAD University, Portugal
Paulo Novais	Universidade do Minho, Portugal
Pedro Larranaga	Technical University of Madrid, Spain
Quang Minh Tran	Ho Chi Minh City University of Technology, Vietnam
Rabel Guharoy	Bharati Vidyapeeth Deemed University, India

Radhika S.	Sathyabama Institute of Science and Technology, India
Radu-Emil Precup	Politehnica University of Timisoara, Romania
RajaniKanth Aluvalu	Vardhaman College of Engineering, India
Richard Weber	University of Chile, Chile
Roberto Iglesias Rodriguez	Universidade de Santiago de Compostela, Spain
Rohit Anand	GB Pant Govt. Engineering College, India
Ronald Yager	Machine Learning Institute, Iona College, USA
Rudolf Kruse	University of Magdeburg, Germany
Rutuparna Panda	VSS University of Technology, India
Saman Halgamuge	University of Melbourne, Australia
Sancho Salcedo-Sanz	Universidad de Alcalá, Spain
Sandeep Sreekumar	IndsutryApps, Singapore
Santosh Kumar Srivastava	Galgotias University, India
Sarina Sulaiman	Universiti Teknologi Malaysia, Malaysia
Sebastian Basterrech	VSB-Technical University of Ostrava, Czech Republic
Sebastián Ventura Soto	University of Cordoba, Spain
Shabeer Puthukkudi	British Telecom, United Kingdom
Shali Rani	Chitkara University, India
Shankru Guggari	KLE Technological University, India
Shilpa Gupta	Maharaja Agrasen University, India
Shing Chiang Tan	Multimedia University, Malaysia
Silviu Ionita	University of Pitesti, Romania
Sirine Marrakchi	University of Sfax, Tunisia
Stéphane Grieu	Université de Perpignan, France
Steven D. Prestwich	University College Cork, Ireland
Sujata Dash	Nagaland University, India
Sung-Bae Cho	Yonsei University, Korea
Teresa Ludermir	Federal University of Pernambuco, Brazil
Thillaiarasu Nadesan	Reva University, India
Thinagaran Perumal	Universiti Putra Malaysia, Malaysia
Tho Quan	Ho Chi Minh City University of Technology, Vietnam
Thomas Hanne	University of Applied Sciences Northwestern, Switzerland
Tom Gedeon	The Australian National University, Australia
Tzung-Pei Hong	National University of Kaohsiung, Taiwan
Van Vo	Ho Chi Minh University of Industry, Vietnam
Victor M Rivas	University of Jaen, Spain
Vijay P.	Vellore Institute of Technology, India
Vishwanath N.	Toc H Institute of Science & Technology

Wen-Yang Lin	National University of Kaohsiung, Taiwan
Wilfried Elmenreich	Alpen-Adria-Universitat, Austria
Witold Pedrycz	University of Alberta, Canada
Xiao-Zhi Gao	University of Eastern Finland, Finland
Yaochu Jin	University of Bielefeld, Germany
Yilun Shang	Northumbria University, UK
Yukio Ohsawa	University of Tokyo, Japan

Contents

Communication and Rehabilitation Interface for Human Brain Disease

Aastha Sharma[1], Sandhya Avasthi[1](✉), Kadambri Agarwal[1], Khushboo Jain[2], and Ajith Abraham[3]

[1] Department of CSE, ABES Engineering College, Ghaziabad, India
sandhya_avasthi@yahoo.com
[2] School of Computer Science, UPES, Dehradun, India
[3] Machine Intelligence Research Labs (MIR Labs), Auburn, WA 98071, USA

Abstract. The transmission of commands and signals from the brain to an external device is done using Brain-computer interfaces (BCIs) bypassing the conventional output channels of peripheral muscles. First BCIs fetch signals and then translate brain activity signals coming from conscious and subconscious levels. In a recent report by NIH (National Institute on Health), it is stated that more than 57 million people are suffering from dementia which causes Alzheimer's disease. This problem can be dealt with in the BCI system by learning emitted patterns by the brain and then assisting the patient in daily activity. Dementia typically attacks the brain's functions like communication ability to make the judgments on the situation. Machine learning applications are capable of detecting cases of dementia early by health healthcare professionals, hence enabling the implementation of timely therapeutic interventions to mitigate the progression of cognitive decline. In this research paper, a BCI framework to classify EEG signals is proposed that applies SVM, XG Boost, and Neural Network classification algorithms. For the feature selection, principal component analysis (PCA) is utilized to select the most suitable features. XGBoost algorithm performs best in classification tasks to give 80% accuracy in detecting the mental disorder.

Keywords: Dementia · Brain-Computer Interface. Machine Learning · Blockchain · EEG · XG-Boost

1 Introduction

Brain-computer interface (BCI) technology is capable to convert thoughts of user into a specific action. BCI can interact with external devices through brain signals, a system that enables communication between the brain and external devices. External devices such as prosthetic limbs and computers help a person with paralysis or other kinds of disabilities. The primary objective of Brain-Computer Interfaces (BCIs) is to bypass traditional channels like nerves and muscles. The prevalent notion in popular and scientific discourse surrounding Brain-Computer Interfaces (BCIs) is the analogy of "mind-reading" or "wire-tapping." This analogy posits that BCIs primarily aim to monitor brain activity through electrophysiological signals to discern an individual's desires. As a result, it

D. Kriksciuniene et al. (Eds.): IBICA 2023, LNNS 1228, pp. 1–13, 2025.
https://doi.org/10.1007/978-3-031-78937-3_1

serves as the foundation for a significant portion of both popular and scientific speculation concerning BCIs [1, 2]. Human age increases the risk factors for brain illnesses such as dementia, Alzheimer's, and depression, these disorders have the potential to affect brain structure and function, which can affect cognitive function. Brain-computer interface systems and applications may assist in restoring cognition, memory, focus, and consciousness in the elderly patient [3].

There are hidden layers in the human brain that forecast and interpret each perceptron as it connects one after the other. Neurons are the fundamental units of the brain that transmits information via electrical impulses. In BCIs the electrical activity of neurons is captured using techniques like EEG, ECoG or fMRI. Neural firing corresponds to cognitive states or motor intentions. In this way, the SVM and XG BOOST algorithms for processing EEG signals and classification are applied. These machine learning approaches and methodology train and evaluate the required dataset [4–6]. The learning analysis in the formation is stored in the block of the Blockchain technology and learns the system to fetch and make predictions based on the previous data status.

The BCI system are mainly two types, Invasive BCIs and Non-invasive BCIs. Invasive BCIs works through implanting electrodes in the brain directly to receive high quality brain signals. Whereas in Non-invasive BCIs, external sensors connected to human detect brain signals through electroencephalography (EEG) technique. EEG (electroencephalography) technique is used in BCI to make virtual interface between the brain & external world [7, 8]. A Brain-Computer Interface (BCI) system's efficacy depends on the EEG signal recording quality, the characteristics chosen, the classification algorithm robustness, and the system's ability to adapt to particular users. Electroencephalography (EEG)-based brain-computer interfaces (BCIs) are also improving due to signal processing and machine learning advances. The main objective of this research is to study Brain computer interface and signal processing research area and to identify possible limitations in existing solutions. In this research paper a generic Brain Disease Interface is described to classify dementia patient based on their behavioral data. The process goes through four phases to learn the pattern of neurons using machine learning and deep learning methods. Assessment, communication, prediction and Rehabilitation are main four phases. In assessment phase mainly neuron pattern analysis is done and in communication mood, behavior and emotion is dealt with. Prediction phase provides outcome of anxiety and disease detection.

Assessment phase involves evaluating the patient needs, abilities and compatibility with the BCI system, and also ensures system is suitable for patient requirement. In preliminary evaluation medical and neurological assessment is performed. To complete it Coma Recovery Scale-revised (CRS-R) is utilized that assesses disorder of consciousness (DOC) which is designed to differentiate between states of consciousness such as coma, unresponsiveness, minimal consciousness etc. Figure 1 shows the pre-assessment process of the brain signals from patients. The details can be stored in the patient blocks in the Blockchain technology.

Communication part of a patient is related to behavior, mood, emotions, etc. BCI for communication helps patients with motor disabilities helping them in taking action and making strategies. Users can choose letters, words, or phrases on a screen by using

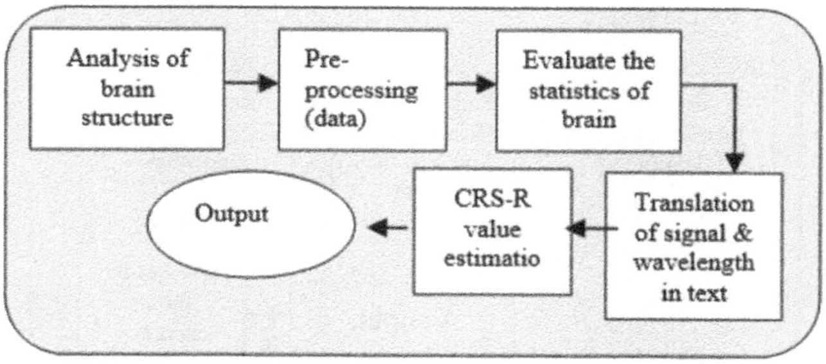

Fig. 1. Analysis and assessment of patient behavior

precise mental commands or focusing their attention. The BCI system enables communication between people by interpreting these signals and producing the appropriate output. Figure 2 describes the basic structure and other steps in analysis. All the necessary details can be stored within the block of the patient data records in the Blockchain in the form of a distributed repository.

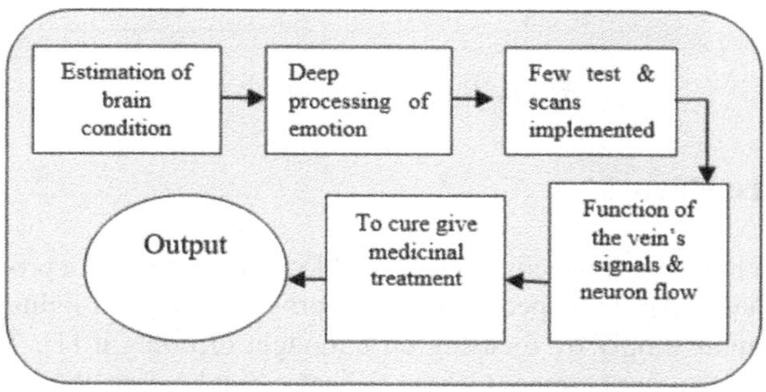

Fig. 2. Analysis of signal during communication

In ***Prediction phase***, Machine Learning processes brain signals, blockchain techniques is employed to verify the authenticity and integrity of the decision-making process. Real-time secure decision-making is possible through blockchain module in BCI systems that ensures transparent communication between BCI systems and external devices (as shown in Fig. 3). Use of deep learning models such as Convolution Neural Network (CNNs) is done to analyses spatial and temporal pattern of brain signals.

Rehabilitation is related to helping patients with neurological conditions like dementia by using BCI devices. As dementia progresses, people may experience difficulties with verbal communication. BCIs have the potential to improve communication by turning brain signals into instructions that guide the operation of communication devices, such as interfaces that convert text to audible voice. Figure 4 illustrates the process of rehabilitation in BCI systems. Cognitive rehabilitation is done through exercises and activities that stimulate attention and memory.

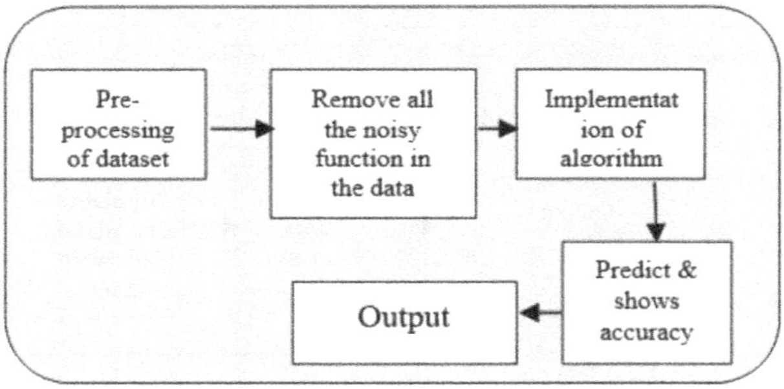

Fig. 3. Prediction Decoding Message Reading

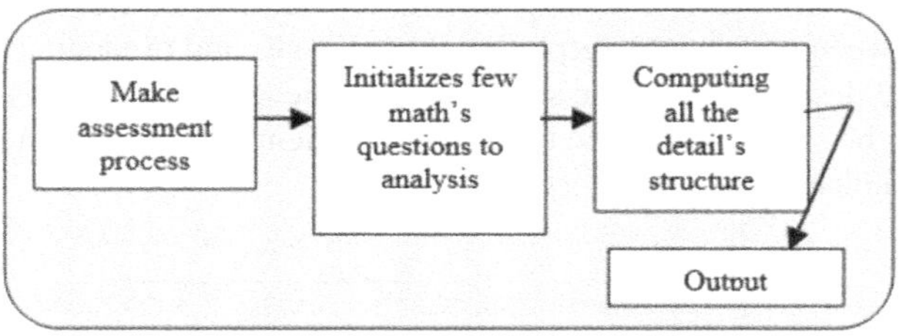

Fig. 4. Rehabilitation Analyses

2 Literature Review

The past researches experimented, where a disabled person through a process can record brain signals that along with specific software program can turn lights on or off, can even send an email simply by focusing on a thought of doing it [1]. The experiment further expanded to a improvement where the same can be done through a neuroprosthetic control like a prosthetic device which can hold a glass of water. The utilization of brain-computer interfaces (BCIs) will enable individuals to manipulate machinery and engage in direct communication with computer systems through neural activity [2, 3]. The research regarding BCI-based therapy [4], shows promising results in patients with multifaceted impairments where patients have motor, cognitive, and emotional deficits. The research supports that BCI-based therapy has the potential to deal with post-stroke cognitive rehabilitation. This study [5] clarifies the architectural framework and underlying principles of the brain-computer interface and virtual reality (BCI-VR) system, as well as their interconnectedness. In this novel hybrid system, Brain-Computer Interface (BCI) uses bio-signals as input, while Virtual Reality (VR) technology shows the output. The CNN model can learn necessary patterns in brain structure and can detect temporal patterns in the BCI system [6]. BCI-enabled patients can control external devices like robotic arms by interpreting motor imagery and fostering motor recovery. This research [7] offers a novel method for creating a graphical user interface (GUI) that may be used to type the 26 English alphabets and carry out tasks including food, drink, and medication. The basic mechanism of the proposed interface is movement imagery (MI).

In addition, the paper discusses the deep learning algorithm to integrate signal generated by the implementation, the recognition of COVID-19 analysis in the brain functionality [9]. Brain age prediction was developed to provide comfort and explain the effects of age-related circumstances [12]. The research [13] surveys key points and challenges in BCI systems mainly in the acquisition of brain signals and processing. Acquisition is mainly done using non-invasive implants which are easy to wear but produce poor signal resolution. An Introduction to Brain-Computer Interfaces and their Applications [14] provides a concise review of the framework architecture and working area for the patient, devices, and signal modulation in the display system. The study aims to develop a methodology for the identification of epileptic seizures by integrating fuzzy-based and traditional machine learning techniques. The method extracts the most salient and discriminative features from electroencephalogram (EEG) data associated with seizures. The proposed approach categorizes unidentified EEG signal segments into two distinct groups: interictal and ictal [15]. In this paper [16], a novel face-computer interface (FCI) was proposed as a means to facilitate human-computer interaction (HCI) for those who have impaired or non-functional limbs which centers around the optimization of facial movements and the selection of collecting locations for facial electromyography (fEMG).

A hands-free human-computer interface was created to assist patients in utilizing computers by using facial motions in this research [17, 18], it includes the detection of five distinct facial movement patterns using four electromyography (EMG) sensors. These patterns were then recognized utilizing myoelectric pattern recognition algorithms. The MRI scan and the deep convolutional analysis of the brain structure to predict and analyze the brain-behavior analysis of the disease prediction techniques. The combination of "Logistic Regression" (LR), "Random Forest" (RF), and "Gradient Boost" (GB) is presented as an ensemble model called Logistic Random Forest Boosting (LRFB) in this research [23], LRFB was utilized in detection of Parkinson disease. The researchers [24] employed a variety of classifier methods, such as Support Vector Machine, Voting, Decision Tree, and Random Forest, to determine the optimal parameters for Alzheimer's disease prediction. The Open Access Series of Imaging Studies (OASIS) data is used to make predictions about Alzheimer's disease, and metrics like Precision, Recall, Accuracy, and F1-score are used to evaluate how well ML models perform [25, 26].

Many earlier research focuses on was improvement of BCI systems, the paper [27] describes potential threats and strategies for improving privacy in BCI systems, it provides insights on how to protect privacy when it comes to protecting information about EFG. A prediction approach for dementia using rehabilitation therapy based on transfer learning is proposed that uses the patient's condition [28, 29]. The main challenge that advances BCI technology pertains to the use of non-invasive or minimally invasive technologies designed for paralyzed individuals [30]. Cost and physical distress associated with a surgical procedure can be significantly diminished through the use of non-invasive equipment. On the other hand, low signal-to-noise ratio (SNR) attenuated signals, diminished source precision and diminished spatial resolution may result from non-invasive methodologies [31–33]. To better understand brain activity, the study [34, 35] proposes an in-home occupational therapy setting that employs a variety of occupational therapy-related activity detection methods, including EEG signals. The therapy data for the purpose is stored and secured through blockchain based distributed repository.

3 Methodology and Materials

The brain-computer interface, or BCI, demonstrates the necessary modulation strategies, as well as the steps for diagnosing and implementing all in-depth brain procedure analysis. Deep learning techniques and machine learning algorithmic approaches are used to demonstrate and communicate movement and mood behavior study of brain anatomy and neuron message transmission signals. The algorithmic method for deep brain survey analysis recommendation:

SVM (Support vector machine) [4]: The SVM model works well with small, high-dimensional datasets and is ideal for data like EEG and other brain signal formats. The kernel functions enable to classify non-linear patterns in brain activity signals from patients. The trained SVM model is applied to new brain signals and the user's intent is predicted or the classification of motor imagery task is done. The training of the model takes as input signals corresponding to specific motor tasks.

XG Boosting: XGBoost is widely used in BCI systems for rehabilitation tasks. The model can classify imagined movement e.g. left or right hand or foot movement from EEG data to control external devices like robotic arms or exoskeletons. The model can distinguish between different mental states to optimize neurofeedback training for cognitive rehabilitation and analyze motor-related brain signals during rehabilitation exercises [8].

Blockchain Technology [35]: Blockchain technology is a fully innovative, collaborative, and developed platform that provides any transaction digitally. It also acts like a database that records all the types of data in the block in the Blockchain. The block can be stored in the hyper ledger or immutable ledger and broadcasted in the peer-to-peer network. It is also called DLT (distributed ledger technology) which creates and maintains all the records of the user within the block. The copy of block information can be updated in multiple locations at the same time as updating the block. Blockchain is an outgoing source that provides the freedom to make a secure environment and transparent data and can be shared in the peer-to-peer network (p2p).

3.1 Secure BCI Framework

Figure 5 shows a layered architecture for a typical BCI system for brain disease data. Each payer performs a specific set of tasks from acquiring brain signals to delivering feedback for rehabilitation. Here blockchain layer logs the signal acquisition sessions and ensures that data integrity is maintained from the source. Figure 6 illustrates the steps in the machine learning process to detect the pattern.

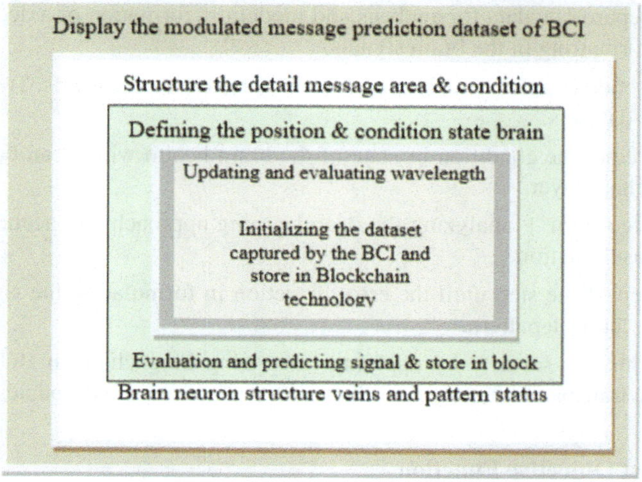

Fig. 5. Brain Data Pattern Structure

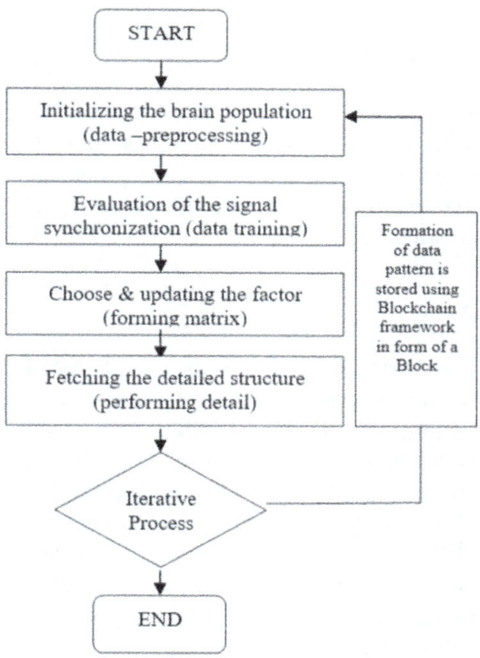

Fig. 6. Process Flow for Learning

The steps are given as follows.

Step 1: Collect raw brain signals that represent neural activity that will be captured by the BCI systems.

Step 2: Prepare raw data for analysis and modeling, formation of a deep detailed structure formatting in the brain signals.

Step 3: Decode features from brain signals to actionable commands. Train models on labeled datasets.

Step 4: Predict the condition or state of the brain health which can be stored in the Blockchain layer.

Step 5: Algorithm + analyzing the deep learning approach and accuracy of the brain disease function.

Step 6: Repeat the step until the error detection in formulating the signals until the system learns a pattern.

Step 7: Enhance security by applying integrating blockchain in BCI systems. Maintain a transparent history of BCI system performance and updates.

3.2 XGBoost Objective Function

The objective functions for the XGBoost classifier are described is given in Eq. 1. The function given is at iteration 't'.

$$\mathcal{L}^{(t)} = \sum_{i=1}^{n} l(y_i, \widehat{y_i}^{(t-1)} + f_t(X_i) + \Omega(f_t) \quad (1)$$

Here 'l' CART learner's function. The function is a sum of the current and previous values. Whether used for classification (e.g., motor imagery tasks) or regression XGBoost provides a powerful framework to decode neural signals while ensuring model interpretability and performance.

3.3 Proposed Model

The model for classifying brain signals entails the completion of multiple stages, which comprise data collection, preprocessing, feature extraction, and algorithm execution. A condensed version of a proposed machine learning-based model for classifying brain signals, specifically electroencephalography (EEG) signals, is provided below. Gather electroencephalography (EEG) data from individuals performing certain mental tasks or activities that are relevant to the intended use, including cognitive or motor imaging tasks. The dataset should be collected from a diverse group of participants and conditions. Support vector machine, CNN, and XGBoost machine learning models are used in the experiments for classification tasks. Figure 7 shows samples of brain signals of specific frequencies. EEG signals between scalp electrodes have several distinct waveforms, the brain signals are unique for every individual.

All the data from the learning process can be stored in the Blockchain framework for the pattern analysis of brain structure in the future.

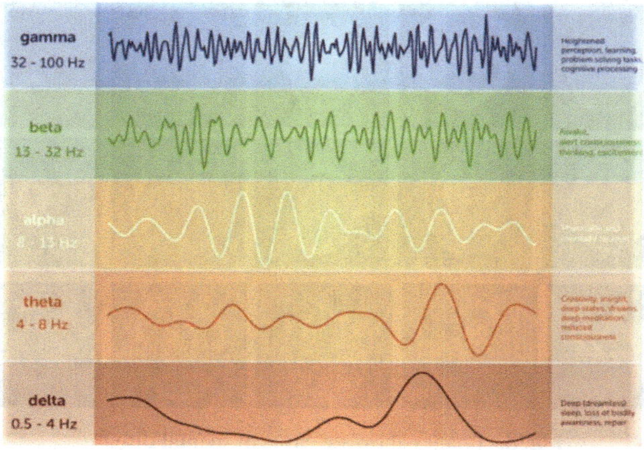

Fig. 7. Brain signals samples with specific frequencies

4 Implementation and Experiments

The BCIs utilize EEG data acquired from the scalp, the surface of the brain, or recorded neural activity within the brain as input. Consequently, in addition to the fundamental distinction between dependent and independent BCIs.The digitalized signals are then subjected to several feature extraction processes, such as single-neuron isolation, voltage amplitude measurements, spectrum analysis, and spatial filtering. The signal properties that, presumably, encode the user's commands or communications are recovered through this analysis.

- **Initial stage dementia**: starting age of the brain disease prediction i.e. remembering the other people's identity, struggles with a task that was once a routine.
- **Middle stage dementia**: stage of predicting the brain disease analysis, where the patient cannot remember their home location, and may not remember family members.
- **Last stage of dementia**: the patient needs help in bathing and feeding, and a profound decline in motor skill.

Dataset1 [21] – To increase the accuracy of brain disease prediction for this dataset, the hidden layer error is reduced. We achieved an accuracy of 95.02% for both the training and testing sets after applying SVM for illness detection. Training the dataset is expected to take up 80% of the time, with the remaining 20% going towards testing and evaluating the data analysis.

Dataset 2 [31] – Every detection iteration shows the disease's symptoms in the brain region. The training and testing analysis of the dataset reveals 70% training, 20% testing, and 10% validation. The accuracy is 82.6% when the dataset is used. The accuracy is about 82.7% when the program is repeated with the same amount of data boosted using the XG Boost technique.

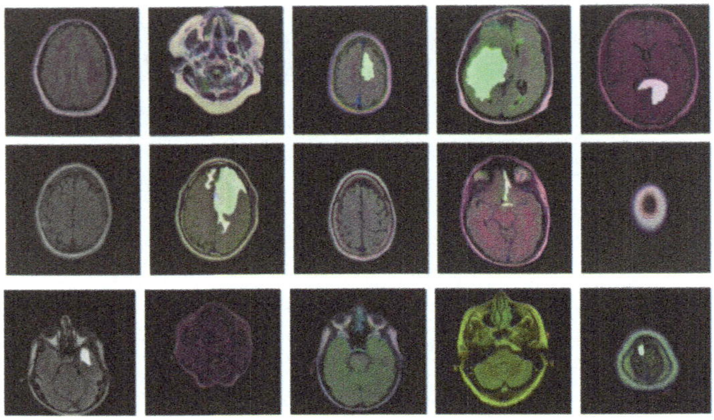

Fig. 8. Images showing cases of mental disease

5 Results and Discussions

The experiment was performed by dividing both dataset 1 and dataset 2 into training and testing set in a ratio of 80:20. The SVM and XG Boost machine learning approaches are used to forecast brain tumors stage-by-stage in Fig. 8, and also for the classification task for dementia patient. The experiment demonstrated the effectiveness of XGBoost and SVM in classifying dementia-related brain activity where XGBoost showed higher accuracy and robustness. Four class labels for classification are "Mild", "Moderate", "Non-dementia" and "VM dementia". Figure 9 shows highest number of the class labels in dataset 1 are " Non-dementia" patients, and Fig. 10 shows the same. Classification accuracy is higher at 93% for XGBoost and 88% for SVM. The XGBoost is better at detecting early stage dementia in patients.

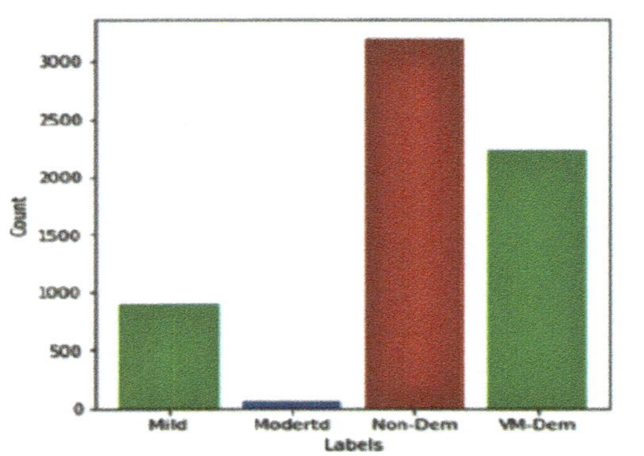

Fig. 9. Dementia patient Dataset classification results

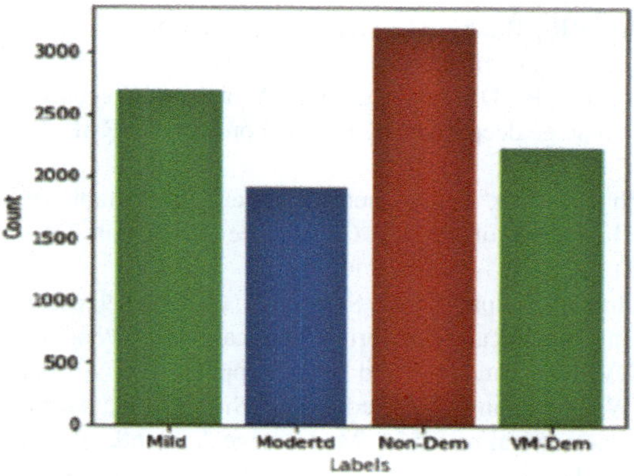

Fig. 10. Alzheimer disease dataset results

6 Conclusion

The brain analysis shows how neurons are interlinked in the brain and can receive and transmit signal during a patient's activity. A BCI allows a person to interact with or manipulate their surroundings without using their brain's typical peripheral nerve and muscle output routes. Electrophysiological events such as cortical neuronal activity or evoked or spontaneous EEG patterns are employed to communicate signals. A BCI system along with machine learning can classify brain signals to detect possible brain disorders. The BCI system is beneficial for patients and doctors for communicating through devices, the system is improved using blockchain to protect privacy as well. While challenges like inter-user variability persist, ongoing innovations promise to further enhance usability, accessibility, and outcomes. In the future one can investigate long-term effects of BCI-assisted rehabilitation on neural plasticity and functional outcomes.

References

1. Pfurtscheller, G., Müller-Putz, G.R., Scherer, R., Neuper, C.: Rehabilitation with brain-computer interface systems. Computer **41**(10), 58–65 (2008)
2. Zander, T.O., Kothe, C., Jatzev, S., Gaertner, M.: Enhancing human-computer interaction with input from active and passive brain-computer interfaces. Brain-Comput. Interfaces: Applying Minds Hum. Comput. Interact. 181–199 (2010)
3. Buch, V.P., et al.: Network brain-computer interface (nBCI): An alternative approach for cognitive prosthetics. Front. Neurosci. **12**, 790 (2018)
4. Mane, R., Chouhan, T., Guan, C.: BCI for stroke rehabilitation: motor and beyond. J. Neural Eng. **17**(4), 041001 (2020)
5. Tan, X., Li, Y., Gao, Y.: Combining brain-computer interface with virtual reality: Review and prospect.: In 2017 3rd IEEE International Conference on Computer and Communications (ICCC), pp. 514–518. IEEE (2017)
6. Sakhavi, S., Guan, C., Yan, S.: Learning temporal information for brain-computer interface using convolutional neural networks. IEEE Trans. Neural Networks Learn. Syst. **29**(11), 5619–5629 (2018)

7. Kumar, S., Verma, P.R., Bharti, M., Agarwal, P.: A CNN based graphical user interface controlled by imagined movements. Int. J. Syst. Assur. Eng. Manage., 1–11 (2021)
8. Bougourzi, F., Contino, R., Distante, C., Taleb-Ahmed, A.: Recognition of COVID-19 from CT scans using two-stage deep-learning-based approach: CNR-IEMN. Sensors **21**(17), 5878 (2021)
9. Gu, Y., Li, H., Zhao, J., Ma, X.:. A photovoltaic charger suitable for brain computer interface equipment.: In Journal of Physics: Conference Series, vol. 2254, no. 1, p. 012012. IOP Publishing (2022)
10. Avasthi, S., Chauhan, R., Tripathi, S.L., Sanwal, T.: COVID-19 research: open data resources and challenges.: In Biomedical Engineering Applications for People with Disabilities and the Elderly in the COVID-19 Pandemic and Beyond, pp. 93–104. Academic Press (2022)
11. Peng, H., Gong, W., Beckmann, C.F., Vedaldi, A., Smith, S.M.: Accurate brain age prediction with lightweight deep neural networks. Med. Image Anal. **68**, 101871 (2021)
12. Kwon, O.Y., Lee, M.H., Guan, C., Lee, S.W.: Subject-independent brain–computer interfaces based on deep convolutional neural networks. IEEE Trans. Neural Networks Learn. Syst. **31**(10), 3839–3852 (2019)
13. Bonci, A., Fiori, S., Higashi, H., Tanaka, T., Verdini, F.: An introductory tutorial on brain–computer interfaces and their applications. Electronics **10**(5), 560 (2021)
14. Lupu, R.G., Ungureanu, F., Cimpanu, C.:. Brain-computer interface: challenges and research perspectives.: In 2019 22nd International Conference on Control Systems and Computer Science (CSCS), pp. 387–394. IEEE (2019)
15. Aayesha Qureshi, M.B., Afzaal, M., Qureshi, M.S., Fayaz, M.: Machine learning-based EEG signals classification model for epileptic seizure detection. Multimedia Tools Appl. **80**, 17849–17877 (2021)
16. Zhu, B., Zhang, D., Chu, Y., Zhao, X.: A novel limbs-free human-computer interface: face-computer interface (FCI) with channels optimization.: In 2021 IEEE International Conference on Robotics and Biomimetics (ROBIO), pp. 1092–1097. IEEE (2021)
17. Lu, Z., Zhou, P.: Hands-free human-computer interface based on facial myoelectric pattern recognition. Front. Neurol. **10**, 444 (2019)
18. Anders, C., Arnrich, B.: Wearable electroencephalography and multi-modal mental state classification: a systematic literature review. Comput. Biol. Med. **150**, 106088 (2022)
19. Castro, M.C.F., Arjunan, S.P., Kumar, D.K.: Selection of suitable hand gestures for reliable myoelectric human computer interface. Biomed. Eng. Online **14**, 1–11 (2015)
20. Longo, L., Wickens, C.D., Hancock, G., Hancock, P.A.: Human mental workload: a survey and a novel inclusive definition. Front. Psychol. **13**, 883321 (2022)
21. https://www.kaggle.com/datasets/sachinkumar413/alzheimer-mri-dataset
22. Suzuki, M., et al.: Mental workload during endoscopic sinus surgery is associated with surgeons' skill levels. Front. Med. **10**, 1090743 (2023)
23. Rad, E.M.: EEG and MRI Processing for Alzheimer's Diseases.: In Vision Sensors-Recent Advances. IntechOpen (2022)
24. Kavitha, C., Mani, V., Srividhya, S.R., Khalaf, O.I., Tavera Romero, C.A.: Early-stage Alzheimer's disease prediction using machine learning models. Front. Public Health **10**, 853294 (2022)
25. Kumar, N., Avasthi, S., Prakash, A.:. Establishing the correlation between parkinson's and heart disease using machine learning algorithm.: In 2023 International Conference on Computational Intelligence, Communication Technology and Networking (CICTN), pp. 581–586. IEEE (2023)
26. Avasthi, S., Chauhan, R., Acharjya, D.P.: Topic modeling techniques for text mining over a large-scale scientific and biomedical text corpus. Int. J. Ambient Comput. Intell. (IJACI) **13**(1), 1–18 (2022)

27. Xia, K., et al.: Privacy-preserving brain–computer interfaces: a systematic review. IEEE Trans. Comput. Soc. Syst. **10**(5), 2312–2324 (2022)
28. Kuo, P.H., Huang, C.T., Yao, T.C.: Optimized transfer learning based dementia prediction system for rehabilitation therapy planning. IEEE Trans. Neural Syst. Rehabil. Eng. **31**, 2047–2059 (2023)
29. Avasthi, S., Chauhan, R., Acharjya, D.P.: Extracting information and inferences from a large text corpus. Int. J. Inf. Technol. **15**(1), 435–445 (2023)
30. Di Benedetto, M., et al.: Deep networks for behavioral variant frontotemporal dementia identification from multiple acquisition sources. Comput. Biol. Med. **148**, 105937 (2022)
31. https://www.kaggle.com/datasets/mateuszbuda/lgg-mri-segmentation
32. Tafuri, B., et al.: Behavioral variant frontotemporal dementia in patients with primary psychiatric disorder: a magnetic resonance imaging study. Brain Behav. **13**(4), e2896 (2023)
33. Belkacem, A.N., Jamil, N., Palmer, J.A., Ouhbi, S., Chen, C.: Brain computer interfaces for improving the quality of life of older adults and elderly patients. Front. Neurosci. **14**, 692 (2020)
34. Kumar, R., Badwal, L., Avasthi, S., Prakash, A.: A secure decentralized e-voting with blockchain & smart contracts. In 2023 13th International Conference on Cloud Computing, Data Science & Engineering (Confluence), pp. 419–424. IEEE (2023)
35. Rahman, M.A., Hossain, M.S., Rashid, M.M., Barnes, S.J., Alhamid, M.F., Guizani, M.: A -based non-invasive cyber-physical occupational therapy framework: BCI perspective. IEEE Access **7**, 34874–34884 (2019)

Home Healthcare Optimization: A Systematic Literature Review

Emerson José de Paiva[1,2](✉) and Ana Maria A. C. Rocha[2]

[1] Federal University of Itajubá UNIFEI, Itajubá, Brazil
emersonpaiva@unifei.edu.br
[2] ALGORITMI Research Centre / LASI, University of Minho, Braga, Portugal
arocha@dps.uminho.pt

Abstract. Currently, healthcare has attracted the attention of administrators, medical teams, and researchers due to the numerous home care programs. These programs facilitate society's access to treatment, minimize hospital costs, alleviate waiting rooms, and speed up care. However, in order to make it operational, complex visit planning is required to optimize travel and care teams, while ensuring that operational costs are minimized. Such planning requires the use of mathematical optimization methodologies, which are not within the expertise of most of the professionals involved. In this study, a Systematic Literature Review was carried out with the aim of identifying the main optimization algorithms used in the complex operationalization of these services, in order to provide professionals with the main methodologies for solving these problems. The results indicate that the studies focus on cost minimization, with most optimization models utilizing solutions based on Simulated Annealing and Genetic Algorithm.

Keyword: Optimization · Scheduling · Home healthcare · Solution algorithms

1 Introduction

Fully meeting customer expectations, maximizing the efficiency of production processes, delivering high-quality products and services, and minimizing costs are the primary objectives of organizations in today's times. This objective also holds for healthcare organizations, such as hospitals or clinics, whether they are public or private [1].

Recently, several authors have pointed out that the number of constraints involved in the scheduling and planning of care teams is quite complex, and that HHC providers face constant pressure regarding operational costs, quality improvement, and productivity [2–4]. This is why there is significant interest in suitable Decision Support Systems (DSS). DSSs are typically generated through mathematical modeling of real-world problems using Operations Research methods. Begur et al. [5] presented the first study involving

This work has been supported by FCT— Fundação para a Ciência e Tecnologia within the R&D Units Project Scope: UIDB/00319/2020 and GEQProd – Grupo de Estudos em Qualidade e Produtividade, Federal University of Itajubá, Campus Itabira, MG, Brazil.

D. Kriksciuniene et al. (Eds.): IBICA 2023, LNNS 1228, pp. 14–23, 2025.
https://doi.org/10.1007/978-3-031-78937-3_2

Operations Research, developing a spatial decision support model for the HHC scheduling and routing problem, aiming to balance workloads and travel times. Since then, several researchers have developed studies focused on HHC intending to obtain better operational results, such as reducing costs [5], optimizing routes [6], assigning teams appropriately [7], maximizing the number of visits [8], ensuring patient satisfaction [9], among others.

In summary, an HHC problem can be described as the daily scheduling of a specific and limited number of caregiver teams with different skill levels, such as doctors, paramedics, and nurses, who must visit a certain number of patients at least once a day, while minimizing costs [4, 10, 11]. In this scheduling process, assignment constraints and narrow time windows for visits must be respected, as well as maximum daily working times and mandatory rest breaks [12]. Among the main objectives are minimizing the total travel cost and caregiver team waiting time [4]. The mode of transportation used for visits should also be considered since it significantly affects the available travel time [13]. In some cases, a combination of different transportation modes may or should be used. Thus, the number of variables and constraints that need to be considered increases the complexity of planning the daily schedule of visits. There are several studies available in the literature, but each one has a different focus, making it challenging for managers to access relevant information.

In this work, a systematic literature review (SLR) is presented with the aim of identifying the main optimization models available in HHC, considering their objectives, the solution methods adopted (exact, heuristics, and/or meta-heuristics), and the implemented algorithms.

2 Methodology

Team scheduling in HHC is a recent field of research that has aroused great interest from the scientific community [4, 14]. More specifically, considering the combined approaches of optimization and team scheduling in HHC, the first publications date back to 1997, experiencing significant growth in subsequent years, especially in recent years. Figure 1 illustrates the volume of publications within the research period from 2018 to 2023.

It is worth noting the significant increase in publications between 2020 and 2021. In fact, this behavior may be the result of the global pandemic that began at the end of 2019. Since that time, social distancing restrictions compelled healthcare facilities to invest in HHC as a way of guaranteeing the population's access to necessary healthcare while minimizing the risk of contagion spreading. Furthermore, Fig. 1 also highlights the number of publications in the year 2023, which in just nine months has already surpassed the volume of publications in previous years.

For the implementation of the SLR, a methodology based on the PRISMA 2020 checklist [15] was chosen and adapted to the needs of this research. A systematic literature review is a study that searches, selects, analyzes, and synthesizes the findings of studies to answer a structured research question and reduce possible methodological biases. Thus, in this study, the SLR aims to evaluate which are the main optimization algorithms and solution methods (exact, heuristics, or meta-heuristics) that have been used to solve home healthcare team scheduling problems. It is important to note that, for

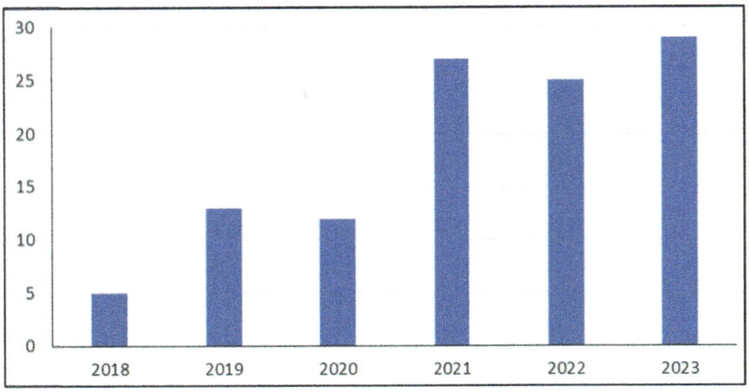

Fig. 1. Articles per year – 2018–2023

this research, it was crucial to outline the current state of the art of the problem. It was therefore decided to look for articles published in peer-reviewed journals between 2018 and 2023, using the Web of Science and Scopus databases. Consequently, publications in conference proceedings, book chapters, technical reports, etc., were excluded from consideration. To achieve the SLR goal, some filters were applied to reduce the number of studies to be analyzed, with a specific focus on this research field. Thus, the search was carried out using only the terms "optimization," "scheduling," and "home health*" found in the title, abstract, or keywords. The term "home health*" specified the type of care desired, which is domiciliary, and the asterisk ("*") was used to indicate variations of the term, such as "health care," "health-care," or "healthcare," for example.

The Web of Science and Scopus databases were unified and processed using functions available in the "Bibliometrix" package in R and the results were analyzed using the "Biblioshiny" tool [16].

A total of 93 studies from 57 different sources were selected from both databases. Figure 2 presents the volume of publications grouped by country. It is evident that China has the highest number of publications from 2018 to 2023, possibly reflecting the impact of the pandemic and the reasons previously mentioned, followed by Iran and France.

However, to further refine the selection, studies that, upon reading, were identified as not belonging to the scope of the research were excluded. Additionally, literature review papers were also disregarded. To ensure the reliability of the analysis, only the most relevant studies were considered. For this metric, it was determined that papers with 20 or more ($> = 20$) citations should make up the final list. As a result, a total of 21 articles were selected. Figure 3 summarizes the evolution of the search process by applying filters to refine the results and the number of studies identified.

Table 1 shows the final list of articles with their general characteristics. For each year of the SLR period, the main studies found, focused on the optimization and scheduling of HHC problems, reflecting different perspectives on the same problem in terms of objectives, methods, and solution algorithms. It can be observed that there is a wide variety of objectives and solution algorithms suggested by the researchers. In the following, a brief discussion of this data will be presented to enhance the understanding of the results.

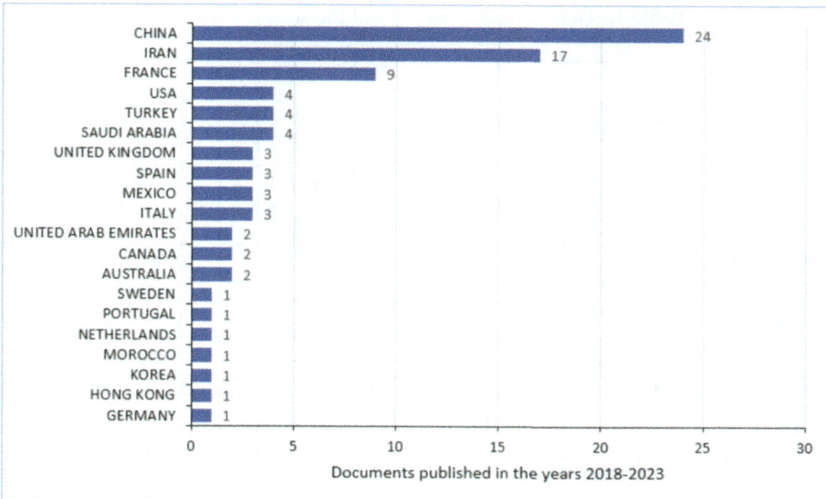

Fig. 2. Documents published by countries

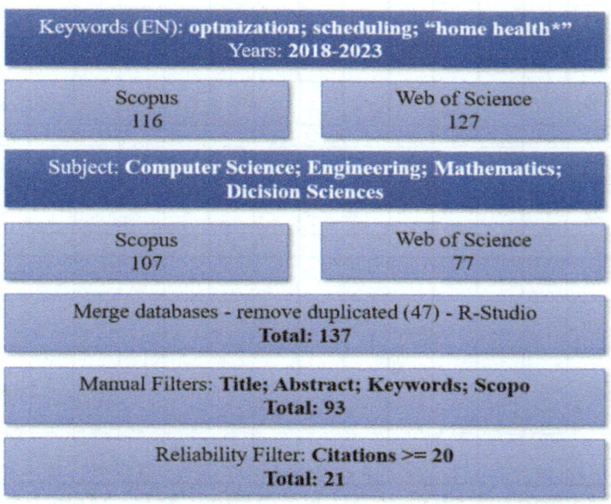

Fig. 3. Bibliography research scheme

3 Discussion

The bibliometric analysis of the results highlights several aspects of the studies that deserve to be emphasized. Initially, it is possible to extract a set of keywords, the results of which indicate the evolution of the HHC research field, as illustrated in Fig. 4.

Furthermore, it is also evident that the search found studies that clearly align with the research focus.

In terms of objectives, Table 1 shows that the majority of articles (66.7% of the evaluated works) aim to minimize costs. It should be noted that, for this study, costs

Table 1. General characteristics of the selected articles.

Year	Authors	Number of citations	Min										Max		Solution Method	Solving Approach
			Costs	Distance travelled	Traveling time	Environmental	Working in time	Time window	Synchronized visits	caregiver inconsistency	workload	unemployment time	Patients visits	Satisfaction		
2018	Fathollahi-Fard et al. [17]	101	✓			✓									MH	MILP, SA, MSA, SSA, MSSA, H-SS
	Nasir & Dang [2]	32	✓	✓											E	MILP
	Liu et al. [14]	24	✓											✓	H	MILP, NSGA-II, MOSA, RAA
2019	Fathollahi-Fard et al. [18]	128	✓			✓									MH	SA, MSA
	Shi et al. [19]	94	✓												H	MIP, SA, TS, VNS
	Grenouilleau et al. [20]	76	✓												H	LNS, SA
	Decerle et al. [21]	55		✓											H	MIP, MA, ACO, HMAC
	Moussavi et al. [22]	48		✓											MH	MIP
	Decerle et al. [23]	39					✓	✓	✓						MH	MIP, NSGA-II, MDLS, MAMO
	Demirbilek et al. [24]	20											✓		H	DH, CH
	Heching et al. [8]	32											✓		E	MILP, LBBD
2020	Fathollahi-Fard et al. [25]	102	✓											✓	MH	NSGA-II, SA, AMSEO
	Nasir & Kuo [26]	32	✓												H	MILP, HGA
2021	Goodarzian et al. [27]	49	✓				✓								MH	ISEO
	Nikzad et al. [28]	41	✓												MH	Math-1
	Cinar et al. [29]	32			✓								✓		MH	MIP, LR, SSPH, ALNS
	Shahnejat-Bushehri et al. [30]	23	✓												MH	SA, GA, MA
	Yang et al. [31]	22	✓							✓	✓				MH	IMOABC
	Fathollahi-Fard et al. [32]	21	✓									✓	✓		H/MH	IMORDA
	Demirbilek et al. [33]	20											✓		H	SBAM
2022	Fathollahi-Fard et al. [34]	51	✓												MH	ICA, WWO, SSA, MICA, MWWO, MSSA, HSIA, HWSA
	Total		14	3	1	2	2	1	1	1	1	1	5	2		

Notes: Solution Method – E: Exact; H: Heuristics; MH: Meta-heuristics.

encompass various types, namely: a) routing costs: travel time or distance; and b) team-related costs: overtime, visit duration or working time, and waiting time. The second most significant objective, regarding minimization objectives, is distance (representing 14.3% of the studies). As for maximization objectives, there are two focuses: a) patient visits, with 23.8% of the works, and b) patient satisfaction, with 9.5%. Among the less frequently mentioned objectives, it is worth noting that "Environment" refers to impacts and pollutant emissions (by the vehicles used to transport the teams), and also "Time Window," to which the researchers attribute characteristics of patient satisfaction (for example, minimizing the "time window" corresponds to maximizing "satisfaction").

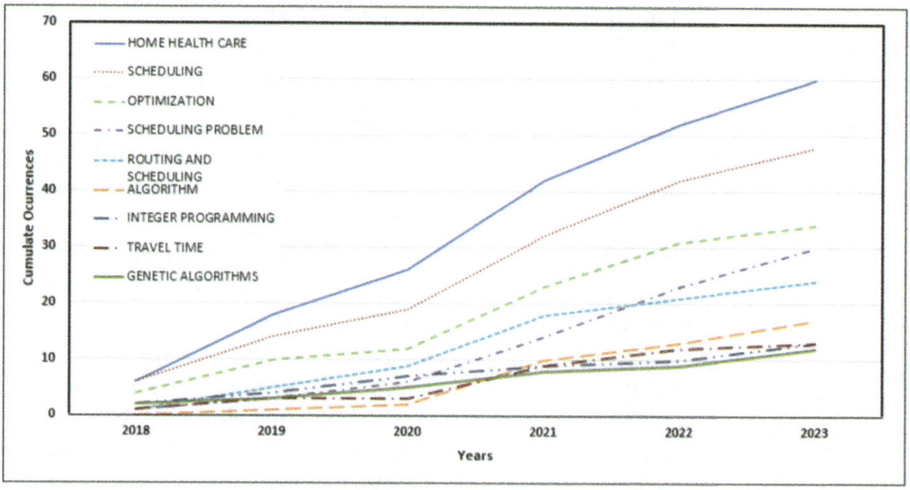

Fig. 4. Historical occurrences by keyword

Regarding solution methods, 52.4% of the studies are related to Meta-Heuristics, while Heuristics are adopted in 33.3% of the works. Exact methods were identified in only 2 studies, representing 9.5%, and finally, a single work presented a hybrid approach between Heuristics and Meta-Heuristics, constituting 4.8% of the works.

The focus of this work is on the algorithms adopted for the different sets of objectives, to identify the most recurrent among researchers and point out the main Decision Support Systems (DSS).

Table 2 lists the different solution approaches found in the analyzed studies.

Initially, it is important to note that optimization models for HHC problems are based on the "Vehicle Routing Problem" (VRP), and their solutions are typically based on Mixed Integer Programming (MIP) or Mixed Integer Linear Programming (MILP). Given this integer programming approach, the analyzed studies present different perspectives due to the specific characteristics of each investigated model. This complexity makes it challenging to define a generalized approach that is closer to reality.

Analyzing Table 2, one can observe a significant diversity of algorithms, sometimes used individually and sometimes in hybrid approaches. Most often, these algorithms are adapted to new situations or incorporate improvements over their predecessors, leading to the development of new algorithmic variations. However, it is observed that the most recurring solution approach is Simulated Annealing (SA), used in 15.8% of the analyzed studies, followed by Non-dominated Sorting Genetic Algorithm (NSGA-II) at 7.9%, Memetic Algorithm (MA), and Salp Swarm Algorithm (SSA), each at 5.3%. Considering that many of the presented algorithms are variations of other classic algorithms, grouping them (e.g., SA, MSA, and MOSA), the result indicates that the most recurring approach is Simulated Annealing, found in 42.9% of the analyzed studies, followed by Salp Swarm Algorithm at 28.6%, and finally, Genetic Algorithm at 19.0%. As mentioned earlier, each analyzed study presents solutions to HHC problems with different

Table 2. List of Adopted Solution Algorithms.

ACO	Ant Colony Optimization	MA	Memetic Algorithm
ALNS	Adaptive Large Neighborhood Search	MAMO	Memetic Algorithm for Multi-objective Optimization
AMSEO	Adaptive Memory Social Eng. Optimizer	Math-1	Multi-phase metaheuristic algorithm
CH	Capacity Heuristic	MDLS	Multi-Directional Local Search
DH	Distance Heuristic	MICA	Modify Imperialist Competitive Algorithm
GA	Genetic Algorithm	MOSA	Multi-objective Simulated Annealing method
HGA	Hybrid Genetic Algorithm	MSA	Modify Simulated Annealing
HMAC	Hybrid Memetic-Ant Colony	MSSA	Modify Salp Swarm Algorithm
HSIA	Hybrid of SSA and ICA	MWWO	Modify Water Wave Optimization
H-SS	Hybrid metaheuristics of SSA & SA	NSGA-II	Non-dominated Sorting Genetic Algorithm
HWSA	Hybrid of WWO and SSA	RAA	Random Assignment Approach
ICA	Imperialist Competitive Algorithm	SA	Simulated Annealing
IMOABC	Improved multi-objective Artificial Bee Colony	SBAM	Scenario-Based Approach for Multiple Nurses
IMORDA	Improved Multi-objective Red Deer Algorithm	SSA	Salp Swarm Algorithm
ISEO	Improved Social Engineering Optimizer	SSPH	Successive Single Period Heuristic
LBBD	Logic-Based Benders decomposition	TS	Tabu Search
LNS	Large Neighborhood Search	VNS	Variable Neighborhood Search
LR	Lagrangean Relaxation	WWO	Water Wave Optimization

characteristics. These characteristics increase the complexity of the solutions, requiring different approaches from the existing solution methods. However, they contribute to enriching the methodological framework of mathematical modeling and operations research (OR).

4 Conclusions

Home Healthcare services have been receiving increasing attention from researchers, managers, and analysts, especially during the pandemic period, due to their social and economic importance. The need for social distancing to contain the spread of the has highlighted the importance of route optimization and scheduling in this context. As a result, various numerical and methodological approaches have been developed and made available in the literature to address this complex assignment problem. These contributions have the potential to significantly improve the efficiency and quality of HHC services, benefiting both patients and healthcare providers.

In this work, a systematic literature review was conducted with the aim of identifying the most recurring solution algorithms among the found studies. Using the terms "optimization," "scheduling" and "home health*", a search was performed in two of the main available databases, "Scopus" and "Web of Science". After applying some refinement filters, twenty-one studies were selected, considered the most relevant due to the number of associated citations.

Analyzing these studies and applying bibliometric techniques, it was possible to verify that: HHC problems are modeled based on the Vehicle Routing Problem; HHC solutions are obtained through methods based on Mixed Integer Programming or Mixed Integer Linear Programming; Meta-heuristic methods are preferred by researchers for solving different insights; Cost minimization is the primary objective indicated by the analysis of selected works, often associated with other secondary objectives; There are different solution algorithms proposed for problems with various characteristics. However, considering the objectives of this study, it can be stated that there is a preference for solutions based on Simulated Annealing (SA) and Genetic Algorithm, which are well-established in the literature, as well as the recent Salp Swarm Algorithm.

Since these methods are non-trivial, it is suggested that future work carry out a more in-depth evaluation of the key algorithms identified and consider implementing them in mobile or web applications. This would make them more accessible to analysts, managers, and decision-makers.

References

1. Wang, L., Demeulemeester, E.: Simulation optimization in healthcare resource planning: a literature review. IISE Trans. **55**(10), 985–1007 (2023)
2. Nasir, J.A., Dang, C.: Solving a more flexible home health care scheduling and routing problem with joint patient and nursing staff selection. Sustainability **10**, 0148 (2018)
3. Fikar, C., Hirsch, P.: Home health care routing and scheduling: a review. Comput. Oper. Res. **77**, 86–95 (2017)
4. Rest, K.D., Hirsch, P.: Daily scheduling of home health care services using time-dependent public transport. Flex. Serv. Manuf. J. **28**, 495–525 (2016)
5. Begur, S.V., Miller, D.M., Weaver, J.R.: An integrated spatial DSS for scheduling and routing home-health-care nurses. Interfaces **27**, 3–48 (1997)
6. Akjiratikarl, C., Yenradee, P., Drake, P.R.: PSO-based algorithm for home care worker scheduling in the UK. Comput. Ind. Eng. **53**(4), 559–583 (2007)
7. Shao, Y., Bard, J.F., Jarrah, A.I.: The therapist routing and scheduling problem. IIE Trans. **44**(10), 868–893 (2012)

8. Heching, A., Hooker, J.N., Kimura, R.: A logic-based benders approach to home healthcare delivery. Transp. Sci. **53**(2), 510–522 (2019)
9. Trautsamwieser, A., Hirsch, P.: Optimization of daily scheduling for home health care services. J. Appl. Oper. Res. **3**(3), 124–136 (2011)
10. Belhor, M., El-Amraoui, A., Jemai, A., Delmotte, F.: Multi-objective evolutionary approach based on K-means clustering for home health care routing and scheduling problem. Expert Syst. Appl. **213**, 119035 (2023)
11. Du, G., Liang, X., Sun, C.: Scheduling optimization of home health care service considering patients' priorities and time windows. Sustainability **9**, 1–22 (2017)
12. Méndez-Fernández, I., Lorenzo-Freire, S., González-Rueda, A.M.: An adaptive large neighbourhood search algorithm for a real-world home care scheduling problem with time windows and dynamic breaks. Comput. Oper. Res. **159**, 106351 (2023)
13. Hiermann, G., Prandstetter, M., Rendl, A., Puchinger, J., Raidl, G.R.: Metaheuristics for solving a multimodal home-healthcare scheduling problem. CEJOR **23**, 89–113 (2015)
14. Liu, M., Yang, D., Su, Q., Xu, L.: Bi-objective approaches for home healthcare medical team planning and scheduling problem. Comp. Appl. Math. **37**, 4443–4474 (2018)
15. BMJ.: The PRISMA 2020 statement: an updated guideline for reporting systematic reviews. 372:n71 (2021)
16. Aria, M., Cuccurullo, C.: Bibliometrix: an R-tool for comprehensive science mapping analysis. J. Informet. **11**(4), 959–975 (2017)
17. Fathollahi-Fard, A.M., Hajiaghaei-Keshteli, M., Tavakkoli-Moghaddam, R.: A bi-objective green home health care routing problem. J. Clean. Prod. **200**, 423–443 (2018)
18. Fathollahi-Fard, A.M., Govindan, K., Hajiaghaei-Keshteli, M., Ahmadi, A.: A green home health care supply chain: new modified simulated annealing algorithms. J. Clean. Prod. **240**, 118200 (2019)
19. Shi, Y., Boudouhb, T., Grundera, O.: A robust optimization for a home health care routing and scheduling problem with consideration of uncertain travel and service times. Transp. Res. Part E **128**, 52–95 (2019)
20. Grenouilleau, F., Legrain, A., Lahrichi, N., Rousseau, L.: A set partitioning heuristic for the home health care routing and scheduling problem. Eur. J. Oper. Res. **275**, 295–303 (2019)
21. Decerle, J., Grunder, O., El Hassani, A.H., Barakat, O.: A memetic algorithm for multi-objective optimization of the home health care problem. Swarm Evol. Comput. **44**, 712–727 (2019)
22. Moussavi, S.E., Mahdjoub, M., Grunder, O.: A matheuristic approach to the integration of worker assignment and vehicle routing problems: Application to home healthcare scheduling. Expert Syst. Appl. **125**, 317–332 (2019)
23. Decerle, J., Grunder, O., El Hassani, A.H., Barakat, O.: A hybrid me-metic-ant colony optimization algorithm for the home health care problem with time window, synchronization and working time balancing. Swarm Evol. Comput. **46**, 171–183 (2019)
24. Demirbilek, M., Branke, J., Strauss, A.: Dynamically accepting and scheduling patients for home healthcare. Health Care Manag. Sci. **22**, 140–155 (2019)
25. Fathollahi-Fard, A.M., Ahmadi, A., Goodarzian, F., Cheikhrouhou, N.: A bi-objective home healthcare routing and scheduling problem considers patients' satisfaction in a fuzzy environment. Appl. Soft Comput. J. **93**, 106385 (2020)
26. Nasir, J.A., Kuo, Y.: A decision support framework for home health care transportation with simultaneous multi-vehicle routing and staff scheduling synchronization. Decis. Support. Syst. **138**, 113361 (2020)
27. Goodarzian, F., Abraham, A., Fathollahi-Fard, A.M.: A biobjective home health care logistics considering the working time and route balancing: a self-adaptive social engineering optimizer. J. Comput. Des. Eng. **8**(1), 452–474 (2021)

28. Nikzad, E., Bashiri, M., Abbasi, B.: A metaheuristic algorithm for stochastic home health care planning. Eur. J. Oper. Res. **288**, 753–774 (2021)
29. Cinar, A., Salman, F.S., Bozkaya, B.: Prioritized single nurse routing and scheduling for home healthcare services. Eur. J. Oper. Res. **289**, 867–878 (2021)
30. Shahnejat-Bushehri, S., Tavakkoli-Moghaddam, R., Boronoos, M., Ghasemkhani, A.: A robust home health care routing-scheduling problem with temporal dependencies under uncertainty. Expert Syst. Appl. **182**, 115209 (2021)
31. Yang, M., Ni, Y., Yang, L.: A multi-objective consistent home healthcare routing and scheduling problem in an uncertain environment. Comput. Ind. Eng. **160**, 107560 (2021)
32. Fathollahi-Fard, A.M., Ahmadi, A., Karimi, B.: Multi-objective optimization of home healthcare with working-time balancing and care continuity. Sustainability **13**(22), 12431 (2021)
33. Demirbilek, M., Branke, J., Strauss, A.K.: Home healthcare routing and scheduling of multiple nurses in a dynamic environment. Flex. Serv. Manuf. J. **33**, 253–280 (2021)
34. Fathollahi-Fard, A.M., Hajiaghaei-Keshteli, M., Tavakkoli-Moghaddam, R.: Bi-level programming for home health care supply chain considering outsourcing. J. Ind. Inf. Integr. **25**, 100246 (2022)

Simulation on Home Healthcare Problem: A Systematic Literature Review

Emerson José de Paiva[1,2](✉) and Ana Maria A. C. Rocha[2]

[1] Federal University of Itajubá – UNIFEI, Itajubá, Brazil
emersonpaiva@unifei.edu.br
[2] ALGORITMI Research Centre / LASI, University of Minho, Braga, Portugal
arocha@dps.uminho.pt

Abstract. The advantages of using the recent Home Health Care model have drawn the attention of experts and managers in recent decades, gaining preference among healthcare institutions and patients. In the literature, various studies can be found, that present complex mathematical optimization models in order to obtain the best scheduling and routing solutions, intending to minimize operational costs, travel or service time, or maximizing the number of visits. To assist in this task, simulation techniques can be used as a way to test different scenarios in search of the best solutions without any intervention in the system until the final decision is made. In this work, we present a Systematic Literature Review focusing on the use of simulation as a tool to assist in the planning of Home Health Care operations. As a result, the application of Monte Carlo Simulation to improve the quality of input data or evaluate the results of the proposed approaches is observed, revealing an important research gap for dynamic simulation models or Discrete Event Simulation applications.

Keyword: Home healthcare · Scheduling · Simulation

1 Introduction

Nowadays, the main objectives of organizations are to fully meet customer expectations, maximize the efficiency of production processes, provide high-quality products and services, and minimize costs. These objectives are also valid for healthcare organizations, such as hospitals or clinics, whether they are public or private [1]. In this context, the Home Health Care (HHC) process has gained prominence in recent decades, and the related literature indicates a substantial increase in research in this area in the coming decades [2]. This is a process of providing healthcare to patients, in which teams composed of professionals of various levels, such as doctors, paramedics and nurses, visit patients in their own homes. They carry out clinical assessments, administer medicines,

This work has been supported by FCT— Fundação para a Ciência e Tecnologia within the R&D Units Project Scope: UIDB/00319/2020, GEQProd – Grupo de Estudos em Qualidade e Produtividade, Unifei, Campus Itabira, and CNPq – Conselho Nacional de Desenvolvimento Científico e Tecnológico, Brazil.

D. Kriksciuniene et al. (Eds.): IBICA 2023, LNNS 1228, pp. 24–36, 2025.
https://doi.org/10.1007/978-3-031-78937-3_3

collect samples for laboratory analysis, provide physiotherapy or nutritional care, in essence delivering healthcare services as if they were in a healthcare institution [1, 3].

This recent model of healthcare has achieved highly satisfactory results, thanks to the comfort it provides to patients who do not need to travel or wait in waiting rooms. It also contributes to a better workload distribution among team members and leads to a significant reduction in operational costs [4, 5]. However, in order to be effective, HHC requires complex planning that involves defining care teams, the number of professionals and their required skills [4, 6], the number of daily visits to be made [5], transportation means, and the routes to be followed [7], among other factors.

There are currently numerous studies in the literature that seek to evaluate these variables and identify the best solutions to the "routing and scheduling" problem [8–11]. All these studies present solutions based on mathematical models which, due to their complex nature, are developed and optimized using Operations Research techniques [12]. Consequently, a good balance between the different input and output variables is expected to obtain the best solutions.

In recent decades, numerous studies have been conducted in this regard and made available to the academic and scientific community. Studies focusing on the HHC routing and scheduling problem (HHCP), can be classified as a variant of the Multiple Travelling Salesman Problem (MTSP) [13], Vehicle Routing Problems (VRP) or Dial-a-Ride Problem (DARP) [14], and their solution is generally obtained from exact, heuristic or meta-heuristic methods based on Mixed Integer Programming (MIP) or Mixed Integer Linear Programming (MILP) models.

In addition to mathematical models and optimization methods, simulation can be employed as a way of verifying that optimal models actually achieve statistically robust results through by evaluating various hypothetical experimental scenarios [13]. By implementing these simulated scenarios, the results can be assessed, ensuring that decisions are made with the confidence of achieving the best possible results in practice.

Classic authors, such as Jun et al. [15], state that simulation techniques allow analysts and managers to select alternatives to improve the quality of input data, reconfigure existing systems to enhance their performance or design, anticipate and evaluate the potential outcomes of the models to be implemented, and test various implementation possibilities before intervening in the system under study.

For example, Bowers et al. [13] used algorithms based on the classic Clarke-Wright heuristic, which were incorporated into a Discrete Event Simulation (DES) model to explore trade-offs between travel time and the continuity of care. Guinet et al. [16] carried out a simulation study to test the ability to evacuate patients from the HHC in the event of flooding. According to these authors, optimization models and computer simulation are the main approaches used in evacuation studies, mainly because computer simulation models are able to capture all the complexity of these real-world systems. As a result, they were able to minimize costs related to evacuation, hospitalization, and home healthcare. Demirbilek et al. [17] presented a Scenario-Based Approach (SBA), simulating various scenarios for scheduling new clients using a simple and fast heuristic, analyzing the results to decide whether to accept the new client at specific dates and times. The idea was to evaluate the number of times a new request would be assigned across all scenarios and at which time interval it would be scheduled most frequently. These cases merely demonstrate some of the potential applications of various simulation techniques in conjunction with decision-making in HHCP.

Thus, in this work a Systematic Literature Review (SLR) is carried out to determine whether simulation techniques have been applied in the optimization of the HHC routing and scheduling problem, how they have been used, in which problem situations they are used and what results or conclusions have been drawn from their application.

2 Implementation

In this study, two databases were chosen, Scopus and Web of Science, to carry out the SLR. Three search criteria were adopted, each representing a specific domain for the research: a) "home health*," spelled like this to make it easier to find possible variations of the term "Home Health Care"; b) "scheduling"; and c) "simulation".

The preliminary search was conducted by locating the keywords concatenated with logical conjunctions in the search string, present in the titles, abstracts, or in the list of keywords of the articles. In addition to the search string, the following filters were applied: a) Period: between 2010 and 2023 – this period was chosen because it is assumed to encompass the main computer simulation tools available in the market; b) Language: English; c) Type: Articles and conference papers. After applying the filters to both databases, their results were combined, duplicates were removed, and a total of 42 papers were selected from 37 different sources. These papers were then analyzed using the Bibliometrix tool in the R statistical package [18].

Figure 1 shows the classification of the selected papers by type of publication. It can be observed that the majority of the papers (52% equivalent to 22 papers) are articles published in journals. The remaining 48% are divided into "conference paper," "conference review," and "proceedings paper", representing 29%, 12%, and 7% of the total, respectively.

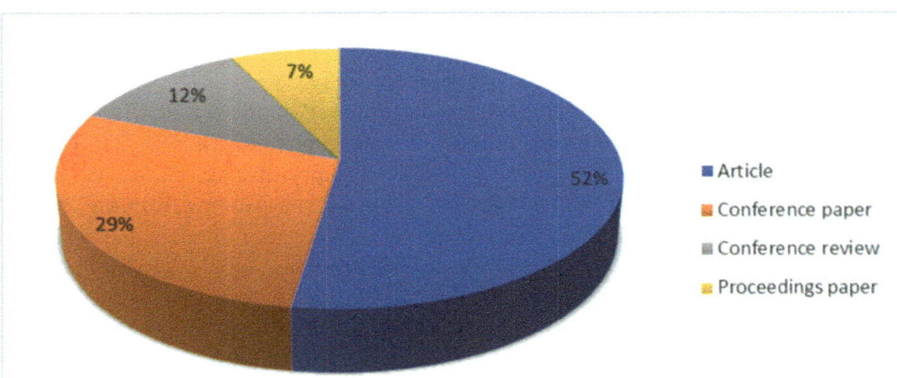

Fig. 1. Papers by type of publication

Among the selected papers, a total of 124 keywords were referenced by the authors. Figure 2 presents the list of the top 10 most referenced keywords, where it can be observed that the three most frequent keywords are keywords of the search criteria used.

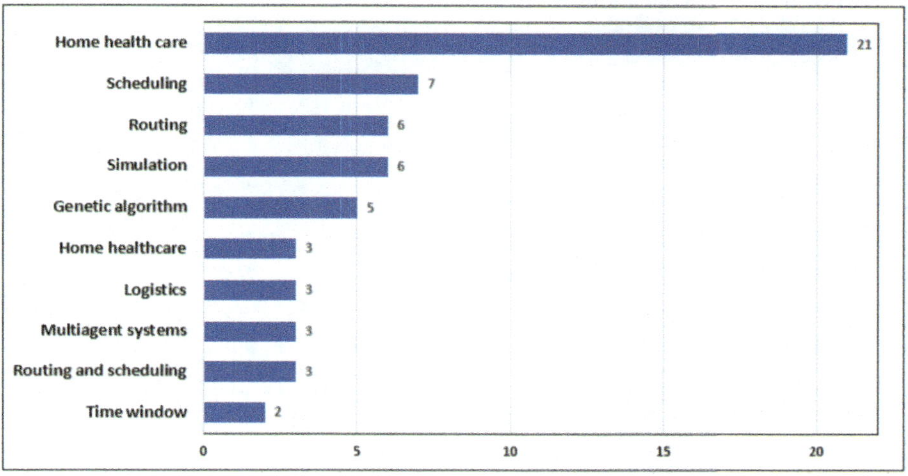

Fig. 2. Most referenced keywords

The string "home health*" was used to locate different possible forms of the term "home health care", resulting in a total of 24 occurrences, including its variations.

The evolution of publications over the defined period from 2010 to 2023 shows significant growth in recent years, as can be seen in Fig. 3.

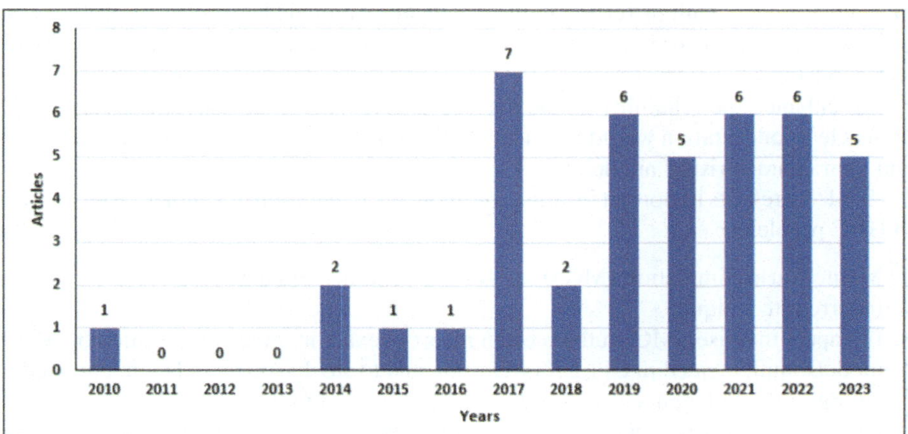

Fig. 3. Annual publication

The productivity of the authors during the period considered can be seen in Fig. 4, where the smaller circles represent just one publication per year, the medium-sized circles are associated with two publications per year, and the larger circles with three publications. Regarding the color variation, the darker the color, the more citations are associated with the publications.

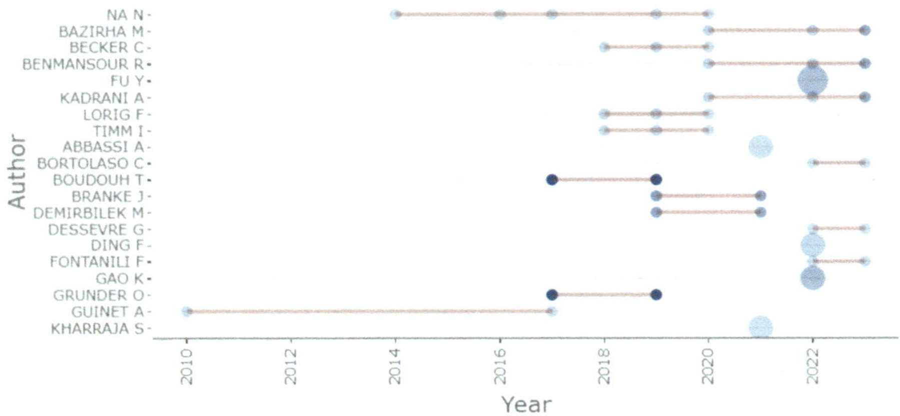

Fig. 4. Authors' production over time

3 Results

In the previous section, metrics related to the selected publications in terms of historical evolution, main authors, and most referenced keywords were presented. With the aim of evaluating publications related to the use of simulation techniques in research on the optimization of HHC processes, a more in-depth content analysis of the selected articles was conducted.

To enhance the reliability of the analyses, it was also established that only works with at least one citation would be selected. As a result, a list of 20 papers was compiled, and their characteristics are described in Table 1.

Table 1 presents important information about the integration of simulation techniques in HHC problems:

a. Monte Carlo Simulation (MCS), used for essentially stochastic models, is the most recurrent technique;
b. In papers that use SMC, authors often resort to exact methods represented by MILP models implemented in CPLEX or Gurobi, and to stochastic models implemented in a programming language such as Java, C + +, or Python;
c. Only 3 papers used specific commercial simulation tools (Simul8, NetLogo, and AnyLogic), with only 1 specific to Discrete Event Simulation (DES), and one paper presented an application developed in a Microsoft Excel spreadsheet (static simulation);

Table 1 Characteristics of the selected papers.

#	Year	References	Citations	Type of simulation	Software	Solver	Algorithms
1	2010	Bachouch et al. [4]	12	Static simulation	Excel	Lingo Solver	Lingo Solver
2	2014	Mankowska et al. [2]	208	Static/stochastic simulation	–	CPLEX, Java	CPLEX, AVNS
3	2015	Bowers et al. [13]	35	Discrete event simulation	Simul8	–	CW
4	2017	Shi et al. [19]	138	Stochastic simulation	–	Gurobi	HGA
5	2017	Guinet et al. [16]	1	Unspecified	–	–	–
6	2017	Marcon et al. [20]	29	Multi-agent system	NetLogo	Lingo Solver	Lingo Solver
7	2017	Xie and Wang [21]	9	Agent-based simulation	AnyLogic	OptQuest	h-a; h-r; h-m
8	2017	Emiliano et al. [14]	23	Review	–	–	–
9	2018	Becker et al. [22]	2	Review	–	–	–
10	2019	Shi et al. [23]	95	Monte Carlo Simulation	–	Gurobi	Gurobi Solver, SA, TS, VNS
11	2019	Demirbilek et al. [17]	37	Scenario Based Approach (SBA)	–	–	DH, CH
12	2020	Bazirha et al. [24]	4	Monte Carlo Simulation	–	CPLEX, C + +	GA
13	2021	Huang et al. [25]	1	Experimental	–	Python	APA, GA, TS
14	2021	Demirbilek et al. [26]	22	Scenario Based Approach (SBA)	–	–	DH
15	2022	Du et al. [27]	3	Monte Carlo Simulation	–	Matlab	TS, SA, GA
16	2022	Fu et al. [28]	2	Monte Carlo Simulation	–	–	GA
17	2022	Ma et al. [29]	6	Monte Carlo Simulation	–	–	MOEAs

(*continued*)

Table 1 (*continued*)

#	Year	References	Citations	Type of simulation	Software	Solver	Algorithms
18	2022	Bazirha et al. [30]	3	Monte Carlo Simulation	–	CPLEX	CPLEX, GA
19	2023	Belhor et al. [31]	2	Unspecified	–	Java IDE Eclipse	LGA-mTSP
20	2023	Bazirha et al. [32]	7	Monte Carlo Simulation	–	CPLEX, C + +	CPLEX, GA, GVNS

Notes: Algorithms: **AVNS**: Adaptive Variable Neighborhood Search; **CW**: Clarke-Wright algorithm; **HGA**: Hybrid GA; **h-a**: updating; **h-r**: rescheduling; **h-m**: merging; **SA**: Simulated Annealing; **TS**: Tabu Search; **VNS**: Variable Neighborhood Search; **DH**: distance heuristic (Euclidian); **CH**: capability heuristic; **GA**: Genetic Algorithm; **APA**: approximation algorithm; **MOEAs**: Multi-objective evolutionary algorithms; **LGA-mTSP**: Learning Genetic Algorithm for mTSP; **GVNS**: General VNS.

d. CPLEX, Java, and C + + are the most commonly used computational tools for implementing solution algorithms, whether they are exact or stochastic;
e. Regarding solution algorithms, there is a prevalence of evolutionary algorithms like the Genetic Algorithm, but various others are used, each with its specific application.

It is possible to notice, from Table 1, that some articles did not specify how the proposed algorithms were implemented and solved (Solver column).

Article 5 does not detail how the simulation was incorporated into the work, making it impossible to identify the type of simulation, the solution algorithm adopted, or how this algorithm was implemented. Obviously, since it is a proposal for the development of a decision support tool (as indicated in Table 2), no commercial simulation software was used.

Finally, two articles (8 and 9) are literature reviews, which were considered because both address aspects of simulation and serve as references in the field.

Table 2 and Table 3 list the selected papers regarding their objectives, the ways simulation techniques are used, and the results achieved.

Table 2 Objectives, application of simulation, and results.

#	Objectives	Referenced simulation technique	Results
1	Balance the workload of the HHC team, ensuring the quality of patient care and inherent activities	Microsoft Excel was used to simulate the scheduling of 10, 15, and 20 tasks assigned to 3 and 5 nurses	An interface generated in a Microsoft Excel spreadsheet was developed for use by managers to define the scheduling of appointments based on the solutions obtained by Lingo

(*continued*)

Table 2 (*continued*)

#	Objectives	Referenced simulation technique	Results
2	Minimize the total daily distance traveled by the staff, and the tardiness of services to avoid unnecessary waiting times for the patients	Random generation of seven datasets to be used in the experimental procedure	For deterministic solutions, CPLEX can be used. However, the proposed heuristic, AVNS, is clearly superior as it managed to solve all simulated instances much more efficiently
3	Define the logistics of midwife allocation and routing, develop an understanding of its logistics implications, and consider team preferences, to explore trade-offs between travel time and care continuity	Simulation experiments: a) Travel & Preferences; b) Travel & Continuity; c) Antenatal clinics; d) Shifts; e) Flexible visits	The results indicate that increases in travel times are necessary but not significant. However, it is not possible to plan allocations and routes perfectly due to shift constraints without the existence of more flexible schedules
4	A model based on HHC (Home Healthcare) routing, in which a vehicle departs from the depot and ends its journey at the laboratory, distinguishing it from the classic VRPTW (Vehicle Routing Problem with Time Windows). The objective is to minimize the total distance	The stochastic simulation method was used to simulate the scheduling indicators while demand changes	After having conducted three series of experiments to test the proposed HGA (Hybrid Genetic Algorithm) approach, the results point out that HGA outperformed all of them
5	A decision support tool is proposed to identify evacuation strategies, resource availability, and the determination of the necessary number of beds for evacuated patients	Simulate a 10-day horizon with the evolution of a hydrological disaster by district for nurses' resources	The proposed tool allowed for managing the hydrological crisis and anticipating the removal of patients before the closure of access points
6	This paper aimed to solve the routing problem in a dynamic context using an MAS-based approach	Simulate caregiver behavior	Four decision rules were developed and tested to help caregivers make decisions according to their level of autonomy and the local context and their performance of them was assessed on a multi-agent platform
7	The paper proposes an algorithm to dynamically adjust the scheduling of home visits, revising it to consider dynamic events occurring within a specific period, to minimize costs for HHC (Home Healthcare) services	To validate the efficiency of the proposal, an agent-based simulation model was developed	The results indicate a 14% reduction in costs with the application of the proposal compared to other approaches

(*continued*)

Table 2 *(continued)*

#	Objectives	Referenced simulation technique	Results
10	The objective is to present a model that considers uncertainties related to travel and service times in HHC problems, minimizing the total travel cost and fixed costs associated with caregivers	MCS is used in an experimental procedure, and the results highlight the importance of considering uncertainty in optimization models	Solutions obtained by the proposed robust optimization model are suitable for effectively obtaining a solution that takes into account uncertainties for all generated scenarios
11	A new heuristic based on various randomly generated scenarios is proposed to maximize the average number of daily visits by a single nurse	Two variants of the Scenario-Based Approach (SBA) are examined, simulating various scenarios from a simple and agile heuristic	The two proposed heuristics showed good results for all simulated scenarios. However, the computational effort is significantly higher for the weekly model compared to the daily model

Note: Due to their nature as literature reviews, articles 8 and 9 were excluded.

Table 3 Objectives, application of simulation, and results (cont.)

#	Objectives	Use of simulation technique	Results
12	The goal is to minimize the transportation costs of caregivers and the additional costs resulting from delays in patient services and caregiver overtime	A GA heuristic, combined with Monte Carlo simulation is used to estimate the expected additional values	Both the deterministic and stochastic approaches produced good results. However, due to the use of simulation to generate the expected additional values in each iteration, the computational effort is much greater in the stochastic approach
13	An approximation algorithm, GA (Genetic Algorithm) and TS (Tabu Search) are used to minimize the total cost of HHC organizations	Twenty instances were simulated in three different set sizes, with 40, 80, and 150 patients, and 5, 10, and 20 caregivers, respectively	The proposed approach (APA), GA, and TS outperformed the results available in the literature in terms of total costs for all simulated scenarios

(continued)

Table 3 (*continued*)

#	Objectives	Use of simulation technique	Results
14	The Distance Heuristic is proposed to make decisions regarding the acceptance of new patients, nurse assignments for these patients, and scheduling their visits in order to maximize the number of patients visited by a group of nurses	The simulation was run for 30 rounds for each experiment. Each experiment covered a working horizon of 360 days, with each working day consisting of 34 teams	The results from different scenarios indicate that the proposed model significantly increased the daily visit average while reducing travel time
15	Minimizing the total costs associated with hiring caregivers, travel, overtime, patient services, and minimizing dissatisfaction among caregivers and patients in the scheduling and routing for a day with stochastic service times	MCS, implemented in MatLab, is used to generate a random set of patient information	To solve the problem, a hybrid SA algorithm, an initial solution generation method and a neighborhood search strategy, and the use of the TS algorithm to improve the algorithm were proposed. MCS was used to evaluate the cost of the uncertain problem. The experimental procedures point out that the total cost was reduced and sensitivity analysis verified the robustness of the proposal
16	A hybrid approach combining the genetic algorithm (GA) and simulation optimization was proposed to minimize the expected total travel time and service time	To reduce the disturbance of random noise, incurred by stochastic travel time and service time, the Monte Carlo simulation approach was implemented to evaluate the performance of searched solutions	The designed approach acquires better solutions for the problem under consideration
17	A multi-objective cooperation evolutionary algorithm with a stochastic simulation method was proposed to minimize the total operation cost and penalty cost incurred by earliness and delay service	The stochastic simulation approach is employed to evaluate the quality and feasibility of the obtained solutions	The results indicate that the proposed model converges to suitable and diverse solutions for the problem under investigation

(*continued*)

Table 3 (*continued*)

#	Objectives	Use of simulation technique	Results
18	The objective function is defined as minimizing the transportation cost and the average number of unvisited patients caused by skipping them if their time windows are violated	The deterministic model is solved using the CPLEX solver and the GA. Monte Carlo simulation is embedded into the GA to solve the stochastic model	The results highlight the efficiency of the GA, show the complexity of the model, and indicate the advantage of using multiple time windows
19	A mixed-integer linear programming model is proposed to minimize the overall service time performed by all caregivers while respecting the workload balancing constraint	The simulation aims to conduct computational experiments to investigate the performance of the proposed solution approach	Computational results show the effectiveness of the LGA-mTSP approach, which is able to minimize service costs. Maximize productivity and balance workload depending on workers' experiences. Simultaneously
20	The objective is to minimize the transportation cost and the expected value of recourse, which is estimated using Monte Carlo simulation	MCS is used to estimate the expected value of recourse caused by patients' delayed services and caregivers' extra working time	GVNS and GA are successfully tested using several instances randomly generated from the literature. The tests prove the high performance of these two heuristics to deal with large instances in a small amount of time

By analyzing the information presented in Tables 2 and 3, we can see the application of simulation to different situations, with significant practical results. However, it's important to note the prevalence of the use of Monte Carlo Simulation, which is usually applied to generate instances for evaluating the proposed optimization models. Furthermore, its application is limited to mathematical implementations for generating random data.

4 Conclusions

This work presented a Systematic Literature Review (SLR) with a focus on the application of simulation techniques in solving Home Health Care problems. Through the analysis of the presented works, it can be observed that there is a relatively small volume of publications, even when considering those without citations, which were excluded from the final list. Among the selected works, the use of Monte Carlo Simulation stands out, primarily applied to improve the quality of input data and evaluate the results of the approaches. These results indicate that the topic may represent an interesting gap in the literature, open to further academic and scientific research, especially when considering other aspects of simulation, such as Discrete Event Simulation and Queueing Theory applied to HHC. Furthermore, exploring existing commercial, free, or paid simulation software can provide valuable insights for future research.

References

1. Belhor, M., El-Amraoui, A., Jemai, A., Delmotte, F.: Multi-objective evolutionary approach based on K-means clustering for home health care routing and scheduling problem. Expert Syst. Appl. **213**, 119035 (2023)
2. Mankowska, D.S., Meisel, F., Bierwirth, C.: The home health care routing and scheduling problem with interdependent services. Health Care Manag. Sci. **17**, 15–30 (2014)
3. Li, Y., Xu, X., Wang, F.: Research on home health care scheduling considering synchronous access of caregivers and vehicles. Sustainability **15**, 6132 (2023)
4. Bachouch, R.B., Liesp, A.G., Hajri-Gabouj, S.: An optimization model for task assignment in home health care. In: IEEE WHCM, Venice, pp. 1-6 (2010)
5. Trautsamwieser, A., Hirsch, P.: Optimization of daily scheduling for home health care services. J. Appl. Oper. Res. **3**(3), 124–136 (2011)
6. Akjiratikarl, C., Yenradee, P., Drake, P.R.: PSO-based algorithm for home care worker scheduling in the UK. Comp. Ind. Eng. **53**(4), 559–583 (2007)
7. Rest, K.D., Hirsch, P.: Daily scheduling of home health care services using time-dependent public transport. Flex. Serv. Manuf. J. **28**, 495–525 (2016)
8. Tsang, M.Y., Shehadeh, K.S.: Stochastic optimization models for a home service routing and appointment scheduling problem with random travel and service times. EJOR **307**(1), 48–63 (2023)
9. Yadav, N., Tanksale, A.: An integrated routing and scheduling problem for home healthcare delivery with limited person-to-person contact. EJOR **303**(3), 1100–1125 (2022)
10. Fathollahi-Fard, A.M., Ahmadi, A., Karimi, B.: Multi-Objective optimization of home healthcare with working-time balancing and care continuity. Sustainability **13**, 12431 (2021)
11. Fathollahi-Fard, A.M., Ahmadi, A., Goodarzian, F., Cheikhrouhou, N.: A bi-objective home healthcare routing and scheduling problem considering patients' satisfaction in a fuzzy environment. Appl. Soft Comput. J. **93**, 106385 (2020)
12. Grieco, L., Utley, M., Crowe, S.: Operational research applied to decisions in home health care: a systematic literature review. J. Oper. Res. Soc. **72**(9), 1960–1991 (2021)
13. Bowers, J.A., Cheyne, H., Mould, G., Page, M.: Continuity of care in community midwifery. Health Care Manag. Sci. **18**, 195–204 (2014)
14. Emiliano, W., Telhada, J., Sameiro Carvalho, M.: Home health care logistics planning: a review and framework. Procedia Manufact. **13**, 948–955 (2017)
15. Jun, J.B., Jacobson, S.H., Swisher, J.R.: Application of discrete-event simulation in health care clinics: a survey. J. Oper. Res. Soc. **50**(2), 109–123 (1999)
16. Guinet, A., Barkaoui, H., Wang, T., Dubost, E.: A mitigation tool to protect a home health care structure facing a hydrological disaster. IFAC-PapersOnLine **50**(1), 4630–4635 (2017)
17. Demirbilek, M., Branke, J., Strauss, A.: Dynamically accepting and scheduling patients for home healthcare. Health Care Manag. Sci. **22**, 140–155 (2019)
18. Aria, M., Cuccurullo, C.: Bibliometrix: an R-tool for comprehensive science mapping analysis. J. Informet. **11**(4), 959–975 (2017)
19. Shi, Y., Boudouh, T., Grunder, O.: A hybrid genetic algorithm for a home health care routing problem with time window and fuzzy demand. Expert Syst. Appl. **72**, 160–176 (2017)

20. Marcon, E., Chaabane, S., Sallez, Y., Bonte, T., Trentesaux, D.: A multi-agent system based on reactive decision rules for solving the caregiver routing problem in home health care. Simul. Model. Pract. Theory **74**, 134–151 (2017)
21. Xie, Z., Wang, C.: A periodic repair algorithm for dynamic scheduling in home health care using agent-based model. In: IEEE 21st Int CSCWD, Wellington, pp. 245–250 (2017)
22. Becker, C.A., Lorig, F., Timm, I.J.: Survey of multiagent systems for improving home health care management. AIH@IJCAI (2018)
23. Shi, Y., Boudouh, T., Grunder, O.: A robust optimization for a home health care routing and scheduling problem with consideration of uncertain travel and service times. Transp. Res. Part E: Logistics Transp. Rev. **128**, 52–95 (2019)
24. Bazirha, M., Kadrani, A., Benmansour, R.: Scheduling Optimization of the Home Health Care Problem with Stochastic Travel and Care Times, 2020 5th Int Conf Log Oper Manag (GOL), pp. 1–8. Rabat, Morocco (2020)
25. Huang, J., Wu, Y., Huang, X., Chen, L., Wu, J.: Efficient algorithms for caregiver routing and scheduling in home health care services. In: 2021 IEEE SmartWorld/SCALCOM/UIC/ATC/IOP/SCI, pp. 139–145 (2021)
26. Demirbilek, M., Branke, J., Strauss, A.K.: Home healthcare routing and scheduling of multiple nurses in a dynamic environment. Flex. Serv. Manuf. J. **33**, 253–280 (2021)
27. Du, G., Zhang, J.: Cross-regional manpower scheduling and routing problem with stochastic service times in home health care. Comp Ind. Eng. **173**, 108668 (2022)
28. Fu, Y., Ding, F., Mu, Z., Sun, C., Gao, K.: Integrating scheduling and routing decisions into home health care operation with skill requirements and uncertainties. J. Simul. **18**(3), 259–282 (2022)
29. Ma, X., Fu, Y., Gao, K., Sadollah, A., Wang, K.: Integration routing and scheduling for multiple home health care centers using a multi-objective cooperation evolutionary algorithm with stochastic simulation. Swarm. Evol. Comp. **75**, 101175 (2022)
30. Bazirha, M., Kadrani, A., Benmansour, R.: Optimization of the stochastic home health care routing and scheduling problem with multiple hard time windows. IJSOM **9**(2), 235–250 (2022)
31. Belhor, M., El-Amraoui, A., Jemai, A., Delmotte, F.: Learning-based metaheuristic approach for home healthcare optimization problem. CSSE **45**(1) (2023)
32. Bazirha, M., Kadrani, A., Benmansour, R.: Stochastic home health care routing and scheduling problem with multiple synchronized services. Ann. Oper. Res. **320**, 573–601 (2023)

Detecting Age Related Macular Degeneration Using Integrated Auto coder and Particle Swarm Optimization

F. Ajesh[1,2](✉) and Ajith Abraham[3,4]

[1] Machine Intelligence Research Labs (MIR Labs) Scientific Network for Innovation and Research Excellence 11, 3rd Street NW P.O. Box 2259, Auburn, Washington 98071, USA
ajeshf@gmail.com

[2] Department of Computer Science and Engineering, Sree Buddha College of Engineering, Alappuzha, Kerala, India

[3] Machine Intelligence Research Labs (MIR Labs), Auburn, WA 98071, USA
ajith.abraham@ieee.org

[4] Center for Artificial Intelligence, Innopolis University, Innopolis, Russia

Abstract. Age-related macular Degeneration (AMD) could become the leading cause of long-lasting vision impairment in individuals aged 50 and above if not properly treated. The paper describes an effective real-time strategy for recognizing and classifying macular problems using deep learning, with the goal of improving AMD care. The procedure entails creating a dataset in conjunction with medical retina specialists, preprocessing photos, extracting critical characteristics, choosing significant features with particle swarm optimization, and detecting bolt loosening using the densenet169 approach. The various steps involved in the process are: a) Generating a dataset requires the collaborative work of a subcommittee operating within The Royal College of Ophthalmologists' Informatics and Audit Sub-committee. This subcommittee is composed of a varied group of medical retina experts hailing from various healthcare institutions across the United Kingdom. b) Preprocessing using image masking, color normalization, illumination normalization, contrast enhancement and identified blood vessels c) Feature extraction of quintessential features from the inputs using an autoencoder, d) selection of features from the extracted version using particle swarm optimization and finally e) detection of bolt loosening using densenet169 technique. According to the results derived from the experimental investigation, the proposed system surpasses existing cutting-edge models across various aspects, including but not restricted to achieving remarkable outcomes in terms of accuracy (0.95), sensitivity (0.97), and specificity (0.97).

Keywords: Age Macular Degeneration · Deep Learning · Autoencoder · Densenet169 · PSO

D. Kriksciuniene et al. (Eds.): IBICA 2023, LNNS 1228, pp. 37–48, 2025.
https://doi.org/10.1007/978-3-031-78937-3_4

1 Introduction

Age-Related Macular Degeneration (AMD) is a condition that affects the central retina and causes vision loss in the elderly. With about 196 million cases in 2020 and a predicted increase to 288 million cases by 2040, it is a developing worldwide concern that poses considerable public health and economic challenges.

Clinical indicators like as drusen and Retinal Pigment Epithelium (RPE) abnormalities contribute in the early detection of AMD. Advanced AMD causes central vision loss, which has a substantial impact on quality of life. AMD prevalence in adults aged 45 to 85 years old worldwide is 8.69% (any type), 8.01% (early), and 0.37% (advanced). At age 49, the chance of early AMD is 22.7%, whereas advanced AMD is 6.8%.Given this rising incidence and the associated risks, there is an urgent need to enhance the efficiency of AMD screening and diagnosis, as depicted in Fig. 1.

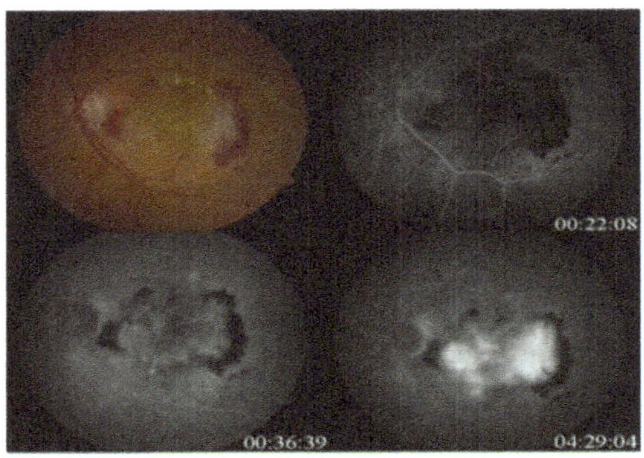

Fig. 1. AMD instances

Deep Learning (DL), a subfield of machine learning, relies on Artificial Neural Networks (ANNs) as its core architecture. ANNs consist of input, hidden, and output layers, with hidden layers learning features from stimuli to handle complex tasks. Convolutional Neural Networks (CNNs) stand out as particularly effective tools for the analysis of image data in the field of deep learning structures. Due to significant developments in recent years, the diagnosis of Age-Related Macular Degeneration (AMD), a common cause of decreasing vision in older adults, has become increasingly challenging. Nevertheless, there remain various obstacles in current approaches to AMD detection that require resolution to enhance the precision and effectiveness of diagnosing and managing this ailment [10].

- Limited Data Availability: Due to privacy issues, AMD identification is hampered by a lack of varied, expert-annotated datasets.
- Interpretable AI Models: While promising, deep learning models lack transparency, making it challenging for clinicians to trust and comprehend their diagnosis.
- High False Positive Rates: Current methods frequently yield significant false positive rates, resulting in needless tests and patient concern.

- Variability in Imaging Various imaging techniques are used for AMD detection, but integrating these data sources remains challenging.
- Longitudinal Tracking: Monitoring AMD development over time is critical, and present technologies may not be capable of doing so.
- Resource Constraints: Deploying sophisticated AMD detection systems may necessitate large computing resources, raising difficulties in resource-constrained environments.
- Patient Acceptance: When AI is integrated into healthcare, patients may have privacy and accuracy concerns, necessitating efforts to earn their trust.

Through the analysis of clinical images such as fundus photos, Fundus Fluorescence Angiography, and Optical Coherence Tomography, Artificial Intelligence has shown promise in diagnosing ocular diseases such as Diabetic Retinopathy, Age-Related Macular Degeneration, Retinopathy of Prematurity, glaucoma, and papilledema.

1.1 Key Highlights

The main objective of this study was to develop an incredibly efficient DL method for the early identification of AMD. To achieve this, the following objectives were pursued:

- Development of AMD classification and detection using densenet 169
- Improving the classifier performance by usage of additional stages with advanced techniques than making it a standalone system
- With optimized deep learning parameters, the model has been trained and tested using real-time AMD images.
- The usage of autoencoder and PSO gives additional angles for the low-complex classifier.
- The results of the experiments show that the proposed system performs better than current state-of-the-art models.

2 Literature Review

In 2023, research by Holomcik et al. [11], achieved noteworthy results with F1 scores of 0.73 for accurately delineating CNV lesion size and 0.65 for identifying leakage area. In both datasets, automated segmentations outperformed human experts, with equivalent proficiency in CNV lesion size and leaking area segmentation. Positive feedback was received from clinical feasibility testing, indicating the potential for automated deep learning approaches to improve precision in objective biomarker analysis for clinical usage.

Wei et al. (2023) [12] proposed a deep learning strategy aiming at automating the identification and quantification of six crucial clinical indicators connected to Macular Atrophy (MA) in patients with wet AMD. In all datasets, expert review favored automated segmentations, showing the model's competency in CNV lesion and leaking region segmentation. According to positive clinical reviews, automated deep learning improves objective biomarker analysis in FA pictures for therapeutic usage. Farag et al. (2022) [13] investigated the extensive use of Optical Coherence Tomography (OCT) as a non-invasive approach to diagnose and manage AMD. Their research revolves around an innovative approach that employs Deep Learning (DL) techniques, which involve intricate Artificial Neural Networks (NN) for the extraction of pertinent features.

Chen et al. (2021) [14] A specific deep learning technique for identifying Choroidal Neovascularization (CNV) in Optical Coherence Tomography Angiography (OCTA) pictures of nAMD patients was created. The model included a ResNeSt module that used group convolution and split-attention approaches to improve feature extraction. A spatial pyramid pooling module was created to accommodate CNVs of varied sizes, allowing the model to incorporate contextual information at several scales. Trial results on a clinical OCTA dataset of 116 pictures revealed remarkable performance, with an AUC of 0.9476 and high specificity (0.9950) and sensitivity (0.7271). Finally, in OCTA pictures of nAMD patients, this model efficiently separates CNV patches from the background.

3 Methodology

Figure 2 illustrates the proposed system's structure, with comprehensive explanations for each stage presented in subsequent sections.

3.1 Proposed System

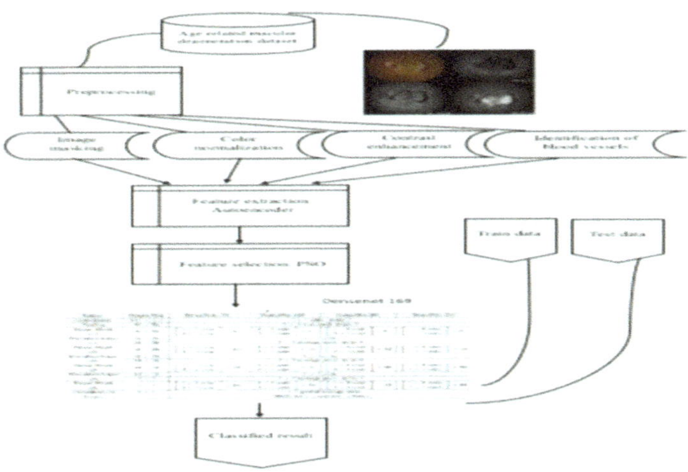

Fig. 2. Proposed architecture overall outlook.

3.2 Data Collection

This study introduces a freshly assembled dataset specifically tailored for investigating Age-Related Macular Degeneration (AMD). The dataset was thoughtfully compiled by a committed team from The Royal College of Ophthalmologists' Informatics and Audit Sub-committee, which comprises a diverse array of medical retina specialists practicing across various healthcare institutions throughout the United Kingdom. By collecting and utilizing this dataset, researchers can facilitate comparisons of outcomes across different healthcare sites. One promising option to consider is the utilization of the International Consortium for Health Outcomes Measurement (ICHOM) Standard Set for Macular Degeneration for the assessment and examination of these discoveries.

Table 1. Dataset description and its datatypes

Item	Description	Datatypes
Patient ID	A unique identifier, such as the NHS number or a local patient identifier, is used to distinguish and identify patients. In anonymized datasets, this identifying information is excluded or omitted	INT & STRING
Age	The patient's age, measured in years, can be determined either at the time of their presentation, an event, or treatment. Since the age is subject to change, it will be calculated based on their date of birth	DATE
Sex	The patient's gender	ENUM
Consultant	Identification for the consulting professional responsible for a patient's care, enabling individual audit capability, through their GMC Number	INT
Ethnic category	The patient's ethnic background according to the categorization employed in the 2011 census	ENUM
Smoke status		ENUM
Route of referral	The method through which the patient reaches the ophthalmic department is determined by their source or origin	ENUM
Date of initial referral		DATE

3.3 Preprocessing

The preprocessing phase aimed to improve the classification system's efficiency by initially enhancing the quality of the images. The subsequent procedures were implemented as fol-lows: The concept of a "mask image" (referred to as IBackground) was initially introduced following Harr's proposal. This was accomplished by applying morphological and intensity thresholding techniques to the original image I, as shown in Fig. 3A. As seen in Fig. 3B, pixels within the fundus' circular region of interest were given the value "1," while those outside of it were given the value "0."

Using a histogram specification technique, the original image, denoted as "I" in Fig. 3C, was employed to create a fresh image named "Icolor." This process aimed to standardize the color distribution in the original image. The methodology suggested by Foracchia et al. in their study from 2018 was subsequently employed on Icolor to mitigate the effects of illumination variations, leading to the generation of Illumination as depicted in Fig. 3D.

A fresh representation, denoted as Fig. 3E, was produced using the Contrast Limited Adaptive Histogram Equalization (CLAHE) technique to improve the contrast in the illumination image (Illumination). This choice was made due to its proven effectiveness compared to similar methods, as illustrated in Fig. 4.

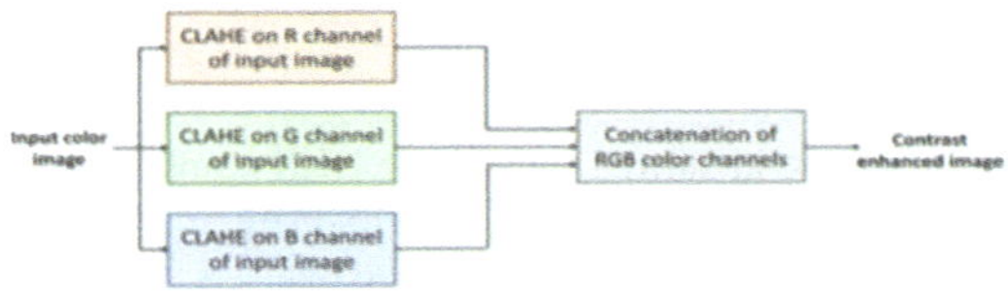

Fig. 3. CLAHE steps

In the image analysis, we utilized wavelet features along with a supervised classification method to detect blood vessels. However, we excluded the pixels related to blood vessels in Ivessel (Fig. 3F) and those designated as "0" (black) in Ibackground (Fig. 3B) from any further analysis. It's worth noting that in this study, we intentionally chose not to focus on locating and eliminating the optic disc. We made our decision relying on our previous experience, which showed that this specific step didn't result in any noticeable enhancement in classification performance.

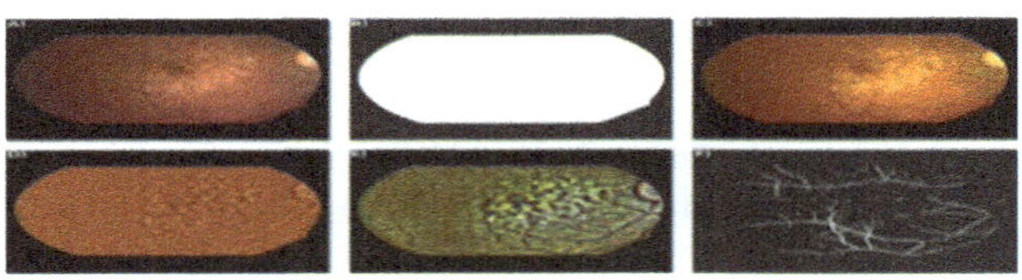

Fig. 4. Demonstration of preprocessing stages: (A) initial image; (B) image masking; (C) image following color standardization; (D) image after illumination equalization; (E) image with enhanced contrast; and (F) the detected blood vessels.

3.4 Feature Extraction: AE

Autoencoders (AEs) are unsupervised learning algorithms that capture data patterns. The AE's encoder captures complicated information from vibration signal data using Long Short-Term Memory (LSTM) units. The decoder then employs these attributes to reassemble the original signal. AEs, on the other hand, focus exclusively on learning regular patterns during training, making them less successful at recognizing substantial abnormalities. The network is made up of an encoder and a decoder, both of which use LSTM units, which excel in incorporating time-dependent information due to their hidden state vector's memory capacity. A) Encoder: We analyze the input data sequence labelled as X, which can be described as (x(1), x(2),..., x(N)), with each x(N) representing an element in the RT^d space. This characterization indicates that, for each index N, there exists a time-series sequence with d dimensions and T time steps. To address the dynamic nature of the input sequence and capture sequential patterns within the time-series data, we utilize an RNN. This process can be described as follows for each sequence.

$$C_t' = tanh(W_C.x_t + R_C h_{t-1} + b_C) \tag{1}$$

Ct0 is the memory state in this case, xt is the input vector, and ht1 is the out-put vector from the previous phase. The input weights, recurrent weights, and bias are designated by the letters WC, RC, and bC. A Tanh activation function is used to generate output values between [–1, 1].

The input sequences are subjected to processing in the encoder section of the LSTM network. The result of this encoding process yields the output of the encoder as follows:

$$h_t = \sigma_{\emptyset}^{e}(x_t, h_{t-1}) \tag{2}$$

The ith LSTM-encoder's output, denoted as ht, is affected by the encoder's parameter set φ. To tackle the issue of vanishing and exploding gradients, the ReLU activation function, denoted as σ, is selected for both the encoder and decoder.

b) Decoder: The vibration signal is first processed by the encoder to transform it into lower-dimensional embeddings. These embeddings are subsequently employed by the decoder to reconstruct the original signal.

$$h_t' = \sigma_{\varphi}^{d}\left(h_t, h_{t-1}'\right); \; x_t' = \sigma\left(h_t'\right) \tag{3}$$

In the context of this explanation, we define ϕ as the parameter set for the de-coder, and x0 represents the symbol denoting the reconstructed input used in computing the Reconstruction Error (RE), commonly referred to as Mean Squared Error (MSE). The fundamental properties of the Autoencoder (AE) model are accurately recreating the original signal and withstanding the effects of training data and noise, guaranteeing resilience in capturing a varied data representation. As a result, every divergence from the norm during the prediction phase increases Reconstruction Error (RE).

3.5 Feature Selection: PSO

PSO operates by employing a group of particles that evolve over successive iterations. In pursuit of the best possible solution, each particle adjusts its movement based on its own previously achieved best position (pbest) and the overall best position within the entire swarm (gbest).

$$pbest(i, t) = \underset{k=1,\ldots,t}{\arg\min} \left[f(P_i(k))\right], \; i \in \{1, 2, \ldots, N_p\}, \tag{4}$$

$$gbest(t) = \underset{\substack{i=1,\ldots,N_p \\ k=1,\ldots,t}}{\arg\min} \left[f(P_i(k))\right], \tag{5}$$

"i" serves as the particle index, "Np" is indicative of the total particle count, "t" designates the current iteration count, "f" symbolizes the fitness function, while "V" and "p" respectively denote particle velocity and position.

$$V_i(t+1) = \omega V_i(t) + c_1 r_1 (pbest(i,t) - P_i(t)) + c_2 r_2 (gbest(t) - P_i(t)), \quad (6)$$

$$P_i(t+1) = P_i(t) + V_i(t+1) \quad (7)$$

In this context, V is velocity, and w is the inertia weight for global vs. local exploration. r1 and r2 are random variables, while c1 and c2 are constants used to manage particle velocities. Techniques like velocity clamping and the constriction coefficient method are employed to keep particles within search space bounds during optimization. Equation (2) consists of the inertia, cognitive, and cooperation elements, guiding particles in their search for the best solution.

3.6 Classifier SVM

The model leverages the DenseNet169 architecture, which is noted for its thick interconnections between layers and has a significantly reduced amount of trainable parameters. This decision is motivated by the need to improve the depth of deep Convolutional Neural Networks (CNNs) while minimizing the danger of information loss that can occur with long connection pathways between input and output layers. DenseNet169 accomplishes parameter reduction by removing duplicate feature maps, leading in fewer filters.

The DenseNet169 architecture includes convolutional, max-pooling, dense, and transition layers, as well as two activation functions: ReLU and SoftMax. Convolutional layers are essential in the processing of X-ray pictures because they provide feature maps that characterize the retrieved feature intensities. When a L N picture is processed using a m m filter, it produces a feature map with dimensions (L–m + 1) (N–m + 1). In DenseNet169, max-pooling layers are employed to reduce feature map dimensions (nh, nw, nc) by applying pooling filters and aggregating features inside specific areas, thus lowering feature map dimensions (nh, nw, nc).

$$MaxPool = \frac{n_c{*}(n_h - f + 1){*}(n_w - f + 1)}{s^2} \quad (8)$$

The dense layer in the DenseNet169 architecture is made up of neurons that receive input from every node in the preceding layer. This input interaction makes use of matrix-vector multiplication. We use the notations h for height, w for width, c for feature map channel, and f for filter size in this procedure. To offer further information, we assume M is a matrix with dimensions x y, p is a vector with a size of (1 y), and the parameter matrix from the preceding layer was obtained by the back-propagation method.

Hence, you can compute the weights (ϕly) and biases (ηly) linked to the layer by using the following method:

$$\varphi^{l_y} = \varphi^{l_y} - \alpha{*}d\varphi^{l_y} \quad (9)$$

$$\eta^{l_y} = \eta^{l_y} - \alpha{*}d\eta^{l_y} \quad (10)$$

The symbols dϕly and dηly are indicative of the partial derivatives of the loss function concerning ϕ and η. Furthermore, the transition layer plays a role in simplifying the model by employing 1×1 convolutions to reduce the number of channels. DenseNet169's tiered architecture is thoroughly de-scribed in Table 1, providing insights into each layer, including data like as kernel size, tensor size, and parameter consumption. A closer look at the table reveals that both Relu and SoftMax serve as activation functions. Furthermore, a stride value of 2 is applied uniformly throughout all convolutional, pooling, and transition layers, defining the pixel shift across the input matrix. In addition, a dropout rate of 0.2 is used consistently for all thick layers to minimize overfitting and reduce variation.

4 Performance Analysis

The proposed approach used specific hardware components, including a GTX NV graphics card and a 1 TB HDD, running on the Windows 10 operating system. The software setup included Python, a machine learning library, and Google Colab for model development. The experiments included a comprehensive evaluation with metrics like accuracy, sensitivity, specificity, precision, recall, F1-score, detection rate, True Positive Rate (TPR), and False Positive Rate (FPR). To address overfitting and data imbalance in the limited dataset, testing was conducted with various ADM images. Table 2 provides a side-by-side comparison between the proposed system and other cutting-edge models, designated as 'L.' The proposed system, known as AMDAED, demonstrated superior performance attributed to the classifier's training and testing, along with improvements in other phases.

Table 2. Comparison analysis of accuracy, sensitivity, specificity

Models	Accuracy	Sensitivity	Specificity
L1	76	85	88
L2	84	87	89
L3	86	81	86
L4	83	79	80
L5	88	92	89
SVM-AE	96	97	97

Table 3 shows a comparative analysis of precision-recall f1-score of various models in which the proposed system outperforms due to non-training of the model as a standalone to improve its knowledge with complex classifiers i.e., Densenet 169. By using of autoencoder with its convolutional approach and improving selection ability with PSO, the efficacy of the classifier increased than expected (Fig. 4).

An analysis comparing the proposed system's performance in terms of detection rate, True Positive Rate (TPR), and False Positive Rate (FPR) is presented. The proposed

Table 3. Comparison analysis of precision, recall, f1-score

Models	Precision	Recall	F1-score
L1	72	77	83
L2	76	82	82
L3	83	84	85
L4	85	84	86
L5	81	86	89
AMDAED	86	90	92

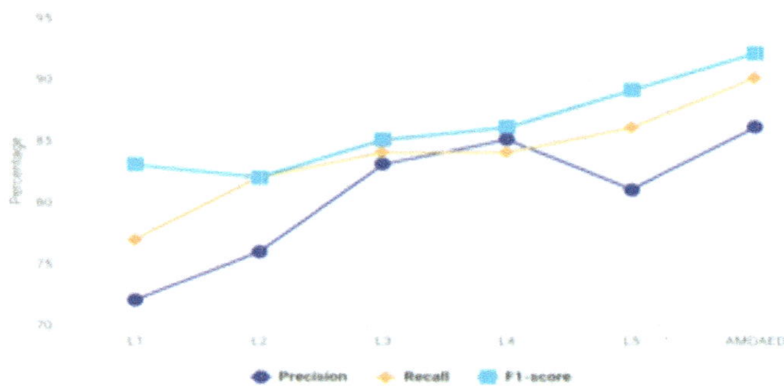

Fig. 5. Models' vs Precision, Recall, F1-score.

system outperforms by overcoming the issue of overfitting and small data imbalance. The system was able to train under greater parameters and also with quality processed inputs (Fig. 5).

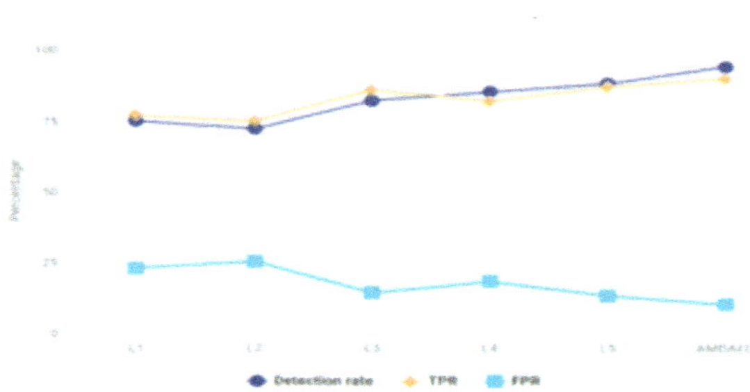

Fig. 6. Models' vs Detection rate, TPR, FPR

5 Conclusion

Employing advanced techniques such as employing DenseNet-169 for the classification of AMD, utilizing Autoencoder for feature extraction, and incorporating Particle Swarm Optimization (PSO) for feature selection has yielded impressive results in AMD classification, attaining an outstanding accuracy rate of 96%. This amalgamation of cutting-edge deep learning architecture, feature extraction, and optimization algorithms has showcased the potential to significantly elevate the diagnosis and comprehension of AMD. These discoveries offer hope for enhancing the early detection and treatment of this debilitating eye condition, ultimately contributing to improved patient care and outcomes in the realm of ophthalmology. In future, the use of transfer learning and also hybrid mode of integration with existing complex architecture will show another dimension of performance in the classification of AMD (Fig. 6).

References

1. Yim, J., et al.: Predicting conversion to wet age-related macular degeneration using deep learning. Nat. Med. **26**(6), 892–899 (2020)
2. Burlina, P., Freund, D.E., Joshi, N., Wolfson, Y., Bressler, N.M.: Detection of age-related macular degeneration via deep learning. In: 2016 IEEE 13th International Symposium on Biomedical Imaging (ISBI), pp. 184–188. IEEE (2016)
3. Peng, Y., et al.: Predicting risk of late age-related macular degeneration using deep learning. NPJ Digit. Med. **3**(1), 111 (2020)
4. Moraes, G., et al.: Quantitative analysis of OCT for neovascular age-related macular degeneration using deep learning. Ophthalmology **128**(5), 693–705 (2021)
5. Motozawa, N., et al.: Optical coherence tomography-based deep-learning models for classifying normal and age-related macular degeneration and exudative and non-exudative age-related macular degeneration changes. Ophthalmol. Ther. **8**, 527–539 (2019)
6. Kaymak, S., Serener, A.: Automated age-related macular degeneration and diabetic macular edema detection on oct images using deep learning. In: 2018 IEEE 14th International Conference on Intelligent Computer Communication and Processing (ICCP), pp. 265–269. IEEE (2018)
7. Fu, D.J., et al.: Predicting incremental and future visual change in neovascular age-related macular degeneration using deep learning. Ophthalmol. Retina **5**(11), 1074–1084 (2021)
8. Liefers, B., et al.: Quantification of key retinal features in early and late age-related macular degeneration using deep learning. Am. J. Ophthalmol. **226**, 1–12 (2021)
9. Kadry, S., Rajinikanth, V., González Crespo, R., Verdú, E.: Automated detection of age-related macular degeneration using a pre-trained deep-learning scheme. J. Supercomput. 1–20 (2022)
10. Banerjee, I., et al.: Prediction of age-related macular degeneration disease using a sequential deep learning approach on longitudinal SD-OCT imaging biomarkers. Sci. Rep. **10**(1), 15434 (2020)
11. Holomcik, D., et al.: Segmentation of macular neovascularization and leakage in fluorescein angiography images in neovascular age-related macular degeneration using deep learning. Eye **37**(7), 1439–1444 (2023)
12. Whitney, J., et al.: Multi-model integration for enhancing deep learning retinal layer segmentation performance in age-related macular degeneration with geographic atrophy on optical coherence tomography. Invest. Ophthalmol. Vis. Sci. **64**(8), 1127 (2023)

13. Farag, M.M., Fouad, M., Abdel-Hamid, A.T.: Automatic severity classification of diabetic retinopathy based on densenet and convolutional block attention module. IEEE Access **10**, 38299–38308 (2022)
14. Chen, Y.M., Huang, W.T., Ho, W.H., Tsai, J.T.: Classification of age-related macular degeneration using convolutional-neural-network-based transfer learning. BMC Bioinform. **22**(5), 1–16 (2021)

A Review of Machine Learning Techniques for Epileptic Seizure Prediction

Akshita Modi, Anu Bajaj(✉), and Vikas Sharma

Thapar Institute of Engineering and Technology, Patiala, India
er.anubajaj@gmail.com

Abstract. More than 50 million people worldwide are affected by the severe brain disorder known as epilepsy. It is a long-term disorder that causes unprovoked seizures repeatedly. It leaves many epileptics resistant to medicines and are not good candidates for surgery. To prevent seizures using medication, seizure prediction before onset is beneficial. It can be a challenging task to predict epileptic seizures. Without compromising on performance, deep learning, and machine learning classifiers can identify EEG data, predict seizures, and reveal sensible patterns. This paper aims to present an overview of different deep learning and machine learning techniques that have been explored over the past few years for epileptic seizure prediction.

Keywords: Epilepsy · Epileptic seizure · Prediction · Machine Learning · Deep Learning

1 Introduction

Epilepsy is a medical disorder characterized by sudden abnormal bursts of brain cells. It affects around 1% of people worldwide which makes it one of the most important challenges in neurological science. One major concern for researchers has been epileptic seizure prediction [2]. Early prediction of epileptic seizures has the potential to warn patients about approaching seizures so that they can get medication rapidly and avoid any potential injuries by taking protective measures in advance. It also paves the way for personalized treatment of epileptic seizures. Imminent seizures can be stopped using systems with seizure intervention [5]. A crucial area of research involving seizure prediction uses EEG (electroencephalography) signal-based feature extraction and classification techniques which involves using small metal discs to measure the electric activities of the scalp [1]. EEG demonstrates the brain's electrical activity using electrodes [2].

Due to the limited scope of variation and analysis in morphology related to interictal spikes, various neurophysiologists have different views regarding the same EEG recording. Additionally, some artifacts in data such as background noise and eye blinking make it difficult to distinguish other patterns from seizure patterns in the EEG signal [3]. Visual assessment of patients' EEG records has its own drawbacks which include dependence on several experts or being time-consuming due to which it is no longer recommended to use [10, 11]. Furthermore, using a computer for the detection of epileptic seizures

D. Kriksciuniene et al. (Eds.): IBICA 2023, LNNS 1228, pp. 49–60, 2025.
https://doi.org/10.1007/978-3-031-78937-3_5

is quite a challenge, and doing so manually makes it nearly impossible [2]. If not kept under check, seizures can also abrupt trivial activities of a seizure patient's life, and prolonged and uncontrolled seizures may cause irrevocable brain damage [3].

The challenges faced in seizure prediction using EEG-based approaches have led to the transition from EEG-based approaches to machine-learning and deep-learning methodologies. Machine learning and deep learning algorithms have emerged as powerful tools for addressing these challenges. Machine Learning includes a range of techniques and algorithms to create models that can generate predictions or decisions by analyzing data learned on the other hand, deep learning is a branch of machine learning focusing on multilayer neural networks. These computational approaches excel in feature extraction, handling of vast datasets, and pattern recognition. These learning models can be trained to provide real-time predictions enabling proactive interventions and improved seizure management.

The goal of this paper is to present a comprehensive analysis of the transition from EEG-based methodologies to Machine Learning and Deep Learning techniques, such as CNN (Convolutional Neural Networks), SVM (Support vector Machine), transfer learning models, Random Forest (RF), Neural architecture search, Support Vector Machine (SVM) with K fold cross-validation in the context of epileptic seizure prediction. The study contributes to throwing some light on underlying trends of Machine Learning and Deep Learning methods in the healthcare field.

Table 1. briefs about the various machine learning and deep learning methodologies used by researchers in the field of epileptic prediction.

Table 1. Machine Learning algorithms [29]

Models	Methods	Description
Baseline Models	SVM (support vector machine)	It is a kind of supervised learning algorithm that can be applied to non-linear or linear regression classification and outlier detection tasks whose main goal is to identify an optimal hyperplane for classifying data or to make continuous output predictions
	RF (random forest)	An Ensemble classifier can be used to form a single solution by combining the output of more than one decision tree
	KNN (k-Nearest Neighbors)	It is one of the most basic classification algorithms in ML which locates the nearest 'k' data points to a given input using majority class to make a prediction

(*continued*)

Table 1. (*continued*)

Models	Methods	Description
	NB (naive bayes)	Supervised learning algorithm used for classification which is based on the Bayes' theorem. If given the class, it makes the assumption that features are conditionally independent of each other
	DT (decision tree)	Non-linear model which uses feature thresholds for decision-making. It is structured like a tree where predicted outcomes are represented by leaf nodes and specific decision-based features are represented by internal nodes
Deep Learning	CNNs (convolutional neural networks)	It is a deep learning algorithm that is capable of learning from patterns of data automatically. It is suited for tasks namely, image generation and classification
	LSTM (long short-term memory)	Enhanced RNN (Recurrent Neural Network) architecture which involves learning in sequential data and is also capable of modeling long-range dependencies in data

We formulated the following research questions to assess the potential of deep learning and machine learning algorithms in solving the problem:

- **RQ1**: What deep learning and machine learning models are employed to predict epileptic seizures?
- **RQ2:** What are the datasets used for validation?
- **RQ3:** Which performance metrics are used for evaluation?
- **RQ4:** What are possible paths for applying machine learning approaches for epileptic seizure prediction in the future?

The structure of this paper is as follows: sect. 2 summarizes existing literature, sect. 3 gives observations and future findings with the conclusion in sect. 4.

2 Literature Review

There are possibilities for early prediction and detection of epileptic seizures using deep learning and machine learning models. In this literature review, we examine various methodologies and findings aiming to provide a comprehensive perspective on the evolving dynamics of this field and enable a deeper understanding of progress made and the challenges that lie ahead.

2.1 Machine Learning Models

Machine learning methodologies have emerged as a revolutionary force in the domain of epileptic seizure prediction. In this section, we dig into various core methodologies, algorithms, and significant studies that employ machine learning to enhance the accuracy and efficiency of seizure prediction.

S. Ra et al. [9] employed (Genetic Algorithm) GA(KNN-GA) combined with k-NN (k Nearest Neighbors) which was used to differentiate PE (Permutation Entropy) values and aimed to select the most informative channels for efficient analysis and seizure prediction using the scalp EEG database from CHB-MIT. They used 3-fold approach: calculated PE values and sampled data, selected channels using KNN-GA, and employed the machine learning algorithm SVM (Support Vector Machine) for modelling.

Meanwhile, Dominique L. Tanner et al. [15] demonstrated that creating individualized treatment plans is essential for epilepsy patients. They utilized statistical modeling techniques and individual seizure diary data, intended to expand their research by highlighting the importance of personalized medicine and establishing more reliable healthcare regimes based on epilepsy. Maurice Ntahobari et al. [17] implemented an extra tree classifier, combined with SHAP (Shapely Additive exPlanations) algorithm for feature selection. This approach helped to maintain the performance of the model's AUC (Area Under Curve) while drastically cutting down prediction and training time which made the model more appropriate to implement it in wearable or small implantable devices to extend battery life and enhance the applicability of the model in a clinical environment. The findings of their study revealed that on the basis of extra tree classifiers, the most essential EEG characteristic features for the prediction of epileptic seizure were maximum frequency and zero crossing value.

Salvador et al. [10] presented an approach to classify interictal and 20 min of preictal data using different channel topologies and employed EEG peripheral channels to predict epileptic seizures. This approach allowed adapting real-time wearable devices and the use of fewer EEG channels rather than entire scalp data. B. R. et al. [18] employed XGBoost (Extreme Gradient Boosting) for epileptic seizure detection on a real-time data set taken from an EEG sensor, demonstrating impressive results.

To overcome the challenges of class imbalance, Cai Chen et al. [19] used the CHB-MIT dataset to propose a feature-enhancing strategy that combined hyperparameter tuning with an automatic machine learning method. They combined CTGAN (Conditional Tabular Generative Adversarial Network), KNN, MLP (multilayer perceptron), LR (Logistic regression), and NB (Naive Bayes) and obtained 0.97, 0.94, 0.87, and 0.95 of average precision. This combination outperformed traditional machine learning algorithms. Meanwhile, in a separate approach, an integrated machine learning approach was presented by Mohammad Mausam et al. [24] for the detection of epilepsy that can learn from unbalanced data efficiently. To extract both empirically customized high and low-variant principal components, PCA (Principal Component Analysis) was used at the first stage in this approach for imbalanced data, and principal components associated with low variances were hypothesized to enhance model performance by capturing the implicit pattern of dataset's minor class.

Meanwhile, Carolina Rebello et al. [20] employed SVM using CHB-MIT dataset for prediction of seizure, resulting in the best outcome for seizure prediction. In a parallel approach, I. Bhattacherjee [23] combined SVM with K fold cross-validation for epileptic seizure prediction on real-time EEG signals to achieve 85.9% accuracy with a 60-min prediction window before the onset of Seizure on real-time epileptic seizure patients' data.

AbdElminaam et al. [11] brought together three ML algorithms namely, DT (Decision Tree), KNN, and RF (Random Forest), and one DL algorithm namely CNN (Convolutional Neural Network). This approach prevented Machine Learning models from inclining towards the majority class of the dataset which resulted in accuracy varying in the range of 94% and 96%. Meanwhile in a separate approach, Purnima P S [13] combined classic machine learning models namely, RF, gradient boosting, and SVM models with classic deep learning models namely, LSTM (Long Short Term Memory) and Convolutional neural network and deep learning models based on graph namely, convolutional and LSTM neural networks on a dataset comprising 18 channel EEG recordings obtained at Helsinki University Central Hospital, Finland for epileptic seizure prediction.

As Researchers continue their hunt for the ideal ML model for the prediction of epileptic seizures, some have also engaged in comparative analysis to find the best model for seizure prediction. Farooq et al. [25] provided an overview of the selected papers in the field of epileptic seizure detection. The main focus was on the feature extraction techniques for epileptic seizure prediction which used wavelet transform and signal decomposition. The major classifiers studied were ANN (Artificial neural networks), SVM, KNN, and RF. Meanwhile, Seifi et al. [21] provided a comparison of the Naïve Bayes classifier with SVM and KNN for seizure prediction and demonstrated Naïve Bayes classifier outperformed the other two classifiers as it ignores data noise and considers that the contribution of every feature equal to the making of a final decision.

Table 2. describes the different Machine learning techniques for the prediction of epileptic seizures.

Table 2. Machine learning methodologies for prediction of epileptic seizure.

Author (Year)	Dataset	Model Used	Model Compared	Performance Metrics
Sook Ra et al. (2021)	CHB MIT Scalp EEG Database	KNN combined with SVM	Using all channels	Accuracy, Sensitivity, and Specificity rates
Salvador et al. (2023)	Dataset from CHB-MIT	Machine learning algorithm		Accuracy Specificity Sensitivity
Abdelminaam et al. (2023)	EEG data	KNN, Decision Control, Random Forest, CNN	Artificial Intelligence and Machine Learning	Accuracy

(*continued*)

Table 2. *(continued)*

Author (Year)	Dataset	Model Used	Model Compared	Performance Metrics
Purnima P S et al. (2023)	EEG recordings obtained at Children Hospital's NICU, Helsinki University Central Hospital, Finland	Classic ML models namely, gradient boosting, SVM and Random forest classic deep learning model LTSM	CNN, ANN, SVM	Sensitivity Accuracy ROC _ AUC
Dominique L. Tanner et al. (2023)	Patient specific morning and evening seizure diaries	Logistic regression		Accuracy
Ntahobari et al. (2022)	A tiny contest dataset and a considerably larger dataset	Extra trees classifier		$p < 0.004$
B.R. et al. (2019)	Data set in real time extracted from EEG sensor	XGBoost		Accuracy
Cai Chen et al. (2022)	CHB MIT EEG dataset	Automatic machine learning (Auto ML) combined with CTGAN	Combination of KNN, CTGAN, RF, DT	Accuracy, Precision
I. Bhattacherjee (2021)	EEG Scalp dataset in real time	K-fold cross validation with SVM (Support vector machine)		Accuracy
Masum et al. (2020)	Epileptic Seizure Recognition dataset	Integrated ML and PCA techniques	Classical ML approaches like SVM, LR, RF, DT	F-beta score
Bruna Carolina Rebello et al. (2022)	CHB MIT	Supported Vector Machine (SVM)		Accuracy

2.2 Deep Learning Algorithms

Deep learning techniques have emerged as a groundbreaking technique for the prediction of epileptic seizures. In this section, we dig into various core methodologies and neural network architectures that utilize the capabilities of deep learning to advance efficiency and precision of epileptic seizure prediction.

Sook Ra et al. [1] took unprocessed EEG signals and transformed them using the STFT (Short-time Fourier transform) and SET (Singular Spectrum Transformation) and SVD (Singular Value Decomposition) was used to reduce the dimensionality of the results acquired by the SET which were further classified using different neural networks namely, 1D-CNN(MLP and 1D-CNN (Convolutional Neural Network), and then their classification performances were compared. Meanwhile in a separate approach. A new feature extraction technique was proposed by Feizbaksh et al. [2] which clustered EEG signals in phase space and treated density of data points within each cluster as a feature and utilized CNN and random forest classifiers to identify different brain states of epilepsy patients. Meanwhile, to overcome the challenge of manual extraction of selected features separately and classifying them subsequently, Eberlin et al. [4] directly applied CNN to the multichannel EEG time series where suitable data representation and classification models were derived from data directly.

S. Hussain [3] used localization for epileptic seizure prediction. This method involved Multivariate Empirical Mode Decomposition for localization. Variations from inter-ictal to preictal periods were observed. Lowest number of channels was chosen on the basis of the source localization results and features were extracted using Empirical Mode Decomposition which resulted in 98.33% specificity and 100% sensitivity with overall prediction accuracy of 99.16%.

Hussein et al. [5] introduced an algorithm called MViT (multi-channel vision transformer) to predict epileptic seizures accurately. They demonstrated that the model proposed outperformed several previous studies on seizure prediction that include recurrent neural network models and sophisticated convolutional models.

LTSM in seizure prediction was introduced by Kostas et al. [7] which was able to provide solid results with seizure prediction and unattainable false alarm rates in each preictal duration test ranging between 15 min and 2 h. Meanwhile, Wang et al. [6] employed GRU (Gated Recurrent Unit) neural network which used EEG signal data for the prediction of epilepsy. They established a random forest and LSTM neural network model for prediction. This study showed exceptional 99.9%, 87.5%, and 99.5%, prediction accuracy rates, respectively. Meanwhile, in a separate approach, Kumar et al. [12] harnessed a model named CNN to differentiate between seizure and non-seizure on the EEG dataset which significantly improved the efficiency of seizure recognition. Nejedly et al. [8] classified each 30-s long iEEG data segment into preictal or interictal by using CNN. This approach used inputs of two different types: spectrogram images of the 16 iEEG channels each and unprocessed iEEG data in the time domain. Meanwhile, Noble Nnakeny et al. [16] combined three deep learning algorithms namely, LSTM, CNN and Sparse Autoencoder (SAE) to construct a single model for epileptic seizure prediction. A superior prediction system was created which incorporated the strengths of each model over their weaknesses. An ensemble model for prediction of epilepsy with 97.4% accuracy rate was presented in this study.

The importance of optimizers in each model was established by Manju et al. [14] who talked about various algorithms including Adagrad, SGD (Stochastic Gradient Descent) and Adam demonstrated that employing the right optimizer might boost performance and produce favorable outcomes. This study observed that when compared to the other two, the performance of Adagrad was poor. Adam is the best optimizer for enhancing neural network training and increasing their speed where stochastic gradient descent is an optimal choice if consumers choose to utilize a gradient descent algorithm. This research will benefit the study of neural networks and deep learning algorithms. These three optimizers offered prediction accurately in this case of prediction of epilepsy whereas others provided inaccurate results.

Mohammad et al. [22] developed SPERTL, a model that used TL approaches for early prediction of ES using EEG data by training ResNet. A group of twenty patients from the CHB-MIT dataset was trained and evaluated to predict seizure five minutes in advance. In terms of accuracy, the proposed model outperformed the state-of-the-art method by 14.7%. Meanwhile, to design a baseline model for the prediction of epileptic seizure Shiqi Zhao et al. [26] introduced a novel approach based on neural architecture search which outperformed the performance of the state-of-the-art method. A new patient-specific epilepsy prediction method based on deep learning was introduced by Hisham G. Draud et al. [27] which took long-term EEG scalp data into consideration. Their proposed four deep learning-based models aimed to provide reliable and early seizure prediction considering the real-time operation.

Seif Soliman et al. [28] reviewed various deep learning algorithms namely, LTSM, CNN, SGD, and SVM for prediction of epileptic seizure taking CHB-MIT ASSEP Freiburg.

Table 3. describes the various deep learning algorithms used for the prediction of epileptic seizures.

Table 3. Deep Learning methodologies for prediction of epileptic seizures.

Author(year)	Dataset	Model Used	Model Compared	Performance Metrics
Sook Ra et al. (2023)	Database from Bonn University and CHB-MIT	MLP (Multilayer perception), 1D-CNN (1-dimensional convolutional neural network)		Accuracy
Feizbakhsh and Omranpour (2023)	CHB-MIT scalp EEG dataset	CNN Random Forest		Sensitivity, Specificity, FPR
S.Hussain(2018)	EEG scalp database from CHB-MIT	EMD		Prediction accuracy
Eberlin (2018)	Public datasets comprising long-term recordings	Convolutional neural network (CNN) topology	GLM (Generalized Linear Models), SVM (Support vector machine), Random forests, decision trees	

(*continued*)

Table 3. (*continued*)

Author(year)	Dataset	Model Used	Model Compared	Performance Metrics
Hussein et al. (2022)	3 benchmark EEG datasets	Multichannel Vision Transformer (MVIT)	State of art methods	Sensitivity
Wang et al. (2021)	EEG signals for prediction of epileptic seizure	GRU (Gated Recurrent Unit)	LSTM, GRU and random forest	Accuracy rate
Tsiouris et al. (2018)	Publicly available EEG dataset from CHB-MIT	Long Short- Term Memory (LSTM) Networks	CNN(Convolutional Neural Networks) and traditional machine learning techniques	Sensitivity and false prediction rates
Nejedly et al. (2019)	iEEG from 4 canines with naturally occurring epilepsy	CNN system which is subject specific and fully automated	Poisson random predictor with monte carlo simulations having equal time for warning	Sensitivity
Suraj Kumar and Shiva Prakash (2023)	EEG dataset CHB – MIT data	Deep learning CNN Model	LSTM-SzHNN, SVM-KNN-GMM, CNN, MLF-CNN, Deep CNN	Accuracy
SJ et al. (2023)	CHB-MIT dataset's EEG recordings	Using 1-D mobile networks	CNN	Accuracy Precision Average F1 recall
P. Manju et al. (2023)	EEG recordings from CHB-MIT	New deep learning algorithm which used EEG signal for seizure prediction	Optimizers	Accuracy Precision Recall F1 score
Umair Mohammad and Fahad Saeed (2022)	20 patients	TL (Transfer learning) and ResNets (Residual neural networks)	State of art methods	Accuracy, Sensitivity and specificity
Zhao et al. (2021)	3 public datasets	Neural architecture search	State of the art method	Area under characteristic operating curve of receiver, sensitivity and false prediction rates
Hisham G.Draud and M.Bayoumi(2019)	Long term scalp electroencephalogram (EEG) recordings	RNN and CNN		Accuracy, false alarm rate

3 Observations and Findings

RQ1: What deep learning and machine learning algorithms are employed to predict epileptic seizures?

In the domain of epileptic seizure prediction, to analyze electroencephalogram (EEG) data and make predictions various deep learning and machine learning methods were used which include:

- Classic machine learning techniques namely, Decision Trees, SVM (Support Vector Machine), Random Forest, and k-NN (k-Nearest Neighbors) have been used widely in the area of epileptic seizure prediction which require careful feature selection to process EEG signals efficiently.
- There has been a notable shift from machine learning models to deep learning methodologies, especially Convolutional Neural Networks (CNNs). Models like Long Short-Term Memory (LTSM) have thrived in handling temporal dependencies in EEG data.
- Researchers have also conducted a comparative analysis of these models. Various Challenges such as the need of real-time prediction persist in domain of epileptic seizure prediction and these are needed to be addressed to enhance the accuracy of these models.

RQ2: What are the datasets used for validation?

Various datasets have been used in the context of validating machine learning and deep learning models in the area of epileptic seizure prediction. Some commonly used datasets include the EEG scalp database from CHB-MIT and Bonn University and researchers have also used publicly available datasets from sources such as Kaggle.

RQ3: What are the performance metrics used for evaluation?

Performance metrics are used to evaluate a model's efficiency and effectiveness. In this review, various performance metrics were analyzed to assess the effectiveness of models used for prediction of epileptic seizures. Some of them include accuracy, sensitivity, specificity, and F1 score.

RQ4: What are possible paths for applying machine learning approaches for epileptic seizure prediction in the future?

Datasets: Many researchers have used a private dataset in their study which is difficult to access. As we have summarized, the most used dataset in epileptic seizure prediction is CHB-MIT. Researchers have advised to use of other datasets as well and also suggest taking larger data sets to enhance the dependability and accuracy of models used to predict seizures. We can also focus on using patient-reported data.

Alternative Algorithms: Deep learning models such as RNN can be further researched for predicting epileptic seizures. Researchers have been trying to explore various methods to identify meaningful patterns in seizure prediction. Feature Extraction from EEG signals can be done by using techniques such as time-domain analysis, and wavelet analysis.

Multimodal Data Integration: Researchers have advised combining data form various sources such as ECG and EEG data to enhance the accuracy of prediction of epileptic seizure prediction and provide a more systematic view of patients' condition.

Personalized Models: The need for personalized models that adapt to an individual's specific EEG patterns which lead to accurate seizure prediction has been highlighted by many studies.

4 Conclusions

We explored a wide variety of ML and DL algorithms in the prediction of epileptic seizures. The reviewed studies demonstrated improved accuracy and performance in detecting seizures using various algorithms such as CNN, KNN, and Random Forest. The insights emphasized the promising capabilities of machine learning algorithms in epileptic seizure prediction by using multimodal data and personalized models. These findings suggest that these algorithms can contribute to the development of more epileptic seizure prediction models, ultimately enhancing patient care and quality of life.

References

1. Ra, J.S., Li, T.: A novel epileptic seizure prediction method based on synchroextracting transform and 1-dimensional convolutional neural network. Comput. Methods Programs Biomed. **240**, 107678 (2023)
2. Feizbakhsh, B., Omranpour, H.: Cluster-based phase space density feature in multichannel scalp EEG for seizure prediction by deep learning. Biomed. Sign. Process. Control **86**, 105276 (2023)
3. Hussain, S.J.: Epileptic seizure prediction based on localization. In: 2018 Fourth International Conference on Biosignals, Images and Instrumentation (ICBSII), pp. 206–212. IEEE (2018)
4. Eberlein, M., et al.: Convolutional neural networks for epileptic seizure prediction. In: 2018 Conference on Bioinformatics and Biomedicines (BIBM), pp.2577–2582. IEEE (2018)
5. Hussein, R., Lee, S., Ward, R.: Multi-channel vision transformer for epileptic seizure prediction. Biomedicines **10**(7), 1551 (2022)
6. Wang, X., Han, Q., Li, J. Jin, Y.: Research on prediction model of epileptic EEG signal based on GRU. In: 2021 International Conference on Electronic Information Engineering and Computer Science (EIECS), pp. 9–12. IEEE, September (2021)
7. Tsiouris, KM, Pezoulas, V.C., Zervakis, M., Konitsiotis, S., Koutsouris, D.D., Fotiadis, D.I.: A long short-term memory deep learning network for the prediction of epileptic seizures using EEG signals. Comput. Biol. Med. **99**, 24–37 (2018)
8. Nejedly, P., et al.: Deep-learning for seizure forecasting in canines with epilepsy. J. Neural Eng. **16**(3), 036031 (2019)
9. Ra, J.S., Li, T., Li, Y.: A novel permutation entropy-based EEG channel selection for improving epileptic seizure prediction. Sensors **21**(23), 7972 (2021)
10. Salvador, C., et al.: Epileptic seizure prediction using EEG peripheral channels. In: 2023 IEEE 7th Portuguese Meeting on Bioengineering (ENBENG), pp. 60–63. IEEE (2023)
11. AbdElminaam, D.S., Fahmy, A.G., Ali, Y.M., El-Din, O.A.D., Aly, A.R., Heidar, M.: ESEEG: an efficient epileptic seizure detection using eeg signals based on machine learning algorithms. In: 2022 2nd International Mobile, Intelligent, and Ubiquitous Computing Conference (MIUCC), pp. 210–215). IEEE (2022)
12. Kumar, S., Prakash, S.: Prediction of epileptic seizures based on EEG signal using CNN Model. In: 2023 8th International Conference on Communication and Electronics Systems (ICCES), pp. 84–89. IEEE (2023)

13. S, P.P., M, S., Kattepura, S.: Machine learning models for epileptic seizure prediction. In: International Conference on Inventive Computation Technologies (ICICT), pp. 135–141. IEEE (2023)
14. Manju, P., Devassy, B.R., Vidyamol, K. Ajith, A.P.: Evaluation of optimizers for predicting epilepsy seizures. In: 2023 Advanced Computing and Communication Technologies for High Performance Applications (ACCTHPA), pp. 1–6. IEEE (2023)
15. T, D.L., Privitera, M., Rao, M.: Feasibility of regression modelling and biomarker analysis for epileptic seizure prediction. In: 57th Annual Conference on Information Sciences and Systems (CISS), pp. 1–4. IEEE (2023)
16. Noble-Nnakenyi, P.C., Olatunji, K.A., Abiola, O.B., Oguntimilehin, A., Adeyemo, O.A., Babalola, G.: Predicting epileptic seizures using ensemble method. In: 2022 5th Information Technology for Education and Development (ITED), pp. 1–7. IEEE (2022)
17. Ntahobari, M., Kuhlmann, L., Boley, M., Hesabi, Z.R.: Enhanced extra trees classifier for epileptic seizure prediction. In: 2022 5th International Conference on Signal Processing and Information Security (ICSPIS), pp. 175–179. IEEE (2022)
18. Begam, B., Bathri, D., Charavanan, V.: Machine learning-based epileptic seizure detection using xgboost algorithm. In: 2023 International Conference on Recent Advances in Electrical, Electronics, Ubiquitous Communication, and Computational Intelligence (RAEEUCCI), pp. 1–5. IEEE (2023)
19. Chen, C., et al.: Epileptic seizure prediction based on EEG by auto-machine learning. In: 2022 IEEE International Conference on Real-time Computing and Robotics (RCAR), pp. 710–715. IEEE (2022)
20. Rebello, B.C., Ramirez, A.R.G., Heredia-Negron, F. Roche-Lima, A.: A machine learning-based approach to epileptic seizure prediction using electro-encephalographic signals. J. Eng. Res. **2**(8) (2022)
21. Seifi, B., Barfi, M., Esmaeilpour, M.: ECG-based prediction of epileptic seizures using machine learning methods. In: 2022 9th Iranian Joint Congress on Fuzzy and Intelligent Systems (CFIS), pp. 1–7. IEEE (2022)
22. Mohammad, U., Saeed, F.: SPERTL: epileptic seizure prediction using EEG with ResNets and transfer learning. In: 2022 IEEE-EMBS International Conference on Biomedical and Health Informatics (BHI), pp. 1–5. IEEE (2022)
23. Bhattacherjee, I.: Epileptic seizure prediction using machine learning techniques on real-time EEG signals. In: 2021 8th International Conference on Computing for Sustainable Global Development (INDIACom), pp. 221–226. IEEE (2021)
24. Masum, M., Shahriar, H., Haddad, H.M.: Epileptic seizure detection for imbalanced datasets using an integrated machine learning approach. In: 2020 42nd Annual International Conference of the IEEE Engineering in Medicine & Biology Society (EMBC), pp. 5416–5419. IEEE (2020)
25. Siddiqui, M.K., Morales-Menendez, R., Huang, X., Hussain, N.: A review of epileptic seizure detection using machine learning classifiers. Brain Inform. **7**(1), 1–18 (2020). https://doi.org/10.1186/s40708-020-00105-1
26. Zhao, S., Yang, J., Sawan, M.: Energy-efficient neural network for epileptic seizure prediction. IEEE Trans. Biomed. Eng. **69**(1), 401–411 (2021)
27. Daoud, H., Bayoumi, M.A.: Efficient epileptic seizure prediction based on deep learning. IEEE Trans. Biomed. Circuits Syst. **13**(5), 804–813 (2019)
28. Soliman, S., et al.: Deep learning approaches for epileptic seizure prediction: a review. In: 2022 4th Novel Intelligent and Leading Emerging Sciences Conference (NILES), p 0.1–06. IEEE (2022)
29. Sharma, V., Bajaj, A., Abraham, A.: Machine learning techniques for electronics health records: review of a decade of research. Int. J. Comput. Inform. Syst. Ind. Manage. Appl. **15**, 595–604 (2023)

Prediction of Epileptic Seizures Based on EEG Dataset Using an Enhanced Extreme Learning Algorithm

Sujata Dash[1](✉) and Sourav Kumar Giri[2,3]

[1] School of Engineering and Technology, Nagaland University, Dimapur 797112, Nagaland, India
sujta238dash@gmail.com

[2] Department of Computer Application, Maharaja Sriram Chandra Bhanja Deo University, Takatpur, Baripada 757003, Odisha, India

[3] School of Computer Engineering, Kalinga Institute of Industrial Technology (KIIT) Deemed to be University, Patia, Bhubaneswar 751024, Odisha, India

Abstract. Epilepsy is a neurological disorder that causes repeated seizures. One of the biggest problems in accurately classifying epilepsy is that there are not enough samples in the minority class, which makes the dataset unbalanced. This can lead to biased model performance, particularly in predicting the minority class. To address the issue of imbalanced datasets, we propose a framework utilizing an Extreme Learning Machine (ELM) enhanced with Principal Component Analysis (PCA) and Synthetic Minority Over-sampling Technique (SMOTE) to improve classification performance for epilepsy. PCA reduces the dataset's dimensionality while retaining essential data, to improve the discriminative ability of the features used by the classification model. SMOTE ensures that the minority class samples are properly represented, allowing the model to learn their distinguishing features effectively. We compare the performance of ELM on the imbalanced dataset with and without balancing techniques and PCA. We evaluate the classification accuracy, precision, recall, and F1-score to assess the effectiveness of our proposed approach. Our experiments demonstrate that the PCA-ELM model, when combined with SMOTE, outperforms other models such as Support Vector Machine (SVM) and ELM kernel based Support Vector Machine (SVM-ELM). The optimized hyperparameters obtained through the grid-search optimization algorithm further enhance the performance of the algorithms. The findings of this study have important implications for developing more effective diagnostic tools and personalized treatment strategies for epilepsy patients.

Keywords: Imbalanced data · Principal Component Analysis · Synthetic Minority oversampling technique · Support vector machine · hyper-parameter optimization

D. Kriksciuniene et al. (Eds.): IBICA 2023, LNNS 1228, pp. 61–70, 2025.
https://doi.org/10.1007/978-3-031-78937-3_6

1 Introduction

Epilepsy is a neurological disorder that causes uncontrollable brain activity leading to repeated seizures. It is prevalent in about 2% of the population, making it one of the most common and severe neurological conditions. To enable timely diagnosis of epilepsy, an automated approach using Computer-Aided Diagnosis (CAD) tools needs to be developed. These tools are fast, cost-effective, and accurate, and developed utilizing artificial intelligence, computer vision and image processing. Various neuroimaging techniques, such as Magnetic Source Imaging (MSI), Magnetic Resonance Imaging (MRI), Magnetoencephalography (MEG), and CT scans, use magnetic fields, radio waves, and electrical currents to produce images that can map brain electrical activities. Computer-aided design (CAD) tools can be used to enhance low-quality images, which is crucial for detecting epileptic seizure activity [1]. This can help in the diagnosis of seizures and speed up the detection process.

Epilepsy treatment involves interpreting complex EEG signals. Manual diagnosis is time-consuming and can lead to incorrect decisions. Studies extract time/frequency-domain features from EEG using DWT [2]. They then use feature selection or dimensionality reduction techniques to identify the best features. Traditional classifiers like LR, SVMs, RF, KNN, NN, NB, DT, PNN, and MLP are used but require domain expertise. Few studies use deep neural networks like CNNs, LSTMs, or BiLSTMs for epilepsy detection. Limitations exist for methods that rely on handcrafted features or majority-voting techniques. LSTMs and BiLSTMs [3, 4] can identify temporal dependencies, but they tend to be complex.

In this study, we propose a framework that combines PCA, ELM and SMOTE [5]. This comprehensive approach improves the model's ability to identify critical features to recognize epilepsy. The contribution of this study is as follows:

1) Developed an integrated framework, using ELM, PCA, and SMOTE, to accurately identify epilepsy.
2) PCA was used to reduce data dimensionality while preserving information and enhancing feature identification for accurate classification. Additionally, SMOTE was applied to rectify data imbalance in epilepsy datasets by generating synthetic samples for minority classes.
3) The proposed framework's effectiveness was evaluated using key metrics, including classification accuracy, precision, recall, AUC, and F1-score, on a publicly available Kaggle dataset.
4) Analyzed the ELM model's [6] performance on imbalanced datasets with and without PCA [7] and SMOTE [5] to determine their impact on classification accuracy and used optimized hyperparameters to improve efficiency, then compared the performance with two established models like SVM and SVM with ELM kernel [8].

This manuscript is divided into several sections. Section 2 covers related work, Sect. 3 describes the proposed framework and its implementation, and Sect. 4 presents simulation results and associated findings. Section 5 concludes the research, and the references used in this work are listed at the end of the manuscript.

2 Literature Review

EEG-based seizure detection involves two phases: feature separation and classification—features extracted from time, frequency, time-frequency, and non-linear domains [9]. Morphological parameters and wavelet transform are used for time and frequency domain analysis, respectively. Other methods include empirical mode decomposition and its variants. Complexity metrics and techniques, such as dynamic pattern decomposition and variational mode decomposition, are used to analyze epileptic signals. Moreover, recurrence plots generate two-dimensional images from EEG signals to preserve the non-linear EEG [10] characteristics.

Researchers have proposed using various machine learning algorithms to extract features from EEG signals, including artificial neural networks and logistic regression [11]. Different classifiers, such as Naïve Bayes, multi-layer perceptron, k-nearest neighbours, and support vector machine, have been applied to detect seizure activity [12]. Although speed and size issues may arise during training and testing, SVM is slower but more accurate than other ML techniques. SVM performance highly depends on appropriate kernel functions, which can be challenging to choose.

DL-based algorithms have made strides in detecting seizures. A deep CNN by Acharya et al. [13] distinguishes normal, preictal, or seizure states. VWCNN, proposed in [14], is a new network architecture that adapts dynamic weights to different EEG characteristics. Sahani et al. [15] combined multiple deep neural networks to detect seizures. DL techniques automatically generate features, resulting in superior performance compared to traditional machine learning systems that rely on manually crafted features [16]. Research focuses on improving epileptic seizure detection through Deep Learning. The Extreme Learning Machine (ELM) model [17] achieved 100% accuracy, promising better patient outcomes.

Our work presents a model that efficiently integrates ELM with PCA and an oversampling technique (SMOTE) to improve the model's ability to identify critical features for recognizing epilepsy. Finally, the model's effectiveness is established by comparing it with a support vector machine (SVM) and an SVM with an ELM kernel in terms of better classification accuracy at a lesser execution time.

3 Proposed Work

3.1 Dataset

The dataset has 500 individuals, with each contributing 4097 EEG recordings taken over 23.5 s [18]. The 4097 data points were divided into 23 chunks, each having 178 data points representing a one-second interval. Each record has 178 data points and a label 'y' {1, 2, 3, 4, 5}. The 'y' variable, which represents the category associated with the 178-dimensional input vector, is located in the 179th column, while the explanatory variables, 'X1,' 'X2,'..., 'X178,' make up the remaining columns. Class 1 denotes 'Seizure activity' and classes 2–5 represent non-seizure activity. The focus is on binary classification, with class 1 classified as epileptic and class 0 as non-epileptic shown in Table 1.

Table 1. Dataset description

Attributes	Description
X1 to X178	The EEG data contains values ranging from -1415 to 2047
y	When the response variable is 0, there is no seizure, and when it's 1, there is an active seizure

3.2 Workflow Diagram

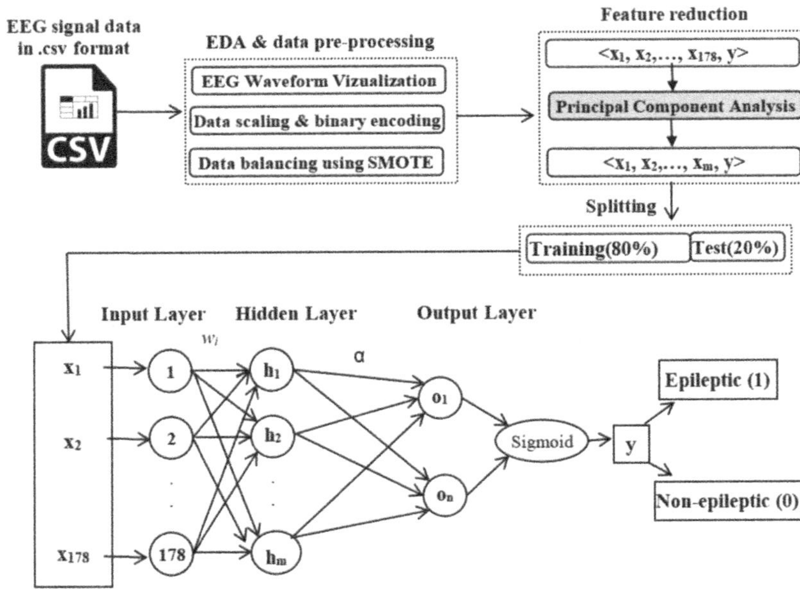

Fig. 1. Workflow diagram

EDA and Data Preprocessing

Data preprocessing is a crucial step in many data-driven applications. Its primary objective is to transform raw data into a suitable format, eliminate any irrelevant noise, handle missing values, and extract essential features. In Fig. 2, we can see the characteristics of five different waveforms. Preprocessing enhances data quality, reliability, and usefulness for analysis by applying data scaling to normalize the data and convert multiclass data to binary. Thus, the preprocessing method aims to prepare the data for analysis and improve its overall quality.

Data Balancing

The provided dataset exhibited class imbalance, comprising 2300 data points (20%) for class 1 (indicating epilepsy cases) and 9200 data points (80%) for class 0 (representing normal cases). To rectify this imbalance and ensure accurate diagnostics, a method of oversampling was employed. Oversampling involves augmenting the instances in the minority class to achieve a more balanced dataset [5]. The Synthetic Minority Oversampling Technique (SMOTE) was utilized for this purpose, creating new samples through interpolation between existing minority class samples. This involves selecting a minority class instance, identifying its k-nearest neighbours, and generating synthetic samples along the line segments connecting the instance with its neighbors. Following the application of SMOTE, the model was equipped with 18,400 records, evenly distributed between epilepsy and normal cases, resulting in a well-balanced dataset.

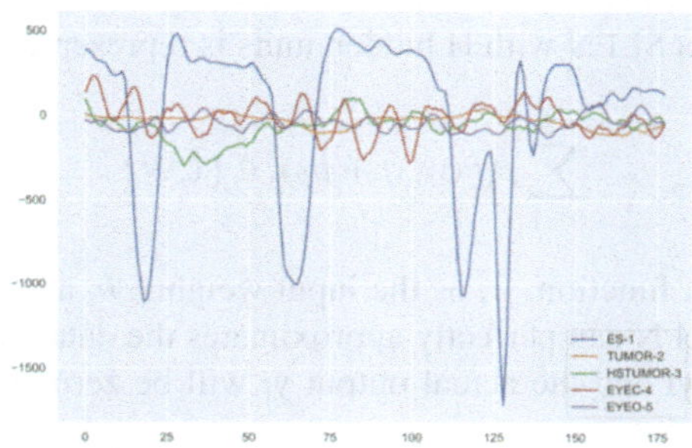

Fig. 2. Visualization of EEG Waveforms (four normal conditions vs one epileptic condition)

Proposed Model

The proposed model combines Principal Component Analysis (PCA) with Extreme Learning Machine (ELM). After feature reduction by PCA, the data is transmitted to the ELM model, which undergoes processing through a weighted SLFN system. Finally, the processed data is directed to the output layer, corresponding to the response variable's prediction. The workflow diagram of the proposed model is depicted in Fig. 1.

Principal Component Analysis (PCA)

PCA, as outlined in the literature [7], serves as a valuable tool in signal processing and machine learning for dimensionality reduction in high-dimensional datasets, preserving crucial information. The principal component and Eigenvalues were calculated for each feature set and the respective instance through the covariance or correlation matrix. Subsequently, we assessed the distance of each feature instance from the mean principal component (0.5 in this instance). Consequently, instances with elevated distances were excluded, while those with lower eigen-distances were retained, under the assumption that they hold significance for classification, as depicted in Fig. 3.

Training Algorithm

The Extreme Learning Machine (ELM) stands out as a contemporary neural network architecture leveraging random projections. Comprising a single hidden layer with random weights and an analytically calculated output layer, ELM demonstrates notably swift training and prediction processes compared to various nonlinear methods. It's noteworthy that ELM was initially conceived as a rapid neural network training method, acquiring a mapping in a fully trained network where the weights encapsulate information about the training data.

The ELM algorithm, rooted in the single-layer feed-forward neural network (SLFN) and introduced in [6], operates on the premise of assigning random weights and biases to the SLFN hidden layer, while analytically determining the output weights. This unique approach allows for the rapid and computationally efficient construction of the network. An eminent advantage of ELM lies in its ability to self-train expeditiously without relying on slow gradient descent-based learning algorithms.

Let us consider a collection of N individual samples (x_i, y_i) where $x_i \in R^{d1}$ and $y_i \in R^{d2}$. In this case, a SLFN with H hidden units is represented as the sum below:

$$\sum_{i=1}^{H} \beta_{i}f(w_i x_i + b_b), \in [1, N] \tag{1}$$

with f *as the* activation function, w_i as the input weights, b_i as the biases and β_i as the output weights, if a SLFN can perfectly approximates the data, then the errors between the predicted outputs yi and the actual output $\mathrm{y_i}$ will be zero, and the relation can be expressed as,

$$\sum_{i=1}^{H} C_{i}f(w_i x_i + b_i) = y_i, j \in [1, N] \tag{2}$$

which writes compactly as $H\beta = Y$, with

$$\mathrm{H} = \begin{cases} f(w_1 x_1 + b_1) \ldots f(w_H x_1 + b_H) \\ \quad\quad\quad\quad\quad . \\ \quad\quad\quad\quad\quad . \\ \quad\quad\quad\quad\quad . \\ f(w_1 x_N + b_1) \ldots f(w_H x_N + b_H) \end{cases} \tag{3}$$

and $\beta = \left(\beta_1^T \ldots \beta_H^T\right)$ and $y = \left(y_1^T \ldots y_H^T\right)^T$

ELM algorithm

Given a training set (x_i, y_i) with $x_i \in \Re^{d1}$ and $y_i \in \Re^{d2}$, an activation function f : $\Re \rightarrow \Re$ and the number of hidden nodes H.

1) Randomly allocate input weights w_i and biases $b_i, foreachi \in [1, H]$.
2) Compute the matrix H representing the output of the hidden layer.
3) Compute the output weights matrix b = HfY

The proper determination of the number of hidden units in ELM holds significance. This involves initiating with a substantial number of hidden neurons and iteratively refining the network by pruning.

Optimized Hyperparameters

Table 2. Optimized value of Hyperparameters by Grid Search

Model	Hyperparameters
SVM	c = 11, kernel = RBF
SVM with ELM kernel	c = 13, kernel = ELM
ELM	$\alpha = 0.1$, activation function = sigmoid, no of hidden neurons = 1000

4 Results and Discussion

The experimental assessments were conducted using Python 3.7, leveraging various libraries. These experiments utilized a core i7 processor operating at 3.4 GHz, paired with 8GB RAM and GPU support. The proposed ELM model underwent testing with multiple activation functions, including tanh, sigmoid, multiquadric, hard limit, and Gaussian. Figure 4 illustrates the comparison of these activation functions on the EEG epileptic dataset for the proposed model. The most effective activation function identified in this comparison was subsequently employed for further experimentation to determine the optimal model for disease detection.

The grid search algorithm is employed to determine optimized hyperparameters, and the corresponding parameter values for different techniques used in the EEG data analysis experiments are documented in Table 2. In Fig. 4, the ELM model's performance is compared across various activation functions using the EEG dataset. This specific experiment aims to explore the behavior of different ELM activation functions and their suitability for our problem. The figure illustrates the overall accuracy of these activation functions for the dataset. Based on these findings, it is concluded that ELM with the optimized parameters of the best logistic activation function will be employed for subsequent experiments.

In this experimental setup, neurons are randomly selected, mitigating concerns related to overfitting and providing advantages such as universal approximation, minimal intervention, globally optimal solutions, and significantly faster convergence with reduced training and testing durations. As depicted in Fig. 5, the accuracy of the ELM model exhibits an upward trend with an increasing number of hidden neurons, reaching nearly 100% accuracy at 1000 hidden neurons for an 80:20 data split. The model's best training and testing accuracy is highlighted in Fig. 7, showcasing a false positive rate of 1.0 and a true positive rate of 1.0 in the combined ROC curves—a key performance evaluation metric.

Table 3 illustrates the comparison of performance of ELM models with SVM and SVM with ELM kernel in three different scenarios. In the first case, the model is tested with imbalanced dataset and the comparison shows except for precision and AUC score SVM-ELM has outperformed the ELM model. In the second case, with balanced dataset, ELM model outperforms the other two models by achieving 96.46% precision and 96.76% AUC score. In the third scenario, with PCA based balanced dataset, the performance of ELM model is outstanding in comparison to the other two models with 99.47% accuracy and 99.42% AUC score. Figure 6 establish the efficiency of PCA based ELM in detecting epilepsy in terms of computing time. The comparison shows the training time of PCA based ELM model is 7.3 times faster than SVM and 1.5 times faster than SVM-ELM model. Similarly, the predicting time of ELM is 4.2 times faster than SVM and almost same predicting time both ELM and SVM-ELM models share. The ELM model performance is so promising that, it outperforms Support Vector Machines (SVM), and SVM with an ELM kernel.

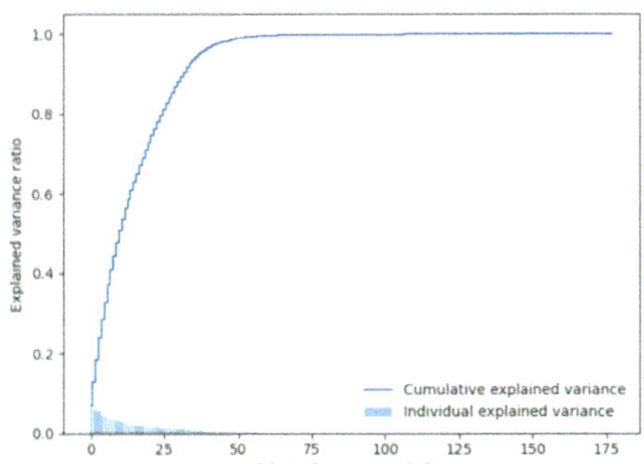

Fig. 3. Principal component index VS explained variance

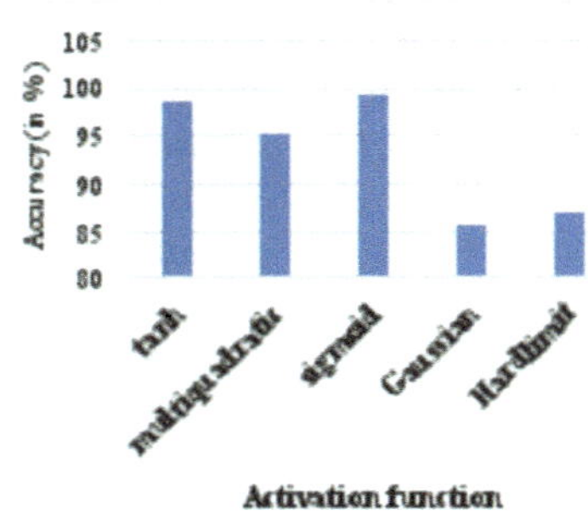

Fig. 4. Performance comparison of different activation functions

Table 3. Performance of various models using Logistic Activation Function

Performance Metric	Models using imbalanced/ balanced/ balanced dataset with PCA		
	SVM	SVM with ELM kernel	ELM
Accuracy	88.87/ 93.7/ 98.01	93.75/ 95.87/ 99.23	93.52/ 96.46/ 99.47
Precision	92.56/94.78/ 96.98	97.83/ 97.54/ 99.87	100.0/ 100.0/ 100.0
F1-Score	79.12/ 91.90/ 97.29	82.79/ 95.61/ 99.38	80.67/ 96.33/ 99.42
AUC Score	87.54/ 94.45/ 98.7	90.23/ 96.53/ 99.79	90.76/ 96.76/ 99.83

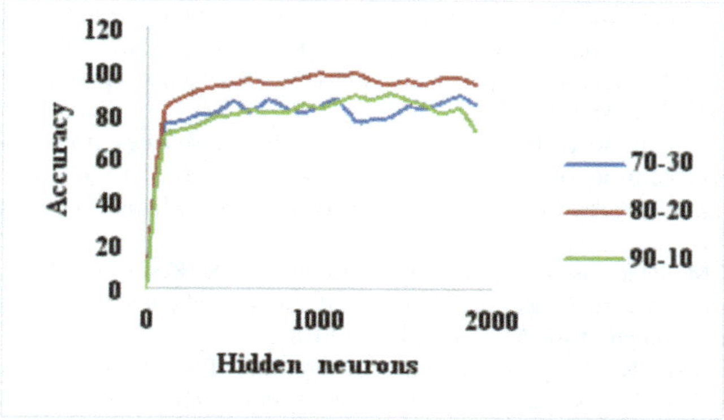

Fig. 5. ELM Performance for identifying Epilepsy

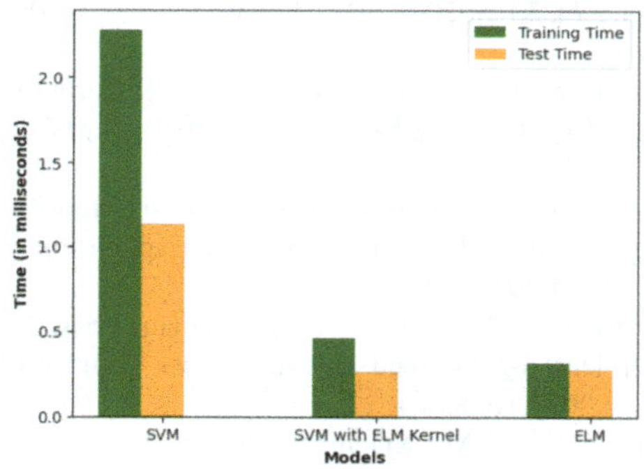

Fig. 6. Training & Test time of PCA based models

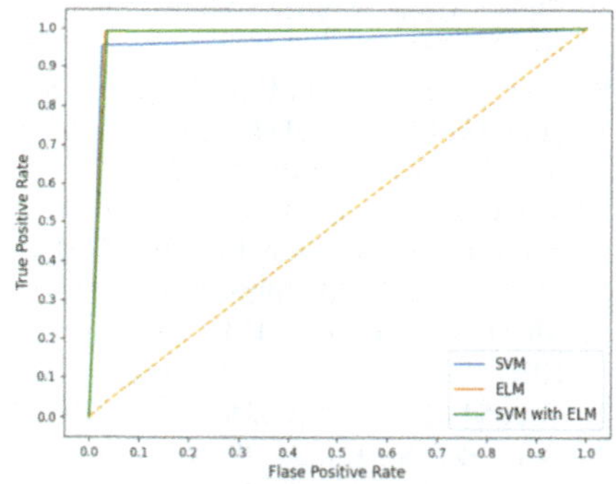

Fig. 7. ROC curve

5 Conclusion

Diagnosing epilepsy through the analysis of EEG datasets poses a considerable medical challenge, necessitating the utilization of artificial intelligence tools. Given the complexity of epilepsy diagnosis, we've devised an advanced machine learning system in this study, leveraging EEG dataset analysis for the automated identification of epileptic seizures. To achieve a precision rate of 100%, we have employed an Extreme Learning Machine equipped with a sigmoid activation function, PCA for feature reduction, and SMOTE for addressing data imbalance.

References

1. Elghamry, H., Ghoneim, M., Haggag, A.A., Darweesh, M.S., Ismail, T.: Comparative analysis of various machine learning techniques for epileptic seizures detection and prediction using EEG data. In: Proceedings of the 2020 2nd Novel Intelligent and Leading Emerging Sciences Conference (NILES), pp. 175–180, Giza, Egypt, 24–26 October 2020. IEEE, New York, NY, USA (2020)

2. Orhan, U., Hekim, M., Ozer, M.: EEG signals classification using the K-means clustering and a multilayer perceptron neural network model. Expert Syst. Appl. **38**(10), 13475–13481 (2011). https://doi.org/10.1016/j.eswa.2011.04.149. https://www.sciencedirect.com/science/article/pii/S0957417411006762
3. Tuncer, E., Doğru Bolat, E.: Classification of epileptic seizures from electroencephalogram (EEG) data using bidirectional short-term memory (Bi-LSTM) network architecture. Biomed. Signal Process. Control, **73**, Article 103462 (2022). https://doi.org/10.1016/j.bspc.2021.103462
4. Tsiouris, K.M., Pezoulas, V.C., Zervakis, M., Konitsiotis, S., Koutsouris, D.D., Fotiadis, D.I.: A long short-term memory deep learning network for the prediction of epileptic seizures using EEG signals. Comput. Biol. Med. **99**, 24–37 (2018)
5. Dash, S., Behera, R.: Sampling based hybrid algorithms for imbalanced data classification. Int. J. Hybrid-Intell. Syst. **13**, 77–86 (2016). https://doi.org/10.3233/HIS-160226.IOSPress
6. Huang, G.B., Zhu, Q.Y., Siew, C.K.: Extreme learning machine: theory and applications. Neurocomputing **70**, 489–501 (2006)
7. Jolliffe, I.T., Cadima, J.: Principal component analysis: a review and recent developments. Philos. Trans. R. Soc. A Math. Phys. Eng. Sci. **374**, 20150202 (2016). https://doi.org/10.1098/rsta.2015.0202
8. Murugavel, A.S.M., Ramakrishnan, S.: Hierarchical multi-class SVM with ELM kernel for epileptic EEG signal classification. Med. Biol. Eng. Comput. **54**, 149–161 (2016). https://doi.org/10.1007/s11517-015-1351-2
9. Wang, Z., Wu, D., Dong, F., Cao, J., Jiang, T., Liu, J.: A novel spike detection algorithm based on multi-channel of BECT EEG signals. IEEE Trans. Circ. Syst. II Exp. Briefs **67**, 3592–3596 (2020). https://doi.org/10.1109/TCSII.2020.2992285
10. Shankar, A., Khaing, H.K., Dandapat, S., Barma, S.: Analysis of epileptic seizures based on EEG using recurrence plot images and deep learning. Biomed. Signal Process. Control **69**, 102854 (2021). https://doi.org/10.1016/j.bspc.2021.102854
11. Beniczky, S., Karoly, P., Nurse, E., Ryvlin, P., Cook, M.: Machine learning and wearable devices of the future. Epilepsia **62**, S116–S124 (2021). https://doi.org/10.1111/epi.16555
12. Anter, A.M., Abd Elaziz, M., Zhang, Z.: Real-time epileptic seizure recognition using bayesian genetic whale optimizer and adaptive machine learning. Future Gen. Comput. Syst. **127**, 426–434 (2022). https://doi.org/10.1016/j.future.2021.09.032
13. Acharya, U.R., Oh, S.L., Hagiwara, Y., Tan, J.H., Adeli, H.: Deep convolutional neural network for the automated detection and diagnosis of seizure using EEG signals. Comput. Biol. Med. **100**, 270–278 (2018). https://doi.org/10.1016/j.compbiomed.2017.09.017
14. Jia, G., Lam, H.K., Althoefer, K.: Variable weight algorithm for convolutional neural networks and its applications to classification of seizure phases and types. Pattern Recogn. **121**, 108226 (2022). https://doi.org/10.1016/j.patcog.2021.108226
15. Sahani, M., Rout, S.K., Dash, P.K.: FPGA implementation of epileptic seizure detection using semisupervised reduced deep convolutional neural network. Appl. Soft Comput. **110**, 107639 (2021). https://doi.org/10.1016/j.asoc.2021.107639
16. Mahmood, D., Riaz, H.N., Nisar, H.: Introduction to non-invasive biomedical signals for healthcare. In: Advances in Non-Invasive Biomedical Signal Sensing and Processing with Machine Learning, pp. 1–24. Springer International Publishing. Cham, Switzerland (2023)
17. Ayman, U., et al.: Epileptic patient activity recognition system using extreme learning machine method. Biomedicines. **11**(3), 816 (2023). https://doi.org/10.3390/biomedicines11030816
18. Andrzejak, R.G., Lehnertz, K., Rieke, C., Mormann, F., David, P., Elger, C.E.: Indications of non-linear deterministic and finite dimensional structures in time series of brain electrical activity: dependence on recording region and brain state. Phys. Rev. E **64**, 061907 (2001)

Self-supervised Learning for Pathological Speech Analysis of Parkinson's Disease

Ashwin Kumar Uniyal, Huizhi Liang(✉), and Varun Ojha

School of Computing, Newcastle University, Newcastle, UK
{a.k.uniyal2,huizhi.liang,varun.ojha}@newcastle.ac.uk

Abstract. We explore self-supervised learning (SSL) techniques to predict Parkinson's Disease (PD) by analysing PD speech signals. PD speech abnormalities can be identified up to five years before a medical diagnosis; however, a large labelled dataset is required to train a machine learning (ML) model to predict the disease. SSL can reduce data labelling requirements by learning data representations without explicit supervision. We focus on examining different SSL techniques. Tabular data with multiple acoustic and mel-spectrogram images are extracted from PD speech data to study the PD classification problem. Various SSL techniques, including TabNet, Encoder, Autoencoder, and SimCLR, are trained to classify the disease and compared with supervised ML methods. We observed that on several experimental scenarios, SSL performed better than supervised ML on the PD speech data, and even when only a third of the labelled data was SSL, accuracy only dropped by 1.86% on average. SSL approaches on tabular data performed much better than on mel-spectrogram image classification tasks, with the highest accuracy recorded at 94.15%.

Keywords: pathological speech analysis · healthcare · Parkinson's disease · self-supervised learning · deep learning

1 Introduction

Parkinson's disease (PD) is a neurodegenerative disease that primarily affects dopamine-producing ("dopaminergic") neurons in the *substantia nigra*, a specific region of the brain. It is the second most frequent kind of neurodegenerative illness, affecting one in every 100 persons over the age of 65 [1,2] and more than 10 million people worldwide. This condition causes involuntary movements such as tremors and stiffness [3], vocal dysfunction that affects speech, motor skills, and other processes such as behaviour, emotion, sensation, and thinking [4]. An early and precise diagnosis of PD is critical for disease management, intervention, and personalised treatment methods. Traditional diagnostic methods often rely on clinical assessments, neuroimaging, and biochemical tests, which can be expensive, time-consuming, and subject to inter-observer variability.

Advancements in machine learning and signal processing have paved the way for innovative disease diagnosis and monitoring approaches. Multiple studies

D. Kriksciuniene et al. (Eds.): IBICA 2023, LNNS 1228, pp. 71–80, 2025.
https://doi.org/10.1007/978-3-031-78937-3_7

have been done to detect PD by examining MRI scans [5], tremors, and movements, performing olfactory and cognitive tests. One such promising avenue is the analysis of speech samples for predicting Parkinson's disease. Speech, as a complex and multidimensional physiological process, offers valuable insights into a person's health condition. Variations in speech patterns, prosody, and vocal characteristics have been observed in individuals with PD, even in its early stages [6], providing a potential avenue for early detection and monitoring. Authors in [7–9] show that speech disorders symptomatic of PD can be detected as early as five years before a clinical diagnosis. This lead time may be essential to predict this disease before it progresses.

Numerous supervised machine learning (ML) and deep learning (DL) algorithms have been developed and used to classify the disease based on speech data [10–12]. However, developing a robust and efficient supervised ML/DL model requires a large labelled dataset that is very expensive, time-consuming, error-prone and exhibits class imbalances. Therefore *our research focuses on developing efficient and resilient self-supervised learning algorithms to classify Parkinson's disease* that do not require extensively labelled data.

Self-supervised learning (SSL) algorithms have gained significant attention in recent years due to their ability to leverage unlabeled data for learning meaningful representations. The use of SSL not only addresses the challenge of limited labelled data but also enhances the model's ability to generalize diverse data (speech) characteristics and variations across different individuals/classes [13].

We present a comprehensive study that includes pre-processing and augmentation of speech data, extraction of features, design and implementation of various SSL frameworks, fine-tuning of predictive models for PD diagnosis, and comparison of different algorithms. We contribute to the area of data-driven healthcare by providing early identification and monitoring of PD by using speech data analysis. Our method reduces the reliance on costly and error-prone labelled data and diagnoses PD at an early stage.

2 Related Work

Literature has multiple supervised ML methods have been developed using speech samples in these studies [10–12]. Most of the work uses supervised ML/DL algorithms on combined acoustic and MFCC features to predict PD, e.g., Toye et al. [14]. However, an obvious drawback is that the majority of the audio chunks obtained after splitting audios were small files of less than one-second length that contained little to no information for training ML models.

Jiu-Cheng et al. [15] performed a study on the Italian Parkinson's Voice and Speech Database (IPVSD) dataset for PD classification in which they extracted three different types of characteristics from audio data: spectrograms, linear and time-frequency, and time-series features. They applied a particle swarm optimization algorithm to build the 1D convolution network for feature classification, and the results showed improvement in the accuracy of voice-based Parkinson's disease detection. Similarly, in [7], authors extracted linear and nonlinear feature parameters for each speech sample in 26 dimensions. The focus was on the sample selection process and a multi-edit-nearest neighbour (MENN) algorithm combined with an ensemble learning algorithm for classification.

On speed analysis work, Meghraoui et al. [10] concluded that harmonicity, pitch, and pulse are the most prominent features in classifying PD. They followed the multinomial and Bernoulli approaches for automatic classification, where the Naive-Bayes multinomial classifier gave an accuracy of 95.23%.

Zhang et al. [7] trained an SSL model to classify PD on MRI images that prioritise developing pre-train tasks to improve learning the representation from unlabelled data. They proposed a 3D network based on SSL strategy and attention mechanism. Successive frames of MRI data were put into a 3D ResNet-18 with the classifier module for PD classification, and an accuracy of 87.50% was achieved. However, Zhang et al. [7] use MRI data instead of speech data. In contrast, our novel approach is to focus on training SSL models on speech data.

3 Parkinson's Disease Data

3.1 Dataset

We use the IPVSD dataset, which has audio recordings of 15 healthy adults aged 19-29 years, 22 healthy people aged 60-77 years, and 28 Parkinson's disease patients aged 40-80 years. All participants are Italian. They are expected to deliver vocals in a variety of modalities and execution times, as listed below:

- two readings of a phonemically balanced text
- reading phonemically balanced words, each with a one-minute pause
- execution of the syllable 'pa' (5 s), pause (20 s), and the syllable 'ta' (5 s)
- two phonations of the vowels 'a' , 'e', 'i', 'o', 'u'

We merged voices of phonetically balanced texts, words, and phrases, totalling 236 audio samples, 130 of which were from healthy people and 106 from PD patients. Multiple pre-processing and augmentation methods were applied to audio data ranging from one minute to one minute and ten seconds.

3.2 Audio Data Quality Evaluation, Data Pre-processing and Data Transformation

Using the 'mutagen' library, we verified the sample rate of audio and found that audio has varied sample rates. This is due to multiple recording equipment used to record speech data. There are 196 audio samples with a sample rate of 16,000 Hz and 40 samples with 44,100 Hz. Many samples have varied amplitudes that need normalisation to assist ML/DL training stability. We applied multiple pre-processing techniques to improve data quality and reliability:

Normalisation: At first, all audio samples are normalised using PyDub's effects module's normalise() function, which finds the maximum volume of an AudioSegment file and then adjusts the rest of the audio to be in proportion, allowing quiet parts of the audio to get a volume boost.

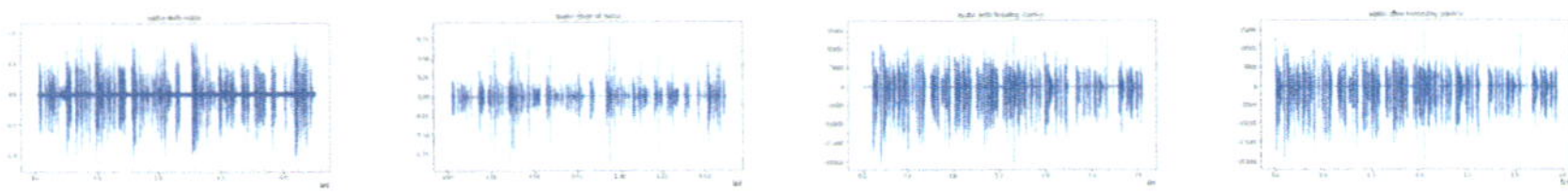

Fig. 1. Left-to-right before and after noise and before and after silent voice removal.

Down Sampling: There are two sample rates found: 16,000 Hz and 44,100 Hz. According to [16], the model's performance varies depending on the sample rate. Hence, all audio recordings are sampled down to a consistent value of 16,000 Hz using the 'librosa' library.

Noise Removal: Some voice samples had unwanted noises and electrical interference. Using 'noisereduce' package, we eliminated noise from audio data as noise can negatively impact model performance [17] (see Fig. 1)

Silence Detection and Removal: Parts of the audio in which the speaker was found less effective in training ML/DL, especially silence at the beginning and the end of audio samples. Thus, we use to truncate the silence segments of audio samples. This was done by applying a specified threshold in decibels (dBFS) iterated over audio chunks of 10 milliseconds duration on the input audio until a non-silent chunk above -25 dBFS (threshold) is found. The same method is applied in reverse to find the silence at silent chunks at the end of the audio (see right two subplots in Fig. 1).

Splitting the audio: We split every audio into chunks of 15 s using *split_audio* function of *librosa* library. This also increases our training data size.

Augmentations. Various transformation strategies were used to improve the performance and generalisation of ML/DL models by increasing the diversity and quantity of audio data and extending the restricted amount of accessible data. However, to avoid data leakage during the training and testing phases, the data is split into training, testing, and validation sets before augmentations, as augmented samples of the same source audio are highly correlated, and having these almost identical samples in both the training and testing sets is not ideal, resulting in poor model performance in the real world. Augmentations such as time stretching, time masking, pitch shifting up and down, and forward and backward shifting were used. Only these methods were used to avoid distorting the meaning and quality of audio data while adding variation to training.

Transformation. After performing pre-processing steps and augmentations, a dataset of 5,976 audio samples was created, 3,180 of which belong to PD patients and 2,796 to healthy individuals. The audio samples were then transformed into two separate data sets to investigate a diverse range of ML/DL algorithms and performances: 1)*mel-spectrogram image dataset* and 2) a *numeric tabular dataset.* Literature suggests that mel-spectrogram (MFCC) is superior to spectrogram images for classification tasks, justifying the reason for transforming the data into MFCC. On the other hand, numerical features in the form of shimmer

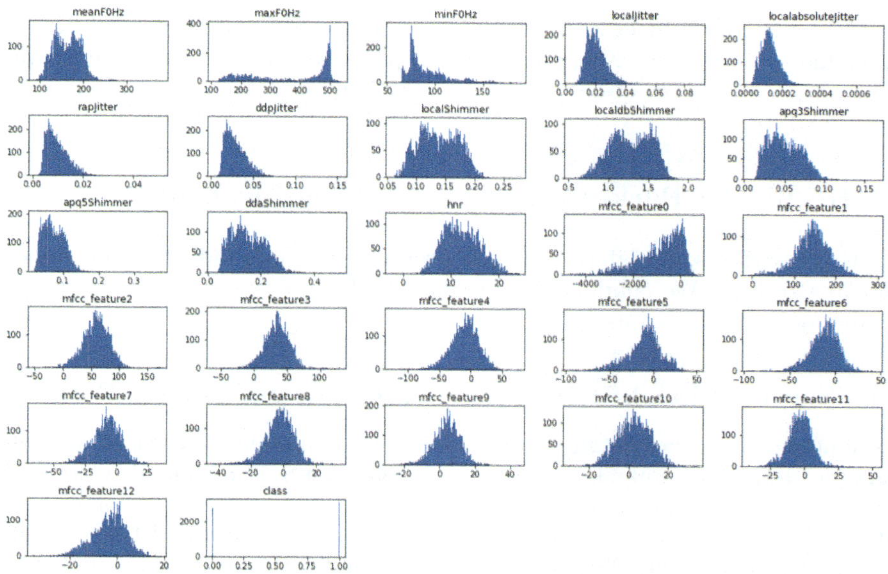

Fig. 2. Histograms of numeric data features. Last subplot is class distribution.

variants, jitter variants, pitch, harmonics-to-noise ratio and MFCC variants are extracted from the pre-processed audio data.

3.3 Feature Extraction and Data Analysis

The acoustic features and MFCC features are extracted from the audio data using 'parselmouth' library. We extracted 13 different MFCC features and 14 acoustic features that include Jitter (absolute), Jitter (relative), Jitter (rap), Jitter (PPQ5), Shimmer(db), Shimmer(relative), Shimmer(APQ3), Shimmer(APQ5), Fundamental Frequency, Harmonics-toNoise Ratio (HNR), and Pitch. These features have been used previously in classifying PD patients [18]. These features, therefore, gave us a numeric tabular dataset containing 5,976 rows and 26 columns. Figure 2 shows histograms of 15 of 26 features. The features are generally seen normally distributed with some skewness of both positive (right) and negative(left). In Fig. 2, the MFCC features are relatively similar in distribution.

The features have been extracted from audio samples recorded for each individual. Hence, the skewness is not considered outliers. Some features contained null values and were removed: four observations in apq3Shimmer, 8 in apq5Shimmer, and 4 in 1ddaShimmer.' The standard deviation shows a deviation from as low as 0.000052 for 'localabsoluteJitter' feature to as high as 988.0 for 'mfcc_feature0' feature. Hence, *a standard scaler was applied to normalise data features to produce scaled and unscaled datasets.* We also observe highly positive and highly negative correlations between some features: a correlation of 0.72 between rapJitter and localShimmer and a correlation of -0.83 between 'hnr' and 'localShimmer.' Finally, the pre-processed data was split into three sets, where 70% (4170 observations) for training, 15% (900) for validation and 15% (906) for testing.

4 Results of Parkinson's Disease Classification

We have various sets of data from our preprocessing methods, such as numeric and image data. We conducted experiments of both types of data, and their results are summarised as follows:

4.1 Supervised and Self-supervised Methods on Numeric Data

From numeric data, we obtained four different sets: unscaled numeric data, scaled numeric data, 30% only labelled numeric data, and 100% labelled data. On unscaled numeric data and scaled numeric data, we apply traditional supervised ML algorithms, and on 30% only labelled numeric data and 100% labelled numeric data, we apply SSL algorithms. The results are summarized in Table 1. Initially, **traditional supervised ML algorithms** such as KNN, SVM, Random Forest, Decision Tree, and Nave Bayes were used [19] for preparing baseline performance and understanding the classification task and data characteristics. In Table 1, the performance of KNN, SVM, Random Forest, Decision Tree, and Nave Bayes algorithms are reported under the supervised learning (SL) block. On unscaled data, the supervised ML approaches were less effective than their performance on scaled versions of data. On scaled data, the accuracy of the KNN classifier went from 84.32% to 97.13%, while the accuracy of the SVM classifier grew from 82.56% to 98.23%. The performance of the Decision tree and Random forest classifiers were similar on both scaled and unscaled data, and the Nave Bayes classifier accuracy dropped from 82.33% to 76.49% when using scaled data. This performance aligns with how these algorithms work [19].

Among **self-supervised learning approaches**, we used TabNet, Encoder, Autoencoder based architectures [13]. We adopted self-supervised pre-training and supervised fine-tuning to perform classification on tabular data, both on only 30% of labelled data and 100% of labelled data.

TabNet. This model is specifically designed for tabular data. It incorporates attention mechanisms that allow the model to focus on relevant features and interactions between them along with decision trees as part of its architecture. The algorithm involves a two-step process. First, a TabNet model is pre-trained in a self-supervised manner using the TabNetPretrainer. Then, the pre-trained model is fine-tuned for a supervised classification task using the TabNetClassifier. This approach leverages the knowledge captured by the self-supervised pre-training to enhance the classification performance on a smaller labelled dataset. The 'mask_type' parameter of the TabNetPretrainer specifies the type of attention mask that is applied within the architecture. It was found that 'entmax' (Entropy Maximization) mask, which is an attention mechanism that

Table 1. Results of supervised ML and self-supervised DL approaches

Methods	Classifier	Unscaled*/SSL (30% labeled data)	Scaled*/ Supervised (100% labeled data)
SL	KNN	84.32	**97.13**
	SVM	82.56	98.23
	Random Forest	95.69	94.15
	DST	86.31	87.08
	Naive Bayes	82.33	76.49
SSL	TabNet	93.1	94.43
	Autoencoder (MSE)	85.65	87.08
	Encoder (CPC)	**94.15**	97.01
(image)	SimCLR (ViT)	71.27	93.43
(image)	CNN Triplate loss	69.34%	72.14%

Note: *these dataset versions only applied on SL approaches.

falls between softmax and sparsemax, performed better than 'sparsemax'. The architecture of this approach is shown in Fig. 3 (left-top). The pre-training of TabNet involved 70% of unlabeled data, and then models were fine-tuned for the PD binary classification task. On testing this model on the unseen test set, TabNet offered an accuracy of 93.01%. However, when TabNet was trained on 100% labelled data, it offered 94.43% accuracy (see Table 1).

AutoEncoder. We perform a two-stage process of SSL pre-training using an autoencoder and then fine-tuning the pre-trained model for a classification task (see Fig. 3 (top-right)). The autoencoder class is defined as containing an encoder which maps the input data to a lower-dimensional latent space, and the decoder attempts to reconstruct the original input from the encoded representation. The mean squared error (MSE) loss and the Adam optimizer were used during the pre-training phase, which is followed by fine-tuning on a smaller set of labelled data for the classification task (see Table 1).

Encoder. An encoder network is designed using the Contrastive Predictive Coding (CPC) for the pre-training (see Fig. 3 (bottom-left)). This CPC Encoder network takes an input, passes it through a hidden layer, and outputs a lower-dimensional representation. A CPC loss function is then used that computes the normalized dot product as a similarity for two representations, calculates cross-entropy loss based on this similarity, and returns the loss. A classifier head is defined by combining the encoder with an additional linear layer that outputs class scores for the downstream classification task. Encoder gave the best accuracy of 94.15% on 30% unlabeled data among SSL approaches (see Table 1).

4.2 Self-supervised Methods on Image Data

We also applied SSL techniques directly on mel-spectrogram images to classify the PD disease. We used SimCLA and CNN with Triplet Margin Loss methods.

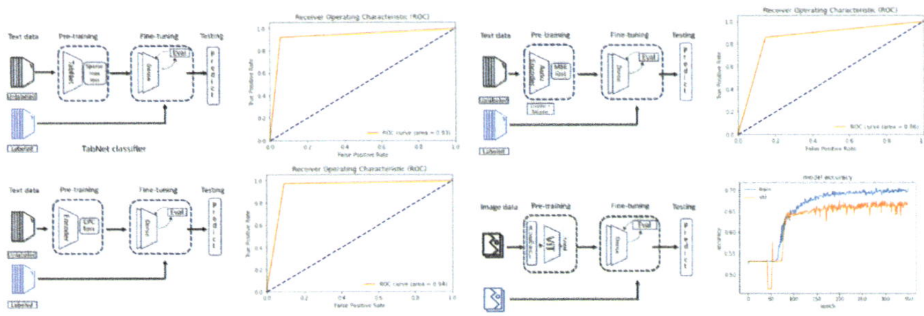

Fig. 3. Self-supervised Learning Models: TabNet classifier, Autoencoder with MSE loss, Encoder with CPC loss, and SimCLR with Vision Transformer.

SimCLR using Vision Transformer. Vision Transformer (ViT) architecture was used as a backbone to create embeddings, and then a SimCLR framework was applied to pre-train a model on an unlabelled dataset. The SimCLR model was created by adding a projection head to the ViT model's output. The projection head consists of a dense layer followed by a normalization layer to create normalized embeddings. Image data augmentations were used for the pre-text task, including random flips, rotations, zooms, translations, contrast adjustments, and resizing. Pre-training was performed based on cosine similarity loss between the embeddings. After pre-training, embeddings were extracted from the projection layer for both the training and validation datasets for the fine-tuning process. A classification head was then added on top of the pre-training as a sequential model with a flattened layer followed by fully connected layers to execute supervised learning on labelled data (see Fig. 3 (bottom-right)).

However, the performance (accuracy of 71.27%) of SSL on the image data was significantly lower than the tabular data (see Table 1). One explanation could be that learning the representations of spectrogram images compared to other images, such as objects, animals, human faces, etc., was a challenging and tricky task as they lack distinguishable patterns and edges. However, ViT was efficient enough to classify the disease with an accuracy of 93.43% when used in a supervised manner (see 100% data column in Table 1).

CNN with Triplet Margin Loss. This algorithm involves a CNN network with triplet margin loss (TML) based SSL pre-training, which is then followed by fine-tuning using supervised learning with a pre-trained backbone. In each epoch, positive and negative pairs are created from the input images, and the forward passes are performed for the anchor, positive, and negative pairs through the model. Triplet margin loss is calculated and used for backpropagation and optimization. This model, too, underperformed compared to models trained on tabular data and had a classification accuracy of 69.34%. The performance of all SSL techniques is summarised in Fig. 4.

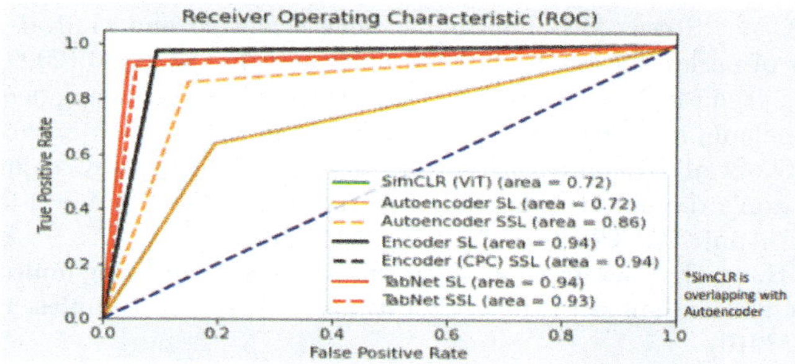

Fig. 4. Performance (ROC curves) of self-supervised learning methods.

5 Conclusions

In this study, we investigated the efficacy of several self-supervised learning (SSL) approaches for classifying Parkinson's disease without relying too much on the labelled dataset. We observed that, with insufficient labelled data, many SSL techniques, along with supervised fine-tuning, could be used to build an effective Parkinson's disease classification model. Also, we found that SSL models with similar architectures could also be used to achieve fully supervised classification for more robust performance. The findings imply that better-performing classification models may be trained in the absence of sufficient labelled data, gathering which is costly and time-intensive. Tabular numerical data (features also extracted from spectrogram images) was found to be substantially easier to deal with than the spectrogram image data. The numeric data was simple and quick to train, outperforming the image dataset. With a classification accuracy of **94.15%**, the encoder SSL model with CPC loss emerged as the best SSL model among all SSL techniques. This study shows that when the data is similar, SSL approaches outperform supervised procedures, and the performance did not deteriorate significantly even after utilising roughly a third of the whole data. The effectiveness of deep learning techniques directly depends on a number of different pre-text activities and hyperparameter tuning [20]. Therefore, future studies will involve investigating these tuning.

References

1. Sriram, T.V., Rao, M.V., Narayana, G.S., Kaladhar, D.: Diagnosis of parkinson disease using machine learning and data mining systems from voice dataset. In: Proceedings of the 3rd International Conference on Frontiers of Intelligent Computing: Theory and Applications 2014: vol. 1, pp. 151–157 (2015)
2. Sakar, B.E., et al.: Collection and analysis of a parkinson speech dataset with multiple types of sound recordings. IEEE J. Biomed. Health Inform. **17**(4), 828–834 (2013)
3. Gugliandolo, G., et al.: A movement-tremors recorder for patients of neurodegenerative diseases. IEEE Trans. Instrum. Meas. **68**(5), 1451–1457 (2019)

4. Olanow, C.W., Stern, M.B., Sethi, K.: The scientific and clinical basis for the treatment of parkinson disease (2009). Neurology **72**, S1–S136 (2009)
5. Shinde, S., et al.: Predictive markers for parkinson's disease using deep neural nets on neuromelanin sensitive mri. NeuroImage: Clinical **22**, 101748 (2019)
6. Sakar, C.O., et al.: A comparative analysis of speech signal processing algorithms for parkinson's disease classification and the use of the tunable q-factor wavelet transform. Appl. Soft Comput. **74**, 255–263 (2019)
7. Zhang, H.H., et al.: Classification of parkinson's disease utilizing multi-edit nearest-neighbour and ensemble learning algorithms with speech samples. Biomed. Eng. **15**, 1–22 (2016)
8. Tsanas, A., Little, M.A., McSharry, P.E., Spielman, J., Ramig, L.O.: Novel speech signal processing algorithms for high-accuracy classification of parkinson's disease. IEEE Trans. Biomed. Eng. **59**(5), 1264–1271 (2012)
9. Harel, B., Cannizzaro, M., Snyder, P.J.: Variability in fundamental frequency during speech in prodromal and incipient parkinson's disease: a longitudinal case study. Brain Cogn. **56**(1), 24–29 (2004)
10. Meghraoui, D., Boudraa, B., Merazi-Meksen, T., Boudraa, M.: Parkinson's disease recognition by speech acoustic parameters classification. In: Proceedings of the 4th International Symposium on Modelling and Implementation of Complex Systems: MISC. pp. 165–173 (2016)
11. Karaman, O., Çakın, H., Alhudhaif, A., Polat, K.: Robust automated parkinson disease detection based on voice signals with transfer learning. Expert Syst. Appl. **178**, 115013 (2021)
12. Gürüler, H.: A novel diagnosis system for parkinson's disease using complex-valued artificial neural network with k-means clustering feature weighting method. Neural Comput. Appl. **28**, 1657–1666 (2017)
13. Ericsson, L., Gouk, H., Loy, C.C., Hospedales, T.M.: Self-supervised representation learning: Introduction, advances, and challenges. IEEE Signal Process. Mag. **39**(3), 42–62 (2022)
14. Toye, A.A., Kompalli, S.: Comparative study of speech analysis methods to predict parkinson's disease. arXiv:2111.10207 (2021)
15. Xie, J.C., Gan, Y., Liang, P., Lan, R., Gao, H.: Exploring robust computer-aided diagnosis of parkinson's disease based on various voice signals. Frontiers Phys. **10**, 1128 (2022)
16. Ssnderson, C., Paliwal, K.K.: Effect of different sampling rates and feature vector sizes on speech recognition performance. In: Proceedings of IEEE TENCON'97. IEEE Region 10 Annual Conference on Speech and Image Technologies for Computing and Telecommunications. vol. 1, pp. 161–164 (1997)
17. McLoughlin, I., Zhang, H., Xie, Z., Song, Y., Xiao, W., Phan, H.: Continuous robust sound event classification using time-frequency features and deep learning. PLoS One **12**(9) (2017)
18. Berus, L., Klancnik, S., Brezocnik, M., Ficko, M.: Classifying parkinson's disease based on acoustic measures using artificial neural networks. Sensors **19**(1), 16 (2018)
19. Bishop, C.M., Nasrabadi, N.M.: Pattern recognition and machine learning, vol. 4. Springer (2006)
20. Taylor, R., Ojha, V., Martino, I., Nicosia, G.: Sensitivity analysis for deep learning: ranking hyper-parameter influence. In: 2021 IEEE 33rd International Conference on Tools with Artificial Intelligence (ICTAI). pp. 512–516. IEEE (2021)

HCU: A Multi-lingual Health Claim Dataset and Its NLP Applications

Huizhi Liang[1(✉)], Zhiling Zhou[1], Zhenyu Hou[1], Chris Ryder[2], and Rodney Jones[2]

[1] Newcastle University, Newcastle, UK
{huizhi.liang,zhiling.zhou,zhenyu.hou}@newcastle.ac.uk
[2] University of Reading, Reading, UK
{c.s.ryder,r.h.jones}@reading.ac.uk

Abstract. We propose a multi-lingual health claim dataset for NLP applications and cross-discipline study. It has six languages including English, French, German, Poland, Romanian, and Hungarian. It has three sub datasets: EFSA authorized, manufacturer reworded, and consumer created Health Claims. We conducted basic data analysis of the dataset, described the data collection process, and discussed two NLP applications including text style transfer and machine translation. We applied recent machine learning models and conducted empirical evaluations of these models using this dataset. The experimental results show that this dataset can be used to investigate the design and evaluation of novel algorithms of NLP tasks. This dataset and its NLP applications can be used for food industry to generate multi-lingual health claims with different styles easily and automatically. This dataset bridges the gap of lacking good quality data for these applications in Health Claim area.

Keywords: health claims · multi-lingual · text style transfer · machine translation

1 Introduction

Health claims (HCs) are sentences on the food product packages to claim the nutrition and the benefits of the nutrition. Often, it relates to the benefits of a nutrient or food for overall health or health related well being [1]. Figure 1 shows an example health claim *DHA Omega-3 supports Brain Health* on a food package. HCs are popularly used by food manufacturers to attract consumers and promote their products.

To protect consumer rights and address the quality of potentially healthy foods, the European Food Safety Authority has developed food labelling regulations[1]. According to the regulation, all HCs on food labels must be supported by scientific data and not be deceptive, and are evaluated by the European Food Safety Authority (EFSA). Clear legal requirements and regulatory systems reduce the likelihood that consumers will be misled by exaggerated HCs on food products, but at the same time, EFSA health claims contain too many technical terms, such as cholesterol and homocysteine, making

[1] https://www.efsa.europa.eu/en/topics/topic/health-claims.

D. Kriksciuniene et al. (Eds.): IBICA 2023, LNNS 1228, pp. 81–90, 2025.
https://doi.org/10.1007/978-3-031-78937-3_8

Fig. 1. An example health claim on a food package [2].

it difficult for consumers to understand the health claims , leading to increased confusion and decreased trust in the foods they purchase. Consumers in different European contexts often have difficulties understanding Health claims on food packages, leading to increased confusion about and decreased trust in the food they buy [3]. European Commission (EC) Regulation 1924/2006 was designed to increase consumer trust and promote healthy food choices by regulating the kinds of HCs. Although, Regulation (EC) 432/2012 encourages manufacturer presents more acceptable HCs to consumers that has "the same meaning" as permitted HCs, consumer confusion and mistrust continuously persist, as documented in a range of academic studies [4–6].

Our EU funded project HealthClaimUnpacked [7] aims to solve this problem by engaging consumers, manufacturers and food safety regulators in determining the different linguistic and cultural barriers to the effective communication of HCs and formulating solutions that can be implemented on multiple levels, including regulation, enforcement, marketing, and consumer education. This project adopted user survey, focus group, online gamified toolkit [7] to engage users and collect multi-lingual health claim related data. This project was under the policies applied to all EU countries (while US shares the similar regulations), involving combined expertise in linguistics, computer science, design, behavioural economics, and dietitians.

In this paper, we describe a multi-lingual health claim dataset that collected by HealthClaimUnpacked project. This dataset contains English and other five EU languages including French, German, Polish, Romanian, and Hungarian. It bridges the gap of lacking good quality multi-lingual health claim data in this domain. From the computer science point of view, this dataset can be used to design and evaluate some NLP tasks. For example, automatic text style transfer task between authorized health claims and manufacture revised health claims. It also can be used for the training and evaluating multi-lingual machine translation algorithms in health claim area. This dataset also can support new NLP tasks such as attractiveness analysis of health claims in our previous work [8]. Moreover, this dataset can be used for other multi-disciplinary researchers such as linguistics, behavioural economists, and dietitians for research purposes. In this paper, we only focus on the example applications of NLP tasks based on this dataset. In the following section, we describe the dataset in Sect. 2. The example NLP applications including style transfer and machine translations are discussed in Sect. 3. A conclusion is given at Sect. 4.

2 Dataset Description

We first describe the data collection process. Then, we describe the dataset in details.[2]

2.1 Data Collection Process

We collected data from the viewpoints of both industry and consumers. To get the industry perspective, interviews were undertaken with chief scientists, nutritionists and/or dietitians and representative from the regulatory and legal departments from five large and mostly global food manufacturers in order to gain an understanding of the practices and processes by which EFSA claims are selected and transformed, and their thoughts and opinions on using HCs on food packages. To get the consumer perspectives, focus groups from each selected country were conducted with consumers from a range of demographics: students (aged 18–25), health-conscious men (aged 30–45), mothers with children aged 7–8, and members of the "aging well" population (aged 50–65). Participants were given tasks about their understanding of healthy or unhealthy products and HCs.

Based on the findings of focus groups, we designed and developed a digital toolkit to engage consumers in a variety of activities designed to inform/educate consumers. These activities include Activities like *What's good for what?*, *What is a health claim*, *Similar or different*, *What would you buy?* and *Design your own food pack*. Users can design their own food pack as well as their own HCs. These activities also contribute to collecting data from consumers about their understanding and preferences of HCs for research purpose.[3]

Data from all stages of the project are pulled together to derive the best practice guidelines aimed at food manufacturers across the UK and Europe in the employment of health claims on their food product packaging, revised from those outlined in 2019 in order to account for cross-cultural and cross-linguistic factors that have emerged from this research. The research findings can be found in URL. From all the data that we collected, we constructed a health claim dataset that is suitable for public usage for research purpose, which will be described in the following subsection.

2.2 Health Claim Dataset

We name this dataset as **HealthClaimsUnpacked** and the abbreviation is named as **HCU**. The whole dataset has the following three kinds of sub datasets: 1) HCU-A-2022-6: Authorized Health Claims; 2) HCU-M-2022-7: Manufacturer Produced Health claims; 3) HCU-C-2022-6: Consumer Designed Health claims. We will discuss the three sub datasets in the sub sections.

---

[2] The dataset is publicly accessible with this link https://doi.org/10.5281/zenodo.7993709.

[3] Note: The collection, storage, and usages of data follow the guidance of the General Data Protection Regulation (GDPR).

HCU-A-2022-6: Authorized Health Claims. This is the sub dataset of Authorized English Health Claims approved by Europe Food Safety Association (EFSA) and their corresponding French, German, Polish, Romanian, and Hungarian translations that are translated by professional translators. The English health claims was downloaded from EFSA website[4] The related regulation article is EC 1924/2006. This sub dataset contains 261 health claims in six languages including English (EN), French (FR), German(DE), Polish (PL), Romanian (RO), and Hungarian (HU). An example English health claim is *Calcium is needed for normal growth and development of bone in children.* In this sub dataset, we manually labelled the nutrient for each claim. For example, the corresponding nutrient of this health claim is *Calcium*. Table 1 shows an example authorized English health claim and their multi-lingual translations. This sub dataset also contains the corresponding EC Regulation name and the section of Article of EC 1924/2006. For example, the corresponding EC regulation of the selected health claims of Table 1 is *"432/2012"* and the corresponding section of Article of EC 1924/2006 is *"Artical 13(1): function health claim"*.

Table 1. Example claims in sub dataset HCU-A-2022-6 (Authorized Health Claims)

	Example Claim
English	Alpha-linolenic acid (ALA) ALA contributes to the maintenance of normal blood cholesterol levels
French	L'acide alpha-linolénique (ALA) contribue au maintien d'une cholestérolémie normale
German	ALA trägt zur Aufrechterhaltung eines normalen Cholesterinspiegels im Blut bei
Polish	Kwas alfa-linolenowy (ALA) pomaga w utrzymaniu prawidłowego poziomu cholesterolu we krwi
Romainan	AAL contribuie la menținerea concentrațiilor normale ale colesterolului din sânge
Hungarian	Az ALA hozzájárul a vér normál koleszterinszintjének fenntartásához.

HCU-M-2022-7: Manufacturer Produced Health Claims. This is the sub set of health claims shown in food packages that produced by food manufactures. It includes six country-specific sub sets collected from six countries (HCU-M-2022-UK, HCU-M-2022-FR, HCU-M-2022-DE, HCU-M-2022-PO, HCU-M-2022-RO, and HCU-M-2022-HU) and one English sub set that not specific for UK consumers (HCU-M-2022-EN). HCU-M-2022-EN was collected with an online tool provided by industry collaborator Foodmaestro (now a part of Warlmart).

To better analysis the consumer behaviour and understanding of a specific region, we manually collected a set of health claims of food packages from either offline or online supermarkets of UK, France, Germany, Poland, Romanian, and Hungary. We try to collect the same set of nutrients and food categories, but cover different brands and manufactures. For each country specific sub set, it records *Substance (i.e., nutrient)*, *Claim on Pack*, *Authorized EFSA Claim*, *Product Category (e.g., nuts, cereal, water)*, *Brand (e.g., Weetabix, Nestle)*, *Product (e.g., Additions Red Berries 582g)*, and *Source (e.g., Ocado, Morrisons)*.

[4] https://ec.europa.eu/food/food-feed-portal/screen/home.

Table 2. Example claims in sub dataset HCU-M-2022-7 (Manufacturer Produced Health claims)

Language	**Example Claims on Food Packages**
English	Omega 3 alpha-linolenic acid (ALA) is found naturally in linseeds. Thanks to these seeds, this bread provides a good source of Omega-3 ALA which helps to maintain healthy levels of cholesterol in your body.
French	L'acide alpha-linolénique (oméga 3 - ALA) naturellement présent dans l'huile de colza contribue au maintien d'une taux normal de cholestérol dans le sang. L'effet bénéfique est obtenu par la consommation journalière de 2g d'ALA (soit 3 culières à sole de ce mélange), dans le cadre d'une alimentation équilibrée.
Polish	Czy wiesz co sprawia, że Flora jest dobra dla Ciebie?. Kompozycja wysokiej jakości olejów:. olej słonecznikowy. olej lniany. olej rzepakowy. ...naturalnie bogatych w Omega 3 i 6, co sprawia, że Flora jest dobra dla Ciebie i Twojego serca. Dlaczego Flora jest dobra dla Twojego serca?. Dzięki cennym Omega 3 i 6 pochodzącym z kompozycji wysokiej jakości olejów. Ich spożywanie, w ramach zbilansowanej diety i zdrowego stylu życia, pomaga utrzymać prawidłowy poziom cholesterolu we krwi. 20 g Flory Gold zawiera istotną część dziennej porcji kwasów Omega 3 (ALA) i Omega 6 (LA). Korzystne działanie występujące w przypadku spożywania 2 g ALA i 10 g LA dziennie.
German	Gesättigte Fettsäuren in der Ernährung durch ungesättigte Fettsäuren (wie. z.B. Omega 3 & 6) zu ersetzen, trägt zur eines normalen Cholesterinspiegels bei. Daneben ist eine Aufrechterhaltung abwechslungsreiche und ausgewogene Ernährung und gesunde Lebensweise wichtig.
Romanian	Omega 3 contribuie la menţinerea concentraţiilor normale de colesterol
Hungarian	A vas legfontosabb feladata az oxigén és a szén-dioxid szállítása a szervezetben. Nélkülözhetetlen továbbá a sejtek növekedésében, regenerációjában és a védekező rendszer működésében.
	Corresponding EFSA Authorized Health Claims:
English	ALA contributes to the maintenance of normal blood cholesterol level
Hungarian	A vas a szervezetben részt vesz a normál oxigénszállításban

Table 3. Basic Statistics Summary of HCU-M-2022-7

FileNaume	Country	Nutrient	Claims	Authorized Claims	Product Category	Brand	Product	Sources
HCU-M-2022-UK	UK	39	352	117	24	50	152	4
HCU-M-2022-FR	France	41	147	74	18	40	69	3
HCU-M-2022-DE	Germany	27	74	52	8	16	32	4
HCU-M-2022-PL	Poland	33	163	76	14	24	58	2
HCU-M-2022-RO	Romanian	29	85	68	12	28	44	4
HCU-M-2022-HU	Hungarian	25	89	55	15	22	36	3

Table 2 shows example claims in the six country-specific sub sets of HCU-M-2022-7 (Manufacturer Produced Health claims). Please note they are not necessary from the same brand or food category. The corresponding English EFSA authorized health claim is *"ALA contributes to the maintenance of normal blood cholesterol level"* for the English, French, German, Polish, Romanian example health claims. As the Hungarian sub set does not contain health claims for the nutrient ALA, an example Health claim about the nutrient Iron and the corresponding Hungarian EFSA Authorized health claim is shown in this table. We can see that many manufactures reworded their health claims and the health claims of manufacturers are usually different from the authorized health claims.

Table 3 summarizes the basic information about the six country-specific sub sets from the aspects of country, total number of claims, total number of referred nutrients, total number of product categories, total number of Brands, total number of Products and total number of sources. For the NON country-specific English sub set HCU-M-2022-EN, it has 4608 health claims with *Product Name*, *Category*, and *Sub Category*. They are general English health claims, not specifically for UK consumers. Please see the data file for detailed information.

HCU-C-2022-6: Consumer Designed Health Claims. collected with our website. Not free text, Semi-free text (give some options and ask them to select). Platform breif introduction, collected information. 6 languages. demographic information. The consumer platform aims to popularise the role, constitution, and importance of HCs to reduce consumer confusion and build up their trust in food products. It provides multiple interactive online *activities* including educating activities (Activity 1A, Activity 1B), and practice activities (Activity 1C, Activity 2, Activity 3 in Figure ??) to teach HCs knowledge to consumers. While the toolkit conveys the knowledge to the users, it tests the users' understanding as well as letting the users practice the knowledge. There are multiple perspectives to the users to understand HCs. The user need to register to the website and fill in forms about their profile such as country, age group, gender and some information related to health such as health condition and health goal.

In Activity 3 (*Design your own food pack*), users can design their own food pack as well as their own HCs. First, users need to choose one product (i.e. milk, yoghurt, etc.), then designing a HC for the product. The designing process is governed by a template. It covers a group of words which are sufficient to write usual HCs for the option products. Based on the template, users only need to literally select the words for the HC sentences they are creating. Figure 2 shows one template of some options of selecting the words of a HC. The template always lead users to create legal HCs. This activity also gives users the chance to manually design the product package by changing the icon, colour, and layout. We collected all the user created health claims in six languages with their user information. Researchers can conduct research based on this sub dataset. For example, conducting consumer preferences analysis based on different consumer groups. We conducted statistical analysis of consumer designed health claims. Table ?? shows the example most popular consumer designed health claims about Omega-3 in six languages of sub dataset HCU-C-2022-6.

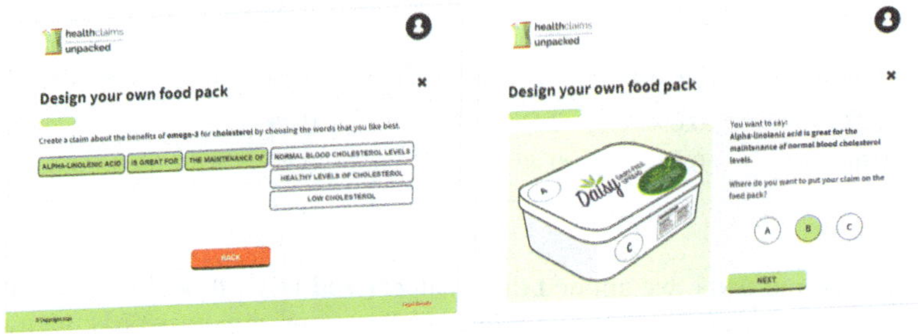

Fig. 2. Activity 3:'Design your own food pack'

3 NLP Applications

The dataset can be used to conduct NLP tasks and other cross-discipline research for example, cross-country and cross-cultural linguistic analysis, marketing analysis, economic behaviour analysis. We focus on NLP applications in this paper. Example applications include automatic attractiveness analysis [8], text style transfer, and machine translations. In our published work [8], we conducted automatic attractiveness analysis and consumer preference prediction of health claims for different consumer groups. This helps develop an AI-powered decision making support tool for food manufacturers in designing attractive health claims for consumers. We mainly focus on the example applications of automatic text style transfer and machine translation in this paper.

3.1 Text Style Transfer

The task of automatic style transfer is to defined as: the text style of the health claim that satisfies the EFSA assessment can be converted to the text style of the health claim reworded by manufactures. Note the health claims produced by manufactures can be further divided into different sub-groups of styles. For easy explanation and processing, in this paper, we only consider them as one big group. For example, the text style transfer algorithms can convert the difficult-to-understand EFSA authorized health claim *"ALA contributes to the maintenance of normal blood cholesterol levels."* of the sub set HCU-A-2022-6 to the reworded easily understandable health claim *"Omega 3 alpha-linolenic acid (ALA) is found naturally in linseeds. Thanks to these seeds, this bread provides a good source of Omega-3 ALA which helps to maintain healthy levels of cholesterol in your body."* in the sub dataset HCU-M-2022-7. The trained style transfer models is expect to help manufacture companies to speed up the process of designing health claims for their food products and increases their competitiveness in the market. It is also expect to enable consumers to have an easy and better understanding of health claims on the food packs.

We applied two popularly used style transfer models to our dataset. The HCU-A-2022-6 sub dataset and the HCU-M-2022-7 sub dataset is used for this task. For the Supervised model, we applied the BILSTM model [9] to our dataset. Based on the fact that some of the biological and medical words in the health declaration do not change in

Table 4. Results of Text Style Transfer

Selected Model	Naturalness	Content Preservation	Style Transfer Intensity
DRG	0.69	0.58	0.74
BILSTM	0.81	0.42	0.90

the input and output texts, we applied the unsupervised DRG model [10]. It can retain the keywords in the sentence, remove relevant phrase enrichment words, then retrieve new phrases matching the target by keywords, and finally output the new combination using a neural network model. The specific text style transfer process of the model is shown in Fig. 3.

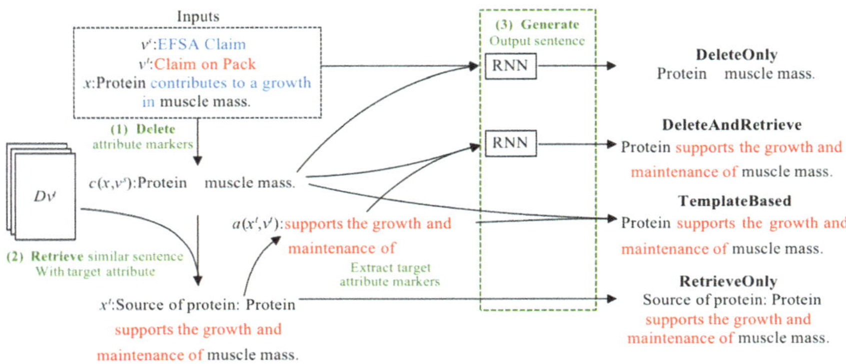

Fig. 3. We selected a sentence from the EFSA authorized HC as an example. Its text style transfer is achieved in the model mainly through the following processes: 1) Deleting the vocabulary marking the attributes in the source sentence, thus separating the content and attributes of the source sentence; 2) RetrieveOnly: Retrieving sentences with the same content as the source sentence among other sentences; 3) TemplateBased: Combining the desired target attributes with the retrieved sentences to generate a new style of sentence. In addition to this, DeleteAndRetrive will use the output attributes generated by the RNN marker DeleteOnly.

In this study, content preservation, style transfer intensity, and naturalness [11] are used to evaluate the effectiveness of text style transfer. The results are shown in Table 4. We can see that the BILSTM supervised model has a higher quality of text style transfer. However, the text content preservation is lower with a value of 0.42. The DRG unsupervised model has a lower style transfer and naturalness of 0.47 and 0.69 respectively, but its content preservation is higher than BILST Model.

3.2 Machine Translation

It is challenging to translate one European language completely into another, which is an important issue for the wording of health claims across the EU. For example, the

Table 5. Results of Machine Translation (BLEU score)

Model	Translation to English					
	FR	DE	PL	RO	HU	Overall
mBART50	37.2	40.1	22.3	27	/	31.9
M2M-100	36.7	36.9	19.4	33.5	26.1	29.0
Model	Translation from English					
mBART50	42.7	27.3	22.9	25.8	/	31.0
M2M-100	41.2	29.5	25.8	47.5	25.2	36.8

translated health statement would employ the word *'prawidowy,'* which means *'proper'* or *'right,'* even though the word *'normal'* already exists in Polish. As an illustration, the official English health statement uses the word *'normal'* several times. As laws and regulations regarding health claims vary from country to country in Europe, cultural and linguistic differences can lead to different wording of the same health claim in different countries, which can lead to confusion and mistrust among consumers about the food products they buy [12]. Machine translation facilitates cross-language communication and helps people overcome language barriers. The lack of gold standard datasets makes machine translation challenging. To ensure the quality of translation, professional translators were hired to translate health claims from English to other five selected languages. The constructed dataset filled the gap of lacking high quality gold standard dataset for machine translation task in the domain of health claims.

Based on the constructed dataset, we compare the translation quality of the task of translating from English to other languages and the task of translating from other languages to English. We applied two multilingual translation models mBART50 [13] and M2M-100 [14] for the health claim machine translation task. The HCU-A-2022-6 sub dataset is used for this task. The results of the mBART50 and M2M-100 models, as evaluated by BLUE score, are shown in the Table 5. Please note that the MBART50 model does not support Hungarian language. There are no evaluation results of MBART50 for HU in this table. The M2M-100 model supports all the other five languages. We can see that overall the mBART50 model performed better in the task of translating from other languages to English with an average BLUE score of 31.9, while that of the M2M-100 model has a value of 29.0 for all the supported languages. On the other hand, overall the M2M-100 model performs better in the task of translating from English to other five languages with an average BLUE score of 36.8. The mBART50 model has an average BLUE score of 31.0 for the all supported languages of our dataset.

4 Conclusions

In this paper, we proposed a multi-lingual Health Claim dataset **HCU** for NLP applications. It has three sub datasets in six languages. We conducted basic data analysis of the dataset, described the data collection process, and conducted two typical NLP applications including text style transfer and machine translation based on this dataset.

The experiments show that this dataset can be used as good quality dataset for NLP applications in health claim area. In the future, we will explore novel NLP application such as multi-lingual style transfer. This dataset also can be used for cross-discipline study for researchers in areas including linguistics, marketing, and economics.

Acknowledgements. This project is funded by European Institute of Innovation and Technology (EIT) Food/EU Horizon 2020 (project number 19098).

References

1. Geiger, C.J.: Health claims: history, current regulatory status, and consumer research. J. Am. Diet. Assoc. **98**(11), 1312–1322 (1998). https://www.sciencedirect.com/science/article/pii/S0002822398002946
2. Factual food labels: health claims. https://he.utexas.edu/ntr-news-list/factual-food-labels-health-claims Accessed 18 Nov 2023
3. Wansink, B., Sonka, S.T., Hasler, C.M.: Front-label health claims: when less is more. Food Policy **29**(6), 659–667 (2004)
4. Nocella, G., Kennedy, O.: Food health claims-what consumers understand. Food Policy **37**(5), 571–580 (2012)
5. Orquin, J.L., Scholderer, J.: Consumer judgments of explicit and implied health claims on foods: misguided but not misled. Food Policy **51**, 144–157 (2015)
6. Annunziata, A., Mariani, A.: Do consumers care about nutrition and health claims? some evidence from italy. Nutrients **11**(11), 2735 (2019)
7. Li, X., Liang, H., Liu, Z.: Health claims unpacked: a toolkit to enhance the communication of health claims for food, pp. 4744–4748. Association for Computing Machinery, New York, NY, USA (2021). https://doi.org/10.1145/3459637.3481984
8. Li, X., Liang, H., Ryder, C., Jones, R., Liu, Z.: Attractiveness analysis for health claims on food packages. In: Data Mining, pp. 217–232. Springer Nature Singapore, Singapore (2022)
9. Siami-Namini, S., Tavakoli, N., Namin, A.S.: The performance of lstm and bilstm in forecasting time series. In: 2019 IEEE International Conference on Big Data (Big Data), pp. 3285–3292 (2019)
10. Li, J., Jia, R., He, H., Liang, P.: Delete, retrieve, generate: a simple approach to sentiment and style transfer. CoRR abs/1804.06437 (2018). http://arxiv.org/abs/1804.06437
11. Mir, R., Felbo, B., Obradovich, N., Rahwan, I.: Evaluating style transfer for text. CoRR abs/1904.02295 (2019). http://arxiv.org/abs/1904.02295
12. van Trijp, H.C., van der Lans, I.A.: Consumer perceptions of nutrition and health claims. Appetite **48**(3), 305–324 (2007). https://www.sciencedirect.com/science/article/pii/S019566630600612X
13. Tang, Y., et al.: Multilingual translation with extensible multilingual pretraining and finetuning. CoRR abs/2008.00401 (2020). https://arxiv.org/abs/2008.00401
14. Fan, A., et al.: Beyond english-centric multilingual machine translation. CoRR abs/2010.11125 (2020). https://arxiv.org/abs/2010.11125

HealthTech Horizons: Promoting Sustainable Healthcare in Developing Nations

Vijay Bhargav Bhamidi[1], Yash Kanani[1], Yajnaseni Dash[2](✉), and Ajith Abraham[3]

[1] School of Computer Science Engineering and Technology, Bennett University, Greater Noida, UP, India

[2] School of Artificial Intelligence, Bennett University, Greater Noida, UP, India
yajnasenidash@gmail.com

[3] School of Artificial Intelligence, Sai University, Chennai, India
abraham.ajith@gmail.com

Abstract. The integration of Artificial Intelligence (AI) in the healthcare sector has transformed traditional practices, addressing inefficiencies and manual data processing challenges. This paper explores the advancements and applications of AI in pharmaceuticals, disease prognosis, and healthcare decision-making. It highlights the benefits of AI in improving resource allocation, personalized care, and data analysis while addressing concerns about data privacy and confidentiality. The discussion extends to the impact of AI on reducing avoidable mortality, particularly in underdeveloped and developing countries. The paper also examines the manipulation of datasets for accurate prognoses and the balance between local and global data in AI applications. Ultimately, the study underscores the need for innovative strategies, international collaboration, and tailored approaches to enhance healthcare outcomes globally.

Keywords: Artificial Intelligence · Machine Learning · Healthcare · Mortality · Developing nations

1 Challenges and Limitations Preceding AI Integration

In the area of healthcare, Artificial Intelligence (AI) has become an effective strategy that could completely transform patient care, diagnosis, and therapy. However, due to resource limitations and technological aspirations, its extensive implementation in developing nations confronts various hurdles. The particular factors and challenges associated with the usage of AI techniques in developing country healthcare systems are explored in this study. The availability of different and high-quality datasets is the key limitation. The efficiency of AI is restricted in developing nations due to the frequent absence of comprehensive and standardized healthcare data. In this paper, we have analyzed the issues that the healthcare sector experienced before the introduction of AI and why it is essential to use these strategies in the healthcare sectors of developing nations. The healthcare sector has worked under several paradigms and practices before the incorporation of AI. Manual data processing, limited customization, and inefficiencies were

D. Kriksciuniene et al. (Eds.): IBICA 2023, LNNS 1228, pp. 91–98, 2025.
https://doi.org/10.1007/978-3-031-78937-3_9

widespread, causing difficulties for both healthcare personnel and patients. For sensitive information such as patient records and medical images, healthcare personnel completely rely on manual data storing methods. The application of AI in healthcare has significantly improved resource allocation, disease outbreak prediction, personalization of care, and data analysis. As these technologies solve the problems caused by manual processes and inefficiency, the healthcare sector has undergone a revolution. Healthcare providers may now provide more prompt interventions, individualized therapies, personalized medications, and accurate diagnoses due to AI which will ultimately improve patient outcomes and healthcare efficacy [1].

2 Utilizing AI in Pharmaceuticals

The application of artificial intelligence techniques in underdeveloped countries' healthcare systems necessitates a careful balance between technology aspirations and resource limitations. To solve challenges, collaboration, innovative thinking, and an international viewpoint are required due to the interconnection of illnesses, limited data, and resource limits. Machine learning approaches can have a big influence on healthcare results and lessen the gap between wealthy and developing nations. We need innovative strategies such as working with international organizations and transferring expertise. Tailoring to individual nations can enhance accuracy and lessen prejudice, but it can also result in care being ignored. It's necessary to achieve equilibrium between regional databases and worldwide expertise to guarantee the efficient establishment of healthcare services.

The authors [2] underscored the significance of adapting to the advancements in technological innovation and the possible integration of AI in the therapeutic or healthcare sector for the mutual advantage of all healthcare personnel. They have focused on using AI to eliminate tedious activities so they can concentrate on quiet doctor-patient relationships and foster the growth of empathy and emotional intelligence. In this field, machine learning (ML) can assist medical professionals in discovering, treating, diagnosing, and stopping an individual from getting the disease. Yet, it is imperative to address both the loss of confidentiality and the possibility of patient data being misused. Precision medicine has made much of the concept of subtyping [3]. It has shown how different subtypes within the population might be found using ML techniques, opening up new possibilities for more specialized and individualized treatment plans. The work conducted by the authors [4] aims to enhance scalability and accuracy in the healthcare industry by integrating deep learning techniques using electronic health records (EHRs). The use of deep learning to analyze and interpret medical images was discussed [5]. Improvement in monitoring and treatment through telemedicine shows the power of AI in telemedicine [6]. The application of big data and ML was discussed [7, 8]. Authors used ML techniques for activity recognition using sensor-based data [9, 10]. For the analysis of large-scale genomics data, Engelhardt and Stephens [11] illustrated the usefulness of ML models in identifying genetic patterns and connections at the population level, which may have consequences for customized treatment. In health informatics, Ravi et al. [12] presented an outline of deep learning applications in healthcare. To improve decision-making and patient outcomes, it addresses how algorithms using deep learning can be used for a variety of healthcare tasks, including identifying diseases, prognosis, and healthcare administration. Authors [13] focused on creating probabilistic graphical models for conveying uncertain knowledge and drawing probabilistic conclusions. This study provides instances of how to create graphical models, which might be helpful in healthcare applications. The prospects and challenges of digital health across the African continent are explored [14]. A description of the present situation of ML for healthcare may be found in this comprehensive mapping study [15]. It provides insights into the current state of research and potential future paths by identifying and classifying different machine-learning approaches, applications, and challenges in the healthcare industry.

3 Understanding Prognosis: Impact on Normal Patients and Healthcare Decision-Making

Prognosis plays a crucial role in healthcare as it provides valuable insights into the likely outcome of a medical condition or disease. It allows healthcare professionals to make informed decisions, develop appropriate treatment plans, and support patients in understanding and managing their health. Over the last two decades, there has been a significant advancement in the application of ML techniques and feature selection algorithms in disease prognosis and prediction [16]. Though ML approaches have demonstrated encouraging outcomes in terms of disease prognosis, it is crucial to remember that they ought to constantly be utilized in combination with clinical judgment and knowledge. With the help of this development, medical practitioners can now give patients better treatment plans, more precise diagnoses, and more efficient use of their resources. By leveraging ML-based prognostic models, healthcare systems can optimize patient care and empower individuals to actively participate in their healthcare journey. ML approaches and feature selection algorithms have been widely applied to disease prognosis and prediction over the last two decades [16]. This paper has utilized ML approaches to simulate cancer progression, find valuable features, and later use those in classification [16].

The distribution of cancer susceptibility, recurrence, and survival using ML approaches is shown in Fig. 1 based on published research. The data was gathered using a range of searches conducted in the Scopus biomedical database. Healthcare providers consider all factors when providing patients with information about their prognosis. It's important to note that prognostic information is not always certain, and individual outcomes can vary widely. It's often given as a range or statistical likelihood based on the available data and research.

Prognosis allows the patient to get the nearest hospitalization available to a certain problem in an extremely short time, though it may not be 100% accurate it is near best given a certain given condition. In underdeveloped and developing countries there is a lack of infrastructure in many places, and in many cases, hospitals are just empty buildings. Approximately 50% of the global population still faces challenges in accessing essential healthcare services, while a staggering 100 million individuals are forced into extreme poverty annually due to the financial burden of healthcare expenses [17]. In 2015, there was an index developed to measure the extent of coverage provided by Universal Health Care (UHC) services [17]. This index aimed to assess the level of accessibility and availability of healthcare services for all individuals within a given

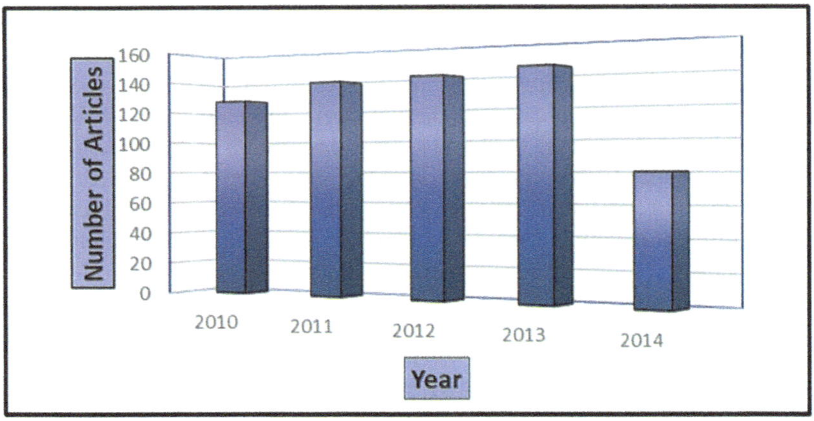

Fig. 1. Distribution in published studies employing machine learning approaches to forecast survival, recurrence, and susceptibility to cancer.

population. Its purpose was to provide a comprehensive measure of the degree to which a country's healthcare system was able to achieve universal coverage. The UHC coverage revolves around ensuring that all individuals have access to the necessary healthcare services without facing financial difficulties. However, the extent of service coverage can significantly differ from one country to another.

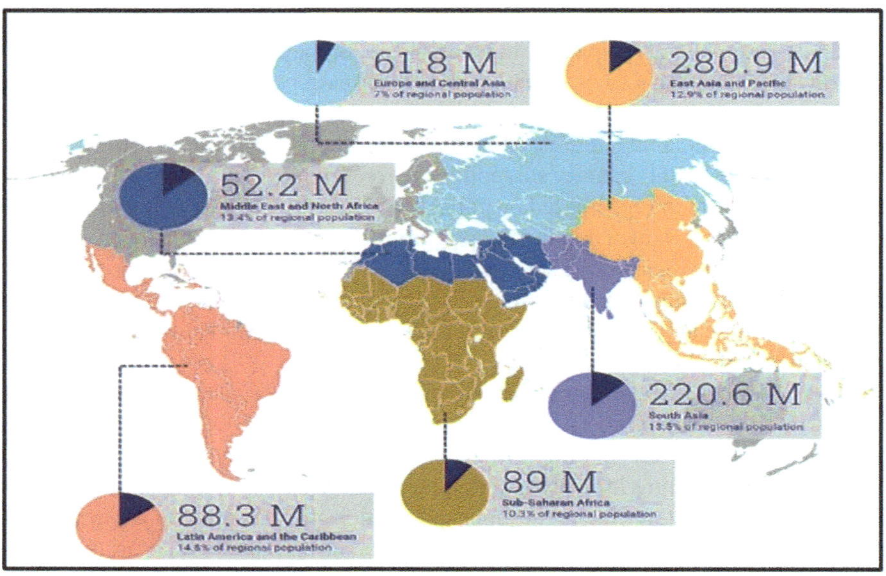

Fig. 2. A significant proportion of individuals are allocating more than 10% of their household budget towards healthcare expenditures [17]

In situations where individuals residing in such areas are unable to access healthcare facilities or receive timely medical attention, even for common ailments that can be effectively treated with proper medication, rest, or minor procedures, the consequences can be severe. A significant proportion of individuals are allocating more than 10% of their household budget towards healthcare expenditures as shown in Fig. 2. They are not only unable to reach hospitals or treatment centers promptly but also even when they do, the available medical personnel may lack the necessary skills and knowledge to recommend basic medications or interventions that could prevent the deterioration of the condition. This critical delay in receiving appropriate care not only puts the patient's health at risk but also prolongs the time before a qualified medical professional can arrive at the location to initiate the necessary treatment. In some cases, this delay can last for days or even weeks, significantly impacting the safety and well-being of the patient.

4 Utilizing Prognosis to Reduce Avoidable Mortality: Enhancing Healthcare Outcomes

Prognosis is a proactive approach that allows for timely interventions, potentially preventing complications and adverse outcomes that may lead to avoidable deaths. Furthermore, prognosis helps medical professionals to recognize high-risk patients and effectively distribute resources, guaranteeing that patients in dire need receive the care they require. Healthcare systems can work to reduce the rate of preventable mortality by utilizing prognosis, which will enhance overall healthcare results and save lives. The rate of avoidable mortality will be lowered in part by the prognosis [18]. The total number of deaths that could have been prevented is known as averted mortality.

Fig. 3. Mortality Associated with Inadequate Healthcare Services Across Countries [18]

As Fig. 3 illustrates, mortality linked to insufficient healthcare is a serious problem that differs between nations. Poor health outcomes and avoidable fatalities can arise from inadequate access to high-quality healthcare. Lack of necessary medical resources, insufficient infrastructure, a scarcity of qualified healthcare workers, and a restricted supply of healthcare facilities are some of the factors causing this issue. A comprehensive strategy is needed to address this problem, one that includes strengthening medical training and education, expanding financing for healthcare, and putting in place efficient healthcare systems and policies. By prioritizing and investing in quality healthcare services, countries can work towards reducing mortality rates and ensuring better health outcomes for their populations.

4.1 Different Underlying Manipulation

For instance, in Africa, there is a high number of malaria cases and other waterborne diseases, but these diseases are not as common in Western countries anymore. When collecting data on symptoms and underlying diseases, the symptoms of malaria, such as high fever, shaking chills ranging from moderate to severe, profuse sweating, headache, and abdominal pain, can be easily confused with symptoms of seasonal flu, cold exposure, or other viral infections. In Western and developed countries, the most common cause of symptoms like headache, sweating, and shaking chills is typically attributed to cold and flu. This biased data can lead to false prognoses if patients or doctors rely solely on the model's predictions. If doctors are not consulted or if they are instructed to unquestioningly follow the ML model's diagnosis (thus being subconsciously influenced by the model's prediction), they may prescribe incorrect medication and consider the prediction to be true, inadvertently perpetuating a false feedback loop. To avoid this, the model should be provided with a dataset that includes altered weights, ensuring that even with less data on symptoms specific to malaria, there are similar chances of the correct output prediction. Additionally, considering other factors such as the patient's location and the local temperature/climate in combination with symptom analysis would help ensure accurate prognoses, which can help break the cycle of false predictions.

4.2 Changing the Dataset for Different Locations/countries

Changing the dataset for different countries may appear to be a viable option as it eliminates the need to modify the dataset itself. Instead, we can provide users in specific regions with the respective dataset of their country or similar countries. For instance, in developed countries like the USA, Canada, UK, Germany, etc., if a user interacts with the ML model, only the dataset of that particular country, or possibly similar countries, will be available to the model. This approach not only helps to mitigate bias stemming from countries with larger datasets and more resources, but also eliminates the need for complex models that account for other factors like weather, history, and climate in each request. Consequently, the model becomes easier and faster to understand and train, and the reduced complexity of the dataset increases the likelihood of higher accuracy in prognoses. But we also need to think about the disadvantages of this strategy. It is extremely uncommon to isolate any illness in a particular area. Regardless of how it affects the populace if a virus or bacteria appears in one place, it is expected to spread across the entire nation. This is especially pertinent to COVID-19 cases. It doesn't matter how quickly or slowly the disease spreads; once it spreads to other regions worldwide, the first few dozen individuals who might benefit from current treatments and prognoses will almost certainly be misdiagnosed and treated incorrectly. As was covered in earlier sections, this scenario may result in a rise in avoidable mortality.

5 Conclusion

The healthcare sector faced numerous challenges before the incorporation of AI, including manual data processing and inefficiencies. AI's integration has revolutionized resource allocation, disease prediction, personalized care, and data analysis, significantly improving patient outcomes and healthcare efficiency. AI in pharmaceuticals has reduced tedious tasks, allowing healthcare professionals to focus on doctor-patient relationships, empathy, and emotional intelligence. Machine Learning (ML) techniques have been pivotal in discovering, diagnosing, treating, and preventing diseases, enhancing scalability and accuracy in healthcare. Prognosis, a critical aspect of healthcare, has benefited from ML techniques and feature selection algorithms, aiding in disease prediction and progression. Despite the promising outcomes, these models must complement clinical judgment and knowledge. The study discusses the importance of addressing biased data and the potential for false prognoses. It suggests adapting datasets for different regions to enhance accuracy, though this approach has limitations, particularly in the context of globally spreading diseases like COVID-19. Prognosis also plays a vital role in reducing avoidable mortality by enabling timely interventions and efficient resource distribution. However, challenges such as inadequate healthcare infrastructure and access persist, particularly in underdeveloped regions. The paper concludes that integrating AI in healthcare systems, especially in underdeveloped countries, requires balancing technological aspirations with resource limitations. International collaboration, innovative strategies, and tailored approaches are essential for improving healthcare outcomes and bridging the gap between developed and developing nations.

References

1. Jiang, F., Jiang, Y., Zhi, H., et al.: Artificial intelligence in healthcare: past, present and future. Stroke Vasc. Neurol. **2**(4), 230–243 (2017). https://doi.org/10.1136/svn-2017-000101
2. Alghadier, M., et al.: A study of various applications of artificial intelligence (AI) and machine learning (ML) for healthcare services. Int. J. Appl. Eng. Technol. **5**(1), 87–94 (2023)
3. Saria, S., Goldenberg, A.: Subtyping: What it is and its role in precision medicine. IEEE Intell. Syst. **30**(4), 70–75 (2015)
4. Rajkomar. et al.: Scalable and accurate deep learning with electronic health records. npj Digital Med. **1**(1), 1–10 (2018)
5. Razzak, M.I., Naz, S., Zaib, A.: Deep learning for medical image processing: overview, challenges and the future. In: Dey, N., Ashour, A.S., Borra, S. (eds.) Classification in BioApps, pp. 323–350. Springer International Publishing, Cham (2018). https://doi.org/10.1007/978-3-319-65981-7_12
6. Pacis, D.M.M., Subido, E.D., Bugtai, N.T.: Trends in telemedicine utilizing artificial intelligence. In: AIP Conference Proceedings, vol. 1933, no. 1, AIP Publishing (2018)
7. Obermeyer, Z., Emanuel, E.J.: Predicting the future big data, machine learning, and clinical medicine. N. Engl. J. Med. **375**(13), 1216–1219 (2016)

8. Sarangi, S.C., Dash, Y.: Application of machine learning and big data analytics in pharmacovigilance and drug safety. In: International Conference on Intelligent Computing, Instrumentation and Control Technologies, Kerala, India, ICICICT, IEEE Xplore Digital Library, pp. 555–559 (2019)
9. Sarangi, S.C., Dash, Y.: Sensor-based physical activity monitoring: Application of machine learning approaches. In: International Conference on Intelligent Computing, Instrumentation and Control Technologies, Kerala, India, ICICICT, IEEE Xplore Digital Library, pp. 621–626 (2019)
10. Dash, Y., Kumar, S., Patle, V.K.: A novel data mining scheme for smartphone activity recognition by accelerometer sensor. In: Das, S., Pal, T., Kar, S., Satapathy, S.C., Mandal, J.K. (eds.) Proceedings of the 4th International Conference on Frontiers in Intelligent Computing: Theory and Applications (FICTA) 2015, pp. 131–140. Springer India, New Delhi (2016). https://doi.org/10.1007/978-81-322-2695-6_12
11. Engelhardt, B.E., Stephens, M.: Analysis of population structure: a unifying framework and novel methods based on sparse factor analysis. PLoS Genet. **6**(9), e1001117 (2010)
12. Ravi, D., et al.: Deep learning for health informatics. IEEE J. Biomed. Health Inform. **21**(1), 4–21 (2017). https://doi.org/10.1109/JBHI.2016.2636665
13. Saranti, A., Taraghi, B., Ebner, M., Holzinger, A.: Insights into learning competence through probabilistic graphical models. In: Holzinger, A., Kieseberg, P., Min Tjoa, A., Weippl, E. (eds.) Machine Learning and Knowledge Extraction: Third IFIP TC 5, TC 12, WG 8.4, WG 8.9, WG 12.9 International Cross-Domain Conference, CD-MAKE 2019, Canterbury, UK, August 26–29, 2019, Proceedings, pp. 250–271. Springer International Publishing, Cham (2019). https://doi.org/10.1007/978-3-030-29726-8_16
14. World Health Organization: Maximising digital health technology to improve quality and patient safety in Africa. World Health Organization (2018)
15. Ngayua, E.N., He, J., Agyei-Boahene, K.: Applying advanced technologies to improve clinical trials: a systematic mapping study. Scientometrics **126**(2), 1217–1238 (2021)
16. Kourou, K., et al.: Machine learning applications in cancer prognosis and prediction. Comput. Struct. Biotechnol. J. **13**, 8–17 (2015)
17. Lack of Health Care is a Waste of Human Capital: 5 Ways to Achieve Universal Health Coverage By 2030. https://www.worldbank.org/en/news/immersive-story/2018/12/07/lack-of-health-care-is-a-waste-of-human-capital-5-ways-to-achieve-universal-health-coverage-by-2030. Accessed 9 Nov 2023
18. Kruk, M.E., et al.: Mortality due to low-quality health systems in the universal health coverage era: a systematic analysis of amenable deaths in 137 countries. The Lancet **392**(10160), 2203–2212

Unbalanced Dataset Preprocessing Using Hybrid Combination Algorithm for Arrhythmia Detection

Lufiya George Cheruvathoor(✉) and Jyothi Thomas

Department of Computer Science, CHRIST (Deemed To Be) University, Bangalore, India
lufiyageorgec@gmail.com, j.thomas@christuniversity.in

Abstract. Nowadays timely vaticination of cardiovascular conditions with the aid of a computer-backed opinion system minimizes the mortality. Cardiac arrhythmia discovery is one of the most grueling tasks because the variations of electrocardiogram (ECG) signal are veritably slight, which can not be recognized by mortal eyes. The data set under disquisition in this work is taken from the well-known Arrhythmia Dataset, which is codified into different classes. The correct identification of the health condition can lead to an easier, more effective, and less precious reclamation. This proffered study incorporates times of exploration on arrhythmia discovery exercising coincidental technologies. This project addresses the challenge of class imbalance in the analysis of arrhythmia data, utilizing hybrid combination of three approaches like Tomek Links, RUSBoost Classifier, and ADASYN. Here, used the hybrid combination of these three methods.The study commences with data preprocessing, including the loading and preparation of the dataset. Feature extraction and label separation are performed to enable further analysis. Subsequently, resolve the dataset into groups of testing and training to facilitate robust model valuation. Additionally, this article provides a thorough analysis of new preprocessing model approaches for diagnosing heart disease. By comparing the performance of these methods, this research contributes to the development of robust and accurate arrhythmia classification models, with potential applications in clinical diagnostics and healthcare decision support systems.

Keywords: Arrhythmia · Preprocessing · Tomek Links · RUSBoost Classifier · ADASYN · Hybrid Algorithm

1 Introduction

With the greatest fatality rate worldwide, cardiovascular illnesses are regarded as the most lethal disorders. They have spread over time and are already straining the healthcare systems of several nations. According to a WHO estimate, 70–80% of persons have this condition. Herein lies the relevance of our effort, which focuses on applying deep learning to forecast the forms of cardiovascular illness [1]. Arrhythmia is a term used to describe any irregular change from the regular cardiac rhythms and is a representative form of CVDs. Atrial fibrillation, premature contraction, ventricular fibrillation, and tachycardia, bradycardia are only a few examples of arrhythmia.

D. Kriksciuniene et al. (Eds.): IBICA 2023, LNNS 1228, pp. 99–108, 2025.
https://doi.org/10.1007/978-3-031-78937-3_10

Continuous arrhythmia beats can have lethal consequences, even though An individual palpitation from an arrhythmia might not have a big impact on survival. For instance, persistent premature ventricular contractions (PVCs) may infrequently develop into ventricular tachycardia (VT) [1] or ventricular fibrillation (VF), both of which can cause heart failure to occur suddenly. To manage and prevent CVDs, it is crucial to frequently evaluate cardiac rhythms. A prosperous course of treatment is invariably the outcome of an accurate and thorough assessment. The arrhythmia is typically brought on by plaques of fatty tissue along the vessel walls, which increases the risk of thrombosis [9]. Throughout the day, blood pressure varies based on activity. The assessment of ECG data is thoroughly investigated in this work, in addition to pre-processing, feature extraction, feature selection, application fields, and categorization, the issue is tackled from all sides.

The MIT-BIH Arrhythmia Database stands as a valuable resource for arrhythmia research, providing annotated electrocardiogram (ECG) recordings with a diverse range of cardiac conditions. However, like many real-world medical datasets, it suffers from a severe class imbalance, with certain arrhythmia classes being significantly underrepresented. In this project, we delve into the exploration and mitigation of class imbalance within the Arrhythmia Dataset at MIT-BIH. Specifically, here balance the dataset using a blend of the three forms: Tomek Links, RUSBoost Classifier, and ADASYN. Each technique targets the class imbalance problem from a different angle, offering unique advantages and trade-offs.

Through the application of these three techniques, we aim to enhance the performance of arrhythmia classification models trained on the input data. Precision, recall, and F1-score are some of the typical classification metrics used to evaluate the performance of the models. Additionally, the computational efficiency of each approach is considered. The outcomes of this investigation offer important perspectives into the efficacy of Tomek Links, RUSBoost Classifier, and ADASYN for addressing class imbalance in arrhythmia data analysis By conducting a comparative study, we seek to gain insights into the strengths and weaknesses of each method in the context of arrhythmia detection.

2 Key Contributions

1) Prediction of cardiovascular conditions that is focused on the most effective datasets available based on relevance.
2) Broader implications for imbalanced medical datasets and machine learning applications in healthcare.

The sections of this article are as follows: Sect. 2) focuses on this paper's contributions. Section 3 provides a concise overview of the current methodology and its drawbacks. Methodology is discussed in Sect. 4). Sections 5 and 6 of this result-oriented work explain the result analysis of the proposed approach with another methods and the conclusion.

3 Related Work

Seminal contributions have been made by author, who utilizes a conjunction of the One-Sided Selection (OSS) method, Support Vector Machine (SVM), and K-Nearest Neighbour (KNN) algorithms for arrhythmia detection. By eliminating raucous, moderate, and duplicate samples, the One-Sided Selection (OSS) approach is utilized for resampling to address the imbalance issue and minimize data [3]. The most important frequency range characteristics are extracted using Wavelet Transformation (WT) and Power Spectral Density (PSD). There is no information provided regarding the generalization of the proposed model to different datasets or patient populations. It is unclear whether the proposed model has been tested on a large and diverse enough dataset to ensure its robustness and reliability.

Author explains the ensemble model for the detection of heart diseases, which utilizes wavelet self-adaptive thresholding methods for denoising the ECG signal. For enhanced feature extraction, a 12-layer deep 1D Convolutional Neural Network (CNN) is introduced. Heartbeat categorization is done using SoftMax and machine learning classifiers [2]. Pre-processing, feature extraction, and classification using a hybrid classifier all have been included in the workflow diagram for the proposed work. The paper does not discuss the computational complexity or resource requirements of the proposed method, which could be important considerations for real-world implementation. For classification, the study only makes use of the MIT-BIH arrhythmia dataset, which may limit the generalizability of the results to other datasets or populations.

In this proposed model author conducted a systematic review by analyzing 368 studies that met the eligibility criteria.The review focused on deep learning-based ECG arrhythmia categorization in different aspects, including the ECG data, initial preprocessing techniques, deep learning methodology, evaluation paradigms, performance metrics, and code availability [8]. Based on the analysis of the chosen research, the report identified patterns, obstacles, and possibilities in deep learning based ECG arrhythmias diagnosis. The paper didn't discuss potential biases or limitations in the selection criteria for the included studies such as the representatives of the data or potential biases in the annotations,code availability and reproducibility of the selected studies.

The author examines three deep learning architectures for electrocardiogram (ECG) signal-based arrhythmia detection. The models' performance was assessed using accepted criteria like accuracy, sensitivity, and specificity [4]. The network was trained using backpropagation, which adjusts the values of filters, biases, weights, and biases in the convolutional and fully connected layers to minimise the difference between predicted and actual outputs. Stochastic gradient descent (SGD) or a variant theorem was used for implementing backpropagation. The paper didn't provide any information on the dataset used for training and testing the models, which could impact the generalizability of the results.It didn't mention any comparisons with other existing methods or algorithms for arrhythmia detection, making it difficult to assess the relative performance of the deep learning architectures.

Author proposes a novel application for arrhythmia detection using the You-Only-Look-Once (YOLO)v8 algorithm to classify single-lead ECG signals. It was fine-tuned on the MIT-BIH arrhythmia dataset to detect arrhythmia in real-time. The model was trained on an NVIDIA Tesla V100 and achieved an average accuracy. To address the class imbalance in the dataset, the authors introduced dynamic inverse class frequency (DICF) to emphasise underrepresented classes and improve overall recall results. Using industry-standard criteria like accuracy, specificity, precision, recall (sensitivity), and F1 score, the classification performance of the model was assessed. The paper does not discuss the potential challenges or limitations of using the YOLOv8 algorithm for arrhythmia detection. There is no mention of the specific limitations or drawbacks of the proposed loss-modified YOLOv8 model. It does not address the potential limitations of using the MIT-BIH arrhythmia dataset for training the model.

4 Methodology

Three phases make up the majority of the suggested method: image collection, preliminary processing, and classification. Several deep learning algorithms are beneficial during these phases for categorizing the ECG dataset as having a regular or an irregular rhythm. Figure 1 illustrates the proposed model's block diagram.

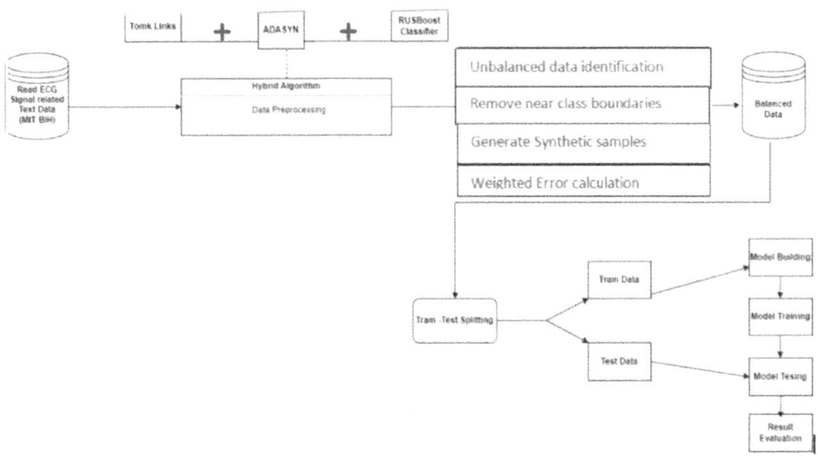

Fig. 1. Proposed Methodology

4.1 Data Acquisition

The MIT-BIH cardiac Dataset is the dataset used in this project. This database is a widely recognized and extensively utilised resource in the field of arrhythmia research. It comprises a collection of neumerical values that is associated to electrocardiogram (ECG) signal recordings, each annotated with specific cardiac conditions. These recordings are obtained from a diverse range of patients, providing a comprehensive representation of cardiac arrhythmias.

The 48 half-hour rhythm recordings in the MIT-BIH Arrhythmia Dataset are recorded at 360 Hz. From a total of 47 subjects. Each recording is labeled with a variety of cardiac arrhythmias, including atrial premature contractions (APC), non-conducted P-waves (PAB) [5], ventricular premature contractions (PVC), and more. Additionally, the database includes (175108, 188) data to serve as a baselin reference. Annotations make it a pivotal resource for researchers and practitioners in the field of cardiology and biomedical engineering. By leveraging this dataset, the project aims to develop and assess effective methods for arrhythmia classification, focusing on techniques like Tomek Links, RUSBoost, and ADALYN for addressing class imbalance.

4.2 Pre-Processing

Even though an database-sourced ECG signal is less likely to have distracting noise than ECG data obtained directly from an individual, It is bound to have an error that needs to be corrected with the goal to boost the activities the system takes after that. Since medical data is constantly non-uniform and low-discrepancy, preprocessing for image improvement is needed to produce accurate segmentation.

4.2.1 Tomek Links

Tomek Links is a technique employed to address class imbalance within the dataset. It focuses on identifying pairs of instances, where one instance is a member of the dominant class, whereas the other is a member of the minority class, and they are particularly close to each other. These pairs form what is known as a Tomek Link [6]. The algorithm then selectively removes the instance from the majority class in each pair. By doing so, Tomek Links aims to improve the quality of the training data by reducing the overlap between classes [12]. This reduction in overlap can be crucial for the effectiveness of classification models, as it promotes clearer and more accurate distinctions between different cardiac conditions. Tomek Links is particularly beneficial then there are regions of overlap between the classes, as it helps in creating a more distinct separation, potentially leading to improved classification accuracy.

4.2.2 RUSBoost Classifier (Random Under-Sampling with Boosting)

RUSBoost is a combination of two powerful techniques - random under-sampling and boosting. It is specifically designed to tackle class imbalance by first applying random under-sampling to the majority class [7]. This step involves reducing the number of samples from the majority class to balance the class distribution. Then, the AdaBoost algorithm is employed, using a base classifier (such as a Decision Tree) trained on the resampled data [9]. During the boosting iterations, more weight is given to the misclassified samples, which are more likely to belong to the minority class. This approach ensures that the model focuses on learning the patterns of the minority class while still benefiting from the boosting ensemble's strength. RUSBoost is particularly effective in cases where both under-sampling and boosting are needed to improve classification performance. By balancing the data and emphasizing the importance of the minority class, RUSBoost equips the model with the necessary tools to make accurate predictions for arrhythmia detection.

4.2.3 ADASYN (Adaptive Synthetic Sampling)

ADASYN is an oversampling technique designed to address class imbalance by generating synthetic tests for the minority group. The unique strengths of ADASYN lies in its adaptability to the local distribution of the data. This makes it particularly effective in regions where the decision boundary is complex and nonlinear. ADASYN utilises a K-nearest neighbors approach to determine where synthetic samples generation should occur. For each sample in the minority class [7], it calculates the K-nearest neighbors and creates synthetic samples in areas with a low proportion of the minority class. This dynamic approach ensures that synthetic samples are generated where they are most needed, capturing the underlying data distribution more accurately [11]. This can lead to a more robust classification model, especially when dealing with intricate patterns associated with different arrhythmias. Nevertheless, it's crucial to remember that the effectiveness of ADASYN may change depending on specific characteristics of dataset, and careful tuning of parameters may be necessary to achieve optimal results.

4.3 Hybrid Algorithm

Proposed preprocessing method is illustrate how to develop a hybrid model to identify the risk level of cardiovascular diseases by analyzing the types of arrhythmia. The following algorithm aids in the development of a highly accurate hybrid model for the preprocessing stage.

Here Fig. 2 shows the hybrid algorithm. After that here explains the newly proposed preprocessing algorithm for the arrhythmia detection and classification. After initialization, read the input data, which is an imbalanced dataset (MIT BIH), with more occurrences of the majority class and fewer examples of the minority class.

```
Hybrid Algorithm

  // Input: MIT BIH Dataset (Imbalanced)

  // Output: Balanced Dataset
Step 1: Initialization
Step 2: Identify Nearest Neighbors
Step 3: Calculate the class imbalance ratio
Step 4: Calculate required number of synthetic samples for each minority class
Step 5: Identify the neighbors from both minority and majority classes
Step 6: Estimate the density distribution difference
Step 7: Calculate the importance of each minority instance
Step 8: Random Undersampling
Step 9: Classifier Weight Calculation
Step 10:Repeat until the boosting process is complete
```

Fig. 2. Hybrid Algorithm

Then identify nearest neighbors for each instance in the dataset by calculating the distance to all other instances in the same dataset. The typical distance metric Euclidean distance is used, but it can be any distance measure suitable for the data. Next procedure is identification of Tomek Links. For the identification firstly a Tomek link is defined as a pair of instances (x, y) such that: X is a member of the minority class, while Y is a member of the majority class. The distance between x and z in the data set is never less than or equal to the distance between x and y, and z corresponds to the majority class. Once Tomek links are identified, remove the instances from the minority class that are part of these links. The removal process aims to eliminate minority instances that are in close proximity to the majority class, as they might be causing confusion to classifiers.

Then apply second methodology to calculate the proportion of the number of instances of the dominant class to those of the minority class is used to compute the class imbalance rate. As per ADASYN methodology, [10] calculate the necessary amount of synthetic samples depending on the class imbalance ratio for every case in the minority group. The goal is to equalize the number of instances between the minority and majority classes. For each minority class instance, identify its k neighbors from both minority and majority classes in the feature space. The number of neighbors (K) is typically a user-defined parameter.Then calculate the density distribution difference between the minority class instance and its neighbors. This difference quantifies show how much similar or dissimilar the instance is to its neighbors based on the density distribution difference. If instances are more dissimilar to their neighbors, that will have higher importance. Moreover, for each minority instance, generate a proportionate number of synthetic samples based on its importance. Sometime more dissimilar instances will generate more synthetic samples. The synthetic samples are created by interpolating (random interpolation factor) across the minority cases and their chosen nearest neighbors' feature vectors.

In last phase Random Undersampling is done by RUSBooster. In this stage firstly Randomly undersample the majority class to create a balanced or nearly balanced subset of the dataset. This step aims to reduce the dominance of the majority class and prevent the classifier from being biased toward it. Secondly train the base classifier and Calculate the weighted error of the base classifier on the entire dataset (both minority and majority class instances). The weighted error accounts for the fact that the dataset is imbalanced, giving more importance to the minority class. Finally calculate the weight of the base classifier in the ensemble based on its weighted error. The formula for calculating the weight is typically:

$$\text{Weight} = 0.5 * \ln((1 - \text{Error})/\text{Error}) \quad (1)$$

where "Error" is the weighted error.

Then update the weights of the training instances in the majority class. Instances that are misclassified by the base classifier receive higher weights to emphasize their importance in subsequent iterations. Repeat until the boosting process is complete and combine the predictions of all the base classifiers in the ensemble using weighted voting. Classifiers with higher weights contribute more to the final prediction. After each iteration, sort the synthetic samples based on their quality, either in descending order (majority-disparity-based) or ascending order (minority-disparity-based). Ultimately return the fully balanced dataset after the desired number of iterations. The outcome contributes to the demonstration of a novel technique for accurate arrhythmia identification.

5 Result and Analysis

The next section initially explains how hybrid technique improved the findings using the MIT-BIH supraventricular dataset as the base dataset.

Table 1. Comparison with Another Methods

Author	Method	Experience	Best performance
Pandey, S. K. et al. [2]	Wavelet self-adaptive thresholding methods, CNN + Softmax	Classification with a denoising process	Accuracy: 84.40%
Zahraa et al. [7]	WT, PSD AND OSS	Classification with multi modal techniques	Accuracy: 85%
Rahman, A.U. et al. [6]	Transfer learning methods, ResNet50, AlexNet, and SqueezeNet are used	Classification with deep learning approaches	Accuracy: 88.8%
Kavya s. et al. [8]	SVM	Classification with a denoising process	Accuracy: 81%
Yang et al. [3]	PCANet,SVM	Classification with a denoising process	Accuracy: 87.77%
Gupta v et al. [5]	FrWT, YWARM,PCA	Classification with a denoising process	Accuracy: 89.89%
R.R Janghel et al. [4]	SMOTE	Classification with DCNN	Accuracy: 78.30%
Newly Proposed method	TomkLink + RUSBoost + ADASYN (Hybrid method)	Classification with preprocessing method	Accuracy: 99.94%

In terms of categorization accuracy, we attained a success rate of more than 99% for each group. In comparison to the findings of the prior approaches in Table 1, our accuracy is excellent. The accuracy of newly balanced data is 99.94%.

In Fig. 3 it shows the class imbalance after PCA, Wavelet Transform, Linear Discriminant Analysis (LDA) applications. Among them The bar graph displays the number of instances or samples in each class after applying the different algorithms like PCA, wavelet transform and Linear Discriminant Analysis (LDA). In a unbalanced dataset, all the bars should be approximately of the different height, indicating an unequal distribution of data across different classes.

Class Distribution Before and After new proposed balancing Method is shown in Fig. 4. After balancing, the bar graph will display bars of nearly equal heights, suggesting a more uniform distribution of instances across all classes. Further more, after balancing the line chart provides statistical insight into the distribution of data within each class. In a balanced dataset, the line would be of similar lengths and located at approximately the same height, indicating similar data distributions across classes. Outliers can also be identified, which may help in further data preprocessing or model tuning. According to the pie chart, it will exhibit nearly equal-sized slices, indicating that the data has been effectively balanced across different classes.

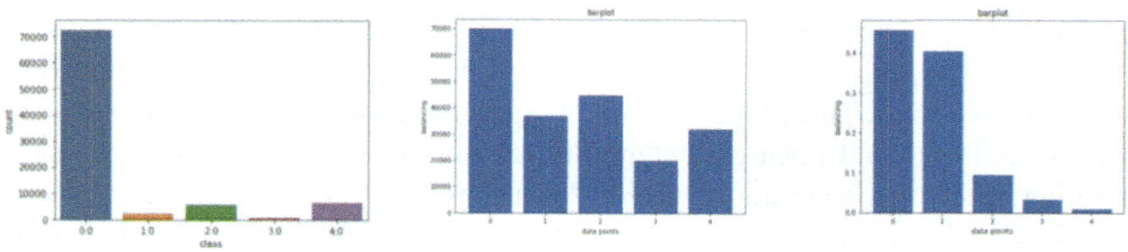

Fig. 3. Class Imbalance After PCA, Wavelet Transform, LDA methods

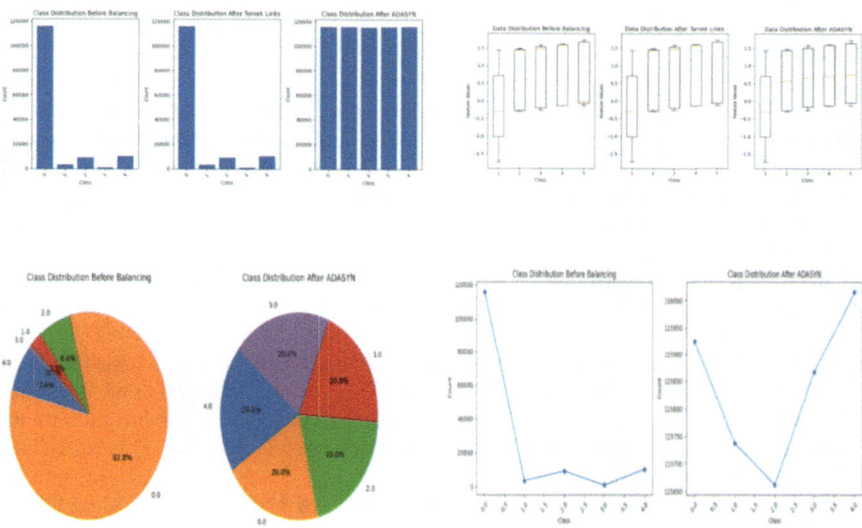

Fig. 4. Class Distribution Before and After New proposed Balancing Method

The hybrid utilization of Tomek Links, RUSBoost, and ADASYN techniques in this study has effectively mitigated the issue of class imbalance, leading to the creation of well-balanced datasets. These methods collectively contributed to a notable enhancement in classification accuracy and overall model performance, culminating in a more dependable arrhythmia detection system. By strategically emphasizing the features of the minority class, these techniques have significantly bolstered the capacity for accurate classification.

This contrast in outcomes marks a significant stride towards the development of more resilient and precise arrhythmia classification models, with promising implications for applications in clinical diagnostics and healthcare decision support systems.

6 Conclusion

In conclusion, the hybrid integration of Tomek Links, RUSBoost, and ADASYN techniques has proven to be instrumental in addressing the challenge of class imbalance within the arrhythmia dataset. These methods collectively led to the creation of balanced datasets, thereby substantially enhancing the accuracy and overall performance of our arrhythmia detection models. By strategically emphasizing the characteristics of the minority class, these approaches have paved the way for more robust and reliable classification systems. This progress holds great promise for the advancement of arrhythmia analysis in clinical settings, potentially revolutionising diagnostic processes and facilitating timely medical interventions. The successful application of these techniques underscores their critical role in improving the accuracy and effectiveness of arrhythmia classification models, and points towards a brighter future for automated cardiac health assessments.

References

1. Lufiya, G.C., Thomas, J., Aswathy, S.U.: A survey on arrhythmia disease detection using deep learning methods. In: Abraham, A., et al. (eds.) Innovations in Bio-Inspired Computing and Applications: Proceedings of the 12th International Conference on Innovations in Bio-Inspired Computing and Applications (IBICA 2021) Held During December 16–18, 2021, pp. 55–64. Springer International Publishing, Cham (2022). https://doi.org/10.1007/978-3-030-96299-9_6
2. Pandey, S.K., et al.: Detection of arrhythmia heartbeats from ECG signal using wavelet transform-based CNN model. Int. J. Comput. Intell. Syst. (2023). https://doi.org/10.1007/s44196-023-00256-z
3. Dong, Y., et al.: An arrhythmia classification model based on vision transformer with deformable attention. Micro Mach. **14**(6), 1155 (2023). https://doi.org/10.3390/mi14061155
4. Hima Vijayan, V.P., et al.: Integrating deep learning architectures for improved arrhythmia detection in electrocardiogram signals. Int. J. Sci. Res. Archiv. **8**(2), 704–715 (2023). https://doi.org/10.30574/ijsra.2023.8.2.0332.
5. Jun, G.. et al.: A novel application for real-time arrhythmia detection using YOLOv8, Arxiv.org. http://arxiv.org/abs/2305.16727 (2023)
6. Rahman, A.-U., et al.: ECG classification for detecting ECG Arrhythmia empowered with deep learning approaches. Comput. Intell. Neurosci. **2022**, 1–12 (2022). https://doi.org/10.1155/2022/6852845
7. Sai Kumar, S., et al.: An IOT framework for detecting cardiac arrhythmias in real-time using deep learning resnet model. Measurement. Sensors **29**(100866), 100866 (2023). https://doi.org/10.1016/j.measen.2023.100866
8. Xiao, Q., et al.: Deep learning-based ECG arrhythmia classification: a systematic review. Appl. Sci. (Basel, Switzerland) **13**(8), 4964 (2023). https://doi.org/10.3390/app13084964
9. Zhang, J., et al.: ECG-based multi-class arrhythmia detection using spatio-temporal attention-based convolutional recurrent neural network. Artif. Intell. Med. **106**, 101856 (2020). https://doi.org/10.1016/j.artmed.2020.101856
10. Ince, T., Kiranyaz, S., Gabbouj, M.: A generic and robust system for automated patient-specific classification of ECG signals. IEEE Trans. Biomed. Eng. **56**(5), 1415–1426 (2009). https://doi.org/10.1109/TBME.2009.2013934
11. Patra, D., Das, M.K., Pradhan, S.: Integration of FCM, PCA and neural networks for classification of ECG arrhythmias. IAENG Int. J. CompuT. Sci.
12. Yanga, W., Sia, Y., Wanga, D., Guoa, B.: Automatic recognition of arrhythmia based on principal component analysis network and linear support vector machine. Comput. Biol. Med. (2018)

Deep Learning Methods for Biomedical Named Entity Recognition: A Comprehensive Review

S. Sabitha[1(✉)] and Anitha S. Pillai[2]

[1] St. Francis College, Koramangala, Bangalore, India
sabithastfrancis@gmail.com
[2] Hindustan Institute of Technology and Science, Chennai, India

Abstract. Biomedical Named Entity Recognition (Bio-NER) contributes significantly to the advancement of biomedical research by facilitating efficient information extraction, text mining, and knowledge discovery. This in-depth analysis examines the applications and significance of Bio-NER within the context of biomedical research. Following an overview of Named Entity Recognition (NER) in natural language processing, this paper discusses the distinctive characteristics and challenges of Bio-NER. This article focuses primarily on deep learning-based approaches for Bio-NER, including single neural network-based methods, multitask learning-based strategies, transfer learning in Bio-NER, and hybrid model-based techniques. Using deep learning, this paper also addresses the obstacles in Bio-NER, evaluation metrics, benchmarks, datasets, and the general framework of Bio-NER. In conclusion, this paper examines the most recent Bio-NER tools and frameworks based on deep learning. This review is a useful resource for researchers and practitioners who wish to comprehend the current state of Bio-NER using deep learning and its potential to revolutionize biomedical research and improve human health.

Keywords: Biomedical Named Entity Recognition (Bio-NER) · Natural Language Processing (NLP) · Deep Learning

1 Introduction

Biomedicine relies on natural language processing, and biomedical named entity recognition (Bio-NER) is essential. The enormous diversity of entity types, including diseases, proteins, and genes, and the complexity of biological nomenclature make Bio-NER harder than Named Entity Recognition [1]. Biomedical research relies on Bio-NER to identify and categorize these items in the text, making it easier to find relevant material in the vast scientific publication collection.

Bio-NER is utilized in biomedical information extraction, text mining, and knowledge discovery [2]. Bio-NER accurately detects and categorizes genes, proteins, and diseases, helping researchers extract relevant information from unstructured text sources like research articles and clinical data. The collected data can be utilized for text-mining to find patterns, linkages, and trends. Bio-NER links entities to their ontologies or databases to provide structured knowledge bases for querying and analysis, aiding knowledge discovery.

D. Kriksciuniene et al. (Eds.): IBICA 2023, LNNS 1228, pp. 109–120, 2025.
https://doi.org/10.1007/978-3-031-78937-3_11

2 Background and Related Work

NLP's named entity recognition (NER) subtask finds and categorizes entities in unstructured text. Entities are words or groups of words that continually refer to the same entity, such as names, organizations, places, currencies, etc. NER helps extract and process textual information for intelligent NLP applications [3, 4]. A subset of NER termed "bio-NER" classifies genes, proteins, medicines, and illnesses. Bio-NER is different from regular NER because of the unique challenges posed by the complexity of biomedical entities, how they are named, and how they are written in text. These challenges make bio-NER harder than normal NER, which handles dates, names, and locations [5].

Deep learning-based techniques have gained identification as valuable tools for Bio-Named Entity Recognition (Bio-NER) due to their capacity to automatically acquire intricate attributes from vast quantities of data. A review of deep learning techniques for Bio-NER challenges is presented by Mishra et al. [6]. There are four categories of deep learning methods: multitask learning-based, single neural network-based, transfer learning-based, and hybrid-based model. These techniques have been used in a variety of fields for BioNER, and their effectiveness depends on the quantity and variety of datasets used. Gao et al., 2021 [17] present a pre-training and self-training approach for Bio-NER. To improve the performance of Bio-NER models, the authors propose a method that leverages large-scale unlabelled corpus pre-training, followed by self-training with a smaller labelled dataset.

2.1 Single Neural Network-Based Methods

These methods integrate recurrent neural networks (RNNs), long short-term memory networks (LSTMs), and convolutional neural networks (CNNs) into a unified model to represent BioNER tasks. In order to enhance the biomedical identification of named entities, Habibi et al. [7] investigate the application of word embeddings in deep learning. DNorm, an advanced named entity recognition system designed specifically for the biomedical domain, is introduced as the system of the future. This approach combines Conditional Random Fields (CRF) and bidirectional Long Short-Term Memory (Bi-LSTM) networks to detect medical entities by integrating conditional random fields, n-grams, as well as bidirectional long short-term memory (BiLSTM), Cho and Lee (2019) [8] present a tailored approach to biological NER using deep neural networks. It is recommended to utilize conditional random fields (CRF) when combined with contextual long short-term memory networks (CLSTM). Based on numerous corpora and word embeddings, the authors examined neural architectures with contextual information for biomedical NER. Zhang & Wu, 2021 [20] present a deep learning-based strategy for medical text entity recognition. An approach is proposed wherein a bidirectional long short-term memory (BiLSTM) model is combined with a conditional random field (CRF) layer to identify and classify items in medical journal articles in an effective way. The authors use medical text datasets to test their strategy. Deep learning showed that the BiLSTM-CRF model outperformed baseline approaches in textual term detection [25]. The recall of biomedical named entity recognition (Bio-NER) was proposed through information extraction as well as label re-correction by Zhou et al., 2021 [16]. Mady et al., 2022 [14], present a method for nested Bio-NER to address the difficulty of identifying entities embedded within other entities. Nested Bio-NER is a difficult problem because it requires recognizing not only the outer entities, but also the inner, overlapping entities. The strategy suggested by the writers is to integrate deep learning techniques, including Bidirectional Long Short-Term Memory (BiLSTM), with Convolutional Neural Networks (CNN). They demonstrate the effectiveness of their proposed method for recognizing nested entities by evaluating it on a benchmark dataset and demonstrating its effectiveness in recognizing nested entities. Asghari, Sierra-Sosa, and Elmaghraby, 2022 [11] study BINER, a low-cost Bio-NER method. Bio-NER's high computational cost and need for extensive annotated data are addressed in the paper. They propose a method that combines word embedding-based feature extraction with a modified Jaccard similarity coefficient. The proposed BINER model exhibited notable computational efficiency when compared to cutting-edge algorithms on four biological entity identification datasets used as benchmarks.

2.2 Multitask Learning-Based

Tong et al., 2021 [15], propose a multi-task approach to improve Bio-NER using multi-granularity information. Due to the complexity of biomedical entities and the lack of diverse, annotated data, current Bio-NER models have limitations, the authors acknowledge. They introduce a multi-task learning framework that combines characters, sub words, and words to address these issues. By jointly learning these levels of representation, the model can better capture complex entity relationships and perform better. The authors test their method on benchmark datasets and compare it to state-of-the-art models. The results show their multi-task approach can improve Bio-NER.

2.3 Transfer Learning in BioNER

Giorgi and Bader, 2018 [9] investigate the use of neural networks and transfer learning in BioNER. Three modifications of the neural network model BiLSTM-CRF are assessed in order to mitigate overfitting and increase generalization. These modifications comprise transfer learning, multi-task learning, and enhanced regularization through variational dropout. In order to enhance the identification of biological terms contained in unstructured text, such as organisms, genes, gene products, diseases, kinds of cells, and substances, transfer learning could potentially be advantageous for the field of BioNER. Chai et al., 2022 [13], propose hierarchical shared transfer learning for biomedical named entity recognition. (Bio-NER). A new transfer learning framework helps Bio-NER overcome limited labeled data and high manual annotation costs. The method improves model generalization by sharing knowledge across tasks and domains. Hierarchical shared transfer learning uses shared and task-specific layers to capture common and unique features for different tasks, improving model performance. Weber et al. 2021 [18] created HunFlair to provide a user-friendly and effective method for identifying biomedical entities in text. The tool is based on the pre-trained language model Flair and its performance has been optimized using multiple biomedical corpora. Genes, proteins, chemicals, diseases, and species are among the entity types supported by HunFlair's NER functionality. Sun et al. [19], implement an innovative approach to biomedical NER by integrating the BERT model into a machine reading comprehension (MRC) framework. In their study, they recast the NER task as a reading comprehension problem, enabling them to utilize the power of pre-trained BERT models for extracting biomedical entities from text. The authors propose a method for converting NER datasets to MRC format, which permits the fine-tuning of BERT for biomedical NER tasks. The recommended technique has been shown successful and demonstrates state-of-the-art efficiency based on experimental results from several benchmark datasets. Syed et al. 2022 [10] introduce the DeIDNER Model, a neural network-based named entity recognition (NER) model for clinical note de-identification. The authors discuss how to protect patient privacy while using EHRs for research in their study. They create a deep learning model to identify and remove PII from clinical notes. The BERT-based DeIDNER model is fine-tuned to perform de-identification on a labeled dataset of considerable scope.

2.4 Hybrid Model-Based Methods

Goyal et al. 2022 propose the utilization of hybrid embedding, a technique that combines bi-directional LSTM-based character embedding and rapid text embedding, in their deep learning-based named entity recognition (NER) system [39]. By encapsulating contextual, syntactic, and semantic characteristics of the text, these embeddings empower deep learning techniques cognitively.

3 Challenges in Bio-NER

Bio-NER faces a number of obstacles that impede the development and deployment of accurate and effective systems.

3.1 Lack of Annotated Data

A shortage of high-quality, annotated datasets is one of the greatest obstacles in Bio-NER. These are essential for training and evaluating machine learning-based NER systems, especially deep learning models that require a substantial amount of data for optimal performance (Cohen, 2005 [22]).

3.2 Ambiguity and Complexity of Biomedical Language

Biomedical terminology is diverse and intricate, which poses challenges for Bio-NER. Ambiguity occurs when a single term refers to multiple entities or a single entity has multiple names. In addition, the use of abbreviations, acronyms, and complex technical language makes NER tasks more challenging (Weeber et al., 2000 [23]).

3.3 Interdisciplinary Nature of Biomedical Research

Biomedical research frequently encompasses multiple disciplines, resulting in a vast array of terminologies and jargon. Due to the interdisciplinarity of this field, it can be difficult for NER systems to adapt and generalise across diverse subdomains and resources (Yeh et al., 2005 [24]).

3.4 Scalability Issues

Biomedical literature is vast and continuously expanding, making scalability of Bio-NER systems essential. However, the complexity of the domain and the requirement for high-quality, annotated data can impede the development of scalable NER systems that can process large volumes of text efficiently (Zhang et al., 2019 [25]).

4 Datasets

For the purpose of facilitating the development, evaluation, and comparison of algorithms in the field of BioNER, researchers have invested significant effort in the creation of standardised data sets. The following is a brief introduction to some of the essential resources:

4.1 GENIA Corpus

For the creation of the GENIA corpus, genes, proteins, and other biological entities were manually annotated in 2,000 PubMed abstracts. Widely employed for training and evaluating Bio-NER systems (Kim et al., 2003 [41]).

4.2 NCBI Disease Corpus

Dogan et al. [47] in 2014 developed the NCBI illness Corpus Dataset, which includes 6,892 illness mentions matched to 790 disease categories. These keywords have MeSH identifiers or OMIM identifiers linked to them in English 88% and 12% respectively, including an additional 6,892 text characters.

4.3 JNLPBA Corpus

Named entities such as proteins, DNA, RNA, cell lines, and cell types are among those comprising the JNLPBA corpus, which is comprised of the GENIA corpus. Kim et al. 2004 [28] developed it for the Bio-Entity Recognition Task at the Joint Workshop on Natural Language Processing in Biomedicine and its applications.

4.4 BioCreative II Gene Mention Recognition (BC2GM) Dataset

The participants were instructed to provide the initial and final characters of a phrase in order to identify a specific gene that was mentioned. The data in question was compiled by Smith et al. in 2008 and is known as the BC2GM Dataset [29]. The training set consists of a collection of sentences accompanied by a catalog of gene annotations (gene mentions) for every phrase.

4.5 CRAFT Dataset

The CRAFT (Colorado Richly Annotated Full-Text) Corpus comprises a compilation of biomedical scientific articles that have been manually annotated. The CRAFT Corpus comprises 67 full-text articles annotated with various commonly employed biomedical ontologies, such as Cell Ontology, Sequence Ontology, and Gene Ontology, which were obtained from the PubMed Central Open Access Subset. Molecular biology and genetics contribute the most to the articles (Bada et al., 2012 [26]).

5 General Framework of Bio-NER Using Deep Learning

A typical framework for a biomedical named entity recognition (Bio-NER) system employing deep learning consists of the steps outlined below (Fig. 1).

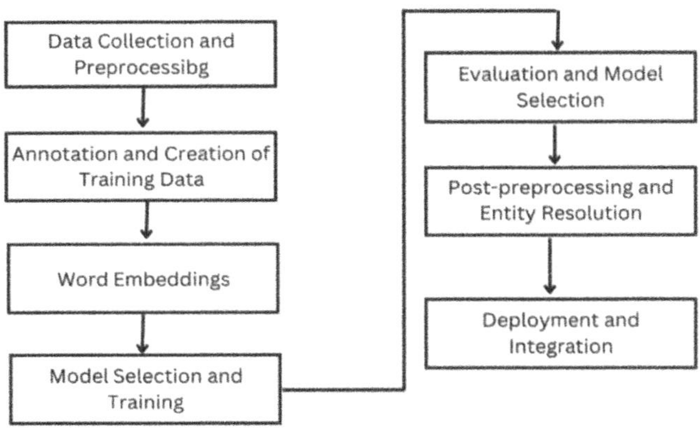

Fig. 1. General framework of Bio-NER using deep learning

5.1 Data Collection and Pre-Processing

- Collect a large corpus of biomedical texts from various sources, including scientific articles, electronic medical records, and patents.
- To prepare the data for further processing, perform pre-processing tasks including tokenization, sentence segmentation, and normalization.

5.2 Annotation and Creation of Training Data

- A subset of the collected corpus is manually annotated with relevant biomedical entity tags (such as genes, proteins, diseases, and chemicals).
- Divide the annotated dataset into training, validation, and test sets for the construction and evaluation of models.

5.3 Word Embeddings

- To convert tokens into continuous vectors that capture their semantic and syntactic properties, use pre-trained word embeddings (e.g., Word2Vec, GloVe, or FastText) or contextualised embeddings (e.g., ELMo).
- Alternately, domain-specific embeddings should be trained on the biomedical corpus to more accurately capture domain-specific terminology and relations.

5.4 Model Selection and Training

- Select an appropriate deep learning model for the task, such as a convolutional neural network (CNN), a recurrent neural network (RNN), or a transformer-based model such as BERT (e.g., BioBERT, SciBERT).
- Apply the annotated training data to the model while fine-tuning its hyperparameters with the validation data to prevent overfitting.

5.5 Evaluation and Model Selection

- Utilize test set metrics like precision, recall, and F1-score to assess the trained model's performance. Compare the performance of various models or model configurations to select the optimal model for the Bio-NER task at hand.

5.6 Post-Processing and Entity Resolution

- Perform post-processing steps, such as merging overlapping entities or resolving entity ambiguity, to enhance the entity recognition quality.
- To facilitate downstream analyses and integration with other resources, map the recognized entities to standard identifiers such as UniProt, MeSH, and ChEBI.

5.7 Deployment and Integration

- Deploy the trained Bio-NER model as a web service, an application programming interface (API), or standalone software for use in a variety of applications.
- Integrate the Bio-NER system with other text mining or analysis instruments to construct comprehensive biomedical text mining pipelines.

6 Deep-Learning Based Bio-NER Tools and Framework

By offering more precise and effective ways to extract and analyze pertinent information from the vast body of biomedical literature, deep learning-based Bio-NER frameworks and tools have revolutionized the area of biomedical text mining. These tools employ cutting-edge deep-learning techniques to improve the extraction of various biomedical components, including genes, diseases, chemicals, proteins, and other pertinent terms (Table 1).

Table 1. Bio-NER tools and frameworks based on deep learning

Sl.No	Tool/Framework	Type	Description	Reference
1	BioBERT	BERT-based	BERT-based models pre-trained on biomedical corpora are powerful biomedical NER tools	Lee et al., 2019 [31]
2	BlueBERT	BERT-based	Pre-trained BERT-based model using PubMed abstracts and MIMIC-III clinical notes for biological NLP tasks like NER	Peng et al., 2019 [32]
3	SciBERT	BERT-based	BERT pre-trained on scientific text	Beltagy et al., 2019 [33]
4	TaggerOne	Semi-Markov Models	Joint learning approach for learning entity types and boundaries, suitable for Bio-NER	Robert et al., 2016 [34]
5	FLAIR	Contextual Embeddings	Contextualized string embeddings for NLP tasks, including Bio-NER in the biomedical domain	Akbik et al. 2018 [35]
6	BiLSTM-CRF	BiLSTM + CRF	Biomedical literature sequence tagging uses BiLSTM networks and a CRF layer	Huang et al., 2015 [36]
7	BC5CDR-BERT	Fine-tuned BERT model for BC5CDR dataset	BC5CDR contains PubMed abstracts annotated for chemicals, diseases, and relationships. The BC5CDR dataset sharpened the BERT model	Lim et al., 2018 [37]

(*continued*)

Table 1. (*continued*)

Sl.No	Tool/Framework	Type	Description	Reference
8	NEEDLE	Multi-stage framework	Includes insufficient label completion, a noise-aware loss function, and fine-tuning	Wang et lt., 2021 [38]
9	BioAug	BART	Low-resource Biomedical Named Entity Recognition data augmentation framework	Ghosh, 2023 [42]
10	Tolerant Pattern Learner (TPL)	Semi-supervised Learning Algorithm	Highlights BioN-ER's vital COVID-19 work. Semi-supervised learning annotates unlabeled data and NLP extracts noun phrases and contextual patterns from huge datasets	Jaura & Ramanna, 2022 [43]
11	AIONER, AIO Scheme	Mult-Task Learning	The report explores BioNER's data scarcity and annotation domain expertise issues. The AIONER tool, developed using an AIO approach, performs well in BioNER tasks and enhances biomedical text processing at scale	Luo et al., 2023 [44]

7 Conclusion

This literature review concludes with a thorough examination of the significance of BioNER in biomedical research, its applications, and the various deep learning techniques used to address its unique challenges. We have obtained insight into the potential of these methods to improve the extraction of significant biomedical information from unstructured text data by evaluating strategies such as single neural network-based, multitask learning-based, transfer learning-based, and hybrid model-based techniques.

This review not only offers a thorough knowledge of the state of deep learning-based Bio-NER at this time, but also emphasizes the need for the continued exploration of novel methods to advance the field, thereby contributing to the advancement of biomedical research and the improvement of human health.

References

1. Pir, D.S., Sanotsh, K., Banbhrani, Arsalan, A.S., Hans, R.: Bio-NER: biomedical named entity recognition using rule-based and statistical learners. Int. J. Adv. Comput. Sci. Appl. **8**(12), 2017. https://doi.org/10.14569/IJACSA.2017.081220.
2. Jansen, S.: Who's Who and What's What: Advances in Biomedical Named Entity Recognition (BioNER). Medium. https://towardsdatascience.com/whos-who-and-what-s-what-advances-in-biomedical-named-entity-recognition-bioner-c42a3f63334 (2021)
3. Nadeau, D., Sekine, S.: A survey of named entity recognition and classification. Lingvisticae Investigationes **30**(1), 3–26 (2007). https://doi.org/10.1075/li.30.1.03nad
4. A Comprehensive Guide to Named Entity Recognition (NER): A Comprehensive Guide to Named Entity Recognition (NER) (n.d.). https://www.turing.com/kb/a-comprehensive-guide-to-named-entity-recognition
5. Sun, P., Yang, X., Zhao, X., Wang, Z.: An overview of named entity recognition. In: 2018 International Conference on Asian Language Processing (IALP), Bandung, Indonesia, pp. 273–278 (2018). https://doi.org/10.1109/IALP.2018.8629225
6. Mishra, P., Biswas, S., Dash, S.: Deep learning based biomedical named entity recognition systems. In: Dash, S., Acharya, B.R., Mittal, M., Abraham, A., Kelemen, A. (eds.) Deep Learning Techniques for Biomedical and Health Informatics, pp. 23–40. Springer International Publishing, Cham (2020). https://doi.org/10.1007/978-3-030-33966-1_2
7. Habibi, M., Weber, L., Neves, M., Wiegandt, D.L., Leser, U.: Deep learning with word embeddings improves biomedical named entity recognition. Bioinformatics **33**(14), i37–i48 (2017). https://doi.org/10.1093/bioinformatics/btx228
8. Cho, H., Lee, H.: Biomedical named entity recognition using deep neural networks with contextual information. BMC Bioinform. (2019). https://doi.org/10.1186/s12859-019-3321-4
9. Giorgi, J.M., Bader, G.D.: Transfer learning for biomedical named entity recognition with neural networks. Bioinformatics **34**(23), 4087–4094 (2018). https://doi.org/10.1093/bioinformatics/bty449
10. Syed, M., et al.: DeIDNER Model: a neural network named entity recognition model for use in the de-identification of clinical notes. In: Proceedings of the 15th International Joint Conference on Biomedical Engineering Systems and Technologies (2022). https://doi.org/10.5220/0010884500003123
11. Asghari, M., Sierra-Sosa, D., Elmaghraby, A.S.: BINER: a low-cost biomedical named entity recognition. Inf. Sci. **602**, 184–200 (2022). https://doi.org/10.1016/j.ins.2022.04.037
12. Kim, H., Kang, J.: How do your biomedical named entity recognition models generalize to novel entities? IEEE Access **10**, 31513–31523 (2022). https://doi.org/10.1109/access.2022.3157854
13. Chai, Z., Jin, H., Shi, S., Zhan, S., Zhuo, L., Yang, Y.: Hierarchical shared transfer learning for biomedical named entity recognition. BMC Bioinformatics (2022). https://doi.org/10.1186/s12859-021-04551-4
14. Mady, L., Afify, Y., Badr, N.: Nested biomedical named entity recognition. Int. J. Intell. Comput. Inform. Sci. **22**(1), 98–107 (2022). https://doi.org/10.21608/ijicis.2022.104170.1134

15. Tong, Y., Chen, Y., Shi, X.: A Multi-Task Approach for Improving Biomedical Named Entity Recognition by Incorporating Multi-Granularity information. Findings of the Association for Computational Linguistics: ACL-IJCNLP 2021 (2021). https://doi.org/10.18653/v1/2021.findings-acl.424
16. Zhou, H., Liu, Z., Lang, C., Xu, Y., Lin, Y., Hou, J.: Improving the recall of biomedical named entity recognition with label re-correction and knowledge distillation. BMC Bioinformatics (2021). https://doi.org/10.1186/s12859-021-04200-w
17. Gao, S., Kotevska, O., Sorokine, A., Christian, J.B.: A pre-training and self-training approach for biomedical named entity recognition. PLOS ONE **16**(2), e0246310 (2021). https://doi.org/10.1371/journal.pone.0246310
18. Weber, L., Sänger, M., Münchmeyer, J., Habibi, M., Leser, U., Akbik, A.: HunFlair: an easy-to-use tool for state-of-the-art biomedical named entity recognition. Bioinformatics **37**(17), 2792–2794 (2021). https://doi.org/10.1093/bioinformatics/btab042
19. Sun, C., Yang, Z., Wang, L., Zhang, Y., Lin, H., Wang, J.: Biomedical named entity recognition using BERT in the machine reading comprehension framework. J. Biomed. Inform. **118**, 103799 (2021). https://doi.org/10.1016/j.jbi.2021.103799
20. Zhang, L., Wu, H.: Medical text entity recognition based on deep learning. J. Phys.: Conf. Ser. **1744**(4), 042209 (2021). https://doi.org/10.1088/1742-6596/1744/4/042209
21. Arutchelvan, K., Ramachandran, R.: TLBIONER: transfer learning based named entity recognition on medical literature documents. Indian J. Comput. Sci. Eng. **12**(5), 1470–1476 (2021). https://doi.org/10.21817/indjcse/2021/v12i5/211205167
22. Cohen, A.M.: A survey of current work in biomedical text mining. Briefings Bioinform. **6**(1), 57–71 (2005). https://doi.org/10.1093/bib/6.1.57
23. Weeber, M., et al.: Text-based discovery in biomedicine: the architecture of the DAD-system. In: Proceedings. AMIA Symposium, pp. 903–907 (2000)
24. Yeh, A., Morgan, A., Colosimo, M., et al.: BioCreAtIvE Task 1A: gene mention finding evaluation. BMC Bioinformatics **6**(Suppl 1), S2 (2005). https://doi.org/10.1186/1471-2105-6-S1-S2
25. Zhang, Y., Chen, Q., Yang, Z., Lin, H., Lu, Z.: BioWordVec, improving biomedical word embeddings with subword information and MeSH. Sci. Data (2019). https://doi.org/10.1038/s41597-019-0055-0
26. Bada, M., et al.: Concept annotation in the CRAFT corpus. BMC Bioinformat. (2012). https://doi.org/10.1186/1471-2105-13-161
27. Doğan, R.I., Leaman, R., Lu, Z.: NCBI disease corpus: a resource for disease name recognition and concept normalization. J. Biomed. Inform. (2014). https://doi.org/10.1016/j.jbi.2013.12.006
28. Kim, J.D., Ohta, T., Tsuruoka, Y., Tateisi, Y., Collier, N.: Introduction to the bio-entity recognition task at JNLPBA. In: Proceedings of the International Joint Workshop on Natural Language Processing in Biomedicine and its Applications, pp. 70–75. Association for Computational Linguistics (2004).
29. Smith, L., et al.: Overview of BioCreative II gene mention recognition. Genome Biol. (2008). https://doi.org/10.1186/gb-2008-9-s2-s2
30. Li, J., et al.: Annotating chemicals, diseases and their interactions in biomedical literature (2015)
31. Lee, J., et al.: BioBERT: a pre-trained biomedical language representation model for biomedical text mining. Bioinformatics **36**(4), 1234–1240 (2019). https://doi.org/10.1093/bioinformatics/btz682
32. Peng, Y., Yan, S., Lu, Z.: Transfer learning in biomedical natural language processing: an evaluation of BERT and ELMo on ten benchmarking datasets. In: Proceedings of the 18th BioNLP Workshop and Shared Task (2019). https://doi.org/10.18653/v1/w19-5006

33. Beltagy, I., Lo, K., Cohan, A.: SciBERT: A Pretrained Language Model for Scientific Text (2019). arXiv preprint arXiv:1903.10676
34. Leaman, R., Zhiyong, L.: TaggerOne: joint named entity recognition and normalization with semi-Markov Models. Bioinformatics **32**(18), 2839–2846 (2016). https://doi.org/10.1093/bioinformatics/btw343
35. Akbik, A., Blythe, D., Vollgraf, R.: Contextual string embeddings for sequence labeling. COLING, pp. 1638–1649 (2018)
36. Huang, Z., Xu, W., Yu, K.: Bidirectional LSTM-CRF models for sequence tagging (2015). arXiv preprint arXiv:1508.01991
37. Lim, S., Lee, K., Kang, J.: Drug drug interaction extraction from the literature using a recursive neural network. PLOS ONE **13**(1), e0190926 (2018). https://doi.org/10.1371/journal.pone.0190926
38. Wang, Z., Zhang, Y., Yu, H., Zhang, M.: NEEDLE: leveraging weak supervision for named entity recognition with distantly labeled data. In: Proceedings of the 2021 Conference on Empirical Methods in Natural Language Processing (EMNLP), pp. 5151–5160 (2021). https://arxiv.org/abs/2106.08977
39. Goyal, A., Gupta, V., Kumar, M.: Deep learning-based named entity recognition system using hybrid embedding. Cybern. Syst. **55**(2), 279–301 (2022). https://doi.org/10.1080/01969722.2022.2111506
40. Dogan, R.I., Leaman, R., Lu, Z.: NCBI disease corpus: a resource for disease name recognition and concept normalization. J. Biomed. Inform. (2014). https://doi.org/10.1016/j.jbi.2013.12.006
41. Kim, J.-D., Ohta, T., Tateisi, Y., Tsujii, J.: GENIA corpus—a semantically annotated corpus for bio-textmining. Bioinformatics **19**(suppl_1), i180–i182 (2003). https://doi.org/10.1093/bioinformatics/btg1023
42. Ghosh, S.: BioAug: Conditional Generation based Data Augmentation for Low-Resource Biomedical NER (2023). arXiv.org. https://doi.org/10.48550/arXiv.2305.10647
43. Jaura, S., Ramanna, S.: Named entity recognition on CORD-19 bio-medical dataset with tolerance rough sets. In: Peters, J.F., Skowron, A., Bhaumik, R.N., Ramanna, S. (eds.) Transactions on Rough Sets XXIII, pp. 23–32. Springer Berlin Heidelberg, Berlin, Heidelberg (2022). https://doi.org/10.1007/978-3-662-66544-2_3
44. Luo, L., Wei, C.H., Lai, P.T., Leaman, R., Chen, Q., Lu, Z.: AIONER: all-in-one scheme-based biomedical named entity recognition using deep learning. Bioinformatics (2023). https://doi.org/10.1093/bioinformatics/btad310

Implementing Deep Learning Models For Identifying And Classifying Infectious Skin Disease In Humans

R. Sandhiya[1(✉)], K. Sruthi[1], S. Suruthi[2], and S. V. Kogilavani[3]

[1] Department of Information Technology, Kongu Engineering College, Perundurai, India
sandhiya.it@kongu.edu

[2] Department is Computer Science and Engineering, K.S.R College of Engineering, Tiruchengode, India

[3] Department of Artificial Intelligence, Kongu Engineering College, Perundurai, India

Abstract. Skin diseases are conditions that affect the skin, including infections, inflammatory conditions, autoimmune disorders, and skin cancers. They can be diagnosed by a dermatologist or a healthcare professional through a physical examination, medical history, and, in some cases, laboratory tests or skin biopsies. This skin disorders may be detected using deep learning algorithms by training on large datasets of healthcare images to learn patterns and features that are indicative of different skin diseases. EfficientNet-B6 and EfficientNet-B7 deep learning models are used to identify different skin diseases like eczema, benign keratosis-like lesions, basal cell carcinoma and melanoma. The proposed neural network model shows high performance on image classification tasks. To identify skin diseases using the different EfficientNet models, the medical images of skin are fed into the model as input. The models extract features and patterns that are indicative of a particular skin disease, and the resulting characteristics are then utilized to categorize the image as belonging to a particular skin disease class. With the right training and testing, the proposed models can provide accurate and efficient identification of skin diseases, potentially aiding in early detection and treatment. By evaluating the performance of all these models using metrics like Precision, Recall, F1-score and accuracy. EfficientNet-B7 achieved an accuracy of 97%.

Keywords: Deep Learning Models · Skin disease Detection · EfficientNetB7 · Feature Extraction

1 Introduction

Health for an individual is the most significant part of their existence. Skin is the biggest organ of the human body. To protect the body from pathogens and to regulate body temperature are the primary functions of the skin. If the skin is not adequately protected, it can develop skin illnesses caused by bacterial infections, fungal infections, and so on. The World Health Organization reports that over 14 million cases were diagnosed in 2018, resulting in 9.6 million deaths worldwide. It is a modification of the skin's

D. Kriksciuniene et al. (Eds.): IBICA 2023, LNNS 1228, pp. 121–128, 2025.
https://doi.org/10.1007/978-3-031-78937-3_12

color or texture. Viruses, bacteria, allergies, or fungal infections can all be the cause of skin conditions. Moreover, skin problems are genetically based. Skin diseases typically affect the epidermis, the skin's thin outer layer, which can be seen by human eyes and can result in both physical harm and psychological sadness [1].

Basal cell carcinoma, benign keratosis-like lesions, melanoma, and eczema are the various types of skin lesions. In terms of symptoms and severity, the lesions differ. Some of them may be mild or painful; some are transitory, and some are permanent. Melanoma is the deadliest and most lethal form of these skin conditions. But if treated early enough, 95% of people with skin diseases can recover. While many people are unaware of the forms, symptoms, and phases of skin diseases, there is a huge knowledge gap between dermatologists and patients. Sometimes the symptoms don't appear for a while [2]. For this reason, it needs to be detected quickly and early. Yet a proper diagnosis of skin conditions that determines the nature and stage of the diseases may be challenging and expensive. Several procedures that were carried out manually using conventional equipment to identify skin diseases have been replaced by a number of deep learning models. Deep learning models such as EfficientNet B6 and B7 are used in the proposed methodology to identify and classify the appropriate type of skin condition in humans [3].

Deep learning is a technique used in machine learning and artificial intelligence that mimics certain aspects of human cognition in order to scare individuals and influence their behavior. It is a key component of data science to build models based on data-driven methods using predictive modeling and statistics. Strong forces, which we commonly refer to as algorithms, are required to propel such a human-like capacity for adaptation, learning, and functioning appropriately. Analyzing the image, extracting pertinent data to highlight the affected skin area, and categorizing it according to the type of skin disease are all steps in the treatment of skin diseases. In order to treat the patient, it uses the EfficientNet B6 and B7 models to identify skin diseases. The main objective is to identify and classify people with skin diseases, thereby preventing infectious diseases in humans using the EfficientNet B6 and B7 models, so that it is easier to get the best medical suggestions. The objective is to correctly identify skin conditions using the EfficientNet models [4].

2 Literature Survey

[5] explains a machine learning-based, automated image-based method for classifying and diagnosing skin issues. The procedure entails utilizing computational methods to analyze and process skin images. These methods extract information from the images, categorize them using sophisticated methods like neural networks with convolution and the softmax classification algorithm, and produce an evaluation report as an output. Before processing the image, the approach filters it to reduce noise and improve its overall quality. It is claimed that the suggested method is faster and more accurate than the previous approaches, resulting in a more effective and reliable system than the traditional method for detecting disorders of the skin.

[6] examined how deep learning techniques are used to diagnose skin conditions. A summary of skin datasets that are openly accessible follows a review of skin conditions

and the many techniques used in dermatology to acquire images. The paper's primary section examines the research on deep learning techniques for diagnosing skin diseases, encompassing a range of tasks such as disease identification, the process of segmentation classification, and diagnosis. The restricted accessibility of large-scale labeled datasets, the uninterpretability of models developed using deep learning, and the possibility of distortions in the training data are only a few of the difficulties addressed in the paper. [7] investigates the advancement of technology that recognizes images for diagnosing skin disorders and how it may complement conventional medical diagnosis. In particular, the paper examines the use of deep learning models—AlexNet, VGG, GoogleNet, and ResNet—for the identification of skin diseases. Multi-model fusion technology is sometimes mentioned as a way to improve the performance of models. The survey aims to encourage more researchers to conduct neural network experiments and apply these models to medical computer vision.

[7] describes the two methods for addressing the special issue of cross-domain disease of the skin. On two sets of skin disease-related data, the network is optimized employing a two-phase progressive learning transfer approach. The network begins with a fully trained, ImageNet-pre-trained convolutional neural network with deep classification. Adversarial learning is then used to convert invariant attributes from the source domains to the target domains, improving recognition performance. [9] discussed about several skin illnesses are categorized using convolutional neural networks. 11 layers make up the CNN, containing convolution, activation, pooling, fully connected, and soft-max classifier layers. Images of various skin conditions like keratosis, eczema herpeticum, urticaria, and acne are included in the DermNet database, which is used to confirm the architecture. The variety in colour of the skin, the exact spot of the disease, and the requirements of the image acquisition system present one of the problems in automating the classification of skin diseases using CNNs. Accordingly, CNN needs to be able to correctly diagnose skin conditions even when the photos it receives differ from one another in these regards.

[10] desires to create an imaging system and computer-based image processing approach for diagnosing skin conditions. This technique involves taking a digital picture of the afflicted area of skin, analyzing it to determine the disease's type, and providing the user with the results, along with details on the disease's severity and spread. The method is intended to be easy, quick, and economical. It entails extracting features from the resized image using a previously trained convolutional neural network, and then categorizing them using a multi-class SVM. All things considered, your study advances the field of skin disease diagnosis by providing a useful and approachable technique for recognizing and interpreting skin illnesses. [11] recommends utilizing transfer learning and deep learning on an Internet of Things platform to assist medical professionals in diagnosing common skin lesions like melanoma and ordinary nodules. The proposed technique uses convolutional neural networks as content extraction methods to analyze images of skin lesions and extract characteristics essential to categorization.

[12] In particular, the authors use several pre-trained CNN models, including VGG, Inception, ResNet, Inception-ResNet, Xception, MobileNet, DenseNet, and NASNet, to extract features from the images. The authors use several classifiers for classifying the skin lesions based on the extracted features. The classifiers used include Bayes,

SVM, RF, MLP, and KNN [13] introduced an AI-based framework is used to describe a system for the real-time and customized detection of skin disorders. The system is made to adapt over time, always learning from fresh data sources and input from users to boost precision and efficiency. The system uses a closed-loop information flow to update the data collections, model parameters, and environmental data between the user and a distant medical data centre [14]. In order to improve the accuracy of the data sets and increase the models' capacity for learning, the study also suggests a data set filter technique based on information entropy. LeNet-5, AlexNet, and VGG16 training on the cloud is another way that the study illustrates the applicability of the procedure's extension interface.

[15] provides a novel methodology that uses six different data mining classification approaches followed by an ensemble approach that employs bagging, adaBoost, and gradient boosting classifier algorithms to forecast categories of skin diseases. Also, the 15 most important traits that will have a significant impact on prediction are chosen using a feature significance technique. The 15 features are chosen, a subset of the original dataset is collected, and an ensemble approach is used on the complete dataset before comparing the output of six machine learning approaches. [16] aimed to increase the precision of classifying dermoscopy images into three types of lesions: seborrheic keratosis, nevus, and melanoma. To obtain high classification accuracy, the authors suggest using an ensemble of deep convolutional neural networks . This method is especially helpful if there aren't many annotated photos available for training [17]. The ensemble approach combines the outputs of various proposed deep neural network models by leveraging their classification layers. According to the authors, these trustworthy CNNs should be combined into a unified framework, resulting in the weighted output of the individual CNNs being used to complete the final classification [18]. All implemented fusion strategies perform better than the individual networks in terms of classification accuracy. This suggests that maximizing the advantages of different networks. In terms of classification accuracy, each deployed fusion technique outperforms the separate networks [19].

3 Proposed System

Pre-processing an image is an approach to improve the quality of the image so that analysis can be done more efficiently. It allows to improve specific qualities and eliminate unwanted distortions that are essential for the application. Deep learning employs the feature extraction of layer by layer technique to acquire the high-level attributes of the image and performs the training of huge data sets using multi-layer network models. By establishing up multiple deep learning models to categorize various diseases of rice leaves. Treatment of skin diseases involves analyzing the image, extracting useful information and classifying the image based on the kind of skin diseases. It is used to determine the skin diseases to be treated by the patient by making use of EfficientNet B6 and B7 deep learning models. In image acquisition, digital cameras are used to capture photos of skin disease. A dataset is taken from Kaggle and it is preprocessed to remove unwanted noise and these datas are augmented by rotating, resizing, cropping and padding which increases the size of the dataset. These augmented dataset is split into training and testing data and the EfficientNet B6 and B7 models are built to classify the skin diseases. These models are implemented to achieve better accuracy and The proposed methodology is illustrated in the Fig. 1.

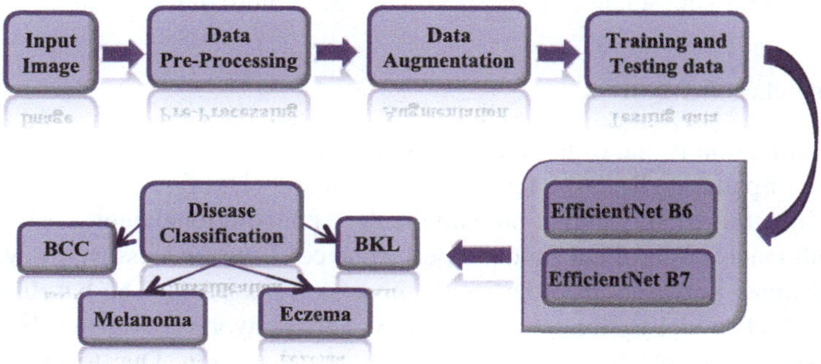

Fig. 1. Model Implementation for skin disease classification.

3.1 Data Augmentation

Convolutional neural networks need a large amount for training the data. One option to enhance the amount of data is to apply augmentation techniques to expand the dataset. Data augmentation comprises already-existing data while maintaining its structural integrity. Mirroring, rotating, shifting, and various photometric adjustments are the most widely used augmentation techniques. The next step of the inquiry is to randomly rotate the dataset by 90 degrees and flip it horizontally and vertically. These techniques, which reduced overfitting during the training phase, are then used to increase the dataset. Fig. 2 shows the different types of skin diseases.

Fig. 2. Skin Disease (a) Eczema (b) Melanoma (c) Benign Keratosis like Lesions (d) Basal Cell Carcinoma

3.2 Preprocessing

Data Preprocessing, a subset of data preparation, is any type of processing applied to raw data in order to prepare it for further data processing operations. Historically, it has been a significant first step in the data extraction process. Recently, techniques for preparing data have been changed to train AI and machine learning algorithms and run inferences on them. Data preprocessing changes the information so that machine learning, data mining, and other data science applications can use it more quickly and efficiently.

3.3 Image classification

Image classification is one of the most fundamental challenges in computer vision and recognition of patterns. Both the feature extraction and classifying networks have been combined and trained in conjunction with the deep convolutional neural network. It has a major influence on the final model's performance. It was necessary to evaluate the efficacy of many networks in order to diagnose skin disorders. The baseline network for EfficientNet model is illustrated in Fig. 3. This network is adjusted for optimum accuracy, but it also suffers if it requires a lot of calculation. Due to the increase in FLOPS, the architecture uses a mobile inverted bottleneck convolution that is larger than MobileNet V2. To create the EfficientNet family, this basic model is scaled up.

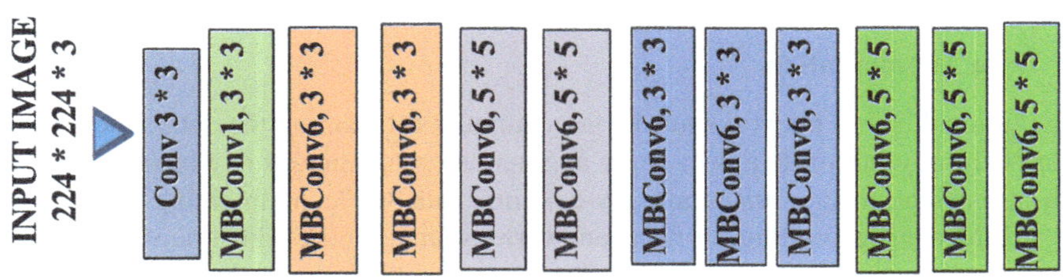

Fig. 3. EfficientNet Network

4 Performance evaluation and analysis

4.1 Validation and results

The performance of the suggested model was determined using the expected loss function and accuracy for the models Efficient-B6 and EfficientNet-B7. EfficientNet-B7 is the best and produces the best results compared with EfficientNet-B6. The classification report for EfficientNet B7 is illustrated in the Table 1.

4.2 Performance evaluation

The testing phase reveals a 97% accuracy rate when using the proposed EfficientNet B6 and B7 models to classify data across four distinct classes. With an accuracy of 97% on the test set, four different forms of skin diseases were accurately identified. During the training of the models EfficientNet B6 and EfficientNet B7 the accuracy and loss obtained for training and validation is illustrated in the Figs. 4 and 5.

Table 1. Evaluation Metrics

	Precision	Recall	F1-score
Basal Cell Carcinoma	94	98	96
Benign keratosis-like Lesions	96	91	93
Eczema	100	100	100
Melanoma	100	99	100

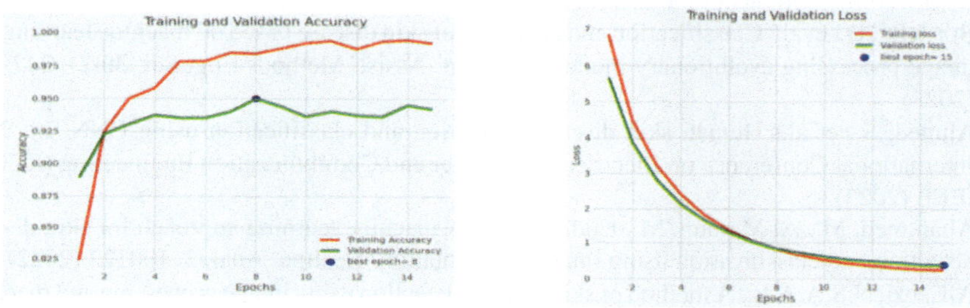

Fig. 4. Accuracy and loss for EfficientNet B7

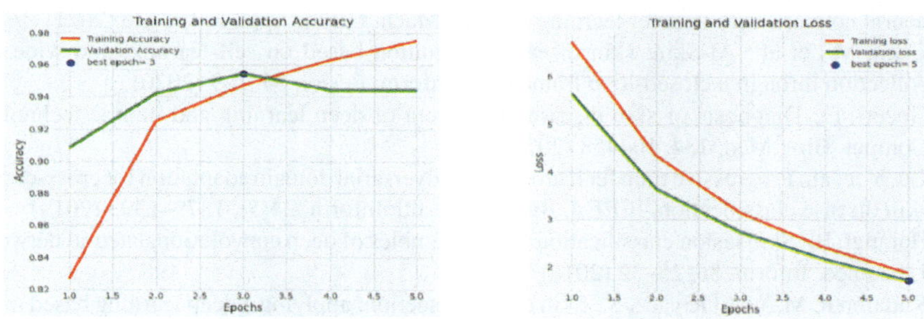

Fig. 5. Accuracy and loss for EfficientNet B6

5 Conclusion

Skin disease is a worldwide concern at this time. Several countries and regions have people who suffer from various skin conditions. By establishing a variety of methods and strategies, we may battle these illnesses. In this research on diagnosing various skin illnesses, deep learning models such as EfficientNet-B6 and EfficientNet-B7 have shown better results in identifying various skin diseases through the analysis of medical images of the skin. The models are trained on large datasets of medical images and can learn patterns and features indicative of different skin diseases, leading to accurate and efficient identification. The early identification and management of skin problems may benefit from this. The reliability and effectiveness of these models can be established by assessing the results using a variety of metrics, including precision, recall, accuracy, and F1 score. EfficientNet-B7 achieved an accuracy of 97%, indicating a high level of accuracy in identifying skin diseases.

References

1. Pai, V.R., et al.: Identification and Classification of Skin Diseases using Deep Learning Techniques (2023).
2. Vajpayee, H., et al.: An effective classification of Skin Disease using Deep Learning Techniques. In: 2023 13th International Conference on Cloud Computing, Data Science & Engineering (Confluence). IEEE (2023).
3. Bordoloi, D., et al.: Classification and detection of skin disease based on machine learning and image processing evolutionary models. Comput. Assist. Methods Eng. Sci. **30**(2), 247–256 (2023)
4. Ahmed, S., et al.: Human skin diseases detection and classification using CNN. In: 2023 International Conference on Electrical, Computer and Communication Engineering (ECCE). IEEE (2023).
5. Ahammed, M., Al Mamun, M., Uddin, M.S.: A machine learning approach for skin disease detection and classification using image segmentation. Healthc. Anal. **2**, 100122 (2022)
6. AlEnezi, N.S.A.A.L.: A method of skin disease detection using image processing and machine learning. Procedia Comput. Sci. **163**, 85–92 (2019). https://doi.org/10.1016/j.procs.2019.12.090
7. Ali, M.S., et al.: An enhanced technique of skin cancer classification using deep convolutional neural network with transfer learning models. Mach. Learn. Appl. **5**, 100036 (2021)
8. Chen, M., et al.: AI-Skin: skin disease recognition based on self-learning and wide data collection through a closed-loop framework. Inform. Fusion **54**, 1–9 (2020)
9. Goceri, E.: Diagnosis of skin diseases in the era of deep learning and mobile technology. Comput. Biol. Med. **134**, 104458 (2021)
10. Gu, Y., et al.: Progressive transfer learning and adversarial domain adaptation for cross-domain skin disease classification. IEEE J. Biomed. Health Inform. **24**(5), 1379–1393 (2019)
11. Harangi, B.: Skin lesion classification with ensembles of deep convolutional neural networks. J. Biomed. Inform. **86**, 25–32 (2018)
12. Kadampur, M.A., Al Riyaee, S.: Skin cancer detection: applying a deep learning based model driven architecture in the cloud for classifying dermal cell images. Inform. Med. Unlocked **18**, 100282 (2020)
13. Li, H., et al.: Skin disease diagnosis with deep learning: a review. Neurocomputing **464**, 364–393 (2021)
14. Li, L.-F., et al.: Deep learning in skin disease image recognition: a review. IEEE Access **8**, 208264–208280 (2020)
15. Murugan, A., et al.: Diagnosis of skin cancer using machine learning techniques. Microprocess. Microsyst. **81**, 103727 (2021)
16. Rodrigues, D.A., et al.: A new approach for classification skin lesion based on transfer learning, deep learning, and IoT system. Pattern Recognit. Lett. **136**, 8–15 (2020). https://doi.org/10.1016/j.patrec.2020.05.019
17. Shanthi, T., Sabeenian, R., Anand, R.: Automatic diagnosis of skin diseases using convolution neural network. Microprocess. Microsyst. **76**, 103074 (2020)
18. Verma, A.K., Pal, S., Kumar, S.: Comparison of skin disease prediction by feature selection using ensemble data mining techniques. Inform. Med. Unlocked **16**, 100202 (2019)

Skin Cancer Detection and Classification Using Convolutional Neural Network

Davinder Paul Singh[1,5](✉), Asheesh Pandey[2,5], Ajay Prakash Pasupulla[3,5], Mandeep Kaur Ghumman[4,5], and Sandeep Gupta[6]

[1] Department of Computer Science and Engineering, School of Technology, Pandit Deendayal Energy University, Gandhinagar, Gujarat, India
davinder.singh@sot.pdpu.ac.in
[2] Department of Computer Applications, ABES Engineering College Ghaziabad, Ghaziabad, India
[3] School of Medicine, College of Health Sciences, Nigisti Elini Comprehensive Specialized Hospital, Wachemo University, Hosanna, East Africa, Ethiopia
[4] Department of Electronics and Communication Engineering, Chandigarh University, Chandigarh, India
mandeep.e10105@cumail.in
[5] College of Health Sciences, Nigisti Elini Comprehensive Specialized Hospital, Wachemo University, Hosanna, East Africa, Ethiopia
[6] Department of Electrical Engineering, Graphic Era (Deemed to Be University), Dehradun, Uttarakhand, India

Abstract. In recent years, the applications of deep learning (DL) techniques has been gained significant attention in the field of skin cancer recognition. In this research article presents a comprehensive study on the utilization of deep learning algorithms for accurate and automated skin cancer detection and classification. By leveraging convolutional neural networks (CNNs) and their ability to extract meaningful features from medical images, our proposed model achieves remarkable performance in identifying malignant skin lesions. The dataset used for training and evaluation comprises a diverse range of skin images, including melanoma and non-melanoma cases. Through extensive experimentation and evaluation, our findings demonstrate the potential of DL technique as a reliable implement for early diagnosis and intervention of skin cancer, contributing to improved patient outcomes and healthcare efficacy.

Keywords: CNN · Skin Cancer · Machine Learning

1 Introduction

Skin cancer is a serious health concern with increasing prevalence worldwide, necessitating the development of efficient and accurate diagnostic tools. In recent years, deep learning has emerged as a promising approach for skin cancer recognition, leveraging its ability to learn intricate patterns and features from large datasets. This paper aims to introduce the applications of DL techniques in skin cancer recognition and highlight

D. Kriksciuniene et al. (Eds.): IBICA 2023, LNNS 1228, pp. 129–136, 2025.
https://doi.org/10.1007/978-3-031-78937-3_13

its potential for advancing diagnostic capabilities. DL is a subset of machine learning, involves training artificial neural networks (ANN) with multiple layers to automatically extract relevant features from input data. In the context of skin cancer recognition, DL algorithms can analyze images of skin lesions and classify them into different categories, such as melanoma, basal cell carcinoma, and squamous cell carcinoma. By utilizing vast amounts of annotated medical images, deep learning models can learn to identify subtle visual cues indicative of malignant or benign lesions [1–3]. The significant advantages of DL is its ability to extract high-level features without explicit manual feature engineering, thereby reducing the dependence on domain expertise. Convolutional Neural Networks (CNNs) are particularly effective in processing image data, as they employ convolutional layers to capture local patterns and hierarchically learn more complex representations. By training CNNs on diverse skin lesion datasets, researchers have achieved remarkable results in terms of accuracy and sensitivity, rivaling or even surpassing human dermatologists [3–5]. Furthermore, DL models can be enhanced through transfer learning, which involves leveraging pre-trained networks on large-scale image datasets and fine-tuning them for skin cancer recognition. This approach addresses the scarcity of annotated medical images by capitalizing on the wealth of available non-medical image datasets, such as ImageNet. Transfer learning enables the models to generalize well and achieve robust performance even with limited labeled medical data. In conclusion, deep learning holds immense potential for advancing skin cancer recognition by automating and improving diagnostic capabilities. Its ability to extract intricate features from medical images and generalize across diverse datasets offers a promising avenue for developing accurate and efficient skin cancer detection systems. By leveraging deep learning techniques, we can make significant strides in early detection, leading to improved patient outcomes and reduced mortality rates associated with skin cancer [6–8].

2 Litrature Review

The skin cancer is a widespread and deadly disease that affects millions of people worldwide. Early time detection of skin cancer are essential for effective treatment, and DL algorithms have shown countless potential in this area. In this literature survey, we will discuss the recent advancements in deep learning for skin cancer recognition. Esteva et al. (2017), demonstrate a DL algorithm was demonstrate and trained on a dataset of more than 129 k clinical images data to identify skin cancer. The algorithm achieved an accuracy level comparable to that of board-certified dermatologists in identifying skin cancer, suggesting that deep learning algorithms can be used as an effective diagnostic tool [9–11]. Brinker et al. (2019) demonstrate the use of DL to identify melanoma from clinical images. They used a dataset of more than 12 k clinical images data and achieved 91.2% accuracy, demonstrating that DL can be a reliable tool for early melanoma detection. In the recent study by Tschandl et al. (2021), the authors developed a deep learning algorithm that could accurately identify skin lesions from clinical images. They used a dataset of over 33,000 images and achieved an accuracy of 90.3%, making it one of the most accurate DL algorithms for skin cancer recognition to date.

A DL has also been used in combination with other techniques, such as dermoscopy, to improve skin cancer recognition accuracy. For instance, in a study by Al-Masni et al. (2021), a deep learning algorithm was combined with a feature extraction method to identify skin cancer using dermoscopy images. The algorithm achieved an accuracy of 91.2%, outperforming existing state-of-the-art techniques.

3 Methodology

In this paper proposes a DL technique based model architecture for the recognition and classification of skin cancer, aiming to enhance the accuracy of diagnosis and provide valuable insights for medical professionals. The proposed model leverages convolutional neural networks (CNNs) to automatically learn discriminative features from dermatological images. The architecture consists of multiple convolutional layers, followed by pooling layers to capture hierarchical features at various scales. Batch normalization is applied to accelerate training and improve model generalization. Dropout is employed to regularize the network and prevent overfitting. The fully connected layers at the end of the architecture perform classification based on the extracted features [15, 16]. To train the model, a large dataset of skin images with corresponding labels is used. Transfer learning is also employed, where pre-trained models on large-scale image datasets, such as ImageNet, are fine-tuned to leverage their learned features. Experimental results demonstrate the effectiveness of the proposed model, achieving high accuracy and outperforming traditional methods. The model holds great potential in assisting dermatologists and improving the early detection of skin cancer, ultimately benefiting patient outcomes [12–15].

3.1 Convolutional Neural Network (CNN)

A CNN is a type of DL model that is particularly active in dispensation and analyzing pictorial data, such as images. It is designed to repeatedly learn and excerpt hierarchical designs and landscapes from input images data, enabling tasks like image classification, object detection, and image segmentation. The key components of a CNN include convolutional layers, pooling layers, and fully connected layers:

Convolutional layers: These layers apply a data set of kernels filters to input image data, performing convolution operations. Convolution helps to detect local patterns and features within the images, such as edges, textures, and shapes. Multiple convolutional filters in a layer extract different features simultaneously.

Pooling layers: These layers downsamples the feature maps generated by the convolutional layers, reducing the spatial dimensions. Pooling helps to retain the important features while discarding unnecessary details, reducing computational complexity and making the model more robust to variations in the input.

Fully connected layers: These layers connect all neurons from the previous layer to the next layer, similar to traditional neural networks. Fully connected layers integrate the learned features from the previous layers and perform the final classification or regression tasks. CNNs often include additional components such as activation functions (e.g., ReLU), normalization layers (e.g., Batch Normalization), and regularization techniques (e.g., Dropout) to improve the model's performance and generalization.

3.2 Methodolody and Results

The deployment of IoT-based library management systems has resulted in considerable gains in library efficiency, security, systems have enhanced the user experience by giving personalised suggestions, enabling remote access to library materials, and boosting the library's accessibility and usability (Fig. 1).

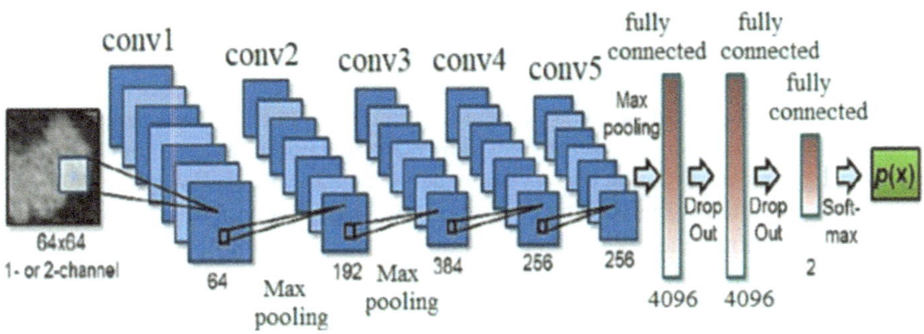

Fig. 1. Image representation of CNN

DATASET: We will be using HAM10000 Dataset in this model. The HAM10000 dataset are widely used data in the area of dermatology and skin cancer detection and classification. It comprises a collection of 10,015 dermoscopic images of skin cancer lesions, which are predominantly benign or malignant melanoma cases. The dataset was curated by the Department of Dermatology at the Medical University of Vienna, Austria.

The HAM10000 dataset encompasses a diverse range of skin scratches, including various types of melanoma, nevi, dermatofibroma, and seborrheic keratosis. Each image is associated with metadata, including lesion characteristics, clinical information, and histopathological features. The availability of such a large and diverse dataset enables researchers to develop and evaluate machine learning and deep learning algorithms for the skin cancer detection and classification. It facilitates training models that can differentiate between malignant and benign skin lesions with high accuracy and sensitivity. Re-searchers often use the HAM10000 dataset for tasks such as skin lesion segmentation, dermoscopy feature extraction, and skin cancer classification. It serves as a benchmark for evaluating the performance of different algorithms and approaches in the field of dermatology. The HAM10000 dataset has significantly contributed to advancing research in skin cancer detection and has played a crucial role in the development of deep learning models for automated diagnosis and decision support systems in dermatology (Fig. 2).

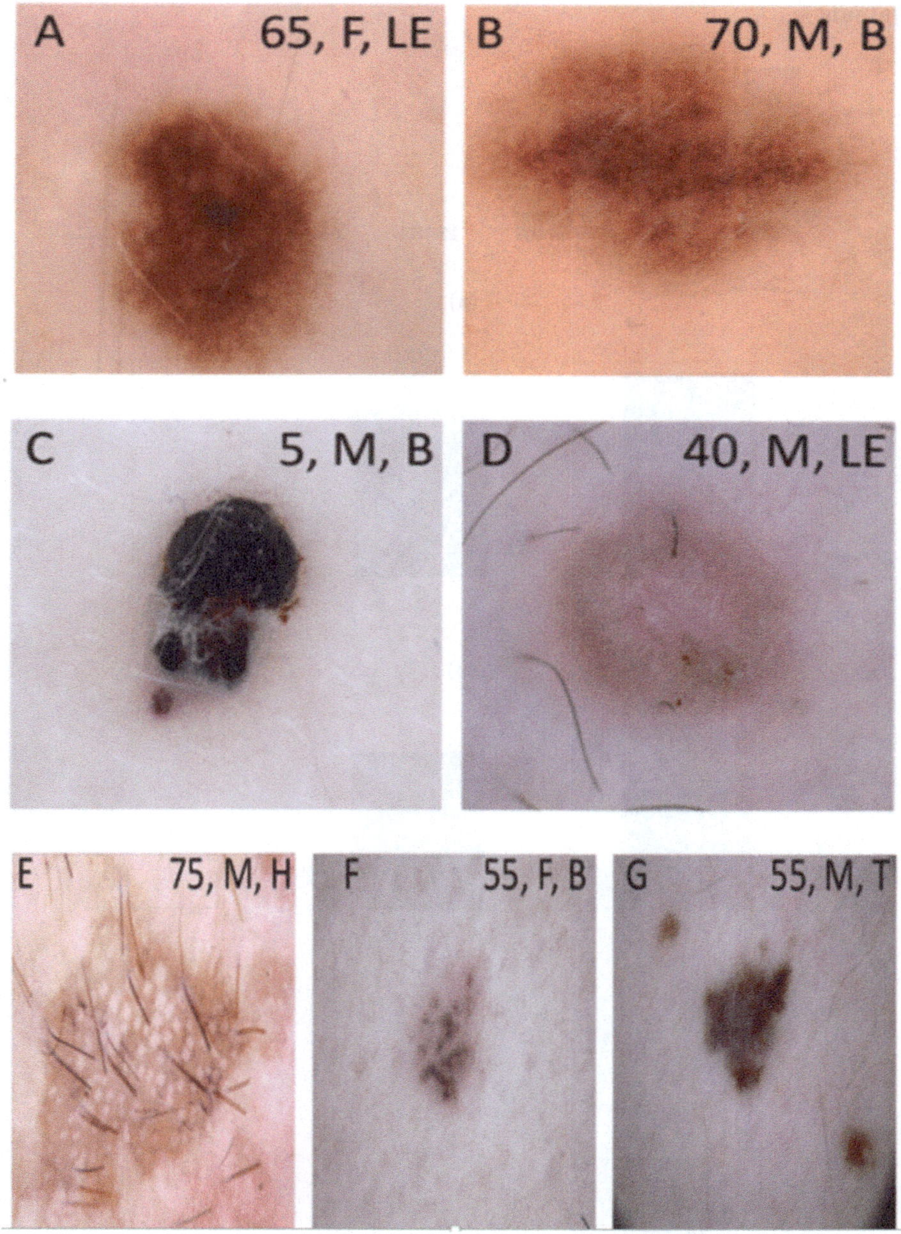

Fig. 2. Sample images from HAM10000 Dataset

Outputs

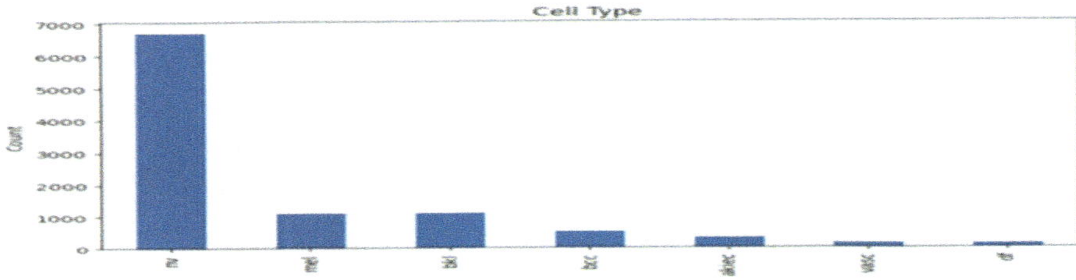

(a)

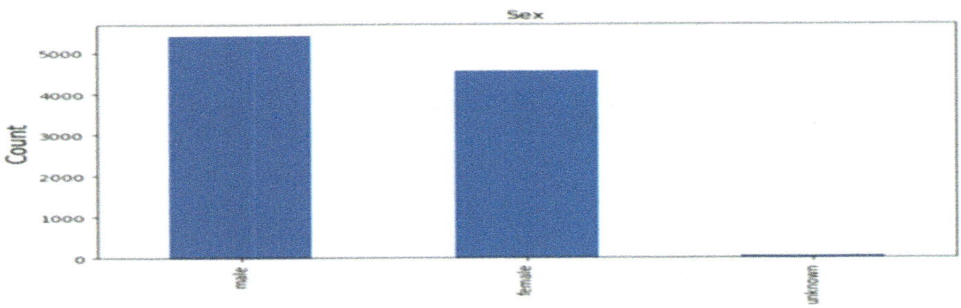

(b)

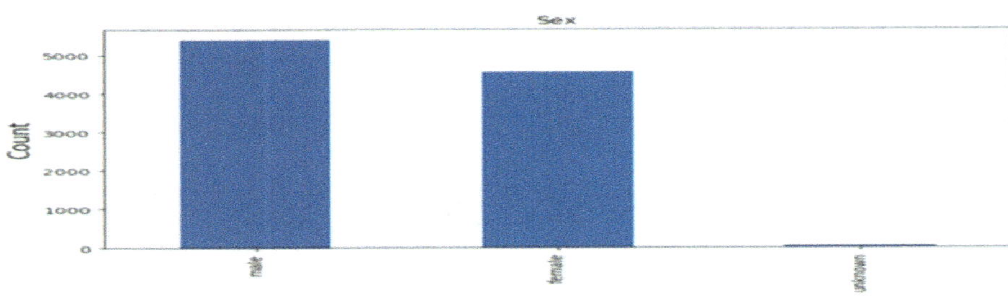

(c)

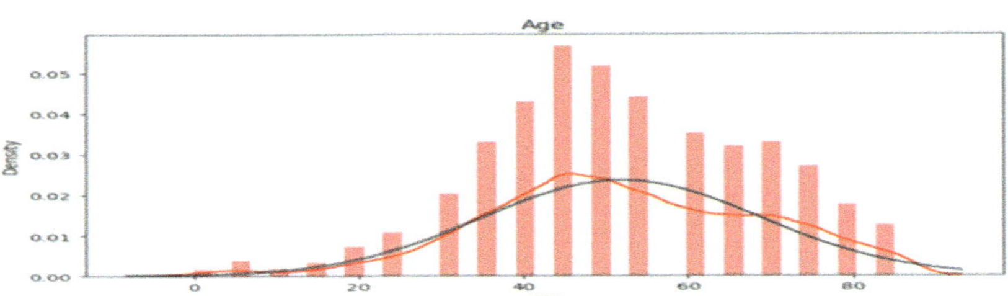

(d)

Fig. 3. Graphical representations of factors regarding skin cancer

4 Identifying Skin Cancer Using Ham10000 Dataset Via Coding Platform

Set up the environment: Create a new Python environment or activate an existing one to work with the dataset. You can use tools like Anaconda or virtualenv to manage your Python environment. Install necessary libraries: Install the required Python libraries for working with images and datasets. Commonly used libraries include numpy, pandas, scikit-learn, and Tensor Flow or Keras for deep learning models. Load and preprocess the dataset: Write code to load the dataset into memory. This may involve reading the image files, extracting features, and preprocessing the data. Libraries like PIL or Open CV can be used to handle image loading and preprocessing tasks such as resizing, normalization, or data augmentation.

Split the dataset: these dataset into training, validation, and testing the data sets. This ensures that your model is trained on one portion of the data, validated on another portion, and tested on a separate portion to evaluate its performance.

Design and train the model: Choose a suitable deep learning model architecture (e.g., CNN) and implement it using libraries like Tensor Flow or Keras. Define the model layers, compile it with appropriate loss and optimization functions, and it train the model using training dataset.

Evaluate the proposed model: After training, to calculate the proposed model's presentation using the authentication set. Determine the evaluation metrics such as precision, accuracy, recall, and F1 score to assess the representation's effectiveness in skin cancer classification.

Test the model: Finally, estimate the trained proposed model on the testing set to get a final assessment of its performance. Calculate and analyze the model's predictions against the ground truth labels (Fig. 3).

5 Conclusion

In conclusion, this research paper utilized the HAM10000 dataset and employed a Convolutional Neural Network (CNN) for skin cancer recognition. The study successfully demonstrated the potential of deep learning in accurately identifying skin cancer lesions. By leveraging the power of CNNs, this approach showcased promising results for automated skin cancer diagnosis. The findings contribute to the growing body of research on deep learning-based methods for dermatological analysis.

References

1. Esteva, S., et al.: Dermatologist-level classification of skin cancer with deep neural networks. Nature **542**(7639), 115–118 (2017)
2. Haenssle, H., et al.: Man against machine: Diagnostic performance of a deep learning convolutional neural network for dermoscopic melanoma recognition in comparison to 58 dermatologists. Ann. Oncol. **29**(8), 1836–1842 (2018)
3. Gessert, A., et al.: Deep neural networks for skin lesion classification in dermoscopy images. In: Proc. IEEE International Symposium on Biomedical Imaging (ISBI), pp. 835–838. Melbourne, Australia (2017)

4. Ranjan, R., Abhishek: Image encryption using discrete orthogonal Stockwell transform with fractional Fourier transform. Multimedia Tools and Applications (2022). https://doi.org/10.1007/s11042-022-14240-w.6
5. Wang, D.S., et al.: Deep learning for identifying metastatic breast cancer. In: Proc. IEEE International Symposium on Biomedical Imaging (ISBI), pp. 837–840. Washington, DC, USA (2016)
6. Esteva, A., et al.: Dermatologist-level classification of skin cancer with deep neural networks." arXiv preprint arXiv: 1703.04977 (2017)
7. Yao, Y., et al.: Accurate classification of malignant melanoma and benign skin lesions using a deep learning approach. J. Dermatol. Sci. **94**(2), 243–245 (2019)
8. Bhopal, M., Ranjan, R., Tripathi, A.: Oral tumor segmentation and detection using clustering and morphological process. Int. J. Electr. Electron. Res. (IJEER) **10**(4), 788–791 (2022). e-ISSN: 2347-470X, https://doi.org/10.37391/ijeer.100401
9. Liang, L., et al.: Skin lesion segmentation using deep learning. In: Proc. IEEE International Conference on Image Processing (ICIP), pp. 2738–2742. Beijing, China (2017)
10. Yu, H., et al.: Automated melanoma detection in dermoscopy images via very deep residual net-works. IEEE Trans. Med. Imaging **37**(11), 2493–2501 (2018)
11. Gaudenzi, S.R., et al.: Automatic lesion boundary segmentation using deep learning for melanoma diagnosis support. Comput. Med. Imaging Graph. **80**, 101679 (2020)
12. Litjens, G., et al.: Deep learning as a tool for increased accuracy and efficiency of histopathological diagnosis. Nat. Sci. Rep. **6**, 26286 (2016)
13. Jain, S.D., et al.: Deep learning algorithms for detection of lymph node metastases from breast cancer: Helping artificial intelligence be seen. Radiology **290**(2), 538–539 (2019)
14. Ranjan, R., Singh, A.K., Jindal, N.: The identities of n-dimensional S-transform. Multimedia Tools and Applications **81**, 16661–16677 (2022). https://doi.org/10.1007/s11042-022-12757-8
15. Sollini, M.L., et al.: Artificial intelligence in nuclear medicine: a step towards clinical implementation. Cancer Biother. Radiopharm. **49**(13), 4452–4463 (2022)
16. Ranjan, R., Jindal, N., Singh, A.K.: A sampling theorem with error estimation for S-transform. Integral Transform. Spec. Funct. **30**(6), 471–491 (2019)

Empowering Patients: Personal Health Clouds for Secure Data Control

V. Hari Priya, N. Samyuktha, A. S. Varunsundar, P. Karthikeyan, V. Vaishnavi(✉), and N. Prakash

Department of Management Studies, Kongu Engineering College, Perundurai, Erode, Tamil Nadu 638060, India

{Vaishnavi.mba,np.mba}@kongu.edu

Abstract. Modern healthcare demands a personalized, efficient system meeting real-time data collection needs. Edge computing, coupled with 5G and advanced computing, satisfies latency requirements. This paper surveys emerging edge computing architectures, emphasizing computing techniques like encryption and authentication for health applications. It delves into edge computing's role in vital sign monitoring, fall detection, and disease symptom monitoring, outlining challenges and proposing future research directions for enhanced healthcare quality. Empowerment is commonly defined through anticipated outcomes rather than its inherent nature, lacking specificity due to a disregard for self-determination principles. The process analysis reveals that educational objectives focus on general psychosocial skills, employing patient-centered experiential learning. A continuous, self-involved provider-patient relationship is crucial. While no comprehensive theory emerges, guiding principles and values are identified.

Keywords: Home hospitalization · Internet of Things · wearable · change management · empowerment · training

1 Introduction

The growing adoption of mobile and wearable devices, such as smartphones, smartwatches, smart bands, and smart glasses, is increasingly evident in health-related applications like remote diagnostics, disease monitoring, and geriatric care. This surge is propelled by rapid advancements in mobile computing, wearable technology, and wireless sensing. These devices generate extensive health data, offering substantial value for academic and commercial healthcare research. Responsible sharing of personal health data can yield benefits for users, patients, researchers, corporations, and the broader public healthcare system. Ideally, personal health data should be viewed and managed as an individual asset, though presently, this perspective is not widely embraced, (Zheng et al., 2018).

D. Kriksciuniene et al. (Eds.): IBICA 2023, LNNS 1228, pp. 137–144, 2025.
https://doi.org/10.1007/978-3-031-78937-3_14

2 Review of Literature

I*ot.* The Internet of Things (IoT) technology is finding applications in diverse fields, spanning industries, smart homes, cities, energy management, vehicular communications, agriculture, healthcare, logistics, and beyond. IoT facilitates the smooth communication and exchange of information among various physical devices via the Internet, establishing an interconnected environment that resembles a global village. Progress in wireless technologies like Wi-Fi, Bluetooth, 6LoWPAN, and ZigBee is instrumental in enabling this interconnected ecosystem (Kadhim et al., 2020).

Hybrid AI The integration of bio-inspired methodologies, genetic algorithms, neuro-fuzzy algorithms, and soft computing approaches in hybrid artificial intelligence (AI) significantly enhances the prediction of critical cardiovascular issues and other healthcare solutions within the ongoing complex research. AI utilizes quantitative methods to enable computer systems to "learn" from new data, discern patterns, and make decisions with minimal human intervention. The potential impact of AI and IoT on healthcare in the future is virtually limitless. This book specifically explores the applications of hybrid AI and IoT in creating an integrated approach for problem -solving in various healthcare domains such as radiology, drug interactions, drug discovery, imaging, electronic health records, disease diagnostics, telemedicine, and mobility- related concerns (De Albuquerque, 2021).

Edge computing Emerging healthcare services, both modern and next-generation, have introduced a fresh set of demands concerning computing power and response time. These advanced devices necessitate faster, more energy-efficient computation, increased storage, and location awareness to operate optimally, needs that conventional cloud computing struggles to meet. Fog computing, alternatively referred to as "edge computing," was initially introduced by Bonomi et al. and stands out as a potentially highly promising technology in addressing these challenges.

Edge Computing Emerging healthcare services, both modern and next-generation, have introduced a fresh set of demands concerning computing power and response time. These advanced devices necessitate faster, more energy-efficient computation, increased storage, and location awareness to operate optimally, needs that conventional cloud computing struggles to meet. Fog computing, alternatively referred to as "edge computing," was initially introduced by Bonomi et al. and stands out as a potentially highly promising technology in addressing these challenges.

Cloud computing. Cloud computing has permeated various industries, including healthcare, indicating a significant shift of healthcare services to the cloud, allowing for a heightened focus on patient care. Cloud computing, a transformative approach to delivering and hosting information technology (IT) services online, has evolved rapidly in the healthcare sector. It provides scalable, multi-tenant, on-demand services with a pay-per-use structure. While there are different interpretations the cloud computing model, the National Institute of Standards and Technology (NIST) offers a comprehensive definition. According to NIST, cloud computing is a model designed to facilitate convenient, ubiquitous, on-demand network access to a shared pool of configurable computing resources (e.g., networks, servers, storage, applications, and services) that can be swiftly provisioned and released with minimal management effort or service provider interaction (Nigam and Bhatia, 2018).

2.1 Methodology

Cloud computing is now so prevalent that it is being used in the health care industry. Because cloud computing in health care is rapidly growing, we can expect a huge number of healthcare services to migrate to the cloud, allowing for more focus on patient care. The phrase "cloud computing" refers to a ground-breaking method of providing and hosting information technology (IT) services online. It offers pay-per-use scalable, multi-tenant, on-demand services.

The cloud computing model has been defined in various ways, with no single description fully encapsulating its complexity. However, the National Institute of Standards and Technology (NIST) provides a comprehensive characterization. According to NIST, it is a model designed to facilitate convenient,

3 Process

Process 1
In the first stage, users employ the User App to gather, compress, and encrypt data. Subsequently, the encrypted data is transmitted to cloud storage, and key shares arekeepers.

Process 2
After uploading the data, a transaction is initiated to notify other participants that the data is prepared for sharing. Upon user confirmation, this transaction is integrated into the block chain through a consensus method, rendering it visible to consumers and data validators.

Process 3
In the process of data validation, data validators employ data-mining and machine-learning techniques to ensure that the data adheres to rules and requirements. Once validated, the data is certified in alignment with the specified criteria.

Process 4
Customers select the data they wish to purchase, ensuring its validation by pertinent data validators. Subsequently, a transaction is generated for the purchase, and upon customer approval, it is appended to the blockchain.

Process 5
The consensus algorithm verifies the transaction for acquiring the data. Approval of the transaction occurs if the customer possesses an adequate amount of PHD coins for the purchase. The corresponding quantity of PHD coins is then transmitted to the smart contract, advancing the process to step 6. In case of insufficient funds, the transaction is declined, and the workflow reverts to step 3.

Process 6
Upon approval of a data purchase transaction, key keepers associated with the dataset receive notifications. Subsequently, they utilize an authenticated communication channel to transmit their keshares for the corresponding dataset to the customer.

Process 7
Following a predefined rule, PHD currencies held within the smart contract are distributed among user and key keeper accounts.

Process 8
Upon approval, the customer receives a link to the encrypted data along with sufficient key shares for decryption. This enables the down loading of data from cloud storage, decryption, and makes it ready for use, thereby concluding the process (Zheng et al., 2018).

3.1 Evolution of Health Care Computing

Healthcare applications can be categorized based on device type, data type, or specific use cases, with key use case types including real-time health monitoring, emergency management systems, health-aware mobile devices, and healthcare information distribution. Real-time health monitoring is adaptable and can operate across various platforms simultaneously, such as smartphones, wearable sensors, or a combination of both. Emergency management systems, similar to real-time health monitoring, generate alerts when a patient's vital signs fall below predefined thresholds. With the advancement of smart mobile devices, patients now have diagnostic capabilities at their fingertips. Websites provide easy access to healthcare information, and personalized apps on mobile devices offer individualized health information and counseling, particularly for chronic disease management. Recent advances in medical monitoring solutions often compare distributed or "fog" architectures with traditional cloud architectures. For real-time emergency situations like fall detection or stroke mitigation, which require rapid medical response times, relying solely on cloud systems for data retrieval is impractical due to prolonged latency. Additionally, transmitting vast sensor-generated data to the cloud results in higher power consumption and associated costs. It has been observed that typical cloud services exhibit excessive latency and poor sustained performance (Hartmann et al., 2022) Real-time health monitoring is a flexible approach that harnesses multiple platforms simultaneously. For instance, the monitoring of vital signs can be conducted using smartphone devices, wearable sensors, or a combination of the two.Emergency management systems exhibit similarities to real-time health monitoring systems, with their primary focus on triggering alarms when a patient's vital signs dip below preset threshold the advancements in mobile technology, patients now possess diagnostic tools at their disposal. Healthcare information is easily attainable through websites, and customized mobile applications provide personalized health information and guidance, particularly for the management of chronic conditions. Contemporary healthcare applications come in various formats, encompassing.

Compromises in Health Systems Arising from Edge Computing
Edge computing plays a pivotal role in advancing healthcare standards, offering faster and more comprehensive treatment on a global scale. The widespread deployment of health sensors reduces the need for patients to make frequent visits to hospitals and clinics. Devices equipped with edge computing capabilities aid in disease diagnosis and patient monitoring, thereby enhancing healthcare accessibility. Patients can easily manage these edge sensor devices, ensuring continuous monitoring of vital signs and the generation of valuable healthcare insights. Edge computing also contributes to cost reduction by minimizing data transmission expenses through the migration of essential data to the edge. The proximity of data minimizes latency issues associated with cloud platforms.

Despite the significant benefits of edge computing in healthcare, the distributed approach comes with trade-offs and challenges. Using different platforms and servers brings challenges such as connectivity, scalability, resource and data management, and node reliability. Integrating diverse sensors and nodes demands extra resources and data processing methods at the edge nodes, in contrast to cloud-based computing, which depends on centralized management. As the demands of the healthcare system grow, managing the complexity of workflows becomes increasingly difficult. Identifying and managing potential bottlenecks and constraints in real time is crucial. Healthcare devices face the challenges of limited computing and storage resources, which makes resource allocation difficult. The concept of Edge Mesh, introduced in a recent study, distributes resources and computing management decision between edge devices. Edge Mesh includes built-in discovery capabilities that address IoT devices' limited knowledge of nearby operating systems. However, the optimization of control systems remains a key area for further research. Tampering with the security of personal data is a critical challenge in the large-scale implementation of IoT in healthcare. The reliability of the communication protocol when using IoT is not exceptionally high, which can lead to network problems. Fault detection in an IoT network is difficult without centralized error reporting. In addition, extensive research in healthcare management is needed to establish a predefined protocol for prioritizing services in a distributed edge network, especially for urgent medical calls (Hartmann and Hashmi et al., 2022).

4 Technical Tools

A grounded theory model for predictive health care in the community: a case study in Beijing, China. Open coding and selective code analysis identified key categories that encompassed all key elements of the "Community-Based Preventive Health Management Model". This marked the beginning of the final stage of developing the theoretical framework. In this phase, a conceptual model of a community-based predictive health care model was formulated as the interrelationships of key areas became clearer. This conceptual model consists of six dimensions: awareness and preparedness, base population, health behavior paradigm, management mechanism, information technology applications, and facilities and services. The main understanding of this model is that the establishment of community-based preventive health management models is based on increasing residents' awareness, attitudes, and readiness for preventive health practices.

4.1 Discussion

The convergence of emerging technological paradigms in Industry 4.0 is prompting a fundamental change in perspective and approach to conventional applications. This transformation is actively taking place in the healthcare sector, marking the beginning of a gradual shift towards e-health. This transition is anticipated to gather momentum and accelerate within the framework of healthcare 4.0 (HC4.0). This shift in perspective necessitates an understanding of the new possibilities, opportunities, as well as the challenges and risks that come with it. Both aspects are explored in this section and summarized in Table II. The objective is to promote a deliberate and efficient integration of new methods, technologies, and tools into healthcare processes.. Malware and intrusion represent significant security threats to IoT systems, including IoMT. Initially, these aspects received limited research attention, but their criticality has surged due to associated life risks and device reliability concerns. Machine learning-based techniques are increasingly employed for malware and intrusion detection in IoT devices. However, given the memory and processing constraints, there is a pressing need for efficient approaches and algorithms to ensure successful implementation.

The primary challenge lies in guaranteeing the security and confidentiality of personal health information stored in the cloud. Patients require assurance that their sensitive data is safeguarded against unauthorized access or potential data breaches. Data Standardization: the lack of standardized formats for health data poses a challenge. Achieving consistency in data formats and structures is essential for effective collaboration and data sharing among different healthcare stakeholders.

5 Future Research Work

Malware and intrusion represent significant security threats to IoT systems, including IoMT. Initially, these aspects received limited research attention, but their criticality has surged due to associated life risks and device reliability concerns. Machine learning-based techniques are increasingly employed for malware and intrusion detection in IoT devices. However, given the memory and processing constraints, there is a pressing need for efficient approaches and algorithms to ensure successful implementation.

While the discussion section has addressed various aspects of trust-related issues in the context of cloud computing and the Internet of Medical Things (IoMT), certain areas of concern necessitate further research. This is crucial to guarantee the reliable and secure adoption of IoMT built upon cloud computing. These specific aspects are detailed in this section.

6 Challenges

Security Concerns: The primary challenge lies in guaranteeing the security and confidentiality of personal health information stored in the cloud. Patients require assurance that their sensitive data is safeguarded against unauthorized access or potential data breaches.

Data Standardization: The lack of standardized formats for health data poses a challenge. Achieving consistency in data formats and structures is essential for effective collaboration and data sharing among different healthcare stakeholders.

User Adoption and Education: Encouraging patients to actively participate in managing their health data through personal health clouds requires education and awareness.

Many individuals may be unfamiliar with the technology or hesitant to adopt new tools for healthcare management.

Adherence to Regulations: It is crucial to ensure compliance with healthcare regulations and standards, such as the Health Insurance Portability and Accountability Act (HIPAA) in the United States or the General Data Protection Regulation (GDPR) in Europe. A substantial challenge involves ensuring that personal health clouds satisfy these regulatory requirements.

Integration into healthcare workflow. The seamless integration of personal health clouds into the existing healthcare workflow is essential for the success of such systems. Healthcare providers and professionals should have effective access to relevant patient information.

Data Ownership and Governance: It is imperative to adaddress concerns related to data ownership and control. Patients need a clear understanding of the ownership of their health data and the extent of their authority over its utilization and sharing.

Technological Infrastructure: Building and maintaining a robust technological infrastructure to support personal health clouds, including ensuring high-speed and reliable connectivity, can be challenging.

Cost and Affordability: The cost associated with implementing and maintaining personal health clouds may be a barrier, especially for individuals with limited financial resources.

Ethical Considerations: Ethical concerns, such as the responsible use of patient data for research purposes, need to be addressed. Striking a balance between empowering patients and advancing medical research while respecting ethical boundaries is challenging.

7 Conclusion

This research paper introduces a system that leverages blockchain, cloud storage, and machine learning technologies for secure sharing of personal health information. The system empowers users to own, manage, and share their health data securely. The study begins by categorizing an individual's health data on portable devices based on characteristics (dynamic and static data) and collection methods (continuous and instantaneous data). To address the challenge of sharing large and dynamic health data, the paper suggests using hash references to indicate the data's storage location. By integrating blockchain and cloud storage, the system effectively overcomes size limitations associated with continuous and dynamic health data. The proposal advocates for storing large sets of health data in encrypted form in the cloud, with only transaction data and metadata shared on the blockchain.

Additionally, the system incorporates a data quality validation module supported by machine learning techniques to ensure data quality from both hardware and software perspectives. The convergence of the Internet of Medical Things (IoMT) and Healthcare 4.0 presents an exciting opportunity to transform the healthcare industry. The paper delves into the core elements of the Healthcare 4.0 paradigm, focusing on hardware and communication technology, and outlines its key features and potential impact on the future of healthcare. It discusses obstacles facing Healthcare 4.0 systems and highlights key technologies to address these challenges. Furthermore, the paper explores recent advances and outlines the future trajectory of both Healthcare 4.0 and IoMT, serving as a comprehensive reference for researchers. The integration of technology holds the potential to make healthcare systems more active, adaptive, and efficient. However, addressing data security, privacy and enforcement barriers is critical to unlocking the full potential of these technological innovations (Ahajan et al., 2023).

References

Aceto, G., Persico, V., Pescapé, A.: Industry 4.0 and health: Internet of things, big data, and cloud computing for healthcare 4.0. J. Indus. Info. Integrat. **18**, 100129 (2020)

Aujoulat, I., d'Hoore, W., Deccache, A.: Patient empowerment in theory and practice: polysemy or cacophony? Patient Educ. Couns. **66**(1), 13–20 (2007)

Bhushan, B., et al.: Towards a secure and sustainable internet of medical things (IoMT): requirements, design challenges, security techniques, and future trends. Sustainability **15**(7), 6177 (2023)

de Albuquerque, V.H.C.: Hybrid Artificial Intelligence and IoT in Healthcare. In: Bhoi, A.K., Mallick, P.K., Mohanty, M.N. (eds.) Springer (2021)

Hartmann, M., Hashmi, U.S., Imran, A.: Edge computing in smart health care systems: Review, challenges, and research directions. Transactions on Emerging Telecommunications Technologies **33**(3), e3710 (2022)

Kadhim, K.T., Alsahlany, A.M., Wadi, S.M., Kadhum, H.T.: An overview of patient's health status monitoring system based on internet of things (IoT). Wireless Pers. Commun. **114**(3), 2235–2262 (2020)

Mahajan, H.B., et al.: Integration of Healthcare 4.0 and blockchain into secure cloud-based electronic health records systems. Applied Nanoscience **13**(3), 2329–2342 (2023)

Mathad, K., Khang, A.: Hospital 4.0: Capitalization of health and healthcare in Industry 4.0 economy. In: Data-Centric AI Solutions and Emerging Technologies in the Healthcare Ecosystem, pp. 319–330. CRC Press (2023)

Nigam, V.K., Bhatia, S.: Impact of cloud computing on health care. Int Res J Eng Technol **3**(5), 2804–2810 (2016)

Zheng, X., Mukkamala, R.R., Vatrapu, R., Ordieres-Mere, J.: Blockchain-based personal health data sharing system using cloud storage. In: 2018 IEEE 20th international conference on e-health networking, applications and services (Healthcom), pp. 1–6. IEEE (2018)

A Study on Medical Diagnosis with Help of LLM: Advancement in Disease Detection

R. S. Karthikeyan, M. Dharshne(✉), M. Mohanasundari, K. Hariharan, T. V. Prathika, and S. Deepa

Department of Management Studies, Kongu Engineering College, Perundurai, Erode, Tamil Nadu 638060, India
{karthikeyanrs.22mba,dharshne.mba,mohanasundari.mba, hariharank.22mba,prathiksatv.22mba}@kongu.edu

Abstract. The incorporation of Large Language Models (LLMs) in medical diagnosis marks a significant advancement in healthcare, harnessing the capabilities of natural language processing and machine learning to transform disease detection. These intelligent software programs demonstrate exceptional proficiency in rapidly analyzing extensive medical data, ranging from electronic health records to intricate medical images, thereby providing healthcare professionals with more precise and efficient diagnostic capabilities. Notably, LLMs contribute to improved accessibility, proving particularly valuable in remote or underserved areas where specialized medical expertise is limited.

These models continually elevate diagnostic accuracy and play a pivotal role in patient education by translating complex medical terminology into easily understandable language. Additionally, LLMs offer crucial support to healthcare professionals, mitigating the risks of human error and fatigue in the diagnostic process. However, the integration of LLMs into healthcare also introduces ethical and privacy concerns, emphasizing the need for a judicious balance between the advantages they offer and the security of patient data.

Despite these challenges, LLMs hold the promise of enhancing disease detection, patient care, and overall healthcare quality. Their implementation suggests a future healthcare landscape characterized by heightened accuracy, personalized treatment approaches, and increased accessibility, portraying a transformative shift toward a more advanced and inclusive healthcare system.

1 Introduction

This paper delves into the crucial intersection of technology and healthcare, where LLM serves as a potent instrument for enhancing disease detection to levels of precision and efficiency previously unattainable. In a healthcare landscape grappling with the challenges posed by the escalating volume and intricacy of medical data, LLM presents a promising solution by tapping into the capabilities of natural language processing and machine learning for the interpretation of extensive clinical information. The primary objective of this study is to investigate the inventive uses of LLM in medical diagnosis, evaluate its influence on disease detection, and appraise the ramifications of its integration into the healthcare system.

D. Kriksciuniene et al. (Eds.): IBICA 2023, LNNS 1228, pp. 145–152, 2025.
https://doi.org/10.1007/978-3-031-78937-3_15

Our research aims to thoroughly explore how medical experts working together with artificial intelligence have made significant advancements. In doing so, we want to answer important questions such as: How does LLM help detect diseases early and accurately? What are the ethical and privacy issues when using LLM in healthcare? How does using LLM for medical diagnosis affect patients' well-being and healthcare expenses? These questions are the basis for our study on how LLM can improve medical diagnosis and healthcare practices in the future.

2 Objectives

1. To assess the effectiveness of utilizing LLM (Large Language Models) in improving the accuracy and efficiency of medical diagnosis.
2. To investigate the potential of LLM in assisting healthcare professionals in the early detection of diseases.
3. To explore the use of LLM in analyzing medical data, such as electronic health records, medical images, and patient history, for disease detection.
4. To evaluate the impact of LLM-based medical diagnosis on patient outcomes, including reduced misdiagnosis rates, earlier disease detection, and improved treatment recommendations.
5. To compare the performance of LLM-based diagnosis with traditional methods and highlight the advantages and limitations of each approach.
6. To present case studies or empirical evidence demonstrating the practical application of LLM in real-world medical diagnosis scenarios.
7. To discuss the ethical and privacy considerations associated with using LLM in healthcare, including data security and patient consent.
8. To provide recommendations for healthcare institutions, policymakers, and researchers on how to integrate LLM into medical diagnosis processes effectively and ethically.

3 Review of Literature

Aleksa Bisercic et al., 2023, The paper presents TEMED-LLM, a novel approach for extracting structured tabular data from textual medical reports using large language models (LLMs). This method leverages the reasoning capabilities of LLMs to accurately infer tabular features, even when not explicitly mentioned in the text. It combines domain-specific reasoning guidelines with a data validation and reasoning correction feedback loop. The extracted data is then analyzed using interpretable ML models like decision trees and logistic regression, resulting in interpretable predictions. TEMED-LLM outperforms state-of-the-art text classification models in medical diagnostics, offering simplicity and interpretability, thus enhancing trust in ML models for medical applications. Multiple real-world medical datasets were used for evaluation, and data privacy was ensured.

Hanyao Huang et al., 2023, The paper discusses the potential applications of ChatGPT, a conversational variant of the GPT-4 language model, in the field of dentistry. It highlights two primary deployment methods for ChatGPT in dentistry: automated dental diagnosis and cross-modal dental diagnosis, emphasizing the model's ability to handle multi-source data and perform complex clinical operations. While ChatGPT offers promising advancements in dental diagnosis and treatment planning, it also addresses challenges such as data privacy, data quality, and model bias. The authors envision that ChatGPT could revolutionize dental healthcare, paving the way for precision medicine and personalized healthcare in dentistry.

Loredana Caruccio et al., 2023, this paper compares the performance of ChatGPT with traditional machine learning models for intelligent medical diagnosis. It explores the benefits and limitations of ChatGPT, how it compares to other language models, and its implications for healthcare. While ChatGPT shows promise in diagnosing low-risk diseases based on symptoms, it also has limitations, highlighting the importance of clinician involvement in the diagnostic process. The study suggests that ChatGPT and traditional models have their own advantages, and further research and development are needed to improve their accuracy and applicability in healthcare.

Jiaxiang Liu et al., 2023, The paper introduces an innovative framework for zero-shot medical image classification, which combines the power of vision-language models (VLMs), such as CLIP, and large language models (LLMs), like ChatGPT, to provide explainable medical diagnoses. In this approach, medical images are classified based on similarity scores computed between the image and possible disease categories, with additional cues and knowledge generated by querying LLMs with category names. The framework demonstrates promising results in improving zero-shot medical image classification accuracy and offers explainable diagnoses, showcasing the potential of VLMs and LLMs in the medical field. However, challenges related to hallucinations and inaccuracies in disease diagnosis by ChatGPT are acknowledged and require further investigation.

Juexiao Zhou et al., 2023, The paper introduces MedAGI, a paradigm for unifying domain-specific medical large language models (LLMs) with minimal cost. It addresses the challenge of integrating LLMs for medical applications and automatically selects appropriate models based on user questions, improving user experience and reducing complexity. MedAGI's versatility and scalability were demonstrated across different medical domains, making it a promising tool for the future of medical artificial general intelligence (AGI). This approach simplifies the integration of various LLMs in the medical field, advancing the goal of AGI for medical applications.

Sheng Wang et al., 2023, The paper introduces ChatCAD, a novel approach that integrates Large Language Models (LLMs) into Computer-Aided Diagnosis (CAD) systems for medical images. ChatCAD combines the strengths of LLMs' medical knowledge and reasoning abilities with CAD's image analysis to provide more interactive and understandable medical reports. This approach enhances patient understanding, streamlines consultations, and has the potential to improve online healthcare services. However, limitations include the quality of LLM-generated reports and the need for more comprehensive datasets and clinical validation. Larger LLMs may further enhance diagnostic accuracy. Collaboration with clinical professionals and future research are essential for refinement.

Samuel Yang and Richard E Rothman (2004), PCR-based diagnostics have transformed infectious disease management in acute-care settings by enabling rapid and accurate pathogen detection. This technology has numerous applications, including identifying specific or broad-spectrum pathogens, tracking emerging infections, and profiling antimicrobial resistance. It has the potential to improve patient outcomes, infection control, and resource allocation. However, challenges like test interpretation, quality assurance, and education must be addressed. Universal PCR assays could revolutionize clinical care by allowing quick triage, targeted therapy, isolation of contagious patients, and precise antibiotic intervention, benefiting both patients and society by reducing hospital stays and enhancing epidemiological surveillance.

4 Methodology

The paper adopts the literature study methodology which have used the secondary data to collected from the relevant articles, research paper and publication made about the opportunities and implications of LLM in healthcare. The secondary data are identified from the data sources which have consistency and quality information.

5 LLM's in Medical Diagnosis

Using Large Language Models (LLMs) in medical diagnosis is a significant advancement in the field of healthcare. These smart computer programs, powered by natural language processing and machine learning, have the capacity to assist doctors in understanding complex medical data. They can swiftly analyze a wide array of information, such as electronic health records, medical images, doctors' notes, and research papers, making it easier for healthcare professionals to provide more accurate and efficient diagnoses. This not only speeds up the diagnostic process but can also reduce healthcare costs by preventing unnecessary tests and treatments. One notable advantage of LLMs is their accessibility, particularly in remote or underserved areas where specialized medical expertise may be scarce. LLMs can serve as a valuable resource for healthcare professionals, offering continuous learning capabilities that improve their diagnostic accuracy over time. Moreover, these programs play a crucial role in patient education, translating complex medical jargon into easily understandable language, empowering patients to become more informed about their health.

Another significant benefit is the support they provide to doctors in complex cases, offering additional information and suggestions to assist in decision-making. LLMs operate without fatigue or emotional bias, reducing the risk of human error in diagnosis and ensuring consistent results regardless of workload or time constraints. They can also contribute to medical research by rapidly processing vast amounts of medical literature, potentially accelerating new discoveries and breakthroughs. However, integrating LLMs into medical diagnosis also presents challenges, including ethical and privacy considerations. Striking the right balance between maximizing the benefits of LLMs and safeguarding patient data is a crucial aspect of their responsible use in healthcare. Despite these challenges, the use of LLMs in medical diagnosis holds tremendous promise in improving disease detection, patient care, and the overall quality of healthcare.

6 LLM's in Disease Deduction

Utilizing Large Language Models (LLMs), which are like smart computer programs, to assist in identifying diseases is a major step forward in the world of healthcare. These programs are equipped to sift through vast amounts of medical information, such as a patient's health records and test results, with remarkable speed and accuracy. Their ability to work tirelessly without fatigue makes them incredibly valuable, particularly in cases where doctors are faced with complex or puzzling health issues. The reliability of LLMs is noteworthy - they consistently produce accurate results, regardless of how busy the healthcare setting is or what time of day it might be. Additionally, they serve as a safeguard against the kinds of mistakes that can happen when humans are tired or under stress.

However, when employing LLMs in the healthcare field, there is an imperative need to address the protection of patient data. Ensuring that sensitive medical information remains private and secure is a top priority. It is crucial to find a balance between making the most of LLMs to enhance disease detection and maintaining the confidentiality of patients.The integration of LLMs into disease detection procedures represents a substantial advancement in healthcare. These intelligent computer programs offer the potential for more accurate and faster disease diagnoses, which can lead to improved outcomes for patients. Nonetheless, a careful and responsible approach is essential, with particular attention to ethical considerations and the security of patient information. Ultimately, LLMs have the potential to enhance healthcare by assisting doctors in pinpointing diseases more precisely and promptly.

7 Challenge and Ethical Considerations

7.1 Challenges

One big challenge is keeping patient information private. LLMs need lots of data, but we must make sure it's safe from unauthorized access. Another challenge is making sure the data used to teach LLMs is accurate. If it's not, LLMs might give wrong answers. Getting LLMs to work smoothly with the systems hospitals already use can be tough. Also, following rules and laws about patient data, like HIPAA, can be complicated. We have to be careful about LLMs inheriting any unfairness from the data they learn from. Fixing this is an ongoing challenge. Training healthcare professionals to use LLMs properly is also a big task.

7.2 Ethical Considerations

One important ethical rule is that patients need to know and agree to LLMs being used in their healthcare. Transparency is key, so patients and doctors understand how LLMs make decisions. Keeping patient data safe and private is not just a challenge; it's an ethical duty.

If something goes wrong with a diagnosis made by an LLM, we must know who's responsible - whether it's the healthcare provider, the LLM developer, or both. LMMs should help all patients equally, no matter where they come from or how much money they have. To make sure LLMs are used fairly, we need rules and oversight. We also have to keep checking for unfairness or mistakes in LLM diagnoses. Balancing the potential benefits of LLMs with these challenges and ethical rules is crucial. It ensures that using these smart computer programs in healthcare is fair, safe, and good for patients and healthcare systems.

8 Future Direction and Research Opportunities

In the future, we can make LLMs even better at predicting diseases and understanding them. They can be like a personal health advisor, looking at your specific factors to suggest the best treatments. We could also create systems that keep an eye on your health all the time and quickly spot any problems. By teaching LLMs to work with pictures and sensor data, we can get a fuller picture of your health. And it's important that the information LLMs give us is easy to understand for both doctors and patients. We also need to keep working on making sure LLMs are fair and don't make any mistakes. Researchers are also thinking about how LLMs can work well with doctors and be part of remote healthcare. Finally, we need rules and regulations to make sure LLMs are used safely and responsibly in healthcare. All of these steps will make healthcare more accurate, faster, and available to more people.

9 Result and Discussion

9.1 Result

As research in the field of LLMs and disease detection continues to progress, we anticipate a future where healthcare is more accurate, personalized, and accessible than ever before. With LLMs fine-tuned to predict diseases and identify health risks more effectively, patients may receive early warnings, leading to quicker interventions and improved health outcomes. Personalized medicine, facilitated by LLMs, has the potential to offer tailor-made treatments based on individual factors, enhancing the effectiveness of healthcare. Continuous health monitoring could transform the way we approach preventive care, leading to more proactive and efficient health management. Multimodal data analysis may provide a more comprehensive understanding of an individual's health, helping doctors make more informed decisions. Moreover, transparent and easy-to-understand explanations from LLMs can foster trust between patients and healthcare providers, improving communication. With ongoing efforts to mitigate biases and ensure fairness, LLMs have the potential to make healthcare more equitable. Collaboration between LLMs and healthcare professionals could lead to more effective medical practices, and integration with telemedicine promises expanded access to healthcare services. Clear regulatory frameworks will ensure the responsible and ethical use of LLMs in healthcare. The future holds the promise of a healthcare system that is not only more efficient and precise but also more inclusive, providing improved healthcare for people around the world.

9.2 Discussion

The anticipated results of advancing research in the use of Large Language Models (LLMs) in disease detection paint an exciting picture of the future of healthcare. As LLMs become increasingly sophisticated, we can expect a notable improvement in disease detection. These models are poised to identify complex disease patterns and risk factors with a high degree of accuracy, leading to earlier and more precise diagnoses. This, in turn, means patients are likely to receive treatment at the earliest possible stage, resulting in better health outcomes, and in some cases, even saving lives. Personalized medicine, another promising outcome, could revolutionize healthcare practices. LLMs, with their ability to process vast amounts of data, have the potential to create tailored treatment plans based on an individual's unique genetic and lifestyle factors. This level of personalization may significantly increase the effectiveness of medical treatments and healthcare interventions. The prospect of real-time health monitoring through LLMs presents an opportunity to shift the healthcare paradigm from a reactive to a proactive approach. With continuous health data analysis, LLMs can identify health issues at their inception, potentially reducing the burden on the healthcare system, cutting costs, and improving overall well-being.

Additionally, the integration of medical images and sensor data into LLMs allows for a more comprehensive health assessment. This can lead to more accurate diagnoses and better-informed treatment decisions, ultimately improving the quality of patient care. Furthermore, the emphasis on creating LLMs that provide clear, transparent, and easily understandable explanations is set to enhance communication between healthcare professionals and patients. Patients will have a deeper understanding of their health and be more actively engaged in their care, ultimately resulting in better healthcare outcomes. Equity and ethics are at the forefront of LLM development. Researchers are diligently working to reduce biases and errors in LLM-generated results, ensuring that healthcare is delivered fairly to individuals from all backgrounds.

With LLMs collaborating with healthcare professionals, we can anticipate a more streamlined and efficient healthcare system. The burden of administrative tasks may be reduced, enabling healthcare providers to focus more on patient care and decision-making. The integration of LLMs into telemedicine platforms could lead to a remote healthcare revolution, allowing patients in remote or underserved areas to access timely and quality healthcare services without geographical constraints. Finally, the establishment of clear and well-defined regulatory frameworks will ensure that LLMs are used responsibly and ethically in healthcare. These safeguards will protect patient data and maintain the highest standards of patient care.

10 Conclusion

Large Language Models (LLMs) like the one we're using are text-based and lack the capability to directly detect diseases. They provide information based on pre-existing knowledge up to their last update in September 2021 but are not a substitute for medical professionals or diagnostic tools. Disease detection requires medical expertise, clinical examinations, and specialized tests conducted by healthcare professionals. AI and machine learning models, specifically designed for medical purposes, can assist in disease detection, such as analyzing medical images or identifying patterns in data. However, these models require extensive training, testing, and validation before being used in clinical settings. In short, LLMs offer general information but aren't suitable for diagnostic purposes, which demand specialized medical knowledge and expertise.

References

Eysenbach, G.: The role of ChatGPT, generative language models, and artificial intelligence in medical education: a conversation with ChatGPT and a call for papers. JMIR Med Educ **9**, e46885 (2023). [https://mededu.jmir.org/2023//e46885/] [CrossRef] [Medline]

Lee, H.: The rise of ChatGPT: exploring its potential in medical education. Anat. Sci. Educ. **1** (2023). [CrossRef] [Medline]

Xue, V.W., Lei, P., Cho, W.C.: The potential impact of ChatGPT in clinical and translational medicine. Clin. Transl. Med.**13**(3), e1216 (2023). [https://europepmc.org/abstract/MED/36856370] [CrossRef] [Medline]

Lee, P., Bubeck, S., Petro, J.: Benefits, limits, and risks of GPT-4 as an AI chatbot for medicine. N Engl. J. Med. **388**(13), 1233–1239 (2023). [CrossRef] [Medline]

Sallam, M.: ChatGPT utility in healthcare education, research, and practice: systematic review on the promising perspectives and valid concerns. Healthcare (Basel) **11**(6), 887 (2023). [https://www.mdpi.com/resolver?pii=healthcare11060887] [CrossRef] [Medline]

Gilson, A., et al.: How does ChatGPT perform on the United States Medical Licensing Examination? The implications of large language models for medical education and knowledge assessment. JMIR Med. Educ. **9**, e45312 (2023). [https://mededu.jmir.org/2023//e45312/] [CrossRef] [Medline]

Kung, T.H., et al.: Performance of ChatGPT on USMLE: potential for AI-assisted medical education using large language models. PLOS Digit Health **2**(2), e0000198 (2023). [https://europepmc.org/abstract/MED/36812645] [CrossRef] [Medline]

Nori, H., King, N., McKinney, S.M., Carignan, D., Horvitz, E.: Capabilities of GPT-4 on medical challenge problems. arXiv. Preprint posted online on March 20 (2023). [https://tinyurl.com/mr3sckcd]

Kurian, N., Cherian, J.M., Sudharson, N.A., Varghese, K.G., Wadhwa, S.: AI is now everywhere. Br. Dent. J. **234**, 72 (2023)

Johnson, S.B., et al.: Using ChatGPT to evaluate cancer myths and misconceptions: artificial intelligence and cancer information. JNCI Cancer Spectr. **7** (2023). https://doi.org/10.1093/jncics/pkad015

Gilson, A., et al.: How does CHATGPT perform on the United States Medical Licensing Examination? The implications of large language models for medical education and knowledge assessment. JMIR Med. Educ. **9**, e45312 (2023)

Li, J., Li, D., Savarese, S., Hoi, S.: Blip-2: bootstrapping language-image pretraining with frozen image encoders and large language models (2023). Preprint at arXiv:2301.12597

Machine Learning Based Clinical Decision Support System for the Diagnosis of Knee Injuries

Parul Chhabra(✉) and Pradeep Kumar Bhatia

CSE Department, GJUS&T, Hisar, India
parul15march@gmail.com

Abstract. Healthcare domain deals with different types of medical data that is available in the form of text or image format. It is very time consuming to correlate the test data with medical images as the text data can be directly produced analyzing the symptoms and on other hand, medical experts are required to interpret the medical imagery as well as it is quite complex to match the symptoms with medical images. In this paper, a machine learning approach will be introduced to correlate the text based medical data with medical imagery using knee injury dataset. Its performance will be also analyzed under the constraints of accuracy, precision value and F-1 score etc.

Keywords: clinical decision support system · machine learning · healthcare · knee injury

1 Introduction

In medical domain, huge volume of medical data is produced on daily basis and this data may be categorized in different types as shown in Fig. 1:

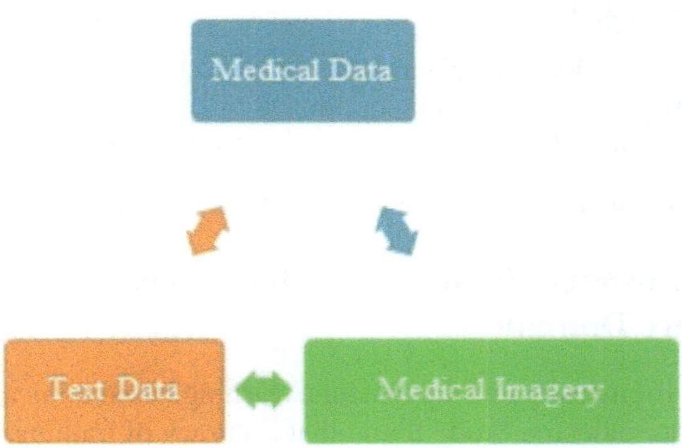

Fig. 1. Medical data forms

D. Kriksciuniene et al. (Eds.): IBICA 2023, LNNS 1228, pp. 153–164, 2025.
https://doi.org/10.1007/978-3-031-78937-3_16

Text data can be easily interpreted by practitioners however, experienced man power is required to analyze the medical imagery for accurate decision making e.g. in case of knee injury, practitioner can interact with the patient to collect the symptoms of knee injury but for the accurate examination of knee MRI images, expert's advice is mandatory. For the accurate treatment of knee injury, there is a need to automate the process of symptom recognition along with MRI image analysis. Machine learning algorithms can be used for the detection and diagnosis of knee injuries and following are the barriers for these methods:

- It is quite difficult to validate the medical dataset
- It is quite complex to correlate the medical text with medical imagery
- Different algorithms are required to interpret the text data and medical imagery.
- Experts are required to interpret medial data
- A precise data is required to build the training dataset
- Expensive medical devices are required to produce the medical images
- It is very time consuming to maintain the patient history, diagnosis status and treatment recommendations.

Recognition of knee injury is quite complex because different patients may have common symptoms with different knee injury types e.g. pain/swelling/limited knee movement/popping sounds are the common symptoms for knee injuries like ACL tear, …… etc.

On the basis of symptoms, practitioners cannot even recognize the knee injury precisely and this may lead to inaccurate diagnosis/treatment. On other hand, there is need to analyze/correlate the MRI images of a knee injuries accurately with the symptoms (text data).

Following are the common machine learning algorithms those can be used to automate the decision support system for knee injury detection/diagnosis:

- Decision Tree:
- Support Vector Machine
- K-Nearest Neighbor

Following are the advantages to integrate the machine learning support with traditional decision support:

- It can analyze the medical data accurately
- Medical data Processing time can be optimized
- Excessive computational overhead can be minimized [1–5]

As per the current health status of the patient, diagnosis plan may be updated.

2 Recent Advancements in Automated Decision Support in Healthcare Domain

N. P. Smadja et al. [6] investigated the machine learning based clinical decision support (CDS) for healthcare industry. Study found that accuracy of training model may varies w.r.t. data validation and the number of datasets used for training as well as it will also affect the accuracy of disease prediction. Study found that CDS automation can reduce the overall diagnosis time/accuracy/operational cost etc.

O. R. Shahin et al. [7] proposed a COVID-19 detection method that uses Support Vector Machine (SVM) classifiers over lung samples for the infected regions. Analysis shows that it is more efficient and predicts the infection level in given samples thus results in higher accuracy of decision making/diagnosis etc.

S. S. Esfahlani et al. [8] used K-Nearest Neighbors (KNN) and SVM classifiers to predict the motor function of patients to facilitate the limb rehabilitation. Analysis shows that accuracy of classifiers can be refined through regression process and outcomes shows that prediction accuracy of SVM classifier is higher as compared to KNN and it is suitable for CDS.

D. Rengasamy et al. [9] used fuzzy logic based feature classifier to distinguish between different features w.r.t. input samples on the basis of weights. Decision making is done using a fusion method. Analysis recommends the fusion method for decision making instead of crisp method.

Y. Ketkar et al. [10] developed a machine learning approach that defines the test cases over Parkinson disease samples using SVM and random forest classifiers. Practitioners can customize the input samples as per requirements to vary the outcomes. Analysis shows that decision making may vary w.r.t. test results.

J. Meng et al. [11] used clinical knowledgebase with machine learning classifier for the disease diagnosis. It uses SVM classifier for attribute selection in given samples. Analysis shows that its performance (accuracy/sensitivity) depends over the attribute density.

M. Unger [12] investigated the machine learning based CDS for healthcare industry. Study found that automated collection and acquisition of medical data can improve the accuracy of decision making as well as it can also reduce the diagnosis time. Study indicates that there is need to develop different datasets/training models w.r.t. each disease, in order to ensure the accuracy of CDS.

F. Shamout et al. [13] explored the impact of prediction models over the process of clinical decision making. Study found that for diseases related to heart/ kidney/liver, large volume of dataset is required thus may degrade the efficiency of the predication model. Validation of samples is another major issue that can affect the decision making.

X. Lin et al. [14] proposed a machine learning based CDS to identify the high risk patient's samples from a large scale medical dataset. Experiments show that filtering of critical samples can be used to set the priority of diagnosis w.r.t. critical patients. With the support of machine learning classifiers, processing time/prediction accuracy/decision making can be optimized.

R. Rai et al. [15] used different classifiers (Decision tree/SVM/ KNN) to detect the neurological disorders. Each classifier recognize the Electroencephalogram (ECG) features (sample entropy/mean/energy/ standard deviation/highest magnitude) to detect the disease. Experiments indicate that SVM outperforms in terms of disease detection in contrast of other classifiers and can be used enhance the accuracy of CDS.

M. Mahbub et al. [16] introduced an expert system for the automaton of CDS. It builds training model using existing medical data and then training model is updated through transfer learning method. Analysis shows that it can assist the practitioners for accurate diagnosis/decision making.

C. Comito et al. [17] developed a framework for CDS. It collects the heterogeneous medical data from various sources (i.e. patient's history/lab test reports/electronic health records) to build a dataset. As per requirements, it can adapt supervised/un- supervised methods for training models. Experiments indicate that it can predict the current health status accurately as well as it can also recommends the guidelines for diagnosis.

P. V. Rao et al. [18] proposed an artificial intelligent (AI) based CDS for healthcare services. It splits the medical dataset in to text and image data and for text data processing, It uses different classifiers (random forest/SVM/Native Bayes). Image data is processed through ResNet. Analysis shows that practitioners can easily interpret the medical data with the help of AI-CDS as compared to traditional CDS.

D. Swain et al. [19] developed a CDS for the diagnosis of kidney disease. It uses SVM classifier with logistic regression method to classify the features from input samples, in order to identify disease. Analysis shows that its prediction accuracy depends over the feature set selection and it can be further improved using cross validation.

Divyashree et al. [20] developed a predictive analysis method for clinical decision making/diagnosis. It uses supervised learning approach over input dataset for disease detection. Comparison with existing methods (Gradient Descent/XGBoost/Logistic Regression/Gaussian Naive Bayes, Random Forest/Decision trees/Linear SVM/Artificial Neural Networks) shows that it outperforms in terms of higher prediction accuracy.

H. Yu et al. [21] proposed an automated decision support scheme for the detection of COVID-19 infection. It uses a knowledgebase to identify the symptoms in input samples. Analysis indicates that it can be implemented in real-time environment.

Y. F. Hsu et al. [22] developed an automated tool to estimate the current health status of patient w.r.t. medical history. It combines the various parameters (i.e. clinical data/symptoms/input samples) for decision making. Analysis shows that it can be utilized to generate health alerts to guide the diagnosis process.

E. Ćojbašić et al. [23] presented a fuzzy logic based prediction model to analyze the current clinical data w.r.t. medical history, in order to classify the patient's health during diagnosis process. As compared to regression/ANN models, it offers higher accuracy for disease classification.

K. Rezaee et al. [24] developed a scheme to classify the symptoms of chronic diseases. It extracts the multiple features and performs SVM based classification for the disease detection. To enhance the accuracy of classification, it also performs component analysis. Analysis shows that it has optimal disease detection ratio as compared to traditional methods.

L. M. Pitti et al. [25] proposed a model to predict the impact of hypertension over patients using clinical data. It performs cross validation over input symptoms and current health status and outcomes show that it can accurately supervise the diagnosis process using XgBoost.

Z. Lv et al. [26] integrated the deep learning methods with big data platform for the recognition of diseases. Disease classification is performed through random forest classifier and finally, large volume medical data is analyzed for decision making. Analysis shows that it is highly efficient as compared to traditional machine learning methods.

Anitha et al. [27] used neural network based multi-perception method to identify the cardio disease. Different classifiers detect the abnormalities in heart vessels to estimate the probability of heart attack. Analysis shows that it can efficiently detect the cardiac disorders and accuracy of decision making can be extended using data mining algorithms.

J. Lu et al. [28] proposed a machine learning method to detect brain disorders. It uses a SVM classifier to distinguish the features of a healthy/ un-healthy brain samples. K-means clustering and correlation analysis is enforced to detect the connectivity of neuron network. Experiments show that it can assist the practitioners for disease detection/diagnosis process accurately.

U. Anda et al. [29] presented a solution for the detection of cardiac disorders using machine learning approach. It build training model by extracting the cardiac features from video streams. On the basis of cardiac training model, input samples are analyzed for the disease detection and decision making. Experiments show that its performance depends over the quality of scanned samples/volume/data acquisition etc.

Y. Wang et al. [30] investigated the multiple disorders caused by thyroid deficiency and proposed an automated diagnosis tool to assist radiologist w.r.t. disease stages. Experiments indicate that radiologists can rectify the diagnosis as per the current disease symptoms.

3 Knee Injuries Recognition Using Clinical Decision Support System

Text based dataset formation as per following:

- CDS dataset {normal, knee injuries, patient's text data}
- CDS dataset {normal, knee injuries, patient's MRI image data}

For training dataset preparation:
Patient's text data = {symptoms, knee injury type}
Patient's MRI data = {MRI images of knee}
Test dataset

Step 1: Prepare knee injury symptom's data
Step 2: Prepare default training dataset using knee injury symptom's data
Step 3: Prepare MRI knee injury training dataset using existing MRI images w.r.t. Injury types
Step 4: Take input symptoms from patients as text dataset
Step 5: Prepare test dataset using above text data
Step 6: Detected the type of knee injury
Step 7: As per the detected injury, use validate dataset provided by MRINET (knee injuries dataset)
Step 8: Recommend the samples of MRI image dataset, in order to train the machine learning algorithm to test the MRI images.
Step 9: perform test over input sample MRI images
Step 10: Recommend the treatment
Step 11: Repeat the above steps, as required.

4 Flow Chart

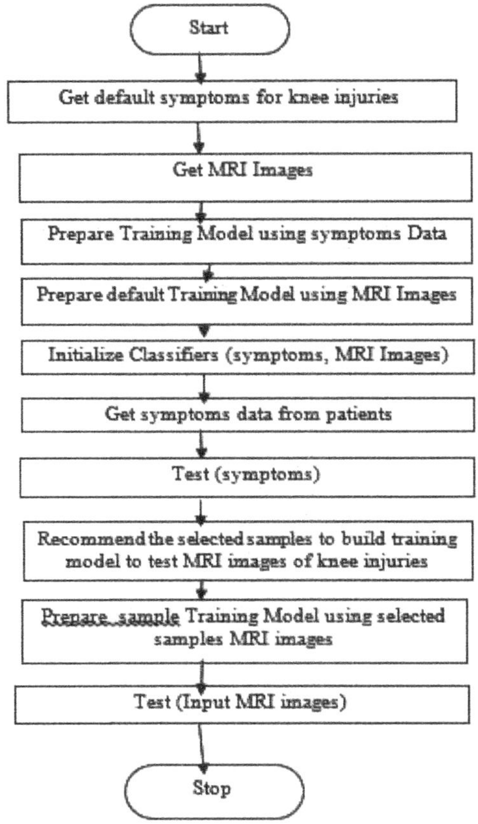

Flow Chart: 1 knee injuries recognition

For experiments, (Decision tree (DT), K-Nearest-Neighbor (KNN), Random Forest (RDF) and AdaBoost (ADAB) were used and following are the sample images from MRINET knee injury datset [31] as shown below (Figs. 2 and 3):

Sample symptoms data for following knee injuries. 0 means no symptom and 1 means there is a symptom of the knee injury and so on.

List of knee injury

Anteriorcruciate-igament
Medialcollateral-ligament
PCL-Knee-Injury
Ligament-Sprains
Quads-Strain
Pulled-Quad
Torn-Quad
Calf-Strains

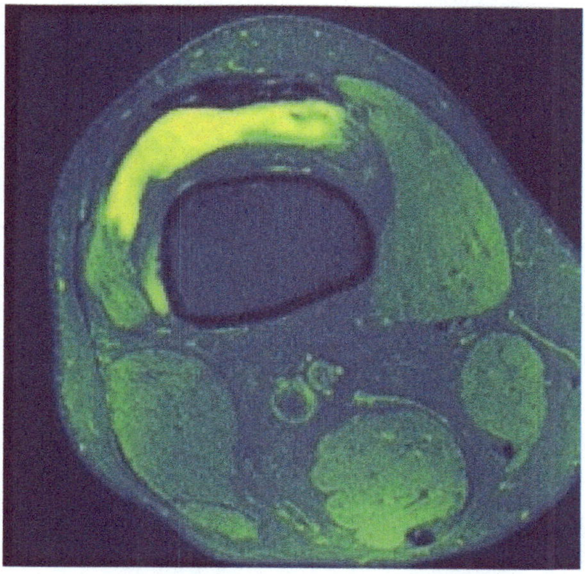

Fig. 2. Axial-MRI of KNEE

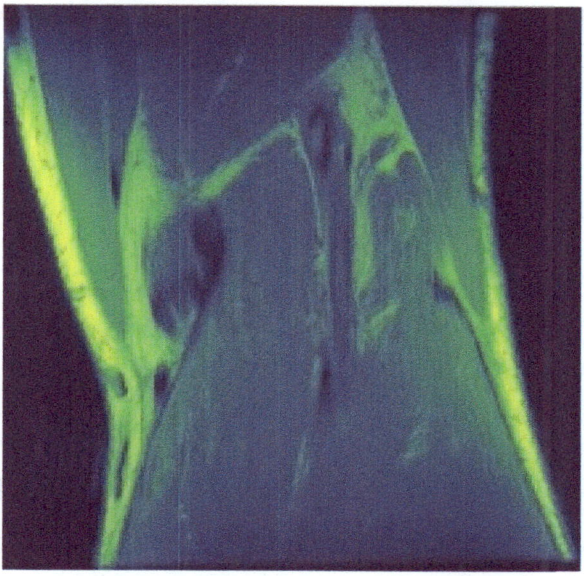

Fig. 3. Coronal - -MRI of KNEE

Dislocated-Patella
Patella-Fractures
MedialImbrication

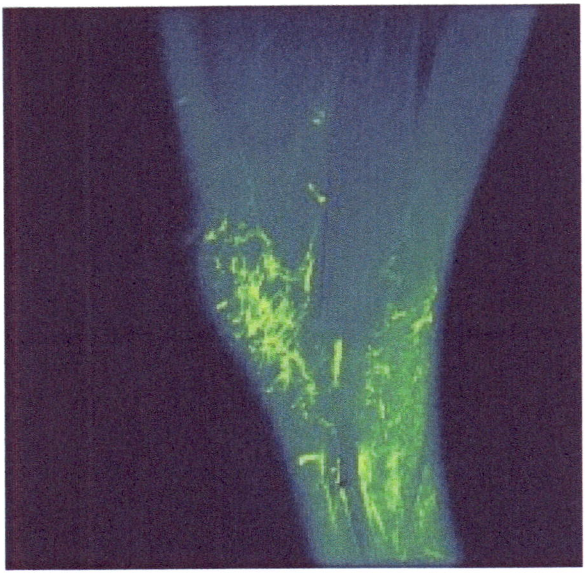

Fig. 4. Sagittal MRI of KNEE

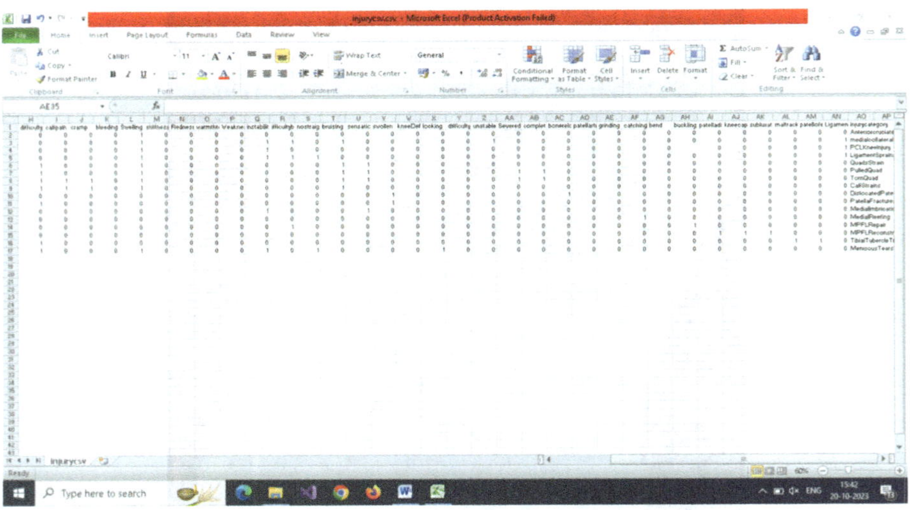

Fig. 5. Sample symptoms data

Medial Reefing
MPFL Repair
MPFL Reconstruction
Tibial-Tubercle-Transfer
Meniscus-Tears

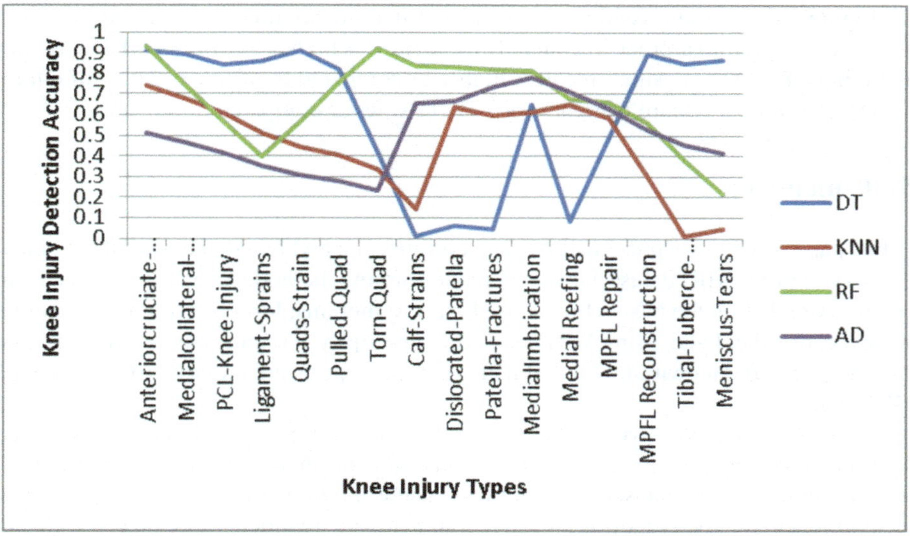

Fig. 6. Knee injury detection accuracy of different classifiers

Figure 4 shows the Knee injury detection accuracy of different classifiers (DT/ KNN/ RF/ AD). As per results, it can be analyzed that decision tree and random forest classifiers delivered the higher accuracy for knee injury detection followed by KNN and AdaBoost (Figs. 6 and 7).

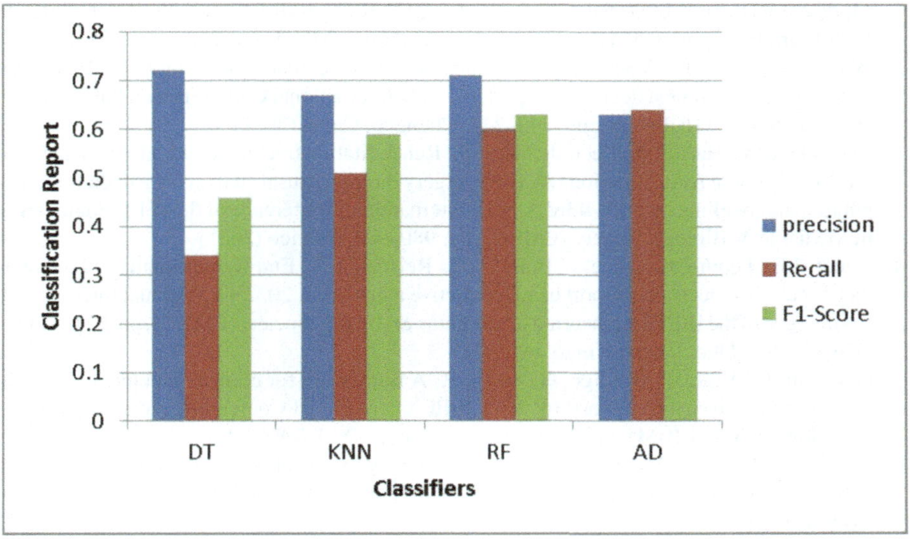

Fig. 7. Classification report of different classifiers

Figure 5 shows the classification report of different classifiers (DT/ KNN/ RF/ AD), it can be analyzed that precision value DT classifier is higher followed by RF classifier and others. For AD classifier, recall is higher as compared to others and it is minimum for DT classifier. F1-score is higher for AD classifier as compared.

5 Conclusion

In this paper, a CDS wa presented that can correlate the text based data with the medical imagery of (knee MRI dataset) using machine learning method. As per the results it can be observed that on the basis of text based input symptoms, level of accuracy cannot be achieved, on other hand, using MRI images of knee injury, it is quite complex to correlate the image based data with text based data, medical experts are required to examine the MRI images.

In proposed scheme, two different training models were prepared, one for text based data, other for image base data. Experiments indicate that using the combination of text and image based datasets together can ensure the higher accuracy of knee injury recognition as well as Practitioners can recommend the treatment more precisely. It can also be observed that DT classifier outperforms as compared to others and AD classifier could not perform as per expectations.

In future, it will be extended to support the other injuries, as its scope is limited up to knee injury, as per the scope of MRINET Knee injury dataset.

References

1. Tang, R., et al.: Integrating Medical Code Descriptions and Building Text Classification Models for Diagnostic Decision Support. In: 2022 IEEE 10th International Conference on Healthcare Informatics (ICHI), pp. 612–613. Rochester, MN, USA (2022)
2. Wu, C., Xiao, L.: Evidence based on patient's experience data and clinical guidelines-for patient-oriented clinical decision support. In: 2021 International Conference on Public Health and Data Science (ICPHDS), pp. 240–247. Chengdu, China (2021)
3. Mandalapu, V., Hart, J.M., Lach, J., Gong, J.: Rehabilitation tracking of athletes post anterior cruciate ligament reconstruction (ACL-R) surgery through causal analysis of gait data & computational modeling. In: 2021 43rd Annual International Conference of the IEEE Engineering in Medicine & Biology Society (EMBC), pp. 980–984. Mexico (2021)
4. Kehoe, I.E., Pepino, J.A., Lee, J., Hahn, J.-O., Reisner, A.T.: End-user evaluation of an interface for clinical decision support using predictive algorithms. 2022 44th Annual International Conference of the IEEE Engineering in Medicine & Biology Society (EMBC), pp. 1149–1151. Glasgow, Scotland, United Kingdom (2022)
5. Giancotti, R., Vizza, P., Tradigo, G., Veltri, P.: A framework for clinical data integration and annotation for decision support. In: 2021 IEEE International Conference on Bioinformatics and Biomedicine (BIBM), pp. 3018–3020. Houston, TX, USA (2021)
6. Smadja, N.P., et al.: Machine learning for clinical decision support in infectious diseases: a narrative review of current applications. Clinical Microbiology and Infection **26**(5), 584–595. Elsevier (2020)
7. Shahin, O.R., Alshammari, H.H., Taloba, A.I., Abd El-Aziz, R.M.: Machine learning approach for autonomous detection and classification of COVID-19 Virus. Comp. Electr. Eng. **101**, 1–15. Elsevier (2022)

8. Esfahlani, S.S., Shirvani, H., Butt, J., Mirzaee, I., Esfahlani, K.S.: Machine Learning role in clinical decision-making: Neuro-rehabilitation video game. Expert Systems with Applications 201, 1–11. Elsevier (2022)
9. Rengasamy, D., et al.: Feature importance in machine learning models: a fuzzy information fusion approach, Neurocomputing. **511**, 162–174. Elsevier (2022)
10. Ketkar, Y., Gawade, S.: A decision support system for selecting the most suitable machine learning in healthcare using user parameters and requirements. Healthcare Analytics 2, 1–7. Elsevier (2022)
11. Meng, J., Xing, R.: Inside the "black box": Embedding clinical knowledge in data-driven machine learning for heart disease diagnosis. Cardiovascular Digital Health Journal **3**(6), 276–288. Elsevier (2022)
12. Unger, M.: Data acquisition and the implications of machine learning in the development of a Clinical Decision Support system. IEEE/ACM 1st Workshop on AI Engineering - Software Engineering for AI, pp.101–104. IEEE (2021)
13. Shamout, F., Zhu, T., Clifton, D.A.: Machine learning for clinical outcome prediction. IEEE Reviews in Biomedical Engineering **14**, 116–126. IEEE (2021)
14. Lin, X., et al.: A case-finding clinical decision support system to identify subjects with chronic obstructive pulmonary disease based on public health data. Tsinghua Science and Technology **28**(3), 525–540 (2023)
15. Rai, R., Martis, R.J.: Automated decision support system for focal epilepsy detection using electroencephalogram. IEEE Bombay Section Signature Conference (IBSSC), pp. 1–5. IEEE, Gwalior, India (2021)
16. Mahbub, M., et al.: cpgQA: a benchmark dataset for machine reading comprehension tasks on clinical practice guidelines and a case study using transfer learning. IEEE Access **11**, 3691–3705. IEEE (2023)
17. Comito, C., Falcone, D., Forestiero, A.: AI-driven clinical decision support: enhancing disease diagnosis exploiting patients similarity. IEEE Access **10**, 6878–6888. IEEE (2022)
18. Rao, P.V., Ashwija, Kanmani, S.K.T., Kulal, B., Jyothika, R.: AI-enabled clinical decision support system. In: International Conference on Distributed Computing, VLSI, Electrical Circuits and Robotics, pp. 1–6. IEEE (2022)
19. Swain, D., Patel, H., Patel, K., Sakariya, V., Chaudhari, N.: An intelligent clinical support system for the early diagnosis of the chronic kidney disease. In: 2nd International Symposium on Sustainable Energy, Signal Processing and Cyber Security, pp. 1–5. IEEE (2022)
20. Divyashree, N.P.: Improved clinical diagnosis using predictive analytics. IEEE Access **10**, 75158–75175. IEEE (2022)
21. Yu, H., Xiao, L.: A flexible clinical decision support model and its application to COVID-19". In: 2nd International Seminar on Artificial Intelligence, Networking and Information Technology, pp. 271–276. IEEE (2021)
22. Hsu, Y.F., Matsuoka, M.: An early warning system for patients in emergency department based on machine learning. International Conference on Computational Science and Computational Intelligence, pp. 1196–1201. IEEE (2021)
23. Ćojbašić, E.: Machine learning as an aid to predicting clinical outcome after stroke. IEEE 21st Mediterranean Electrotechnical Conference (MELECON) pp. 24–27. IEEE, . Palermo, Italy (2022)
24. Rezaee, K., Khosravi, M.R., Attar, H., Alghanim, M.: A decision-making system for medical diagnosis based on iterative nearest component analysis and optimized learning. International Engineering Conference on Electrical, Energy, and Artificial Intelligence, pp. 1–6. IEEE (2022)
25. Pitti, L.M., et al.: Machine-learning model to predict the intradialytic hypotension based on clinical-analytical data. IEEE Access **10**, 72065–72079. IEEE (2022)

26. Lv, Z., Guo, J., Lv, H.: Deep learning-empowered clinical big data analytics in healthcare digital twins. IEEE/ACM Transactions on Computational Biology and Bioinformatics, pp. 1–11. IEEE (2023)
27. Anitha, A., Kumar, D., Gokul, G.: Cardio Vascular Disease Prediction based on ANN Algorithm. 2nd International Conference on Electronics and Renewable Systems, pp. 919–925. IEEE (2023)
28. Lu, J., et al.: An fNIRS-based dynamic functional connectivity analysis method to signify functional neurodegeneration of parkinson's disease. IEEE Transactions on Neural Systems and Rehabilitation Engineering **31**, 1199–1207. IEEE (2023)
29. Anda, U., et al.: Learning deep architectures for the interpretation of first-trimester fetal echocardiography (LIFE) - a study protocol for developing an automated intelligent decision support system for early fetal echocardiography. BMC Pregnancy Childbirth **23**(20), 1–10 (2023)
30. Wang, Y., et al.: Clinical evaluation of malignancy diagnosis of rare thyroid carcinomas by an artificial intelligent automatic diagnosis system. Endocrine **80**, 93–99. Springer (2023)
31. https://stanfordmlgroup.github.io/competitions/mrnet/

Deep Learning Based Early Diagnosis of Malnutrition Among Tribal Pregnant Women in Tamil Nadu

R. Devi Priya[1(✉)], R. Sivaraj[2], T. Rajasekaran[1], M. Sivasankari[1], T. Shanmuga Priya[1], S. M. Nandhagopal[3], and G. Kaushik[1]

[1] KPR Institute of Engineering and Technology, Coimbatore, India
scrpriya@gmail.com
[2] Nandha Engineering College, Erode, India
[3] HKBK College of Engineering, Bengaluru, India

Abstract. In today's world, women have made significant strides across diverse fields, yet the persisting issue of malnutrition among them poses multifaceted challenges. Beyond the immediate health implications, malnutrition in women exacts a heavy toll on the financial stability of families, the vitality of communities, and the overall prosperity of nations. Improved nutrition doesn't just enhance their physical health; it also bolsters their ability to contribute effectively to the workforce, igniting an upward trajectory for entire generations. Breastfeeding women are paramount. They require an additional 50% more calories than usual and a diet rich in balanced nutrition. In India, the challenge of malnutrition persists partly due to the absence of standardized methods to forecast and preempt it before it escalates. With limited access to medical facilities in many regions, by recognizing these patterns beforehand, targeted preventive measures can be implemented, potentially averting the escalation of malnutrition-related issues. This paper proposes a two staged LSTM based deep learning approach called Improved LSTM (i-LSTM) for predicting the occurrence of malnutritional defects by analyzing the health pattern over the past history in pregnant women. The proposed prediction system will be very beneficial for tribal pregnant women to undertake preventive measures and clinicians can easily diagnose the risks of pregnancy complications easily.

Keywords: Malnutrition · deep learning · LSTM · pregnant women · tribal areas

1 Introduction

According to tribal statistics results of 2018, Tamil Nadu holds 7.21 lakhs individuals across 36 tribes and sub-tribes, primarily resides in rural areas of 13 districts. Majority of them are in Dharmapuri, Nilgiris, Salem, Namakkal and Tiruvannamalai districts. However, access to essential healthcare remains a significant challenge. While efforts such as the Integrated Tribal Development Programme and tribal outreach vans aim to support these communities, gaps persist, especially in healthcare provisions. Despite government initiatives, the shortage of healthcare providers and facilities is stark. The existing

D. Kriksciuniene et al. (Eds.): IBICA 2023, LNNS 1228, pp. 165–175, 2025.
https://doi.org/10.1007/978-3-031-78937-3_17

Primary Health Centers (PHCs) fall short of meeting the needs of tribal populations, lacking sufficient medical staff and resources. The diet habits of tribal women in particular are inadequate for their nutritional requirements, and comprehensive health data collection is lacking, hindering effective analysis and prediction of health emergencies, especially during pregnancy (Biswas et al., 2022).

Recognizing the urgency of addressing these gaps, researchers globally, including those in India, are exploring the potential of Artificial Intelligence (AI) techniques to predict and improve nutritional status of tribal pregnant women and children. However, in India, especially within tribal communities, the application of such technologies to forecast the adverse effects of malnutrition is still at a nascent stage.

In Tamil Nadu, there's a pressing concern regarding the healthcare accessibility and nutritional well-being of tribal communities residing in approximately 17,000 villages. While government initiatives like the Integrated Tribal Development Programme aim to support these communities through various welfare schemes, significant gaps persist, particularly in healthcare provisions. The shortfall in healthcare providers exacerbates the challenges faced by these tribal communities. The identified Integrated Tribal Development Programme areas, encompassing regions with over 50% tribal residents, are in need of more robust healthcare infrastructure. As of March 2017, there's a substantial deficit in the number of Primary Health Centers (PHCs) required to adequately cater to the health needs of the tribal population. Although 629 PHCs are deemed necessary, only 512 are operational, leaving a significant gap in healthcare accessibility.

The shortage of doctors further compounds this issue. While 132 doctors are essential to serve these communities effectively, only 86 are available, underscoring the critical shortfall in healthcare professionals. Moreover, the dietary patterns followed by tribal pregnant women in few districts of Tamil Nadu fail to meet their nutritional needs. Insufficient measures have been implemented to address this nutritional deficit, posing long-term health risks. Another crucial gap lies in the absence of a comprehensive database to assess the health conditions of tribal people. Without adequate data, it becomes challenging to analyze their health status effectively and implement targeted interventions.

Of particular concern is the lack of a structured system to predict and address emergencies, especially during pregnancy. This absence of proactive measures exposes tribal women to severe health risks during critical periods, requiring urgent attention. Addressing these systemic deficiencies is crucial to ensure the well-being of Tamil Nadu's tribal communities. By bolstering healthcare infrastructure, enhancing nutritional support, and establishing comprehensive health data systems, we can take significant strides towards safeguarding the health and future of these marginalized populations.

Researchers worldwide are increasingly leveraging Artificial Intelligence (AI) to enhance the healthcare outcomes of rural pregnant women and children. In India, while strides have been made in this direction, the utilization of machine learning techniques for predicting malnutrition risks before and during pregnancy remains in its infancy. Recognizing this critical gap, this paper endeavors to bridge this divide by developing an automated framework to predict the adverse effects of malnutrition among tribal community. Beyond prediction, it will also offer actionable insights into preventive measures. These measures are designed not only to improve physical health but also to bolster psychological well-being among tribal populations, equipping them to combat various serious illnesses, including the challenges posed by the COVID-19 pandemic.

Creating a healthy generation capable of combatting infectious diseases requires a holistic approach that includes access to proper healthcare, education on nutrition, and the implementation of strategies to ensure that vulnerable communities, such as those in tribal regions, receive adequate support.

The significant novelties in the proposed work are:

- The proposed work predicts adverse effects of malnutrition among pregnant women in tribal regions of Tamil Nadu
- Deep learning-based application is used to monitor the growth and health conditions of fetus before birth which can greatly help in avoiding risks
- Mechanisms that are suitable for reaching information to the corresponding tribal regions will be identified and adopted. The framework developed can be used by health care providers for screening, alerting the risks, analyzing the current status and forecasting the future traits of pregnancy and child birth.

2 Literature Survey

Results of various studies regarding malnutritional status and research works carried out are briefly given below.

In a survey conducted by Padmanabhan and Mukherjee (2016) about breast feeding habits of tribal mothers in Yercaud, it is found that most of the tribal women are not adequately feeding their children since they themselves are facing malnutritional problems and make the children also under-nourished. It is also noted that about one fourth of tribal women residing in the state of Kerala remain do not get enough nutritious food.

Mariyam and Dibaba (2018) have done a comprehensive epidemiological study focusing on malnutrition in pregnant women within the Central Rift Valley of Ethiopia, a region characterized by tribal rural areas. The study identified various key factors associated with malnutrition among pregnant women in this context. Factors such as anemia, short stature, MUAC (Mid-Upper Arm Circumference) measurements below 21 cm, dissatisfaction with Antenatal Care (ANC) services, lack of utilization of family planning, and teenage pregnancies emerged as prominent contributors to malnutrition. Mu et al. (2018) undertook an extensive collection of pre-pregnancy health data from a staggering 75,542 couples in China, encompassing a comprehensive array of 308 different health features. Employing a sophisticated deep learning framework, they aimed to predict adverse pregnancy outcomes even before conception. Their innovative approach involved the implementation of a multi-layer neural network structure tailored for the prediction of six distinct pregnancy categories: normal births, premature births, low birth weight, birth defects, spontaneous abortions, and stillbirths. The outcomes revealed superior accuracy, recall rates, and F1 scores compared to existing algorithms, showcasing the potential of their model in effectively predicting and mitigating adverse pregnancy outcomes.

Rigdon and Basu's (2019) research underscores the pivotal role of nutrition in the incidence of Cardio vascular Disease (CVD). Their study highlights the potential of machine learning algorithms, particularly regression models, in predicting CVD risks using nutrition data. This predictive capacity allows for early intervention and appropriate measures to mitigate CVD risks. Parallel to this, Biswas et al. (2019) shed light on the concerning phenomenon known as the Double Burden of Malnutrition (DBM) which is common among South and Southeast Asian women. Their projections indicate that DBM is likely to afflict rural and tribal areas with lower education levels by 2030. Alarming trends indicate that two-thirds of females in wealthier sectors could be overweight or obese, while 10% in poorer rural sectors might remain underweight. This dual challenge requires careful attention as maternal overweight can lead to adverse complications for both mothers and infants. Additionally, Hanieh et al. (2019) have made strides in the field by developing a predictive tool that identifies the consequences caused due to stunting in early childhood, particularly in low-income areas with limited resources. This tool provides a valuable resource for early intervention and targeted support for at-risk children, potentially mitigating the long-term effects of stunting. Research conducted by Sathiyanarayanan et al. (2019) in the Jawadhu hills of Vellore involved a comprehensive house-to-house survey among tribal communities. Their study employed chi-square and ANOVA tests to estimate the prevalence of hypertension and diabetes within these communities. Alarmingly, their findings revealed that the incidence of hypertension and diabetes among females was twice that of males, directly impacting the nutritional well-being of their children. In Adegbosin et al. (2019), deep learning based methods are used to predict under 5 mortality rate in developing nations.

Similarly, Puri et al. (2020) delved into analyzing the correlation between anemia, stunting, incomplete immunization, and maternal factors. They utilized spatial analysis and a binary logit model using data from India's National Family Health Survey (2015-16). Their study sought to uncover crucial associations, shedding light on how maternal parameters influence these health indicators among children. Furthermore, Striessnig et al. (2020) engaged in geographical clustering of malnutrition hotspots across Indian districts. Employing the Principal Component Analysis method, they examined deviations in the growth of under-five children. Their research uncovered strong correlations between factors such as poverty, low female education, inadequate maternal BMI, and specific demographics like high parity and teenage pregnancies, indicating their profound impact on child growth. Moreover, How and Chan's (2020) human-centric AI-thinking approach marks a significant advancement. Their methodology elucidates the interconnected dynamics between malnutrition, health, and population indicators across 180 nations. This approach provides a comprehensive understanding of the complex interactions between these factors, facilitating informed decision-making and policy formulation on a global scale. Santhosh (2020) have also insisted for the acute need of AI based tools to train/model such multinominal data for healthcare analysis.

Allahem and Sampalli (2022) have developed automated machine learning and deep learning techniques specifically aimed at detecting signs of labor pain. Their work focuses on early identification to mitigate the risks associated with premature labor, providing an opportunity for timely interventions and care for pregnant women. In a separate study, Daydulo et al. (2022) have made strides in leveraging technology, employing a pre-trained ResNet 50 model for transfer learning. Their focus lies in detecting fetal distress by analyzing cardiotocogram signals. This approach holds promise in offering a more precise and timely assessment of fetal well-being, potentially improving health conditions of infant and mother. Jin et al. (2022) have also suggested using deep learning algorithms for predicting the malnutritional effects from longitudinal patient trajectories.

Khudri et al. (2023) have explored the efficacy of Support Vector Machines (SVM) and K-Nearest Neighbors (KNN) algorithms in predicting Body Mass Index (BMI). Their findings suggest that these machine learning models outperform others in estimating BMI, providing valuable insights into potential health risks associated with weight during pregnancy.

Additionally, their research sheds light on the multifaceted nature of factors influencing the nutritional status of pregnant women. Observations highlight the significant roles played by age, socioeconomic status, and breastfeeding conditions in determining the nutritional well-being of expectant mothers. Understanding these factors is crucial in developing comprehensive strategies to support and improve the nutritional health of pregnant women. These studies collectively underscore the advancements in leveraging technology and data-driven approaches to address various aspects of maternal health. By harnessing machine learning, deep learning, and insightful analyses of influential factors, researchers are contributing toward more informed and targeted interventions to enhance the health and well-being of pregnant women and their unborn children.

The impact of child immunization emerges as a pivotal factor in the postnatal care of reproductive women, emphasizing its role in shaping reproductive health practices. However, several socioeconomic factors intricately interplay with these practices, shaping the overall health outcomes of women within these communities. The pressing need for a comprehensive approach to address these challenges is evident. Bridging the gap in predicting and preventing malnutrition among women and children demands concerted efforts on both global and local fronts. It requires not just the utilization of advanced technological tools like machine learning but also the integration of social, economic, and cultural factors into preventive strategies. Deep learning algorithms can be used for prediction of malnutrition status effectively. Recognizing this urgency, the proposed system aims to develop an efficient mechanism leveraging improved Long Short-Term Memory (LSTM) technology. This system is envisioned to evaluate the malnutrition status of tribal women and children comprehensively. It aims not only to predict the adverse effects but also to suggest precautionary measures. By harnessing advanced technologies like LSTM, the goal is to empower tribal mothers and children with targeted interventions to combat malnutrition and foster a healthier generation.

3 The Proposed Methodology

The proposed methodology utilizes improved LSTM (iLSTM) which is depicted in Fig. 1 and follows the below steps:

3.1 Survey questionnaire Preparation

A comprehensive survey questionnaire has been meticulously crafted, taking into account distinct parameters relevant to three key categories: health service providers at Primary Health Centers (PHCs), pregnant women, and mothers post-pregnancy. This questionnaire compilation has been guided by inputs from collaborating clinicians, PHC personnel, and insights gleaned from credible journal publications. The paramount objective behind this questionnaire is to encapsulate a holistic spectrum of factors directly or indirectly linked to malnutrition. Expert advice has been instrumental in shaping these questions, ensuring the inclusion of all variables that influence or are influenced by malnutrition. The medical data, pertaining specifically to health service providers and clinical aspects, is designed for direct collection and testing by qualified medical professionals. However, for tribal women and children, a broader set of information encompassing dietary patterns, hygiene practices, familial backgrounds, and other pertinent factors necessitates inclusion in the questionnaire.

By structuring questions that delve into food habits, hygiene practices, familial context, and various influencing factors, this questionnaire aims to paint a comprehensive picture of the circumstances surrounding malnutrition among tribal women and children. Through the amalgamation of these diverse parameters, the survey endeavors to pave the way for a more nuanced understanding and effective interventions to combat malnutrition within these communities.

3.2 Data Collection

The framed survey questionnaire will be given tp health service providers (PHCs), pregnant women and mother after pregnancy for data collection. Scanned images will be developed to collect neonatal details. The android application will make use of the images of womb captured in smart phones to measure the womb's height, weight, head, arm and waist circumference which results in correct data collection quickly. In another way, some date might be available in recorded documents like papers, health records which are not digitally maintained in tribal areas. In order to get those details, images of the records are captured and using CNN classifier, the content are extracted and accordingly the initial data is extracted.

3.3 Data Preprocessing

The collected data will be most of the times incomplete with some missing data. It needs to be preprocessed for accurate data analysis and the steps are explained below.

3.3.1 Missing Values Imputation

Data Cleaning stands as the foundational stage in the preprocessing journey. Often, during surveys, certain data points might be missing due to non-responses from participants. Addressing these missing values within the collected dataset necessitates a thoughtful preprocessing approach. The method proposed in Devi Priya et al. (2017) to handle missing values, Bayesian Genetic Algorithm (BGA) is used to fill the missing values in the dataset collected. This algorithm offers a systematic way to preprocess missing data, ensuring a more comprehensive dataset for subsequent analysis.

Considering the data gathered from tribal communities, a significant portion of the attributes will likely be categorical. This format is more accessible for tribal individuals to respond to effectively. Conversely, data obtained from Primary Health Centers (PHCs) might present a combination of both categorical and numerical attributes, given the diverse nature of information gathered from these sources.

3.3.2 Data Integration

Post-cleaning, the integration of data from three distinct sets involves a meticulous process aimed at enabling comprehensive analysis. This integration phase focuses on harmonizing the information by removing redundant attributes and standardizing data formats across multiple sources. Eliminating redundancy ensures that the integrated dataset remains concise and efficient, devoid of unnecessary duplicate information. Additionally, standardizing the data formats aligns attributes from diverse sources, facilitating a seamless analysis process. Inconsistencies within the data are systematically addressed and rectified during this phase, enhancing the overall quality and reliability of the integrated dataset. The key objective behind this integration is to create a unified dataset that enables meaningful analysis and robust insights. By streamlining the information from multiple sources and ensuring uniformity in attributes, this integrated dataset becomes a valuable resource for conducting comprehensive analyses and deriving accurate conclusions.

3.3.3 Feature Selection

Employing a feature selection mechanism becomes imperative, aiming to identify and isolate the most pertinent attributes. The application of an efficient feature selection mechanism holds the promise of reducing the computational complexity inherent in the process, thereby enhancing overall execution time. Among various methods, the Random Forest algorithm stands out as a robust tool for feature selection. Its efficacy in singling out significant features relevant to a particular analysis has earned it recognition as the standard method in many contexts. Hence, for the proposed methodology in this report, the Random Forest algorithm has been chosen as the preferred mechanism for identifying and isolating the key features necessary for the task at hand. By leveraging the capabilities of the Random Forest algorithm, the aim is to streamline the dataset, focusing only on those attributes that hold the most relevance to the specific analysis. This strategic selection process not only optimizes computational efficiency but also ensures that the subsequent analysis is grounded in the most critical and impactful features, potentially enhancing the overall quality and accuracy of the report's findings.

3.4 Classification/Regression

In the process of gathering information from tribal communities, the attributes collected are often structured as categorical data. This approach is chosen for its simplicity, allowing for easier and more accessible responses from tribal individuals. On the other hand, attributes gathered from Primary Health Centers (PHCs) present a combination of categorical and numerical data types. Long Short Term Memory (LSTM) method is used for predicting the values of occurrence of malnutritional defects. The data usually contains details of certain intervals of time period and hence LSTM algorithm is best suited. The outcomes derived from this data hold immense potential for informed decision-making, particularly in the realm of enhancing maternal and child health. They serve as a valuable resource, guiding policymakers and healthcare providers in devising strategies aimed at improving the well-being of mothers and children. Moreover, these insights streamline the process for health service providers, facilitating easier maintenance of health standards and ensuring more effective delivery of care to those in need.

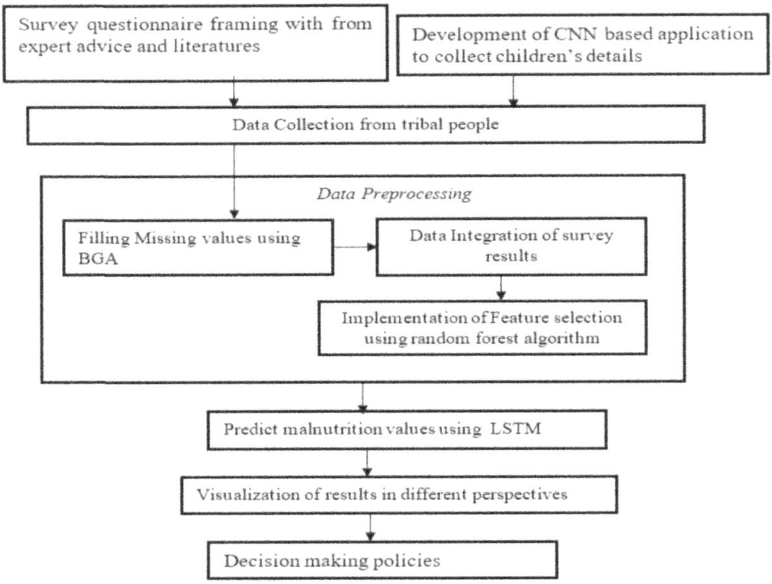

Fig. 1. Process flow of the proposed methodology

3.5 Visualization of Analysis Results

Merely producing numerical outcomes might not resonate well with users, especially those without statistical expertise. Therefore, the visualization of results becomes a crucial component of the process. Various forms of visualization, such as tables, charts, graphs, and histograms, offer distinct advantages in conveying complex data in a digestible format. These visual representations serve as a bridge between raw numbers and actionable insights, aiding not only experts but also casual users with limited statistical knowledge to comprehend and derive meaning from the analysis.

Matplotlib, a powerful Python library, serves as a valuable tool in creating these visualizations. Its flexibility and diverse range of capabilities enable the generation of various types of visual representations based on the analysis outcomes. Additionally, other efficient tools will be leveraged as needed, depending on the specific requirements and outcomes of the analysis. The ultimate goal is to transform complex analysis results into easily interpretable visual formats, empowering users at all levels to grasp the insights derived from the data, thereby enhancing informed decision-making and driving positive outcomes.

4 Experimental Results and Discussion

The experiments are conducted using the malnutrition-across-the-globe and global nutrition report (2021) datasets available in KAGGLE. They contain country-wise malnutrition data. The datasets contain major details like stunting, wasting, lbw values from the year 2000 to 2020. Based on all these values, the model for occurrence of malnutrition can be estimated.

Table 1. Performance comparison on Malnutrition-across-the-globe dataset

Method	Accuracy (%)	Sensitivity (%)	Specificity (%)	F1 Measure	G Mean
Bayesian model	85.2	84.1	81.9	0.84	0.83
Random Forest classifier	83.6	87.9	84.7	0.82	0.86
SVM	87.1	86.0	83.8	0.87	0.81
LSTM	88.3	85.2	87.1	0.88	0.88
i-LSTM	90.7	91.8	92.5	0.91	0.92

The model included advanced age, indicators of poor overall health, nutritional intake etc. Performance of the existing methods like Bayesian model, random forest, support vector machine, LSTM. The results of malnutrition-across-the-globe and global nutrition report (2021) datasets are compared in Tables 1 and 2 respectively. It is found that the proposed method is better than other methods in terms of accuracy, sensitivity, specificity, G Mean and F1 Measure as given in both Tables 1 and 2. The superior performance of the proposed LSTM is due to the fact that it first preprocesses the data and then perform classification. The preprocessing step removes all the inconsistencies and produces more accurate data and model for further processing.

Table 2. Performance comparison on Global Nutrition report dataset

Method	Accuracy (%)	Sensitivity (%)	Specificity (%)	F1 Measure	G Mean
Bayesian model	82.0	82.3	82.4	0.81	0.82
Random Forest classifier	81.6	87.5	81.4	0.80	0.85
SVM	85.9	85.9	82.6	0.84	0.82
LSTM	87.4	86.4	85.4	0.87	0.86
i-LSTM	90.9	92.7	92.6	0.90	0.94

5 Conclusion

Malnutrition, once identified, can be treated with evidence-based and low-cost preventive measures. Identifying the adverse effects of malnutrition early on within the tribal regions holds immense societal and practical significance. i-LSTM is designed precisely for this purpose, leveraging automated computational intelligence. This innovation aims to streamline access to healthcare services, a critical need exacerbated by the challenges posed by the COVID-19 pandemic. The current scenario has made it increasingly arduous for people to access essential health services such as vaccinations and diagnosis. This proposed system not only benefits the society by ensuring early detection of malnutrition-related issues but also significantly assists the government in its efforts to address health disparities among tribal populations.

The implementation of such a system holds the potential to revolutionize healthcare access, especially in regions where traditional avenues face significant limitations. By leveraging technological advancements, we can pave the way for a more inclusive and efficient healthcare system, catering to the needs of every individual, regardless of their geographical or social constraints.

References

Adegbosin, A.E., Stantic, B., Sun, J.: Predicting Under-five mortality across 21 Low and Middle-Income Countries using Deep Learning Methods. medRxiv (2019)

Allahem, H., Sampalli, S.: Automated labour detection framework to monitor pregnant women with a high risk of premature labour using machine learning and deep learning. Informatics in Medicine Unlocked **28**, 100771 (2022)

Biswas, T., Townsend, N., Magalhaes, R.S., Islam, M.S., Hasan, M.M., Mamun, A.: Current Progress and Future Directions in the Double Burden of Malnutrition among Women in South and Southeast Asian Countries. Nutritional Epidemiology and Public Health, Current Developments In Nutrition (2019)

Biswas, S., Hansda, C.K., Singh, N., Malla, V.R., Pal, A.K.: Determinants of nutritional status among scheduled tribe women in India. Clinical Epidemiology and Global Health **17**, 101119 (2022)

DeviPriya, R., Sivaraj, R., Sasipriyaa, N.: Heuristically repopulated Bayesian ant colony optimization for treating missing values in large databases. Knowl.-Based Syst. **133**, 107–121 (2017)

Hanieh, S., et al.: The stunting tool for early prevention: development and external validation of a novel tool to predict risk of stunting in children at 3 years of age. BMJ Global Health **4** (2019)

How, M.L., Chan, Y.J.: Artificial Intelligence-Enabled Predictive Insights for Ameliorating Global Malnutrition: A Human-Centric AI-Thinking Approach. AI **1**(1), 68–91 (2020)

Mariyam, A.F., Dibaba, B.: Epidemiology of malnutrition among pregnant women and associated factors in central refit valley of Ethiopia. J. Nutrit. Disord. Thera. **8**(1), 1–8 (2018)

Mu, Y., Feng, K., Yang, Y., Wang, J.: Applying deep learning for adverse pregnancy outcome detection with pre-pregnancy health data. MATEC Web of Conferences **189** (2018)

Padmanabhan, P.S., Mukherjee, K.: Nutrition in Tribal Children of Yercaud Region. Tamil Nadu Indian J. Nutri. **3**(2), 1–18 (2016)

Puri, P., Khan, J., Shil, A., Ali. M.: A cross-sectional study on selected child health outcomes in India: Quantifying the spatial variations and identification of the parental risk factors. Sci. Rep. **10**(1), 6645–53 (2020)

Rigdon, J., Basu, S.: Machine learning with sparse nutrition data to improve cardiovascular mortality risk prediction in the USA using nationally randomly sampled data. BMJ Open (2019)

Santosh, K.C.: AI-driven tools for coronavirus outbreak: need of active learning and cross-population train/test models on multitudinal/multimodal data. J. Med. Syst. **44**(93), 1–5 (2020)

Sathiyanarayanan, S., Muthunarayanan, L., Devaparthasarathy, T.A.: Changing perspectives in tribal health: rising prevalence of lifestyle diseases among tribal population in India. Indian J. Comm. Med. **44**, 342–346 (2019)

Striessnig, E., Bora, J.K.: Under-five child growth and nutrition status: spatial clustering of Indian districts. Spatial Demography **8**, 63–84 (2020)

Daydulo, Y.D., Thamineni, B.L., Dasari, H.K., et al.: Deep learning based fetal distress detection from time frequency representation of cardiotocogram signal using Morse wavelet: research study. BMC Med. Inform. Decis. Mak. **22**, 329 (2022)

Khudri, M.M., Rhee, K.K., Hasan, M.S., Ahsan, K.Z.: Predicting nutritional status for women of childbearing age from their economic, health, and demographic features: a supervised machine learning approach. PLoS ONE **8**(5), e0277738 (2023)

Jin, B.T., Choi, M.H., Moyer, M.F., Kim, D.A.: Predicting malnutrition from longitudinal patient trajectories with deep learning. PLoS ONE **17**(7), e0271487 (2022)

EEG Signals Classification for Motor Imagery Task Using Different KNN Algorithms

Yogendra Narayan(✉), Riya Khera, Rajeev Ranjan, Rohit katyal, and Neeshu Sharma

Chandigarh University Mohali, Mohali, India
narayan.yogendra1986@gmail.com, neeshusharma.cse@cumail.in

Abstract. Over the past few decades, researchers have utilized motor imagery (MI)-based electroencephalogram (EEG) signals classification for the development of assistive human-robotic interaction. These EEG signals are crucial in the field of brain-computer interfaces (BCIs) and find various applications in the biomedical domain. BCIs enable the acquisition of brain signals, feature extraction, noise reduction, and generation of control signals. The primary aim of this study was to evaluate and compare the performance of Frequency-domain (FD) features using three distinct classifiers: Fine KNN, Medium KNN, and Coarse KNN. The objective was to differentiate between the right and left hand movements using EEG signals. The EEG dataset used in this research was obtained from twenty healthy human subjects during two separate sessions. To minimize noise, band-pass Butterworth filtering was applied to the acquired data. Here employed with Fine KNN, Medium KNN, and Coarse KNN classifiers. Subsequently, the ten features were combined to form a final feature vector, which yielded the best performance when used with the Fine KNN classifier, achieving an accuracy of 93%. Comparatively, the Medium KNN classifier achieved an accuracy of 92.4%. These findings have significant implications for the design of EEG-based technologies such as wheelchairs and prosthetic limbs. In summary, this study aimed to compare the effectiveness of different FD features and KNN classifiers in discerning between right and left hand movements using EEG signals. The results demonstrated that the combination of ten features, when utilized with the Fine KNN classifier, yielded the highest accuracy rate of 93%. These findings have valuable implications for the development of EEG-based assistive devices, including prosthetic limbs and wheelchairs.

Keywords: EEG · BCI · KNN · accuracy · classification

1 Introduction

The central organ of the human nervous system is the human brain which controls most of the activities of our body. It process, integrate and coordinate the information received from sense organs and send instructions to the body. There are billions of neurons present in our brain. These neurons use tiny electrical signals as a mode of communication [1]. Brain activity is measured by electrical signals known as electroencephalogram (EEG) In 1924, Hans Berger, a German neuropsychiatries, recorded first human EEG [2]. The

D. Kriksciuniene et al. (Eds.): IBICA 2023, LNNS 1228, pp. 176–188, 2025.
https://doi.org/10.1007/978-3-031-78937-3_18

main purpose of using EEG signal is to diagnose brain diseases Electroencephalography (EEG) involves the placement of metal electrodes on the scalp to continuously record the brain's electrical activity [2, 3]. The selection and placement of electrodes on the scalp for EEG recording is determined by the specific goal or intended use of the EEG. This may result in variability in both the number and position of the electrodes used. Neuronal cells in the brain communicate with each other by generating electrical currents, which persist even during periods of rest or sleep. Electroencephalography (EEG) is a highly advantageous brain imaging technique due to its low cost, high flexibility, high temporal resolution, non-invasiveness, ease of use, portability, and safety. Compared to other functional neuroimaging methods like PET, MEG, fMRI, and TMS, EEG is a powerful tool for studying brain activity [4]. Due to the dynamic and non-linear nature of EEG signals, they are employed in a variety of tasks, including the classification of a pilot's mental states, detection of drowsiness levels, epilepsy detection, identification of autism spectrum disorder, detection of Alzheimer's disease, stroke analysis, analysis of consciousness and unconsciousness, classification of motor imagery, identification of sleep-related disorders, and diagnosis of schizophrenia-related disorders [5]. EEG is a widely used tool for research in neuroscience, cognitive science, and cognitive psychology. Its non-invasive nature and ability to measure real-time changes in brain activity make it a valuable method for investigating brain function during various cognitive and behavioral tasks.

Surface Electromyographic (sEMG) signals offer a non-invasive approach for acquiring Electromyography (EMG) data, utilizing surface electrodes strategically placed on acupressure points. Widely employed in the realms of rehabilitation robotics and neuromuscular disease classification, sEMG signals play a pivotal role in the control of robotic devices. A significant application of sEMG signals is evident in the development of prosthetic limbs for the rehabilitation of amputees and individuals with physical disabilities. This is owing to the rich information on muscle movement encapsulated within sEMG signals.Despite the profound utility of sEMG signals, a universal model elucidating the intricate relationship between human gestures and recorded sEMG signals is currently lacking. Consequently, the identification of the user's intention becomes a crucial precursor in the development of devices aimed at interpreting and responding to sEMG signals. Addressing this challenge requires leveraging the distinctive features embedded within sEMG signals, thereby paving the way for the advancement of devices that can seamlessly translate human gestures into actionable commands [6]. Brain-Computer Interface (BCI) technology enables individuals to communicate with the external world without using their muscles, by translating their brain signals into commands that can control various devices. This technology is particularly beneficial for individuals with severe disabilities. EEG-based BCIs are commonly used because they can detect brain activity, which is a crucial prerequisite for this type of interface. Brain-Computer Interface (BCI) technology establishes a direct communication pathway between the human brain and an external device, by receiving and converting signals collected from the brain into one or several actions. This technology is especially important for individuals with disabilities or cerebral injuries, as it allows them to communicate with their environment and exert control over their living facilities, thereby enhancing their quality of life. EEG-based BCI systems have been extensively researched and implemented in recent years [7]. Visual analysis of EEG signals may be subject to errors in decision making, especially when identifying epileptic zones for neurosurgical resection [8]. However, advanced signal processing and machine learning techniques can aid in the development of automated classification methods that accurately identify focal onset (FO) EEG signals, thereby overcoming this problem.

2 Materials and Methods

In this section, the process of EEG data recording, feature extraction, and classification is described. Figure 1 demonstrates the list of required materials for simultaneous EEG monitoring: neoprene cap, Control Box, cables, electrodes, measurement tape, saline solution and Bluetooth USB.

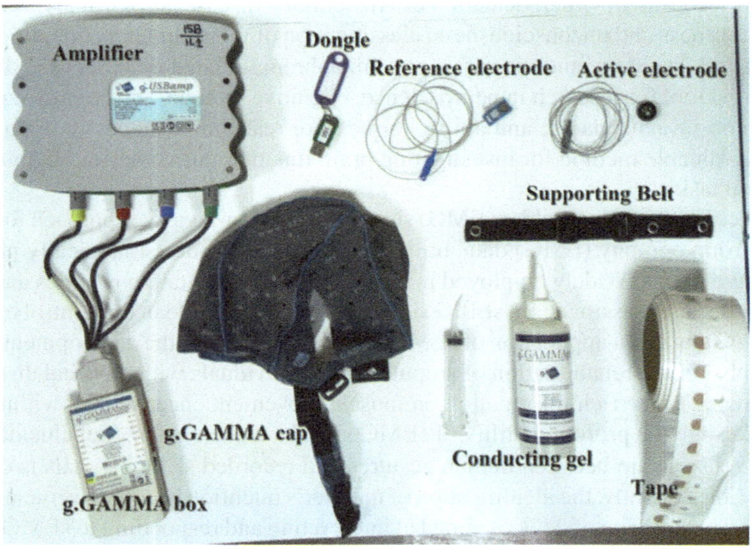

Fig. 1. List of required materials for simultaneous EEG acquisition

The k-Nearest Neighbors (KNN) algorithm is a simple and widely used supervised machine learning algorithm for classification and regression tasks. It falls under the category of instance-based learning, where the algorithm doesn't explicitly learn a model during the training phase but memorizes the training instances and uses them during the testing phase.

A confusion matrix is a table used in classification to describe the performance of a classification algorithm. Each row of the matrix represents the instances in a predicted class, while each column represents the instances in an actual class (or vice versa). The entries in the matrix provide a summary of how many instances were correctly or incorrectly classified.

$$\text{Accuracy} = ((\text{Tp} + \text{Tn}/\text{Tp} + \text{Tn} + \text{Fp} + \text{Fn}))$$

A Receiver Operating Characteristic (ROC) curve is a graphical representation used in binary classification to assess the performance of a classification model at different classification thresholds. It illustrates the trade-off between true positive rate (sensitivity) and false positive rate (1 - specificity) across various threshold settings.

2.1 EEG Data Acquisition and Preprocessing

In this study, EEG data was collected from 20 healthy participants who underwent two separate recording sessions. Prior to recording, participants received instructions on how to perform the correct movements, the concept of Motor Imagery (MI), and the BCI setup. During recording, participants were asked to imagine performing hand movements for both the left and right hand, twenty times per session. Participants were seated comfortably in an armchair at a distance of 150 cm from the computer monitor and instructed to keep their bodies relaxed, while also avoiding any arm or foot movement. Electrodes were placed according to the 10–20 international system, specifically near the C3 and C4 positions over the motor cortex, with a sampling rate of 256 Hz [9]. EEG data was acquired using a total of 16 active electrodes. A block diagram representation of the study's complete workflow can be seen in Fig. 2.

In order to process EEG signals for analysis, the analog EEG signals are converted into digital form using analog-to-digital converters (ADC). Since EEG signals have a bandwidth of around 100 Hz, a minimum sampling frequency of 200 samples per second is required for proper sampling of the signals. The quantization process is then performed on the EEG signals to preserve the diagnostic information [10]. Each sample of the signal is represented by up to 16 bits to ensure accurate recording. This allows for a sufficient amount of memory to store the EEG signals, which can be particularly important for the monitoring of epileptic seizures. Generally, the memory required to store EEG signals is much smaller than that required for storing radiological images.

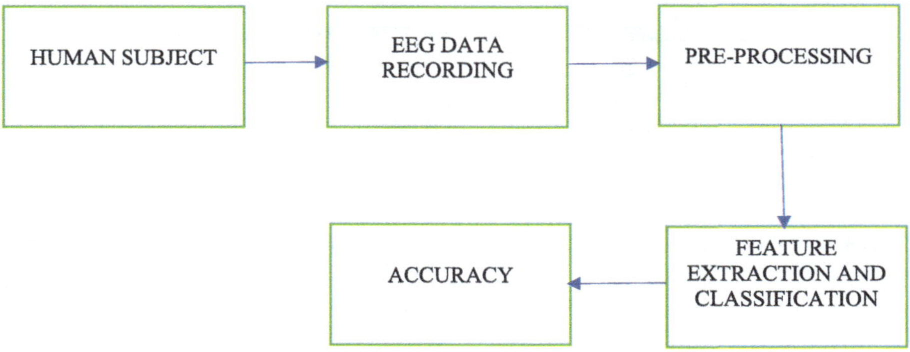

Fig. 2. Block Diagram of EEG Classification Method

After the initial preprocessing of the EEG signals, the feature extraction method known as CSP was utilized to extract features. In order to remove various types of biomedical artifacts, such as EOG and ECG, the ICA technique was used. Additionally, to reduce the dimensionality of the feature vector, the PCA method was employed. It is important to note that while brain-computer interfaces (BCIs) can identify and recognize patterns of brain activity related to specific events or tasks, they are not capable of reading minds or decoding thoughts. Therefore, a BCI relies on the user producing these patterns through mental strategies, which also dictate the constraints on the software and hardware of the BCI, as well as the signal processing techniques to be applied later. Common mental strategies include motor imagery and selective attention (Figs. 3, 4 and 5).

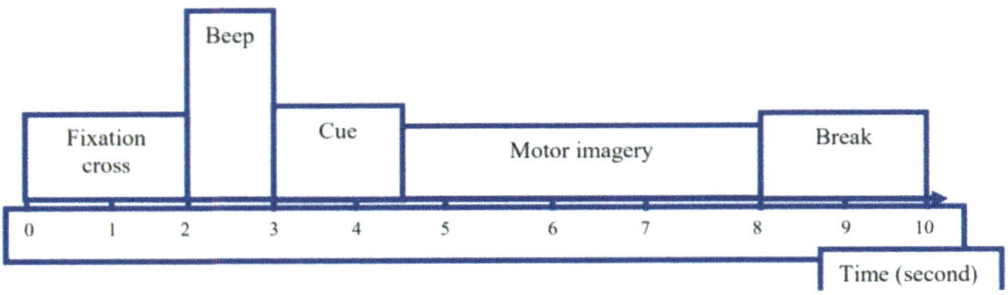

Fig. 3. Timing Schemes of EEG Data Recording

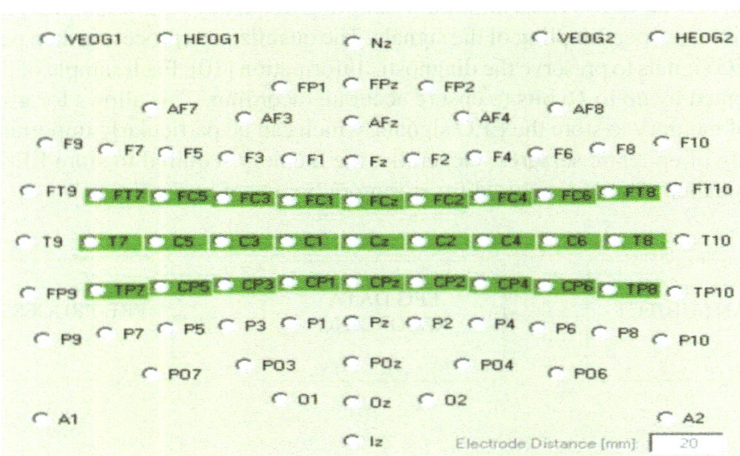

Fig. 4. Location of Different Electrodes Shown in Green Shade for EEG Signals Acquisition

2.2 EEG Feature Extractions

Extracting unique features is a critical step in identifying common patterns within signals. This process is essential for recognizing and categorizing these patterns accurately [11]. Various techniques are employed for extracting features from signals, which include time domain, frequency domain, and time-frequency domain analysis methods. These

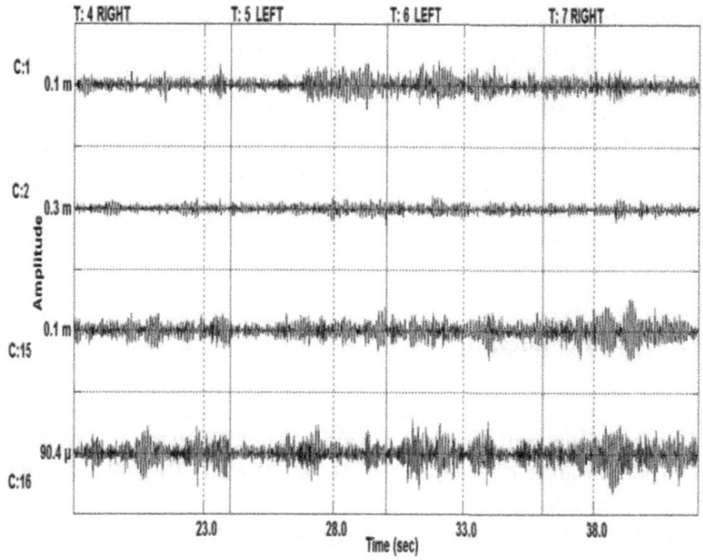

Fig. 5. Temporal Pattern Obtained by CSP Filtering of EEG Signals

methods allow for a comprehensive analysis of the signals, enabling the extraction of relevant features for further analysis or classification [4]. Proper feature selection is vital for achieving high classification performance, as the extraction of features plays a crucial role in this process. Failure to select appropriate features can negatively impact the classification performance [12]. The EMG and EEG signals were first segmented, and each resulting template was windowed to an epoch length of 200 ms with no overlap. To classify the EMG signals, four time-domain features (mean absolute value, zero crossing, waveform length, and slope sign change) were extracted from each template. The power change corresponding to a movement relative to a baseline was used as the feature for EEG classification, which determines the event-related desynchronization/synchronization (ERD/ERS). The baseline signal used for each epoch of the movement trial was the 0.5 s signal template before the movement cue of the same trial. The classification was performed based on different combinations of EEG and EMG channels. Before fusion, the EEG and EMG signals were analyzed independently to determine the effect of reducing the number of channels on accuracy. The channels were optimally selected using a heuristic approach based on the calculated feature sets, and the resulting performance helped in the fusion of different EEG and EMG channels for movement task classification [13].

2.3 Classifiers

Classification is the act of recognizing and understanding objects and concepts by identifying their defining characteristics. It involves categorizing people or objects based on a set of qualities or attributes. For instance, galaxies can be classified based on their shapes, emails can be categorized as spam based on their content, and students can be grouped based on their grades. The expert responsible for this task is called a classifier. In this particular study, the performance of SVM and LDA classifiers was compared to determine their effectiveness for classification purposes [9]. The process of selecting a classifier is crucial in achieving an accurate and satisfactory classification result during the validation of a proposed method. The choice of an appropriate classifier is essential for the success of the classification task, and it can significantly impact the performance of the overall system. Therefore, careful consideration and evaluation of different classifiers are necessary to ensure the selection of the most suitable one for the task at hand.

3 Results and Discussions

To assess the performance of various classification methods, frequency-domain features were computed using the mathematical formulas. The calculation of frequency-domain features involved Variance of Central Frequency (VCF), Total Power (PT), spectral moments (SMk; where $k = 0$ to 5) up to the 5th order, Mean Power (MNP), Central Frequency (F_C) and Root Mean Square Frequency (RMSF). These frequency-domain features were used to evaluate the effectiveness of different classification approaches, providing valuable insights into the performance of the system [10]. The performance of the frequency domain features was evaluated using the KNN classifier with varying fold cross-validation techniques (ranging from two to ten), and the average percentage classification accuracy was recorded for each feature. Table 2 summarizes the results of this analysis. The highest average classification accuracy (57.5%) was obtained using the SM_k features of the 5th order (SM1, SM2, SM3, SM4, and SM5). This was followed by the PT, MNP, VCF, RMSF, and FC features, respectively, using the KNN classifier. These results provide valuable insights into the performance of different frequency-domain features and can aid in the selection of the most effective feature for a given classification task.

Table 1. Average Percentage Classification Accuracies of Frequency Domain Features Using Fine KNN Classifier

Sr No	Features	2FCV	3FCV	4FCV	5FCV	6FCV	7FCV	8FCV	9FCV	10FCV	Mean
1	VCF	31.5	32.1	32.7	33	34.8	31.8	33.3	33.9	34.8	26.79
2	PT	62.1	59.7	60.3	60.6	60	58.5	58.8	59.1	59.1	50.82
3	SM1	63	61.2	61.5	61.5	62.1	61.8	63.3	62.4	62.1	52.89
4	SM2	60	59.7	59.1	59.1	58.8	59.1	59.7	59.1	60	50.46
5	SM3	54.8	54.2	55.5	57.6	55.8	55.8	56.1	54.8	56.7	47.13
6	SM4	55.5	56.1	55.2	51.5	53.3	53	53.6	54.5	54.8	45.75

(*continued*)

Table 1. *(continued)*

Sr No	Features	2FCV	3FCV	4FCV	5FCV	6FCV	7FCV	8FCV	9FCV	10FCV	Mean
7	SM5	55.8	53	54.8	56.4	56.4	55.5	57.3	56.4	55.8	47.14
8	MNP	62.1	59.7	60.3	60.6	60	58.5	58.8	59.1	59.1	50.82
9	FC	30	20.3	22.1	23	26.7	23	25.2	23	24.8	18.81
10	RMSF	26.1	28.5	27.9	29.7	32.1	30.6	30.3	30	33	23.82

As shown in Table 1, the mean classification accuracies of various features namely variance of central frequency, total power, SM_k features of the 5th order (SM1, SM2, SM3, SM4, and SM5), mean power, central frequency and root mean square frequency with Fine KNN classifier were found to be 26.79%, 50.82%, 52.89%, 50.46%, 47.13%, 45.75%, 47.14%, 50.82%, 18.81% and 23.82% respectively. SM_1 yielded the highest mean classification accuracy.

Table 2. Average Percentage Classification Accuracies of Frequency Domain Features Using Medium KNN Classifier

Sr No	Features	2FCV	3FCV	4FCV	5FCV	6FCV	7FCV	8FCV	9FCV	10FCV	Mean
1	VCF	29.7	29.7	32.7	27.6	30	27.6	32.1	30.9	27	23.73
2	PT	63.3	63.3	63.6	63.3	63.3	63.3	63.3	63.3	63.3	54
3	SM1	61.2	60	62.1	61.2	61.2	60.9	59.7	61.2	60.6	51.81
4	SM2	59.4	60.3	59.1	60	60.3	59.7	58.8	60	60.3	50.79
5	SM3	58.5	58.5	57.6	59.4	58.8	59.4	59.4	58.5	58.5	49.86
6	SM4	58.2	57.3	58.5	60	58.2	58.8	59.1	58.8	59.1	49.8
7	SM5	55.8	55.5	55.8	57	56.7	56.1	57.6	56.1	55.8	47.64
8	MNP	63.3	63.3	63.6	63.3	63.3	63.3	63.3	63.3	63.3	54
9	FC	34.2	29.4	32.4	34.5	33.3	30.9	30.9	32.1	33.3	26.1
10	RMSF	28.5	32.1	27.6	30.6	29.4	31.5	29.1	30.6	26.7	23.61

As shown in Table 2, the mean classification accuracies of various features namely variance of central frequency, total power, SMk features of the 5th order (SM1, SM2, SM3, SM4, and SM5), mean power, central frequency and root mean square frequency with Medium KNN classifier were found to be 23.73%, 54.0%, 51.81%, 50.79%, 49.86%, 49.80%, 47.64%, 54.0%, 26.1% and 23.61% respectively. MNP yielded the highest mean classification accuracy.

As shown in Table 3, the mean classification accuracies of various features namely variance of central frequency, total power, SM_k features of the 5th order (SM1, SM2,

Table 3. Average Percentage Classification Accuracies of Frequency Domain Features Using Coarse KNN Classifier

Sr No	Features	2FCV	3FCV	4FCV	5FCV	6FCV	7FCV	8FCV	9FCV	10FCV	Mean
1	VCF	31.5	34.2	33.9	35.2	33.9	34.8	34.2	33.6	33.6	27.49
2	PT	54.2	55.8	55.8	56.1	56.4	56.4	56.4	56.4	56.7	47.42
3	SM1	53	54.8	55.5	55.8	55.8	56.4	56.1	56.4	56.4	47.02
4	SM2	53.3	55.2	55.5	55.5	55.8	56.1	55.8	56.1	56.1	46.94
5	SM3	50.6	54.5	54.5	55.8	55.8	56.7	55.2	56.4	56.4	46.59
6	SM4	47	53.9	54.2	55.8	57	57.6	57.3	57.9	57.3	46.8
7	SM5	45.8	54.8	55.8	55.8	57.3	57.3	57.6	57	57.6	46.9
8	MNP	54.2	55.8	55.8	56.1	56.4	56.4	56.4	56.4	57.7	47.52
9	FC	32.4	33.3	33.9	33.3	33.3	34.2	34.2	33.9	34.2	27.27
10	RMSF	30.6	33.3	33.6	32.4	31.8	33.3	33	31.8	32.7	26.25

SM3, SM4, and SM5), mean power, central frequency and root mean square frequency with Coarse KNN classifier were found to be 27.49%, 47.42%, 47.02%, 46.94%, 46.59%, 46.80%, 46.90%, 47.52%, 27.27% and 26.25%. MNP yielded the highest mean classification accuracy.

The results confirm that the Fine KNN classifier produced the best classification accuracy and Medium KNN was found to be a second best method. The accuracy was found to be 93% when the features were increased. The EEG data from 20 healthy subjects was merged together. The data was collected in four sessions. The individual features were combined to form a complete feature vector of 10 features. As the Fine KNN produced the best performance, the confusion matrix for Fine KNN is also shown in Table 6. Other performance measuring parameters like False Discovery Rate (FDR), Positive Predictive Value (PPV), specificity (true negative ratio), and sensitivity (true positive ratio) can also be calculated from Fig. 6.

Figure 7 shows the ROC of Fine KNN classifier thereby revealing the best performance of the classifier. The results revealed that the Fine KNN method, employing a feature vector consisting of the 10 selected features, outperformed the Medium KNN and Coarse KNN methods. The top five features, ranked in ascending order, were SM5, SM2, PT, MNP, and SM1. These features hold promise for enhancing the system's performance. Remarkably, the Fine KNN classifier achieved an accuracy rate of 93% and operated within a swift 5.58-s timeframe. Frequency-domain features have been calculated by using the mathematical formulae depicted in Table 4, they are also computed for evaluating the performance of different classification approaches. For calculating the frequency domain features, first of all, FFT of SEMG signals has been computed followed by the estimation of Power Spectral Density (PSD). Here, all spectral moments (SM_k; where $k = 0$ to 5) up to 5th order are combined as single features.

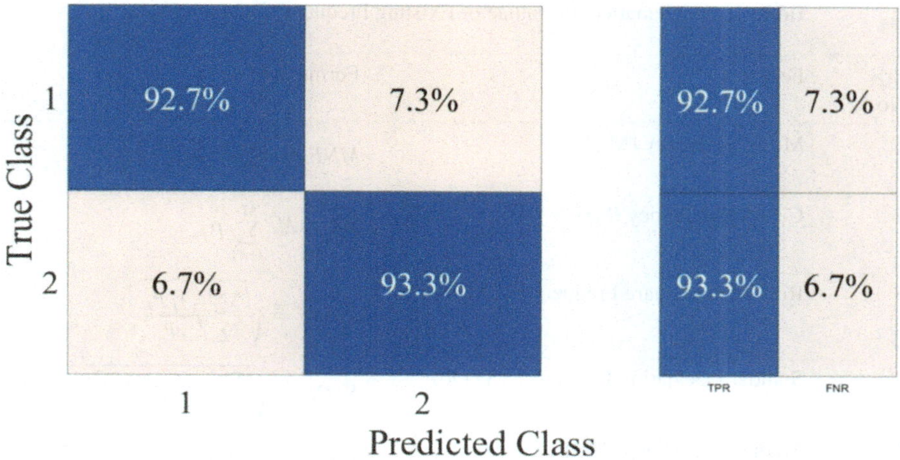

Fig. 6. Confusion matrix for Fine KNN

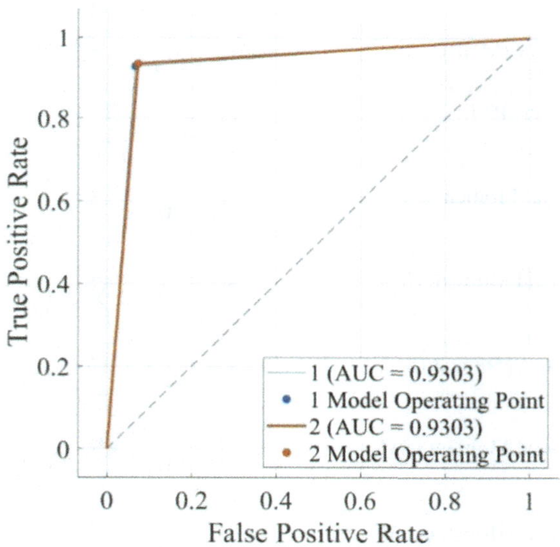

Fig. 7. ROC curve of Fine KNN classifier

Table 4. Mathematical Formulae of Existing Frequency Domain Features

Sr No	Features	Formulae
1	Mean Frequency (MNF)	$MNF = \frac{\sum_{J=1}^{M} f_j P_j}{\sum_{j=1}^{M} P_j}$
2	Central Frequency (F_C)	$F_C = \frac{M}{2} \sum_{J=1}^{M} P_J$
3	Root Mean Square Frequency (RMSF)	$RMSF = \sqrt{\frac{\sum_{J=1}^{M} f_J^2 P_J}{\sum_{J=1}^{M} P_J}}$
4	Standard Deviation Frequency (STDF)	$STDF = \sqrt{\frac{\sum_{J=1}^{M} (f_J - MNF)^2 P_J}{\sum_{J=1}^{M} P_J}}$
5	Median Frequency (MDF)	$MDF = \frac{1}{2} \sum_{j=1}^{M} P_j$
6	Peak Frequency (PKF)	$PKF = max(P_j)$, $j = 1, 2, 3, \ldots\ldots.M$
7	Mean Power (MNP)	$MNP = \frac{\sum_{j=1}^{M} P_j}{M}$
8	Total Power (PT)	$P_T = \sum_{j=1}^{M} P_j$
9	1st Spectral Moment (SM_1)	$SM_1 = \sum_{j=1}^{M} P_j f_j$
10	2nd Spectral Moment (SM_2)	$SM_2 = \sum_{j=1}^{M} P_j (f_j)^2$
11	3rd Spectral Moment (SM3)	$SM_3 = \sum_{j=1}^{M} P_j (f_j)^3$
12	4nd Spectral Moment (SM_4)	$SM_4 = \sum_{j=1}^{M} P_j (f_j)^4$
13	5nd Spectral Moment (SM_5)	$SM_5 = \sum_{j=1}^{M} P_j (f_j)^5$
14	Variance of Central Frequency (VCF)	$VCF = \frac{1}{SM0} \sum_{j=M}^{M} P_j (f_j - F_C)^2$

4 Conclusion

In this study, the performance of three classifiers was compared to assess the ability of 10 time-domain features in distinguishing between EEG signals associated with right and left hand movements. The classifiers under investigation were Fine KNN, Medium KNN, and Coarse KNN. The implications of this study extend to the design of online classification-based robotic assistive devices, specifically aimed at aiding individuals with disabilities. Additionally, comparing the performance of all three domain features could yield more robust combinations of feature sets. In conclusion, this research involved a comparative analysis of three classifiers, evaluating the effectiveness of FD features in discriminating between EEG signals associated with right and left hand movements. The Fine KNN classifier, utilizing the top-performing feature vector, achieved an impressive 93% accuracy in a mere 5.58 s. These findings offer valuable insights for the development of assistive devices that utilize online classification, catering to the needs of disabled individuals. Moreover, further investigations comparing the performance of all domain features could yield even more robust feature set combinations. Accurate source localization of EEG signals remains challenging due to the complex and nonlinear nature of the brain's electrical activity. Developing more robust algorithms for source localization is an ongoing area of research.EEG signals can be susceptible to various artifacts and noise sources, such as muscle activity, eye movements, and environmental interference. Developing advanced signal processing techniques to enhance signal quality and reduce noise is crucial.

References

1. Aydemir, E., Tuncer, T., Dogan, S.: A tunable-Q wavelet transform and quadruple symmetric pattern based EEG signal classification method. Med. Hypotheses **134**, 109519 (2020). https://doi.org/10.1016/j.mehy.2019.109519
2. Tsekoura, K., Foka, A.: Classification of EEG signals produced by musical notes as stimuli. Expert Syst. Appl. **159**, 113507 (2020). https://doi.org/10.1016/j.eswa.2020.113507
3. Narayan, Y.: Hand motion identification based on EEG signals classification. In: 2021 2nd Global Conference for Advancement in Technology, GCAT 2021. Institute of Electrical and Electronics Engineers Inc., Oct. 2021. https://doi.org/10.1109/GCAT52182.2021.9587556
4. Khosla, A., Khandnor, P., Chand, T.: A comparative analysis of signal processing and classification methods for different applications based on EEG signals. Biocybernetics Biomed. Eng. **40**(2), 649–690 (2020). https://doi.org/10.1016/j.bbe.2020.02.002
5. Narayan, Y., Ranjan, R.: EEG signals classification for right- and left-hand movement discrimination using SVM and LDA classifiers. In: Lecture Notes in Electrical Engineering, Springer Science and Business Media Deutschland GmbH, pp. 133–142 (2022). https://doi.org/10.1007/978-981-16-9885-9_11
6. Narayan, Y., Singh, R.M., Mathew, L., Chatterji, S.: Surface EMG signal classification using ensemble algorithm, PCA and DWT for robot control. In: Communications in Computer and Information Science, Springer Verlag, pp. 218–230 (2019). https://doi.org/10.1007/978-981-13-3140-4_20
7. Bakhshali, M.A., Khademi, M., Ebrahimi-Moghadam, A., Moghimi, S.: EEG signal classification of imagined speech based on Riemannian distance of correntropy spectral density. Biomed. Signal Process. Control **59**, 101899 (2020). https://doi.org/10.1016/j.bspc.2020.101899

8. Narayan, Y.: Analysis of MLP and DSLVQ classifiers for EEG signals based movements identification. In: 2021 2nd Global Conference for Advancement in Technology, GCAT 2021, Institute of Electrical and Electronics Engineers Inc., Oct. 2021. https://doi.org/10.1109/GCAT52182.2021.9587868
9. Narayan, Y.: Eeg signals classification using svm and rf classifier for left and right-hand movement. J. Green Eng. **10**(11), 10691–10701 (2020)
10. Narayan, Y.: Motor-imagery EEG signals classificationusing SVM, MLP and LDA classifiers. Turk. J. Comput. Math. Educ. (TURCOMAT) **12**(2), 3339–3344 (2021)
11. George, S.T., Subathra, M.S.P., Sairamya, N.J., Susmitha, L., Joel Premkumar, M.: Classification of epileptic EEG signals using PSO based artificial neural network and tunable-Q wavelet transform. Biocybernetics Biomed. Eng. **40**(2), 709–728 (2020). https://doi.org/10.1016/j.bbe.2020.02.001
12. Gupta, V., Pachori, R.B.: Classification of focal EEG signals using FBSE based flexible time-frequency coverage wavelet transform. Biomed. Signal Process. Control **62**, 102124 (2020). https://doi.org/10.1016/j.bspc.2020.102124
13. Hooda, N., Das, R., Kumar, N.: Fusion of EEG and EMG signals for classification of unilateral foot movements. Biomed. Signal Process. Control **60**, 101990 (2020). https://doi.org/10.1016/j.bspc.2020.101990

Performance Analysis of DS-CDMA Using GIG Orthogonal Codes Under AWGN and Rayleigh Fading Channel

Meenakshi Munjal(✉)

Department of Electronics and Communications Engineering, Chandigarh University, Mohali, Punjab, India
meenakshi6150008@gmail.com

Abstract. Sparse code multiple access (SCMA) is a novel multiple-access strategy used in 5G wireless networks that combines Orthogonal Frequency Division Multiple Access (OFDMA) and Code Division Multiple Access (CDMA). Recent interest in CDMA has been on the use of multi-level orthogonal coding to facilitate spreading. The orthogonal code is used to disseminate the modulated data symbols of each user. The set of spreading codes employed controls how much the signals interfere with each other, and thus impacts the system's performance at the receiver. Grey Inverse Grey (GIG) orthogonal spreading codes are constructed in this study employing grey and inverse grey code, which provide higher correlation features than traditional binary code families such as Hadamard and gold code. This research assesses the Bit Error Rate (BER) performance of CDMA using the Binary Phase Shift Keying (BPSK) and Quadrature Phase Shift Keying (QPSK) modulation techniques. The performance of a DS-CDMA communication system is examined utilizing GIG orthogonal spreading codes in the Additive White Gaussian Noise (AWGN) and Rayleigh fading channels. The scrambling impact on GIG codes is also investigated. Under the same test settings, the results reveal that DS-CDMA systems using GIG spreading codes outperform binary spreading codes in terms of BER.

Keywords: Additive White Gaussian Noise (AWGN) · Binary Phase Shift Keying (BPSK) · Bit Error Rate (BER) · Direct Sequence-Code Division Multiple Access (DS-CDMA) · Quadrature Phase Shift Keying (QPSK)

1 Introduction

In 5G wireless network, SCMA access technique is used which is a combination of OFDMA and CDMA. SCMA is a type of non-orthogonal multiple access technique which is developed for next generation communications systems [1]. The aim of using SCMA is to improve spectral efficiency. It provides less transmission latency and it is able to provide energy saving [2].

In CDMA, a large number of system users can share spectrum resources simultaneously [3]. In comparison to other traditional multiple access systems like Time

D. Kriksciuniene et al. (Eds.): IBICA 2023, LNNS 1228, pp. 189–200, 2025.
https://doi.org/10.1007/978-3-031-78937-3_19

Division Multiple Access (TDMA) and Frequency Division Multiple Access (FDMA), it has the capacity to accommodate a greater number of users, which results in efficient bandwidth utilisation. The W-CDMA and CDMA2000 wireless communication technologies, among others, extensively utilise the CDMA technique [4]. Spreading codes are employed in the CDMA system to separate the various users and disperse the signal. Mainly, spreading codes can be divided into orthogonal and non-orthogonal codes [5].

Binary Walsh Hadamard, Kasami, and Gold codes of 64 or higher length are utilised as CDMA spreading codes in conventional cellular networks. Hadamard codes are perfectly orthogonal and fixed power binary codes. Using the Kronecker product, higher length Hadamard code sets are created iteratively from the shorter length Hadamard code sets. [6]. Codes made of gold are almost orthogonal. The modulo-two addition of two chosen polynomials yields gold code sequences, which are produced from PN sequences [7]. Gold codes are frequently utilised for asynchronous communications while Hadamard codes are used for synchronous coding. Binary code sets have the chip levels 0 and 1 which is mapped to $+1, -1$. It produces a signal with a constant envelope or fixed power, effectively using the RF power available in the CDMA system [8]. This is true only in the case of a single user system, but it is not true in a multiuser scenario where numerous binary signals are joined together.

A statistical model for the impact of a radio signal's propagation environment is called Rayleigh fading [9]. The Rayleigh fading models says that a signal's magnitude will change or fade over time after passing through such a transmission medium in accordance with the Rayleigh distribution [10]. In CDMA-based systems, the RAKE receiver is employed, and it may aggregate multipath signal components, which are time-delayed versions of the initial signal transmission. Combining is done at the receiver to increase SNR (Signal to Noise Ratio) [11]. When a multipath component's relative propagation delay surpasses a chip time, they are effectively uncorrelated from one another [12].

In this paper, the GIG spreading codes are produced by using the gray and inverse gray codes. The algorithm used for generating gray code and inverse gray codes is discussed in [13, 14] and [15]. The construction methodology of generating multi-level spreading codes is discussed in [16]. The multi-level spreading codes used in this paper is called GIG spreading codes. The contributions of this paper are:

i. Firstly, a standardized algorithm is discussed through which GIG codes of any length can be generated. In this paper, 5 and 6-bit GIG codes are generated. The generated GIG codes are not orthogonal to each other. The orthogonal codes are selected depend upon the orthogonality condition.
ii. The GIG codes performance was analysed in terms of different parameters i.e. Mean Square Aperiodic Auto Correlation (MSAAC), Mean Square Aperiodic Cross Correlation (MSACC) and Figure of Merit (FOM). Its performance is also compared with gold code, Hadamard code, 3 and 4-bit GIG codes.
iii. Thirdly, the performance of DS-CDMA using 5 and 6 level orthogonal codes over AWGN and Rayleigh channel is evaluated.
iv. Finally, effect of scrambling on BER is also evaluated.

The remainder of the paper is structured as follows: Sect. 2 discuss the algorithm for gray and Inverse gray code. In Sect. 3, steps of construction of 5-bit and 6-bit GIG spreading codes are given. Section 4 discussed how the orthogonal codes are selected from all GIG codes. In Sect. 5, comparison of correlation properties of different spreading codes is done. Performance is analyzed over AWGN and Rayleigh fading channel is presented in Sect. 6 and then comparison of these codes with Hadamard and gold code is shown in Sect. 7. In Sect. 8, effect of scrambling is shown. Finally, Sect. 9 gives the conclusions.

2 Algorithm for Gray Code and Inverse Gray Code

The algorithm for gray and inverse gray code is designed by taking the m-bit radix 'n' gray code. The permutation of (1, 2, 3 ...... m) is assumed to be (Q_1, Q_2...... Q_n). The algorithm for n^m integers (0, 1, 2...... (n^m-1)) can be structured as.

Algorithm for (m, r) gray code and inverse gray codes

Set m=number of bits required for GIG codes and n=radix of GIG codes
For j = 0, 1, 2......, m
    For k = 1, 2........., n-1
        If $s \leq n^{m-j-1}$-k/n
            $P_{j,k} = n^j$ {k, k+n, ...k+sn}
      End of if loop
    End of k loop
End of j loop

Where 's' is the biggest positive number so that $s \leq n^{m-j-1}$-k/n.

A. *How Algorithm Used for (M, N) Gray Code* [17]

The permutation of number (1, 2, 3 ... (n-1)) is defined by (0, t1, t2..., tn-1). In this paper this algorithm is used for 5-bit and 6-bit code generation. The ith row is then created by taking the $(i-1)^{th}$ row and replacing the Q_j^{th} bit with its successor bit, if it is in $P_{j-1,k}$. and all elements of 0th row is zero.

B. *How algorithm used for (n,r) inverse gray code.*

For the inverse gray code, s_1, s_2... s_{r-1} defines the permutation of (1, 2, 3 ...(r-1)). All elements of 0th row is given by zero and then the ith row is generated from the $(i-1)^{th}$ row by substituting its successor bit if it is in $Q_{j-1,k}$. for all bits other than the p_j^{th} bit.

3 5-bit and 6-bit GIG Spreading Codes

The procedures needed to create higher-level, 10- and 12-length spreading codes are as follows:

Step 1: Utilize the algorithm with increasing permutations to create 5 and 6-bit gray codes.
Step 2: Create 5 and 6-bit inverse gray code using the same permutation as gray code.

Step 3: Applying inverse gray code to gray code yields 10 and 12-length GIG code.
Step 4: To create a higher-level code set, each row of the GIG code of lengths 10 and 12 is mapped to a different amplitude level.

The creation of binary user codes (2-level) involves mapping the radix-2 elements of 0 and 1 to the chip amplitude levels of -1 and 1. An analogous process is used to create higher-level orthogonal codes. For a 3-level coding, radix-3 (ternary) coding elements 0, 1, 2 are mapped to chip amplitude levels −1, 0, 1, and for a 4-level coding, radix-4 (quaternary) number system coding elements 0, 1, 2, 3 are translated to chip amplitude levels −3, 1, 1 and 3.

Step 5: To obtain an m-level, 2n length user code set, choose the first code word and confirm the orthogonality of the other code words with the first code word. Once you have an m-level, 2n length user code established, select a second code word and validate the orthogonality of the remaining code words with it.

The single 5-bit and 6-bit GIG code is shown in Figs. 1 and 2 respectively.

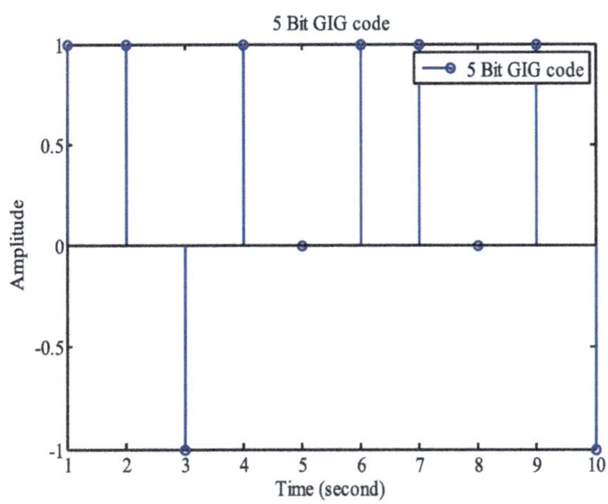

Fig. 1. Plot of 5-bit GIG codes

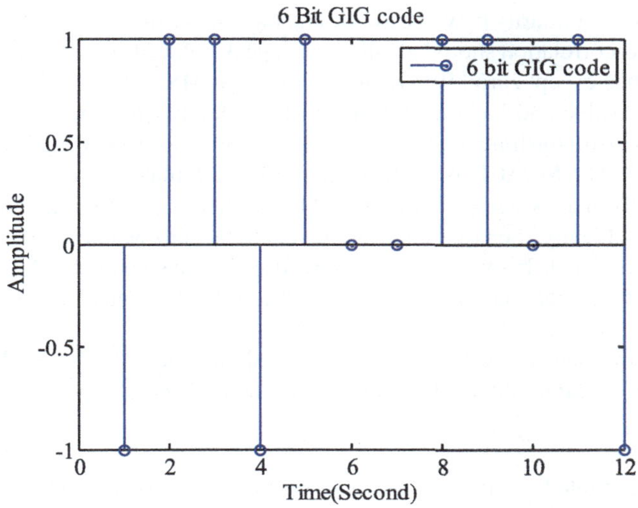

Fig. 2. Plot of 6-bit GIG codes

4 Orthogonal Spreading Codes

All pairwise cross correlations in orthogonal codes are zero. Each mobile user utilizes one of the sequences in the set as a spreading code for CDMA applications. Orthogonal codes ensure that there is no cross correlation among all users. [18]. The orthogonal GIG codes are selected from all GIG codes. The orthogonal codes are selected depend on the Eq. (1).

$$[x].[x]^{T} = I \tag{1}$$

where [x] is the GIG codes matrix, $[x]^{T}$ is the transpose GIG matrix [x] and I is the Identity matrix.

For three level 3 bit, there are total 27 GIG codes but from 27 codes, there are only 8 orthogonal GIG codes which are selected depend on the above equation. For four level 3 bit, there are total 64 GIG codes but not all of them are orthogonal, only 8 codes are orthogonal codes. For four level 4 bit, there are only 37 orthogonal codes out of 256 GIG codes. For three level 5 bit, there are only 56 orthogonal codes from 243 GIG codes. For three level 6 bit, there are 729 GIG codes but from them, there are only 168 orthogonal codes.

Orthogonal sequences provide several benefits, including great spectrum efficiency, various access possibilities, high flexibility, narrow-band interference rejection, easy one tap equalization etc.

5 Comparison of Correlation Properties of Different Codes

The system's performance was evaluated using several metrics i.e. MSAAC, MSACC and FOM. The comparison of different correlation properties of different GIG codes is shown in Table 1.

The degree of similarity between a code and its time-shifted duplicate is measured by MSAAC. Auto-correlation should ideally have a peak value at zero time and zero values at all other times except zero [19]. Table 1 shows that MSAAC with 5 and 6 bit values has the lowest value and has the advantage of correct synchronisation at the receiver side as well as distinguishing the targeted user from other users causing Multiple Access Interference (MAI). MSACC is another metric that calculates the degree of similarity between two distinct code sequences. In CDMA, each user should be allocated a distinct code sequence. Ideally, the cross correlation should be zero which has the advantage of receiver to discriminate between different spread spectrum signals. MSACC of proposed 5-bit and 6-bit GIG codes has the least values, which is basic requirement of orthogonality condition.

The FOM evaluate the performance GIG spreading codes [20].The FOM of 5-bit and 6-bit GIG codes is larger than gold and Walsh Hadamard codes which is also a desirable factor.

Table 1. Correlation Measures of Different Spreading Codes

Different codes	MSAAC	MSACC	FOM
3 bit 3 level GIG codes	0.0280	0.0267	1.2500
4 bit 3 level GIG codes	0.2110	0.0154	1.2829
5 bit 3 level GIG codes	0.0110	0.0100	1.2245
6 bit 3 level GIG codes	0.0085	0.0069	1.2277
Gold code of 8 length	2.6391	3.3442	0.3892
Hadamard code of 8 length	0.0371	0.0103	**0.5056**
3 bit 3 level GIG codes [16]	1.2731	0.7648	**0.7855**
4 bit 4 level GIG codes [16]	1.8641	0.8572	**0.5365**

For orthogonality, it is necessary the all-pairwise cross correlation should be zero. For the CDMA application, it is desired that each mobile user use one of the set's sequences which provides zero cross correlation among all users. Hadamard code is called proper orthogonal codes but as table show that MSACC of 5-bit and 6-bit GIG codes are minimum which is the required condition for orthogonality. So, proposed GIG codes are best among all other codes.

6 Performance Analysis Over AWGN and Rayleigh Fading Channel

In this section, the BER performance of a DS-CDMA system over an AWGN and Rayleigh fading channel is analysed. The capacity of a CDMA system is heavily dependent on spreading codes and the BER performance of the receiver. The number of corrected and destroyed bits as they are sent from transmitter to receiver is defined as BER. As a result, in order to optimize the performance of a CDMA system, BER should be kept minimum.

The performance of DS-CDMA using 5-bit and 6-bit GIG codes is same as shown in Fig. 3. The two lines approximate merge each other that mean BER is same for 5 and 6-bit GIG codes. The AWGN performance for 1 user and 2 users is shown using BPSK is shown in Fig. 4 and using QPSK is shown in Fig. 5. For negative values of SNR, values of BER is larger. However, when the SNR improves from 0 dB, the BER drops. It demonstrates that as the number of users rises, so does BER, which is undesirable. However, due to proper orthogonal sequence, GIG codes are preferred.

The Rayleigh propagation model is most useful when there are many signal paths, none of which is dominant. As a result, all signal routes will vary and have an effect on the total signal at the receiver. The RAKE receiver employs numerous baseband correlators to process various signal multipath components independently [18]. The performance of DS-CDMA using 5-bit and 6-bit GIG codes for 1 user under Rayleigh fading channel is shown in Fig. 6 and for 2 users is shown in Fig. 7. The BER is approximately same for 5-bit and 6-bit GIG code under Rayleigh fading channel. For low value of Eb/N0, BER is larger but as the Eb/N0 increases, BER decreases.

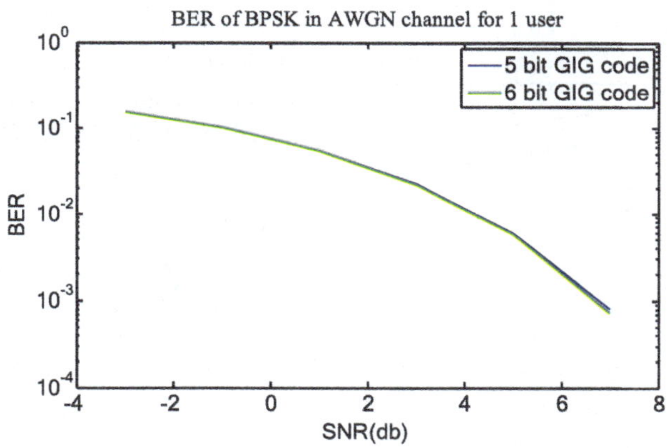

Fig. 3. Performance of DS-CDMA using 5-bit and 6-bit GIG codes for 1 user under AWGN channel

7 Comparison with Walsh Hadamard and Gold Code

There are various types of spreading codes used for spreading of DS-CDMA system. The most widely used binary (2-level) spreading codes are Hadamard and gold codes. The spreading codes can be orthogonal and non-orthogonal. Gold codes are almost orthogonal, whereas Hadamard codes are entirely orthogonal. Researchers have recently focused on developing multiple level spreading codes. The performance comparison of DS-CDMA using gold code, Hadamard code, 5 and 6-bit GIG codes for 1 user under Rayleigh fading channel is shown in Fig. 8 and for 2 users comparison is shown in Fig. 9.

Figure shows that for one user, BER of 5-bit and 6-bit GIG codes is larger than gold and Hadamard code but as the number of users increases, BER performance increases

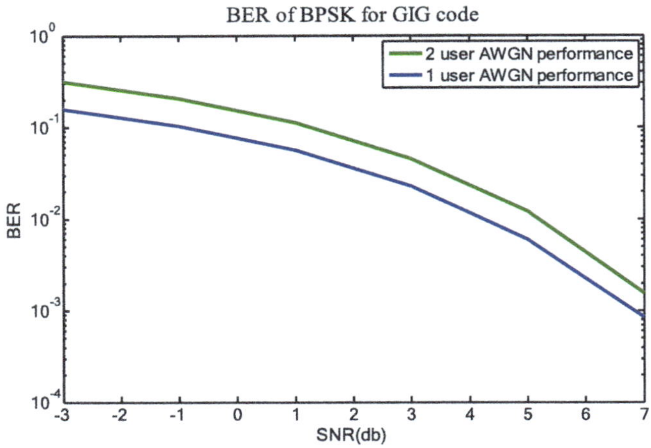

Fig. 4. Performance of DS-CDMA using GIG code for 1and 2 users using BPSK under AWGN channel

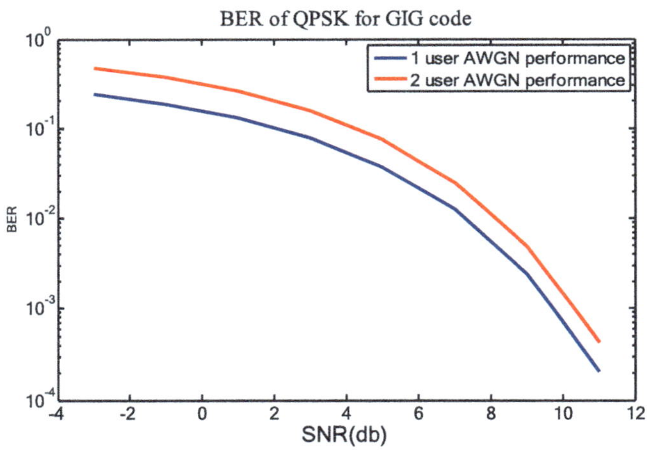

Fig. 5. Performance of DS-CDMA using GIG code for 1and 2 users using QPSK under AWGN channel

for GIG codes. BER performance of 5-bit and 6-bit GIG codes is better than gold code while somewhat less than Hadamard codes. Thus the GIG codes performance increases for larger number of users.

8 Effect of Scrambling

The Scrambling does not affect the signal bandwidth. In scrambling, the downlink is used to differentiate the different cell (one code per base station) while the uplink is used to separate different users per cell (many codes per base station) [21]. Effect of

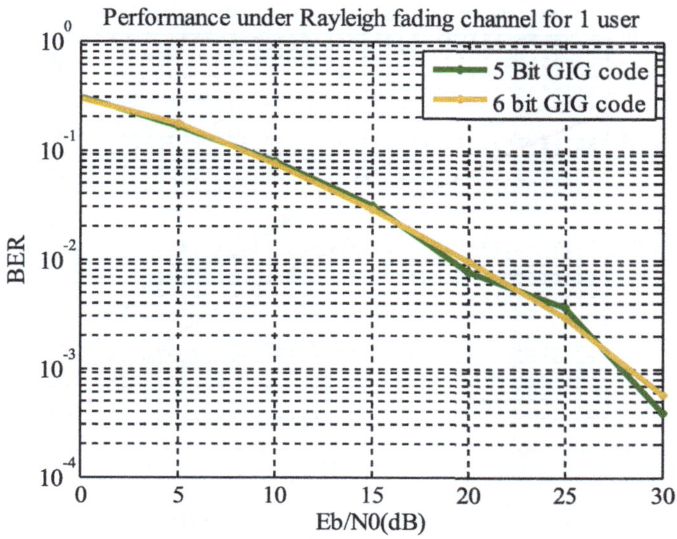

Fig. 6. Performance of DS-CDMA using 5-bit and 6-bit GIG codes for 1 user under Rayleigh fading channel

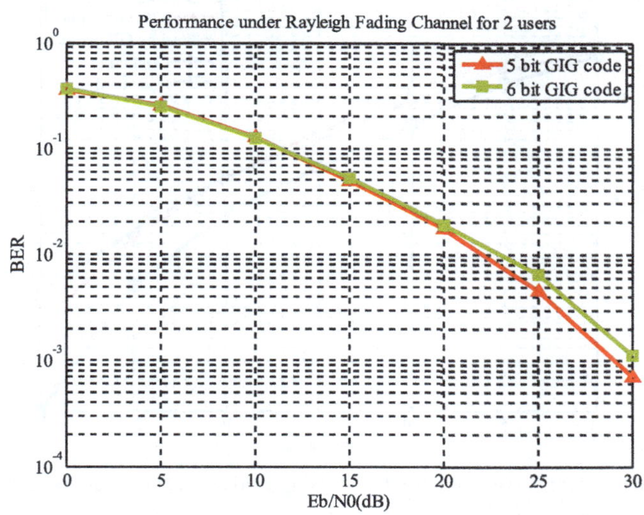

Fig. 7. Performance of DS-CDMA using 5-bit and 6-bit GIG codes for 2 users under Rayleigh fading channel

scrambling on DS-CDMA using GIG code is examined in Fig. 10. The results show that scrambling reduces the BER. For lower values of Eb/N0, BER remain same and there is no effect of scrambling. Scrambling effect appears for larger values of Eb/N0. As the Eb/N0 increases from 10 dB, scrambling effect appears and it reduces the BER.

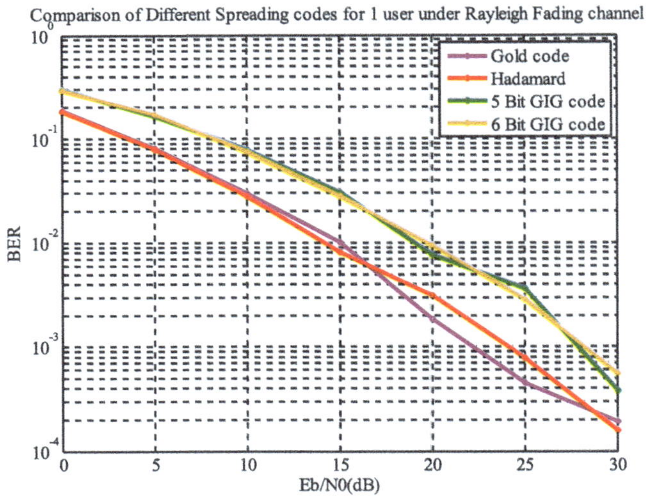

Fig. 8. Performance comparison of DS-CDMA using gold code, Hadamard code, 5 and 6-bit GIG codes for 1 user under Rayleigh fading channel

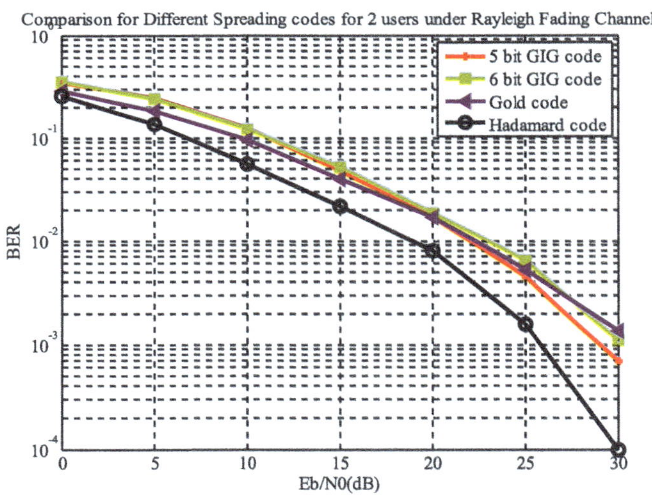

Fig. 9. Performance comparison of DS-CDMA using gold code, Hadamard code, 5 and 6-bit GIG codes for 2 users under Rayleigh fading channel

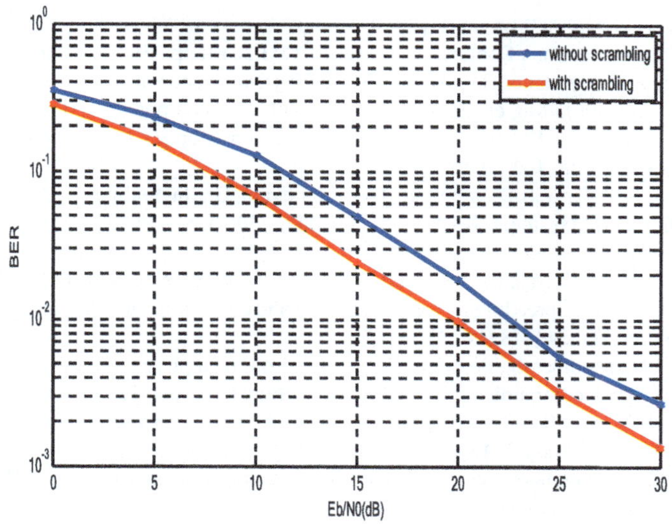

Fig. 10. Performance of DS-CDMA using 5-bit GIG code for 2 users under Rayleigh fading channel with and without scrambling

9 Conclusion

In this paper, a standardized algorithm is proposed through which GIG codes of any length can be generated. Here, 5-bit and 6-bit GIG codes are generated with the algorithm. The generated GIG codes are not orthogonal to each other. Only orthogonal codes are selected depend on the orthogonality condition. Then, The GIG codes performance was analyzed in terms of MSAAC, MSACC and FOM. Its performance is also compared with gold code, Hadamard code, 3 and 4-bit GIG codes and results shows that its MSAAC, MSACC, FOM are best than all codes. After that, the performance of CDMA using 5-bit and 6-bit orthogonal codes over AWGN and Rayleigh fading channel is evaluated. Result shows that for one user, BER of 5-bit and 6-bit GIG codes is larger than gold and Hadamard code however BER performance improves as the user base grows for GIG codes under Rayleigh fading channel. BER performance of 5-bit and 6-bit GIG codes is better than gold code while somewhat less than Hadamard codes. Effect of scrambling on GIG codes is also discussed which show that scrambling reduces the BER. Very important advantage for higher level of spreading codes, the number of orthogonal spreading codes increases. So for lager number of users, higher level of spreading codes can be used. The CDMA system using orthogonal GIG codes can be used in 5G wireless network for SCMA access technique.

References

1. Nikopour, H., Baligh, H.: Sparse code multiple access. In: IEEE International Symposium on Personal, Indoor and Mobile Radio Communications, PIMRC, pp. 332–336 (2013)

2. Nikopour, H., et al.: SCMA for downlink multiple access of 5G wireless networks. In: IEEE Global Communications Conference, GLOBECOM 2014, pp. 3940–3945 (2014)
3. Engineering, C., Sequence, D., Access, M.: BER performance of DS-CDMA system over a multipath rayleigh fading channel considering path gain component and noise variance. In: International Conf. on Electrical Engineering and Information & Communication Technology (ICEEICT) 2015, pp. 1–5 (2015)
4. Derryberry, R.T., Gray, S.D., Ionescu, D.M., Mandyam, G., Raghothaman, B.: Transmit diversity in 3G CDMA systems. IEEE Commun. Mag. **40**(4), 68–75 (2002). https://doi.org/10.1109/35.995853
5. Saini, D., Gupta, A., Joshi, A.: Reduction in code blocking using scattered vacant codes for orthogonal variable spreading factor-based wideband code division multiple access networks. IET Commun. **7**(1), 40–48 (2013)
6. Akansu, A.N., Member, S., Poluri, R.: Walsh-like nonlinear phase orthogonal codes for direct sequence CDMA communications. IEEE Trans. Signal Process. **55**(7), 3800–3806 (2007)
7. Lee, Y.H., Kim, S.J.: Sequence acquisition of DS-CDMA systems employing gold sequences. IEEE Trans. Veh. Technol. **49**(6), 2397–2404 (2000)
8. Amours, C.D., Dahmane, A.O.: Permutation spreading for asynchronous MIMO- CDMA systems using hadamard codes and gold scrambling sequences. In: IEEE Wireless Communications and Networking Conference, pp. 1–6 (2009)
9. Castillo, J.V., Vela-garcia, L., Gutiérrez, C.A., Parra-michel, R.: A reconfigurable hardware architecture for the simulation of Rayleigh fading channels under arbitrary scattering conditions. AEUE Int. J. Electron. Commun. **69**(1), 1–13 (2015)
10. Tan, C.W.: Optimal power control in rayleigh-fading heterogeneous wireless networks. IEEE/ACM Trans. Netw. **24**(2), 940–953 (2015)
11. Kabir, H., Model, A.T.: Performance analysis of a mc-ds-cdma wireless communication system with rake receiver under nakagami-m fading environment. In: IEEE International Conference on Electrical and Electronic Engineering (ICEEE), pp. 181–184 (2015)
12. Kodama, T., Matsumoto, R., Wada, N.: A novel optical code RAKE receiver using a multiport encoder / decoder. J. Light. Technol. **31**(11), 1675–1680 (2013)
13. Sankar, K.J.: Generalized gray codes. In: International Symposium on Intelligent Signal Processing and Communication Systems, pp. 654–659 (2004)
14. Zheng, S.Q., Latifi, S., Member, S.: Optimal simulation of linear multiprocessor architectures on multiply-twisted cube using generalized Gray Codes. IEEE Trans. Parallel Distrib. Syst. **7**(6), 612–619 (1996)
15. Usha, K., Sankar, K.J.: A technique for the construction of inverse gray codes. Int. J. Emerg. trends Eng. Dev. **2**(2), 337–341 (2012)
16. Usha, K., Sankar, K.J.: New multi level spreading codes for DS CDMA communication. In: Conference on Advances in Communication and Control Systems, pp. 154–159 (2013)
17. Meenakshi, M., Gupta, S.: Advanced level cyclic gray codes with application. Int. J. Electron. Commun. Comput. Technol. **4**(3), 619–622 (2014)
18. Zhang, Y., Peng, M., Yang, S.: A clique-based online algorithm for constructing optical orthogonal codes. Appl. Soft Comput. **47**, 21–32 (2016). https://doi.org/10.1016/j.asoc.2016.05.024
19. Koukouvinos, C., Kounias, S., Seberry, J., Yang, C.H., Yang, J.: On sequences with zero autocorrelation. Des. Codes Cryptogr. **4**(3), 327–340 (1994)
20. Verd, S.: Spectral effciency of CDMA with random spreading. IEEE Trans. Inf. Theory **45**(2), 622–640 (1999)
21. Neagu, M., Miclea, L., Manich Bou, S.: Improving security in cache memory by power efficient scrambling technique. IET Comput. Digit. Tech. 1–10 (2015)

Exploring the Performance of Brain-Computer Interfaces in Assistive Technology

Yogendra Narayan(✉), Vishal, Rajeev Ranjan, and Rohit Katyal

Chandigarh University Mohali, Mohali, India
Narayan.yogendra1986@gmall.com

Abstract. Brain-Computer Interface (BCI) has developed a technology with considerable potential, enabling direct communication and control between the human brain and external devices. The present study aims to examine the viability and effectiveness of BCI frequency domain features with the help of three different classification methods for the healthy human person. In this context, electroencephalography (EEG) data was acquired from 10 healthy persons exhibiting diverse levels of control possibility. The participants engaged in training sessions to become familiar with the Brain-Computer Interface (BCI) system and acquire the skill of controlling a robotic arm via the use of motor imagery technique. In this context, ten frequency domain features were extracted to classify the left and right-hand movements for offline robotic arm control. After frequency domain features extraction, four different classification algorithms namely efficient logistic regression (ELR), ensemble boosted tree (EBT), and wide neural network (WNN) were critically compared in terms of classification accuracy. The results demonstrated that the ELR classifier yielded the highest accuracy, followed by the WNN and EBT classifier. Subsequently, these intentions were translated into control instructions that were then executed by the robotic arm.

Keywords: BCI · EEG signals · human-computer interaction · classification

1 Introduction

Brain-computer interfaces, or BCIs, are emerging technology that allows individuals to directly control computers or other electronic devices using their thoughts. BCIs operate by measuring the brain's electrical activity, typically through EEG sensors placed on the scalp [1]. This electrical activity is then analyzed by sophisticated algorithms and translated into commands that can be used to control external devices [2]. The potential applications for BCIs are vast, ranging from medical applications for individuals with disabilities or neurological conditions to gaming and entertainment [3]. For example, BCIs can be used to operate prosthetic limbs, enabling people who have had amputations to carry out difficult jobs with more comfort and flexibility. They can also be used to assist individuals with paralysis or other motor impairments in communicating with others or operating their environment [4].

While BCIs offer exciting possibilities, many technical and ethical challenges still need to be addressed. For instance, current BCI systems have limited accuracy and

D. Kriksciuniene et al. (Eds.): IBICA 2023, LNNS 1228, pp. 201–211, 2025.
https://doi.org/10.1007/978-3-031-78937-3_20

reliability, and there are too many concerns about the security and privacy of the data collected by these devices [5]. Additionally, there are ethical concerns about the potential misuse of BCIs, particularly in the realm of mind reading and manipulation. Despite these challenges, the field of BCI research is advancing rapidly, with breakthroughs and innovations emerging regularly [6]. This technology can fundamentally alter how we interact with the world as it develops and improves the lives of countless individuals [7]. Numerous methods, such as supervised learning, unsupervised learning, and reinforcement learning, can be used to train these algorithms [8]. While in unsupervised learning, the algorithm learns to recognize patterns in the data without being given specific examples, supervised learning involves training the algorithm using a dataset of inputs and outputs that are known [9]. In reinforcement learning, the algorithm learns to associate specific actions with rewards or punishments to maximize a certain outcome.

2 Materials and Methods

The materials and methods used in BCI research can vary depending on the particular approach being used, but some common elements and techniques are often employed [3]. EEG, which includes attaching electrodes to the scalp to detect electrical signals produced by the brain, is a typical technique for gauging brain activity in BCIs. Magnetoencephalography (MEG) and functional magnetic resonance imaging (fMRI), two more techniques for assessing brain activity, are also possible, although they are less frequently utilized due to their high cost and technical complexity [8, 10]. Once brain activity has been measured, Machine learning techniques are often used to analyze it to find patterns that can be utilized to control external devices [11].

In addition to the hardware and software used to measure and analyze brain activity, BCIs may also incorporate a variety of external devices, such as prosthetic limbs or virtual reality systems, that can be controlled using the brain signals detected by the system [12, 13]. As this field continues to evolve, new techniques and technologies will likely emerge that will enable even more sophisticated and precise control of external devices using the power of the human brain [14]. Figure 1 shows the EEG data acquisition system for the motor imagery-based BCI approach.

2.1 EEG Data Acquisition and Pre-Processing

EEG data acquisition and pre-processing are critical steps in BCI research, as they ensure that the recorded data is of high quality and suitable for subsequent analysis [15]. The following are some of the common steps involved in EEG data acquisition and pre-processing [10]:

Electrode Placement: EEG electrodes are placed on the scalp according to a standardized system, such as the International 10–20 system, to ensure consistent and accurate recordings.

Signal Amplification: EEG signals are typically very weak and need to be amplified before they can be recorded. This is usually done using an amplifier that is connected to the electrodes.

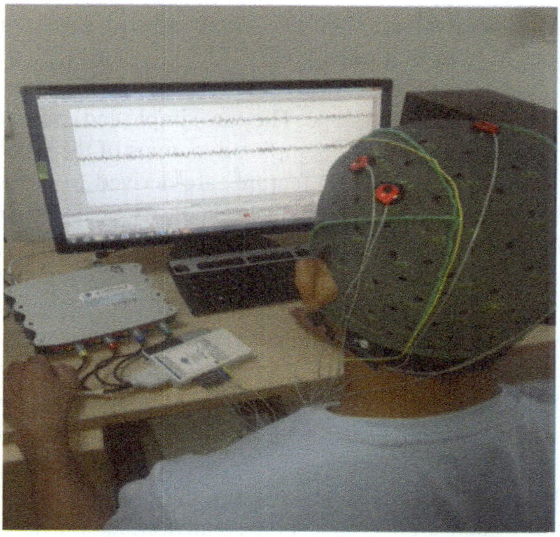

Fig. 1. Experimental setup for data acquisition and classification system

Sampling Rate: EEG signals are typically recorded at a high sampling rate, such as 250–1000 Hz, to capture the fast dynamics of neural activity.

Filtering: Power line interference, muscular activity, ambient noise, and other types of noise frequently taint EEG readings. High-pass, low-pass, and notch filters are used as filtering techniques to remove unwanted noise from the signal.

Artifact Removal: Artefacts like eye blinks and movement artifacts can taint EEG readings as well. These can be eliminated using a variety of methods, including regression-based methods or independent component analysis (ICA).

Epoching: EEG signals are typically divided into epochs, or time windows, of a fixed duration (e.g., 1–2 s) for subsequent analysis. Epochs that contain artifacts or other sources of noise can be removed at this stage.

Reference Selection: EEG signals are typically recorded concerning a reference electrode, which is used as a baseline for comparison. The choice of reference electrode can have a significant impact on the recorded signals, and different reference options are available depending on the research question and experimental design [16].

Overall, EEG data acquisition and pre-processing are critical steps in BCI research, as they ensure that the recorded signals are clean, consistent, and reliable for subsequent analysis. Proper pre-processing techniques can help improve the accuracy and reliability of BCI systems, leading to better performance and more effective use of the technology [17].

2.2 Eeg Feature Extraction

EEG feature extraction is a crucial step in BCI research that involves transforming the raw EEG signal into a set of features that can be used to train a classifier or regression model to control an external device [15]. The following are some common EEG feature extraction techniques used in BCI research [7]:

Time-Domain Features: These features are derived from the time-domain EEG signal, such as amplitude, variance, and slope. They are simple and computationally efficient to compute, but may not capture more complex dynamics of the EEG signal [18].

Frequency-Domain Features: These features are derived from the spectral decomposition of the EEG signal, such as power spectral density (PSD), band power, and peak frequency. They provide information about the distribution of power across different frequency bands, which can be useful for identifying frequency-specific patterns of brain activity.

Time-Frequency Features: These features combine information from both time and frequency domains, such as wavelet coefficients, spectrograms, and event-related spectral perturbation (ERSP). They provide information about the temporal evolution of frequency-specific activity, which can be useful for identifying transient patterns of brain activity [19].

Connectivity Features: These features capture the functional connectivity between different brain regions, such as coherence, phase synchronization, and Granger causality. They provide information about how different brain regions interact with each other, which can be useful for identifying network-level patterns of brain activity.

Higher-Order Statistics: These features capture higher-order statistical properties of the EEG signal, such as skewness, kurtosis, and entropy. They provide information about the non-Gaussian and nonlinear dynamics of the EEG signal, which can be useful for identifying complex patterns of brain activity.

Overall, EEG feature extraction is an important step in BCI research that involves selecting the most informative features from the raw EEG signal to control an external device. By selecting the most relevant and informative features, BCI systems can achieve higher accuracy and reliability, which can ultimately improve the usability and effectiveness of the technology [18].

3 Classifiers

In BCI research, classifiers are used to translate the extracted features from the EEG signal into control signals that can be used to operate an external device [15].

The following are some of the common classifiers used in BCI research:

Linear Classifiers: These classifiers are based on linear models, such as linear discriminant analysis (LDA) and support vector machines (SVM) [16]. They work well for simple BCI paradigms with a small number of classes.

Nonlinear Classifiers: These classifiers are based on nonlinear models, such as neural networks, decision trees, and random forests. They are more flexible than linear classifiers and can handle more complex BCI paradigms with a larger number of classes [19].

Ensemble Classifiers: These classifiers combine multiple base classifiers into an ensemble to improve the accuracy and robustness of the BCI system. Common ensemble techniques include bagging, boosting, and stacking [20].

Deep Learning Classifiers: These classifiers are based on deep neural networks (DNNs), which are capable of learning complex nonlinear relationships between the EEG features and the control signals [21]. DNNs have been shown to achieve state-of-the-art performance in many BCI paradigms, but require large amounts of training data and computational resources [22].

Transfer Learning Classifiers: These classifiers leverage knowledge learned from one BCI paradigm to improve performance in another, related BCI paradigm. This can help overcome the challenge of limited training data in certain BCI paradigms [23].

Overall, the choice of the classifier depends on the specific BCI paradigm and the available resources. While linear classifiers are simple and efficient, they may not be suitable for more complex BCI paradigms [7] Nonlinear and deep learning classifiers can achieve higher accuracy but require larger amounts of training data and computational resources [22]. Ensemble and transfer learning classifiers can improve the performance and generalization of the BCI system but may require additional training and testing data [24] The different frequency domain features used here are Variance of Central Frequency (VCF), 5TH order Spectral Moment (SM_K), Standard Deviation Frequency (STDF), Mean Frequency (MNF), Root Mean Square Frequency (RMSF), Median Frequency (MDF), Peak Frequency (PKF), Mean Power (MNP), Total Power (P_T) and Central Frequency (F_C).

4 Result and Discussion

The present work demonstrated a study conducted on three different classifiers namely ELR, WNN, and EBT. After requiring EEG data, pre-processing and features extraction followed by the classification of two movements using MATLAB2022. Different folds cross-validation methods are applied to average out the accuracy. Feature extraction and Classification both are done in MATLAB using three different classification approaches. Finally, all ten features are merged to form a single feature vector for all classification methods. It was found that ELR classification has the best accuracy with a percentage of 95.2% followed by the WNN with an accuracy percentage of 94.8% while combining all features.

Table 1 shows the average percentage classification accuracy of ten frequency domain features using the ELR classifier. The performance of ten frequency domain features namely SM_k, P_T, MNP, MDF, MNF, F_C, RMSF, STFD, VOCF, and PKF compared with each other's in which PKF is found the best feature with the average classification accuracies of 61.3% followed by P_T, MNP, MDF, MNF, F_C, RMSF, STFD, VOCF, SM_k with 58.8%, 58.8%, 58.8%, 58.68%, 33.9%, 30.3%, 30.93%, 30.4%, 56.7 respectively.

Table 2 shows the average percentage classification accuracy of ten frequency domain features using an EBT classifier. It is clear from Table 2 that SM_K is found the best feature with the average classification accuracies of 56.62% followed by PT, MNP, MDF, MNF, FC, RMSF, STFD, VOCF, and PKF with 54.84%, 54.84%, 54.85%, 54.85%, 51.66%, 28.76%, 34.06%, 33.96%, 51.9% using EBT classification method respectively.

Table 3 shows the average percentage classification accuracy of ten frequency domain features using a WNN classifier. Table 3 shows that MNF is found the best feature with an average classification accuracy of 63.4% followed by PT, MNP, MDF, SKM, FC,

Table 1. Accuracy Rates of Frequency Domain Features using ELR classifier

Sr No	Features	2FCV	3FCV	4FCV	5FCV	6FCV	7FCV	8FCV	9FCV	10FCV	Mean
1	MNF	57.9	58.5	58.5	58.8	59.1	59.1	59.1	58.8	59.1	58.68
2	FC	33	34.2	33.6	33.9	34.2	34.2	33.9	34.2	33.9	33.9
3	RMSF	30.3	30.3	30.9	30.6	30	30.3	30.3	30	30	30.3
4	STDF	31.2	30.9	30.9	30.9	31.2	30.6	30.9	30.6	31.2	30.93
5	PKF	60.3	61.2	61.2	61.2	61.2	61.5	61.8	61.5	61.8	61.3
6	MNP	57.9	58.5	58.8	58.8	59.1	59.1	59.1	58.8	59.1	58.8
7	PT	57.9	58.5	58.8	58.8	59.1	59.1	59.1	58.8	59.1	58.8
8	MDF	57.9	58.5	58.8	58.8	59.1	59.1	59.1	58.8	59.1	58.8
9	SMK	56.4	56.7	56.7	56.7	56.7	56.7	56.7	56.7	57	56.7
10	VOCF	30.3	30.9	30.3	30	30.3	30.6	30.3	30.9	30	30.4

Table 2. Accuracy Rates of Frequency Domain Features using EBT classifier

Sr No	Features	2FCV	3FCV	4FCV	5FCV	6FCV	7FCV	8FCV	9FCV	10FCV	Mean
1	MNF	40.6	46.7	50.9	59.7	58.8	59.7	60	57.9	59.4	54.85
2	F_C	33.6	32.1	32.1	34.2	28.5	31.8	32.1	30	30.6	31.66
3	RMSF	31.5	27	26.4	28.5	29.7	27.9	29.7	27.9	26.7	28.76
4	STDF	32.1	34.2	36.7	34.8	32.1	31.2	31.8	36.1	36.7	34.06
5	PKF	19.7	34.5	43	62.7	61.8	62.7	58.2	62.4	62.1	51.9
6	MNP	40.6	46.7	50.9	59.7	58.8	59.7	60	57.9	59.4	54.85
7	P_T	40.6	46.7	50.9	59.7	58.8	59.7	60	57.9	59.4	54.85
8	MDF	40.6	46.7	50.9	59.7	58.8	59.7	60	57.9	59.4	54.85
9	SM_K	19.7	61.2	60.6	60	61.8	61.8	62.1	61.5	60.9	56.62
10	VOCF	32.1	34.2	36.7	34.8	32.1	31.2	31.8	36.1	36.7	33.96

RMSF, STFD, VOCF, and PKF with 63.1%, 63.1%, 63.1%, 59.53%, 53.06%, 28.63%, 31.4%, 31.73%, 62.2% respectively. Figure 2 shows the confusion matrix in which right and left-hand movement classes are considered as class 1 and class 2 for 95.2% accuracy achieved for the combined features vector. The results indicate that the 95.2% accuracy achieved by the ELR classifier is the best as compared to the other classifiers.

Figure 2 shows the classification accuracy (95.15%) EBT classifier when all features are merged to form a combined dataset and Fig. 3 shows the WNN classification accuracy of 94.25%. Figure 4 shows the offline control of the robotic arm prototype using the motor imagery BCI technique (Fig. 5).

Table 3. Accuracy Rates of Frequency Domain Features using WNN classifier

Sr No	Features	2FCV	3FCV	4FCV	5FCV	6FCV	7FCV	8FCV	9FCV	10FCV	Mean
1	MNF	61.8	63.3	63.6	63.3	63.3	64.2	64.2	63.3	63.6	63.4
2	F_C	31.8	33.9	29.4	34.8	33.3	34.2	33.9	33.3	33	33.06
3	RMSF	31.8	26.1	27	28.5	29.1	29.4	29.1	29.4	27.3	28.63
4	STDF	33.6	30.6	29.1	33	30	32.4	29.4	31.5	33	31.4
5	PKF	62.1	63.3	62.4	62.1	62.1	61.5	61.8	62.7	61.8	62.2
6	MNP	62.4	62.4	63	63.3	62.7	63.6	63.9	63.3	63.9	63.1
7	P_T	62.4	62.4	63	63.3	62.7	63.6	63.9	63.3	63.9	63.1
8	MDF	62.4	62.4	63	63.3	62.7	63.6	63.9	63.3	63.9	63.1
9	SM_K	59.7	60	59.1	59.4	59.4	60.3	59.4	59.1	59.4	59.53
10	VOCF	33.3	30.6	32.4	30.9	29.4	32.7	29.7	33	33.6	31.73

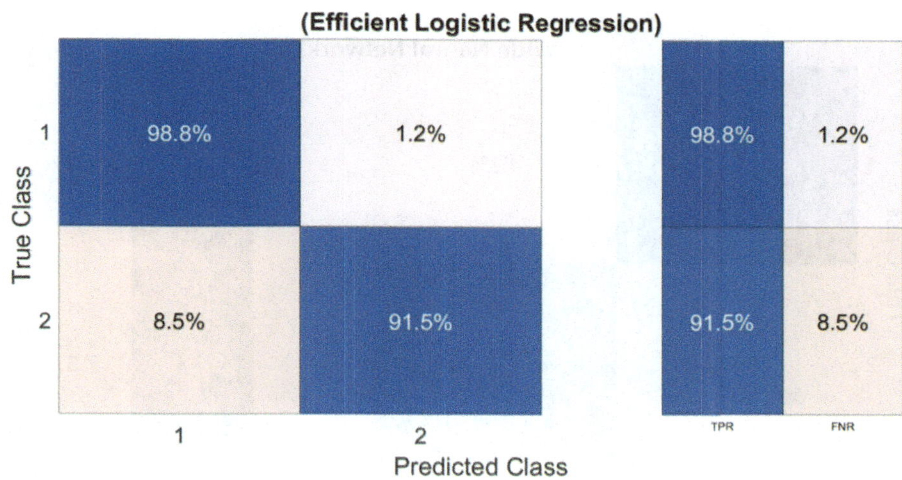

Fig. 2. Confusion matrix for ELR classifier with frequency domain features

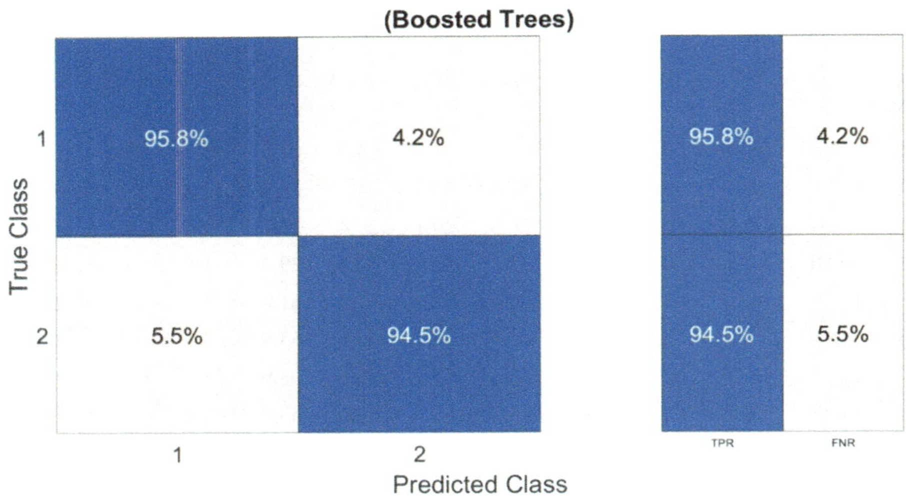

Fig. 3. Confusion matrix for EBT classifier with frequency domain features

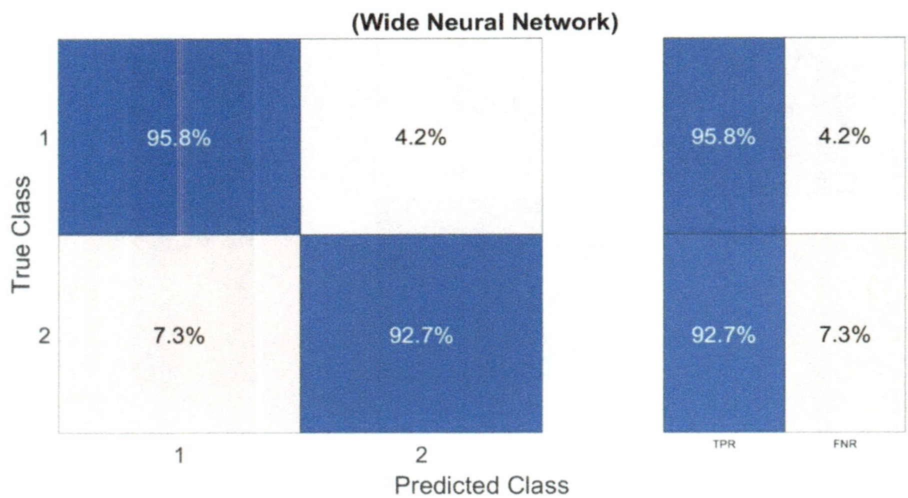

Fig. 4. Confusion matrix for WNN classifier with frequency domain features

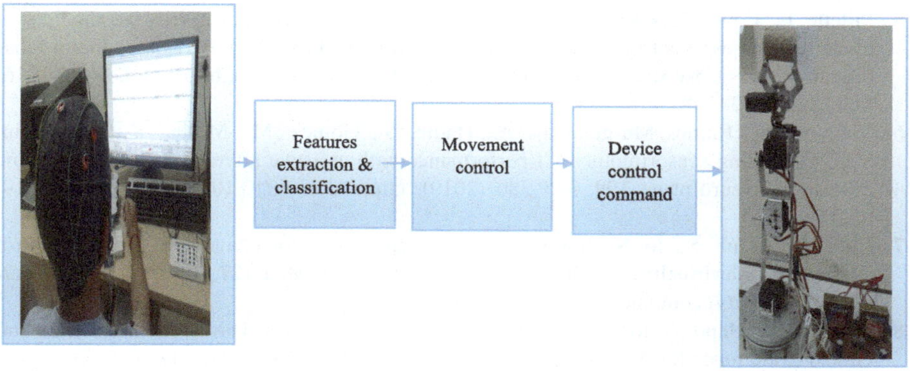

Fig. 5. Controlling of Assistive Devices Using Motor Imagery EEG Signals

5 Conclusion

This study primarily examined the use of EEG data in the context of interacting with a robotic arm. In the present study, a set of ten features, namely MNF, Fc, RMSF, STDF, MDF, PKF, MNP, P_T, SM_K, and VCF, were extracted from the EEG data obtained from a sample of ten individuals who were in good health. These features were then assessed using three distinct classification methods: Efficient logic regression, Ensemble boosted trees, and wide neural network classifiers. The results indicated that ELR yielded the highest performance, followed by the WNN method. The findings validate that the use of the ELR technique, in conjunction with frequency domain features, resulted in a notable improvement in accuracy, reaching a maximum level of 95.2%. Furthermore, there are many other domain feature unexplored that is not discussed here. Additionally, frequency domain features are not compared to time and time-frequency domain with various other classification algorithms. The findings of this study provide support to other researchers to explore more domains and more classification approaches for the betterment of BCI techniques.

References

1. Liu, Y., Liu, Y., Tang, J., Yin, E., Hu, D., Zhou, Z.: A self-paced BCI prototype system based on the incorporation of an intelligent environment-understanding approach for rehabilitation hospital environmental control. Comput. Biol. Med. **118**, 103618 (2020). https://doi.org/10.1016/j.compbiomed.2020.103618
2. Gong, M., Xu, G., Li, M., Lin, F.: An idle state-detecting method based on transient visual evoked potentials for an asynchronous ERP-based BCI. J. Neurosci. Methods **337**, 108670 (2020). https://doi.org/10.1016/j.jneumeth.2020.108670
3. Shi, N., et al.: Steady-state visual evoked potential (SSVEP)-based brain–computer interface (BCI) of Chinese speller for a patient with amyotrophic lateral sclerosis: a case report. J. Neurorestoratology **8**(1), 40–52 (2020). https://doi.org/10.26599/jnr.2020.9040003
4. Lim, H., Ku, J.: Multiple-command single-frequency SSVEP-based BCI system using flickering action video. J. Neurosci. Methods **314**, 21–27 (2019). https://doi.org/10.1016/j.jneumeth.2019.01.005

5. Pillette, L., Lotte, F., N'Kaoua, B., Joseph, P.A., Jeunet, C., Glize, B.: Why we should systematically assess, control and report somatosensory impairments in BCI-based motor rehabilitation after stroke studies. NeuroImage Clin. **28**, 102417 (2020). https://doi.org/10.1016/j.nicl.2020.102417
6. Vidaurre, C., Ramos Murguialday, A., Haufe, S., Gómez, M., Müller, K.-R., Nikulin, V.V.: Enhancing sensorimotor BCI performance with assistive afferent activity: An online evaluation. NeuroImage **199**, 375–386 (2019). https://doi.org/10.1016/j.neuroimage.2019.05.074
7. Choi, J.W., Huh, S., Jo, S.: Improving performance in motor imagery BCI-based control applications via virtually embodied feedback. Comput. Biol. Med. **127**, 104079 (2020). https://doi.org/10.1016/j.compbiomed.2020.104079
8. Narayan, Y.: Hand motion identification based on EEG signals classification. In: 2021 2nd Global Conference for Advancement in Technology, GCAT 2021, Institute of Electrical and Electronics Engineers Inc., Oct. 2021. https://doi.org/10.1109/GCAT52182.2021.9587556
9. Ahlawat, V., Narayan, Y., Kumar, D.: DWT-based hand movement identification of emg signals using SVM. In: Lecture Notes in Networks and Systems, Springer Science and Business Media Deutschland GmbH, pp. 495–505 (2021). https://doi.org/10.1007/978-981-33-6546-9_47
10. Landau, O., Cohen, A., Gordon, S., Nissim, N.: Mind your privacy: Privacy leakage through BCI applications using machine learning methods. Knowl. Based Syst. **198**, 105932 (2020). https://doi.org/10.1016/j.knosys.2020.105932
11. Jeunet, C., Lotte, F., Batail, J.M., Philip, P., MicoulaudFranchi, J.A.: Using recent BCI literature to deepen our understanding of clinical neurofeedback: a short review. Neuroscience **378**, 225–233 (2018). https://doi.org/10.1016/j.neuroscience.2018.03.013
12. Balamurugan, M., Ragavan, C., Balasubramanian, G.: Anti-intuitionistic Fuzzy Soft Ideals in BCK/BCI-algebras. Mater. Today Proc. **16**, 532–539 (2019). https://doi.org/10.1016/j.matpr.2019.05.125
13. Li, Z., Liu, K., Deng, X., Wang, G.: Spatial fusion of maximum signal fraction analysis for frequency recognition in SSVEP-based BCI. Biomed. Signal Process Control **61**, 102042 (2020). https://doi.org/10.1016/j.bspc.2020.102042
14. Petrova, A., Voznenko, T., Dyumin, A., Chepin, E., Cherepanova, A.: The influence of using different mental images as BCI commands on the quality of control. Procedia Comput. Sci. **169**, 235–239 (2020). https://doi.org/10.1016/j.procs.2020.02.141
15. Narayan, D., Kumar, D., kumar, D.: Comparative analysis of sEMG signal classification using different K-NN algorithms. Int. J. Adv. Sci. Technol. **29**(10S), 2257–2266 (2020)
16. Miladinovic, A., et al.: Performance of EEG motor-imagery based spatial filtering methods: a BCI study on Stroke patients. Procedia Computer Science, Elsevier B.V. (ed.). pp. 2840–2848 (2020). https://doi.org/10.1016/j.procs.2020.09.270
17. Sadeghi, S., Maleki, A.: Character encoding based on occurrence probability enhances the performance of SSVEP-based BCI spellers. Biomed. Signal Process. Control **58**, 101888 (2020). https://doi.org/10.1016/j.bspc.2020.101888
18. Khan, M.A., Das, R., Iversen, H.K., Puthusserypady, S.: Review on motor imagery based BCI systems for upper limb post-stroke neurorehabilitation: From designing to application. Computers in Biology and Medicine, vol. 123. Elsevier Ltd, Aug. 01, 2020. https://doi.org/10.1016/j.compbiomed.2020.103843
19. Narayan, Y., Singh, R.M., Mathew, L., Chatterji, S.: Surface EMG signal classification using ensemble algorithm, PCA and DWT for robot control. In: Communications in Computer and Information Science, pp. 218–230. Springer Verlag, Singapore (2019). https://doi.org/10.1007/978-981-13-3140-4_20

20. Muhiuddin, G., MohseniTakallo, M., Borzooei, R.A., Jun, Y.B.: M-polar fuzzy q-ideals in BCI-algebras. J. King Saud Univ. Sci. **32**(6), 2803–2809 (2020). https://doi.org/10.1016/j.jksus.2020.07.001
21. Emanovský, P., Kühr, J.: Some properties of pseudo-BCK- and pseudo-BCI-algebras. Fuzzy Sets Syst. **339**, 1–16 (2018). https://doi.org/10.1016/j.fss.2016.12.014
22. Narayan, Y., Mathew, L., Chatterji, S.: SEMG signal classification with novel feature extraction using different machine learning approaches. J. Intell. Fuzzy Syst. **35**(5), 5099–5109 (2018). https://doi.org/10.3233/JIFS-169794
23. KhaliliArdali, M., Rana, A., Purmohammad, M., Birbaumer, N., Chaudhary, U.: Semantic and BCI-performance in completely paralyzed patients: possibility of language attrition in completely locked in syndrome. Brain Lang. **194**, 93–97 (2019). https://doi.org/10.1016/j.bandl.2019.05.004
24. Zhuralvev, M., Runnova, A., Kiselev, A.: Characteristics of post-stroke patients brain activity with real and imagined movements in the BCI - rehabilitation process. Procedia Comput. Sci. **169**, 677–685 (2020). https://doi.org/10.1016/j.procs.2020.02.184

Cybersecurity Framework for Prediction and Mitigation of Attacks in the Healthcare Systems

C. Syamsundar Reddy(✉) and G. Anjan Babu

SVU College of Commerce, Management and Computer Science, Sri Venkateswara University, Tirupati, India
cssreddi@gmail.com

Abstract. Although the use of digital healthcare technology can enhance clinical outcomes and the standard of care, concerns over security and privacy still exist. Due to increased connectivity, healthcare systems are susceptible to cybersecurity threats. Cybercriminals steal health information and target medical devices, which results in breaches and ransomware attacks. New laws and regulations are required, along with adjustments to technology, methods, and human behaviour, to improve healthcare security. For intelligent healthcare, a study suggests a Bagging ensemble classifier that outperforms Random Forest in every test.

Keywords: cyber-crime · cyber-attack · medical and healthcare systems · artificial intelligence

1 Introduction

Intelligent systems are being developed as a result of the current technological revolution. The world is filled with intelligent systems of all shapes and sizes, used in industries like manufacturing, transportation, medicine, education, and automation. In functional cooperation with smart devices, intelligent systems transform lifestyle into smart healthcare, which has its roots in medical and healthcare systems [9, 10]. In the last ten years, 4G has been used to deploy smart healthcare applications that support legacy systems and backward compatibility [1, 2] with older medical and healthcare equipment. Due to technological limitations, the objective is not satisfactorily accomplished in order to meet the needs and demands of smart health.

By utilising technological advancements like electronic health records (EHRs), wearable technology, and remote services, the healthcare system is changing the way people live. Smart phones are being used more frequently by patients for telehealth and disease management. The system's connectivity provides advantages like increased effectiveness, decreased error rates, and remote patient monitoring. But connectivity also brings with it new risk exposures, including cybersecurity risks and the potential for undetected data breaches [2, 3]. Since health records may be shared among several providers, the interconnected nature of modern records also creates opportunities for attacks.

D. Kriksciuniene et al. (Eds.): IBICA 2023, LNNS 1228, pp. 212–219, 2025.
https://doi.org/10.1007/978-3-031-78937-3_21

Healthcare is experiencing an increase in cyberattacks and medical data theft, with the number of breaches and breaches being reported by the media. Hacking involves illegal access to computing devices [11, 12], while malware, including ransomware and viruses, can cause disruptions. Healthcare must adapt to evolving cybersecurity patterns [5] and risks, investing time and money in maintaining safety and confidentiality of patient data.

Utilising fifth-generation technologies, a greater number of clinical consultations and cases are being handled, including telesurgeries, teleconsultations, telediagnosis, sanitary aids for preventive care [13], and other related medical and contingent considerations. With the fifth generation technologies amassing extensive scientific literatures, healthcare is widely supported. In this work, we present a systematic review that developers and researchers are most interested in because it will motivate them to create complex frameworks for healthcare systems using fifth-generation technologies.

2 Attacks

There have been several attempts to classify cyberattacks. Each of these investigations used a unique set of criteria to categorise cybersecurity attacks according to their purpose, gravity, scope, and legal ramifications. The following are the most prevalent cyberattacks that target the healthcare system that are covered in this article:

2.1 Incidental Data Breach

Both internal and external entities within the organisation have the potential to create data breaches; internal attacks harm the facilities and are significantly more damaging because the attacker has direct access to IT resources and assets. When it comes to dangers to your data, external threats are more frequent but internal threats are more dangerous. Internal threats know the network, know what data is available, and may know ways around your safeguards to avoid detection longer. External threats originate outside the network, including public forums and world events. External cybersecurity involves orchestrating humans and machine intelligence to discover and disrupt threats beyond the corporate perimeter, including data, people, domains, apps, and the brand, through malicious domains, social media, email, and attack chatter. Internal threats, originating within a corporate enterprise, include ransomware, malware, C2 commands, viruses, exploited vulnerabilities, cross-site scripting, and signature-based threats. They can be part of broader cybercriminal schemes and require AI-driven solutions. These threats are detected and fought by enterprise security tools within the company's digital ecosystem.

2.2 Cryptographic Attacks

An attempt is made to decipher encrypted data when a cryptographic invasion is carried out with the goal of revealing concealed information. The process of encoding data into codes that only the sender and intended recipient can decode is known as cryptography. The binary coding is obscured to others since the algorithms used to encrypt the information are only known to its developer. A cryptographic assault is a strategy for getting beyond a system's security by identifying a flaw in a code, cypher, cryptographic protocol, or key management system.

There are various types of cryptographic attacks, including:

Brute Force Attack: In a brute force attack, an attacker tries all possible combinations of keys until they find the correct one to decrypt the data. This is a time-consuming method but can be effective if the key space is not sufficiently large.
Cryptanalysis: Cryptanalysis involves the analysis of the encryption algorithms and ciphertext to discover patterns or weaknesses that can be exploited to recover the plaintext without trying all possible keys.
Man-in-the-Middle (MitM) Attack: In a MitM attack, an attacker intercepts the communication between two parties and can potentially modify or eavesdrop on the data being exchanged. This can compromise the confidentiality and integrity of the data.
Side-Channel Attack: This attack targets vulnerabilities in the physical implementation of a cryptographic system, such as power consumption, electromagnetic emissions, or timing information, to deduce information about the secret key.
Differential Cryptanalysis: This is a specialized form of cryptanalysis that focuses on studying differences between pairs of plaintext-ciphertext pairs to discover information about the encryption key.
Chosen-Plaintext Attack: In this attack, the attacker can choose plaintexts and observe the corresponding ciphertexts to gain insights into the encryption process and potentially discover the key.
Known-Plaintext Attack: In a known-plaintext attack, the attacker has access to both plaintext and ciphertext pairs, which can be used to deduce the key.
Birthday Attack: This attack exploits the birthday paradox to find collisions in hash functions more efficiently than exhaustive search.
Implementation Vulnerabilities: Cryptographic systems can also be compromised due to implementation flaws, software bugs, or configuration errors.

It's crucial to remember that cryptographic assaults can compromise a system's overall security in addition to decrypting data. Using robust encryption algorithms, adhering to recommended practises for key management, and routinely updating and patching systems to fix known vulnerabilities are all essential for defending against cryptographic attacks. In order to respond to new threats and vulnerabilities, security policies and practises should also be continually examined and updated.

2.3 Ransomware

A form of malicious software called ransomware encrypts a victim's files or data and then demands payment in exchange for the decryption key required to decrypt them. Cybercriminals that want to extort money from people, organisations, or enterprises frequently launch ransomware assaults. Ransomware attacks represent serious effects on operations and finances, thus prevention and planning are essential for reducing the risks.

2.4 Denial of Service (DoS)

A Denial of Service (DoS) attack is a deliberate effort to obstruct or reduce a computer system, network, website, or online service's operation. A DoS attack usually involves the attacker flooding the target with an overwhelming amount of traffic, requests, or data, rendering the system unresponsive or useless for the people who are supposed to be utilising it.

2.5 SQL Injection Exploit

An attacker can modify a web application's SQL query via SQL Injection, a sort of security exploit, to get unauthorised access to a database or carry out harmful deeds. When an application neglects to properly validate or sanitise user inputs and inserts those inputs directly in SQL statements, SQL Injection attacks take place. The effects of SQL Injection are:

Data Theft: An attacker has access to the database and can extract, alter, or remove sensitive data. Usernames, passwords, personal data, and financial records can all be stolen, which is known as data exfiltration.

Unauthorised Access: A hacker has the ability to get around authentication processes and obtain access to a web application or its administrative features.

To avoid this kind of vulnerability, developers must be aware of the dangers of SQL Injection and use secure coding techniques. To find and fix SQL Injection vulnerabilities in online applications, regular security testing and in-depth code inspections are essential.

2.6 Escalation of Privilege

A security flaw known as privilege escalation happens when an attacker acquires unauthorised access to a system or application and raises their privileges above those that were initially provided. As a result, the attacker may be able to carry out operations or gain access to resources they shouldn't have been allowed to utilise. Privilege escalation can be caused by vulnerabilities, misconfigurations, weak passwords, social engineering, inadequate access controls, injection attacks, and malicious software. These factors exploit security vulnerabilities, misconfigured settings, weak passwords, social engineering, inadequate access controls, code injection attacks, and malicious software to gain backdoor access to the system.

3 Methods

To protect patient information, guarantee patient safety, preserve the integrity of healthcare systems, and make services available, it is essential to mitigate cyberattacks on medical and healthcare systems.

To mitigate cyberattacks in medical and healthcare systems, regular risk assessments, security training, access control, patch management, network security, endpoint security, an incident response plan, regular backups, third-party vendor security, regulatory compliance, cybersecurity awareness training, zero trust architecture, medical device security, data encryption, regular security audits and penetration testing, and collaboration with cybersecurity experts are essential. These methods help identify vulnerabilities, prioritize and address identified threats, maintain strong access controls, and implement strong encryption for data in transit and at rest. Endpoint security solutions, such as antivirus software and endpoint detection and response, protect devices against malware. Regular backups ensure data recovery in case of ransomware or data breaches. Regular audits and penetration testing help identify vulnerabilities and assess the security of healthcare systems.

3.1 Security with AES

The Rijndael AES algorithm, a modified version of AES, is used to secure electronic healthcare records (EHR) from intruders. It uses a symmetric cryptosystem with varying block and key sizes, overcoming AES limitations. The algorithm uses array transformations, with different rounds based on key and block size, and incorporates linear mix transform, nonlinear, and key added transforms for increased privacy [4, 7] and reduced computational burden.

3.2 Training and Awareness Raising

To avoid data breaches, safeguard patient privacy, and guarantee the reliability of healthcare systems, training and awareness raising for cybersecurity are essential in the medical and healthcare sector.

To establish effective cybersecurity training and awareness programs in healthcare organizations, tailor training to the specific roles and responsibilities of staff, conduct regular training sessions, implement simulated phishing exercises, use interactive methods, teach secure communication practices, and provide training on mobile device security. Educate employees on data handling and classification, incident reporting, physical security awareness, vendor security training, regulatory compliance training, and data breach response. Ensure leadership is well-informed about cybersecurity best practices and actively supports a culture of awareness. Provide easy access to written policies and procedures related to cybersecurity and data protection. Run ongoing awareness campaigns, encourage staff feedback, and recognize and reward employees for their active participation in training and efforts to maintain cybersecurity.

3.3 Ensemble Methods for Security and IDS

A programming paradigm called machine learning is designed to discover a performance criterion using training data or prior knowledge. For the optimisation issue to be solved with massive data samples, an efficient training procedure is needed. The predicted accuracy of the learning algorithm determines how effective it is, despite its spatial and temporal complexity.

A form of ensemble modelling known as bagging uses extremely straightforward techniques to build the foundation model. Base models for training sets are created using bootstrap samples, and unweighted voting is then used to combine them. If the algorithm is unstable, it is better than using just one model created with the algorithm. Prediction quality for stable algorithms does not increase or even slightly decline. There are no prerequisites for the modelling algorithm other than the instability. In comparison to what would often be employed in the event of single model construction, it increases the stability of the model and allows for its simplification. Bagging is based on three fundamental steps: Bootstrapping, parallel training, and aggregation are three steps.

In the context of cryptography, the term "ensemble methods" refers to the combining of several cryptographic techniques or algorithms to increase security and provide more protection for medical and healthcare data. In order to make it more difficult for attackers to violate the confidentiality, integrity, and accessibility of sensitive patient information, ensemble cryptography was developed.

Ensemble cryptographic methods are widely used in medical and healthcare systems to ensure data privacy and security. These methods include Multi-Factor Authentication (MFA), which combines multiple authentication factors, such as passwords, smart cards, and biometric data, to strengthen access control. Hybrid encryption combines symmetric and asymmetric encryption, balancing efficiency with security. Threshold cryptography divides cryptographic keys into multiple parts, making it harder for attackers to access sensitive data. Homomorphic encryption and Secure Multi-Party Computation (SMPC) enable secure data analysis and collaboration without exposing raw data. Blockchain-based cryptography uses decentralized technology to secure data on the blockchain, while Secure Multiparty Computation (SMC) enables collaborative analysis while preserving patient privacy. Attribute-Based Access Control (ABAC) and Quantum Key Distribution (QKD) are emerging technologies that enhance security against quantum-based attacks.

4 Results

A key performance metric in classification tasks is accuracy, which measures how well a model can predict instances in a dataset by counting the percentage of instances that are correctly classified. In classification tasks, precision is a critical metric that evaluates a model's capacity to prevent false positives by calculating the ratio of true positives to false positives. Another key performance indicator in classification tasks is recall, also referred to as sensitivity or true positive rate. It measures how well a model can avoid producing false negatives by calculating the percentage of true positive instances. Time is the length of time it takes a programme or algorithm to complete a particular task or operation. In our study, we used seconds as the unit of time. The following statistics show the performance of the AES and Ensemble methods on the general healthcare data and COVID-19 [6, 8] specific data summarized from WHO.

Table 1 presents the results of implementing machine learning methods for intrusion detection in healthcare, showing performance of AES achieved the highest classification accuracy at 88.78% and by using Ensemble methods with a highest classification accuracy at 89.57%. Table 2 presents the results of implementing machine learning methods for intrusion detection specific to COVID-19 patients data in healthcare, showing performance of AES achieved the highest classification accuracy at 92.34% and by using Ensemble methods with a highest classification accuracy at 94.46%.

Table 1. Performance Measures on general healthcare data

	AES	Ensemble
Accuracy	88.78	89.57
Reliability	90.23	91.22
Specificity	95.72	98.34
Sensitivity	95.72	98.34

Table 2. Performance Measures on COVID-19 patients data

	AES	Ensemble
Accuracy	92.34	94.46
Reliability	93.79	96.11
Specificity	98.28	99.31
Sensitivity	98.28	99.31

5 Conclusion

The healthcare industry faces numerous challenges, including economic, payer policy, and regulatory issues. Cyber-threat response significantly impacts long-term results. This study aims to highlight cybersecurity issues and strategies for managing cyber-attacks in the healthcare sector. An effective anomaly-based intrusion detection system is proposed using the Bagging ensemble classifier, analyzing 20% of the NSL-KDD dataset. As technology advances, more surveillance is needed to safeguard users' private data.

References

1. Vergütz, A., Prates, N.G., Henrique Schwengber, B., Santos, A., Nogueira, M.: An architecture for the performance management of smart healthcare applications. Sensors **20**, 5566 (2020)
2. Jiwani, N., Gupta, K., Whig, P.: Novel healthcare framework for cardiac arrest with the application of AI using ANN. In: 2021 5th International Conference on Information Systems and Computer Networks (ISCON), pp. 1–5 2021. https://doi.org/10.1109/ISCON52037.2021.9702493
3. Sangeetha, S.K.B., Afreen, N., Ahmad, G.: A combined image segmentation and classification approach for COVID-19 infected lungs. J. **8**(3), 71–76 (2021). https://doi.org/10.18280/rces.080302
4. Prema, K., Sriharsha, A.V.: Differential privacy in big data analytics for haptic applications. Technology **8**(3), 11 (2017)
5. Karthik, K., Chandrashekhar, K., Sriharsha, A.V., Siddardha, B., Hitesh Reddy, A.: Adaptable fog computing framework for healthcare 4.0. In: International Conference on Soft Computing and Pattern Recognition, pp. 24–32. Cham, Springer Nature Switzerland (2022)

6. Reddy Madhavi, K., Vijaya Sambhavi, Y., Sudhakara, M., Srujan Kotagiri. R.: COVID-19 isolation monitoring system, Springer series – Lecture Notes on Data Engineering and Communication Technology. Vol. 63, pp. 601–609 (2021)
7. Pustokhin, D.A., et al.: Optimal deep learning approaches and healthcare big data analytics for mobile networks toward 5G. Comput. Electr. Eng. **95**, 107376 (2021)
8. Ye, J.: The role of health technology and informatics in a global public health emergency: practices and implications from the COVID-19 pandemic
9. Sree, S.R., Agarwal, R., Markkandan, S., et al.: Integrated learning algorithms-based epileptologist assistive tool for seizure detection and prediction. Soft. Comput. (2023). https://doi.org/10.1007/s00500-023-07913-7
10. Radhakrishnan, P., Anbarasi, A., Srujan Raju, K., Sai Thrinath, B.V.: Detection of colon cancer using image processing. Cybern. Syst. (2023). https://doi.org/10.1080/01969722.2023.2175131
11. Srujan Raju, K., et al.: Statistical evaluation of network packets in an intrusion detection mechanism using ML and DL techniques, cybernetics and systems (2023). https://doi.org/10.1080/01969722.2023.2175137
12. Gadepaka, L., Lingabathina, V.: Correlation analysis of cognitive regions in automated anatomical labeling atlas using LSTM. In: Kumar, A., Ghinea, G., Merugu, S., Hashimoto, T. (eds.) Proceedings of the International Conference on Cognitive and Intelligent Computing: ICCIC 2021, Volume 2, pp. 575–582. Springer Nature Singapore, Singapore (2023). https://doi.org/10.1007/978-981-19-2358-6_52
13. Lakshmi Narayanan, S., Shunmugavel, S., Prasanth, R., Satheesh Kumar, M., Srujan Raju, K., Suthendran, K.: Design and implementation of cyber threat intelligence data mining model. In: Satapathy, S.C., Lin, J.C.W., Wee, L.K., Bhateja, V., Rajesh, T.M. (eds.) Computer Communication, Networking and IoT: Proceedings of 5th ICICC 2021, Volume 2, pp. 171–185. Springer Nature Singapore, Singapore (2023). https://doi.org/10.1007/978-981-19-1976-3_22

Early Detection and Prediction of Pneumonia Disease Using VGG16 Deep Learning Technique

N. Arfa Taj, Jessin George, and Jissy Thomas(✉)

Department of Computer Applications, Krupanidhi College of Management, Bengaluru, Karnataka, India
mcakric@krupanidhi.edu.in

Abstract. Due to the severity of the disease, immediate and accurate pneumonia diagnostics are urgently required. This study aimed to determine that the VGG16 model could be used with deep learning techniques for diagnosing and predicting pneumonia. The objective is to provide a dependable and precise method to help physicians rapidly identify pneumonia. Chest X-rays were collected of pneumonia (4273) and with other disorders (1583). To improve the model's accuracy, the images were preprocessed and enhanced. Using the VGG16 framework, features were extracted, and labels were assigned. After training in a deep learning framework, the model's performance in various contexts was evaluated. The results show that the VGG16 model is 91% accurate at detecting and predicting pneumonia cases. The model performed admirably overall. Compared to other methodologies, its performance scores are preferable. The proposed system performed well against a database of chest X-ray images for the detection of pneumonia. After an early diagnosis of pneumonia, prompt treatment reduces the risk of complications and increases the likelihood of a positive outcome. This study's findings emphasize the significance of using cutting-edge technology for early disease diagnosis and contribute to the existing corpus of knowledge on applying deep learning to medical imaging.

Keywords: Pneumonia · VGG16 · Deep Learning · chest X-ray images

1 Introduction

Pneumonia is the foremost infectious cause of death as multiple potential pathogens could be the major reason for the infection. Lung alveoli are devastated and filled with fluid and pus, resulting in painful respiration and decreased intake of oxygen [1]. Computer-aided diagnostic (CAD) techniques, such as diagnosis (CADx) and computer-aided detection (CADe), are commonly used to detect abnormalities [2]. Despite their reduced accuracy, CXRAY images are the most common diagnostic tool because many underdeveloped nations need more convenient access to advanced imaging techniques like CT scans and MRIs [3, 4]. This is why deep learning models and other models are used in medical image analysis for classification and detection. We present a CNN pre-trained model that employs transfer learning techniques to classify CXRAY images

D. Kriksciuniene et al. (Eds.): IBICA 2023, LNNS 1228, pp. 220–227, 2025.
https://doi.org/10.1007/978-3-031-78937-3_22

and display symptoms such as fluid and sputum. The contributions of this study include a competitive CNN TL model, a preprocessing approach, and a strategy for addressing dataset imbalances. Previous CNN state-of-the-art designs, such as VGG-16 [5], ResNet-v2 [6], and DenseNet [7], are significantly more accurate than the proposed model. The objective of the models is to classify pneumonic images from normal ones as quickly and accurately as feasible. Several studies have examined deep CNN models, transfer learning methodologies, and ensemble methods to improve the diagnosis of pneumonia [8, 9]. The use of feature extraction, ANNs, and texture analysis with Haralick features can all increase the accuracy of predictions [10]. Transfer learning [11] has accelerated disease detection and diagnosis prediction processes. Some deep learning models, particularly those utilizing lesser datasets, have incorporated attention processes and neutrosophic sets to improve accuracy [12].

The potential of the VGG16 model along with the chest X-rays can be useful to diagnose and predict pneumonia. By training the model on a large dataset of labeled images, we aim to develop a reliable and precise method for the early detection and prediction of pneumonia. The proposed method accelerates pneumonia considerably, aids physicians in making informed decisions, and ultimately enhances patient outcomes.

2 Literature Review

This literature review seeks to provide a concise summary of the studies that have already employed the VGG16 model to detect and predict pneumonia.

Deep learning techniques are indispensable for diagnosing pneumonia, particularly for analyzing lung inflammation [13]. The deep learning models are used, by which the researchers could detect COVID-19 and pneumonia in patients with an accuracy of 80% and 91.46%, respectively, using only X-ray images [14].

On a dataset of pneumonia as well as normal lung X-rays, a comparison of five mainstream deep-learning models demonstrated their ability to recognize pneumonia, with MobileNet exhibiting higher accuracy than other models [15]. The different models used for classification include soft computing methods, deep learning classifiers, and ensemble classifiers. A deep learning architecture utilizing recurrent neural networks (RNN) with long short-term memory (LSTM) has been proposed to accurately detect lung diseases, showcasing robustness and efficiency compared to existing methods [16]. Ensemble learning models incorporating fine-tuned versions of ResNet50, Inception ResNet_V2, and MobileNet_V2 have improved pneumonia classification performance [17]. Another study proposes a deep learning architecture trained with modified images and employing convolutional neural networks and residual network architecture, achieving an accuracy of 78.73% [18]. The combination of "ResNet-based U-Net," and "Efficientnet-B4 based U-Net" provides excellent results for pneumonia detection [19]. For the binary and multi-class classification problems, ensemble learning models and CNN models have been proposed with CNN showing better performance [20]. The chest X-rays for extraction of features using a CNN model have been constructed to classify them for pneumonia detection, achieving an average accuracy of 84% and outperforming other classifiers [21]. CNNs have been used to classify chest X-rays for pneumonia detection, yielding an accuracy of 88.90% [22]. Lastly, a CNN model combined with efficient feature extraction and various classifiers has demonstrated highly accurate results in predicting pneumonia using deep learning, big data, and machine learning techniques [23].

3 Methodology

3.1 Data Collection

The first step in the methodology involves collecting a large dataset of chest X-rays, consisting of cases such as pneumonia-positive and pneumonia-negative. The dataset should be diverse and representative of different populations to ensure the model's generalizability. Several publicly available datasets, such as the Chest X-ray14 dataset, Kaggle, etc., can be used for this purpose. Ethical considerations regarding data privacy and patient consent should be considered during the data collection process.

3.2 Preprocessing

The obtained chest X-ray images may be of superior quality and VGG16-compatible. Normalizing pixel values, reducing to a uniform size, and applying image enhancements such as rotation, scaling, and inverting are typical image preparation procedures. These preparatory procedures reduce noise, improve model robustness, and prevent overfitting.

3.3 Model Training and Validation

Models for detecting and predicting pneumonia frequently employ the VGG16 model, pre-trained on a massive dataset like ImageNet. Images are classified as "pneumonia positive" or "pneumonia negative" by the substitute of a softmax layer for the final fully connected layer in the VGG16 model. The objective function is modified during model training on a labeled data set to mitigate cross-entropy loss.

Stochastic gradient descent (SGD) and Adam frequently enhance model parameters during training. The loss from preprocessed chest X-rays is utilized to modify the model's weight iteratively. Several metrics, such as accuracy, recall, precision, and the F1-score, are the ways to evaluate the efficacy of the model. The validation set, which is not used during training, is used to evaluate the hyperparameters and efficacy of the model. Thus, the model's robustness is maintained and balanced with omitted information.

3.4 Model Evaluations

After the model's training and validation, it is tested on a unique data set to evaluate its performance in laboratory and field settings. The test set consists of chest X-ray images not utilized during training or validation. The performance indicators which include accuracy, specificity, and sensitivity are used to evaluate the model. The VGG16 model has superior accuracy and AUC-ROC for early pneumonia detection and prediction, as shown in Fig. 1 below.

3.5 Architecture

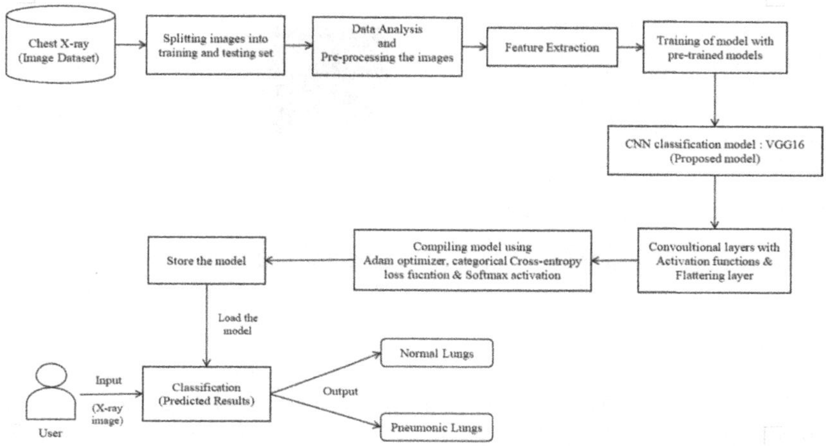

Fig. 1. Architecture Diagram for Pneumonia Detection using Deep Learning

3.6 Algorithms

Transfer Learning: To address a new problem, many machine learning practitioners turn to transfer learning, which entails feeding data to a previously trained model. Transfer learning enables machines to enhance their generalization by applying knowledge acquired in one context to another [24]. This strategy has proven effective when insufficient data exists for the new work or insufficient computational resources are available to train a new model from scratch. It is based on the premise that a trained model can be applied to a different problem that shares similarities with its initial assignment. For example, a model derived from a large dataset of photographs can be fine-tuned using a smaller dataset of related images to achieve a specific categorization objective.

VGG16**:** The architecture VGG16 is a deep convolutional neural network (CNN) model, praised for its deep structure and remarkable accuracy, which makes it ideal for image recognition and classification tasks. VGG16 has 21 layers in total, which include 13 convolutional layers, 3 Dense layers, and 5 Max Pooling layers (hence the "16" in "VGG16") [25]. The maximum number of weight layers (trainable variables) is sixteen. The VGG16 model has already been trained with over a million images from the massive ImageNet dataset, labeled with one thousand categories. This foundational training provides the model with the knowledge and skills to acquire a library of generalizable features applicable to various image recognition tasks. In addition, the pre-trained weights of the VGG16 model can be used to facilitate adaptation to a new task with fewer data points via transfer learning.

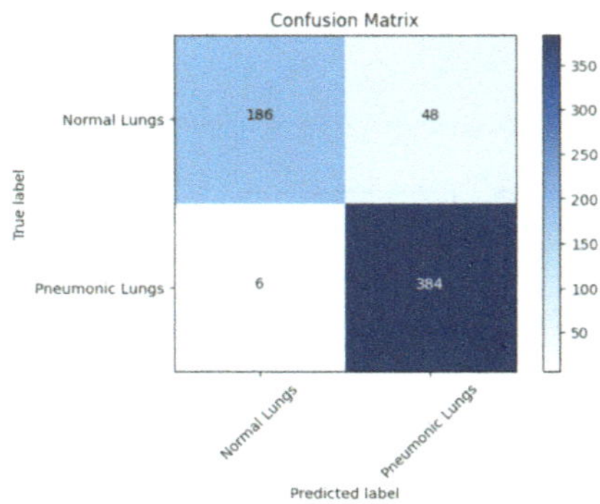

Fig. 2. Confusion Matrix to identify True Positive and True Negative results

4 Result

- Fig. 2 shows that the tested image datasets (624) contain normal and pneumonia lungs to predict the disease by applying CNN techniques for confusion matrix to detect the accuracy of the data, which indicates that there are 186 true positives (N * N), 48 False positives (N*P) for normal lung images (234) 6 False Negatives (N*N) and 384 true negatives (P*P) for pneumonia lung images (390).
- The model correctly classifies 91.3% of the samples as the model's accuracy is 0.913. The model's precision predicts a sample at 88.9% of the time. The recall of the model identified 91.3% of the samples. The F1-measure of the model is 91.3%, which is the mean of recall and precision to predict pneumonia lung disease.
- In addition to training and validation accuracy, total model accuracy can be used as a success metric. This model is trained using a 2-epoch procedure. Initial epoch results revealed a 92.03 percent training accuracy and an 86.70 percent validation accuracy. The accuracy of the training data reached 95.11 percent during the second phase, while the accuracy of the validation data reached 91.35 percent.
- Constructing a confusion matrix, which depicts the actual values on the X-axis and projected values on the Y-axis, is an alternative method for presenting the model's output. According to the confusion matrix, 202 of 234 pneumonia X-ray images and 382 of 390 normal(non-pneumonia) images were correctly identified by our model. Therefore, the model has an aggregate F-measure, recall, and precision of 91.3 percent.

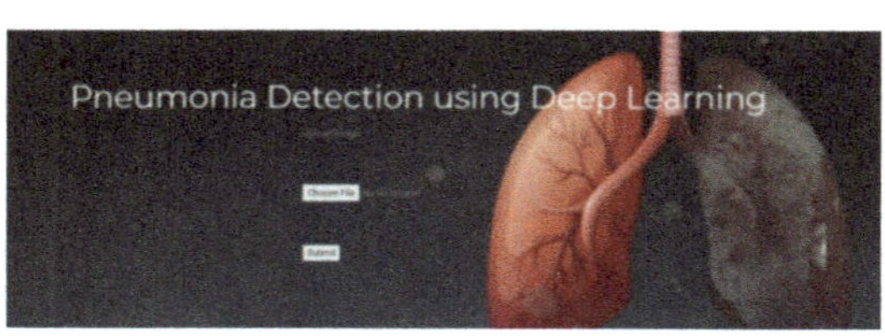

Fig. 3. Prediction for Pneumonia Detection using Deep Learning

Figure 3 is the prediction page which is associated with the input of X-ray images for prediction.

Fig. 4. Result of prediction (Normal) and (Pneumonia)

The VGG16, a pre-trained model tests the uploaded X-ray image, and the results are Normal Lungs and Pneumonic Lungs, as shown in Fig. 4.

5 Conclusion

VGG16, a deep learning technique, has promising results in the early detection and forecasting of pulmonary illness. VGG16 employs convolutional neural networks (CNNs) and transfer learning for the identification of pneumonia from chest X-rays, and it has demonstrated exceptional accuracy, sensitivity, and specificity. The model performed admirably, achieving a sensitivity of 89% and a specificity of 93%. These findings corroborate the model's ability to differentiate pneumonia from other respiratory disorders. The model was dependable and transferable with an AUC-ROC of 0.95. The VGG16 model may be helpful in the early diagnosis and treatment of pneumonia, which would positively impact patient outcomes and healthcare costs. However, additional research is required, particularly substantiating larger, more diverse data sets and evaluations in clinical contexts. Experimenting with ensemble learning approaches, such as combining VGG16 with other transfer learning models, may improve the model's classification performance. The use of VGG16 for early pneumonia detection can potentially improve pneumonia diagnosis and treatment, ultimately saving lives.

References

1. Mahomed, N., et al.: Computer-aided diagnosis for World Health Organization-defined chest radiograph primary-endpoint pneumonia in children. Pediatr. Radiol. **50**(4), 482–491 (2020). https://doi.org/10.1007/s00247-019-04593-0
2. Doi, K.: Computer-aided diagnosis in medical imaging: historical review, current status, and future potential. Comput. Med. Imaging Graph. **31**(4–5), 198–211 (2007)
3. Suetens, P.: Fundamentals of Medical Imaging, 2nd edn., pp. 14–32. Cambridge University Press. New York, NY, USA (2009)
4. Aliyu, G., El-Kamary, S.S., Abimiku, A.L., Hungerford, L., Obasanya, J., Blattner, W.: Cost-effectiveness of point-of-care digital chest-x-ray in HIV patients with pulmonary mycobacterial infections in Nigeria. BMC Infectious Diseases **14**, 1–8 (2014)
5. Geng, L., Zhang, S., Tong, J., Xiao, Z.: Lung segmentation method with dilated convolution based on VGG-16 network. Comput. Assisted Surg. **24**, 27–33 (2019)

6. Reddy Madhavi, K., Jarugula, J., Karuna, G., Kaleru, S., Srujan Raju, K., Sunitha, G.: Bird species recognition using deep transfer learning. In: Ashoka Reddy, K., Rama Devi, B., George, B., Srujan Raju, K., Sellathurai, M. (eds.) Proceedings of Fourth International Conference on Computer and Communication Technologies: IC3T 2022, pp. 473–479. Springer Nature Singapore, Singapore (2023). https://doi.org/10.1007/978-981-19-8563-8_46
7. Narasimharao, J., Laxmaiah, B., Arumalla, R., Unnisa, R., Fatima, T., Nazare, S.S.: Restoration and Deblurring the Images by Using Blind Convolution Method. In: Ashoka Reddy, K., Rama Devi, B., George, B., Srujan Raju, K., Sellathurai, M. (eds.) Proceedings of Fourth International Conference on Computer and Communication Technologies: IC3T 2022, pp. 337–346. Springer Nature Singapore, Singapore (2023). https://doi.org/10.1007/978-981-19-8563-8_32
8. Kumar, V.N., et al.: Anomaly-based hierarchical intrusion detection for black hole attack detection and prevention in WSN. In: Ashoka Reddy, K., Rama Devi, B., George, B., Srujan Raju, K., Sellathurai, M. (eds.) Proceedings of Fourth International Conference on Computer and Communication Technologies: IC3T 2022, pp. 319–327. Springer Nature Singapore, Singapore (2023). https://doi.org/10.1007/978-981-19-8563-8_30
9. Fernandes Dimlo, U.M., et al.: An improved blind deconvolution for restoration of blurred images using ringing removal processing. In: Ashoka Reddy, K., Rama Devi, B., George, B., Srujan Raju, K., Sellathurai, M. (eds.) Proceedings of Fourth International Conference on Computer and Communication Technologies: IC3T 2022, pp. 357–366. Springer Nature Singapore, Singapore (2023). https://doi.org/10.1007/978-981-19-8563-8_34
10. Munawar, S., Geetha, A., Srinivas, K.: Arrhythmia Classification Using BiLSTM with DTCWT and MFCC Features. In: Ashoka Reddy, K., Rama Devi, B., Boby George, K., Raju, S., Sellathurai, M. (eds.) Proceedings of Fourth International Conference on Computer and Communication Technologies: IC3T 2022, pp. 311–318. Springer Nature Singapore, Singapore (2023). https://doi.org/10.1007/978-981-19-8563-8_29
11. Subburaj, T., et al.: Discover Crypto-Jacker from Blockchain Using AFS Method. In: Reddy, K.A., Devi, B.R., George, B., Raju, K.S., Sellathurai, M. (eds.) Proceedings of Fourth International Conference on Computer and Communication Technologies: IC3T 2022, pp. 145–156. Springer Nature Singapore, Singapore (2023). https://doi.org/10.1007/978-981-19-8563-8_15
12. Jagadeesh, S., Shrimali, M., Prasad, P.S., Maheswari, K., Narendrakumar, A., Kotesewara Reddy, G.: Automated arrhythmia classification using harris hawks optimization with deep learning model on IoT environment. In: 2023 Third International Conference on Artificial Intelligence and Smart Energy (ICAIS), pp. 30–35. Coimbatore, India (2023). https://doi.org/10.1109/ICAIS56108.2023.10073736
13. Thaseen, A., Unnisa, R., Sultana, N., Madhavi, K.R., NagaJyothi, G., Kirubakaran, S.: Breast cancer detection using deep learning model. In: Brahmananda Reddy, A., Nagini, S., Balas, Valentina E., Srujan Raju, K. (eds.) Proceedings of Third International Conference on Advances in Computer Engineering and Communication Systems: ICACECS 2022, pp. 669–677. Springer Nature Singapore, Singapore (2023). https://doi.org/10.1007/978-981-19-9228-5_57
14. Madhavi, K.R., Vooradi, S., Mounika, P., Yedlla, S., Tangudu, N.: Emotion analysis using convolutional neural network. In: Reddy, A.B., Nagini, S., Balas, V.E., Raju, K.S. (eds.) Proceedings of Third International Conference on Advances in Computer Engineering and Communication Systems: ICACECS 2022, pp. 523–531. Springer Nature Singapore, Singapore (2023). https://doi.org/10.1007/978-981-19-9228-5_44
15. Nagajyothi, G., Kumar, G.P., Kumar, B.S., Kumar, B.P.D., Damodaram, A.K.: High-Speed Low Area 2D FIR Filter Using Vedic Multiplier. In: Brahmananda Reddy, A., Nagini,

S., Balas, Valentina E., Srujan Raju, K. (eds.) Proceedings of Third International Conference on Advances in Computer Engineering and Communication Systems: ICACECS 2022, pp. 569–578. Springer Nature Singapore, Singapore (2023). https://doi.org/10.1007/978-981-19-9228-5_48

16. Goyal, S., Singh, R.: Detection and classification of lung diseases for pneumonia and COVID-19 using machine and deep learning techniques. J. Ambient Intell. Humanized Comput. 1–21 (2021)
17. El Asnaoui, K.: Design ensemble deep learning model for pneumonia disease classification. Int. J. Multimed. Inform. Retrieval **10**(1), 55–68 (2021). https://doi.org/10.1007/s13735-021-00204-7
18. Jozef Saul, C., Yagmur Urey, D., Doruk Taktakoglu, C.: Early Diagnosis of Pneumonia with Deep Learning.In: arXiv e-prints, arXiv-1904 (2019)
19. Pant, A., Jain, A., Nayak, K.C., Gandhi, D., Prasad, B.G.: Pneumonia detection: An efficient approach using deep learning. In: 11th International Conference on Computing, Communication and Networking Technologies (ICCCNT), pp. 1–6. IEEE (2020)
20. Darici, M.B., Dokur, Z., Olmez, T.: Pneumonia detection and classification using deep learning on chest X-ray images. Int. J. Intell. Syst. Appl. Eng. **8**(4), 177–183 (2020)
21. Shuaib, K.M., Shahid, S.A., Javed, A.A., Zaid, M.: Pneumonia detection through x-ray using deep learning. IOSR J. Comput. Eng. **22**(1), 08–11 (2020)
22. Račić, L., Popović, T., & Šandi, S.: Pneumonia detection using deep learning based on convolutional neural network. In: 25th International Conference on Information Technology (IT), pp. 1–4. IEEE (2021)
23. Swetha, K.R., Niranjanamurthy, M., Amulya, M.P., Manu, Y.M.: Prediction of pneumonia using big data, deep learning, and machine learning techniques. In: 6th International Conference on Communication and Electronics Systems (ICCES), pp. 1697–1700. IEEE (2021)
24. Jessica Powers. What Is Transfer Learning? Exploring the Popular Deep Learning Approach. https://builtin.com/data-science/transfer-learning, (2022)
25. Md Yawar: VGG-16 - CNN Model. Coding ninjas. https://www.codingninjas.com/codestudio/library/vgg-16---cnn-model (2023)

Real Time Face Mask Detector

Ch. Rajyalakshmi[1(✉)], Gampa Sri Varsha[1], Galipalli Vinay Teja[1], Dasi Bhargavi[1], K. Reddy Madhavi[2], and Mohammad Gouse Galety[3]

[1] Department of CSE, B V Raju Institute of Technology, Narsapur, Telangana, India
lakshm.chatragadda@gmail.com, {19211a0585,19211a0584, 19211a0566}@bvrit.ac.in

[2] AI&ML, School of Computing, Mohan Babu University, Tirupati, India

[3] Department of Computer Science, School of Engineering, Samarkand International University of Technology, Samarkand, Uzbekistan

Abstract. Face mask use is crucial for preventing and limiting the development of COVID-19 and other respiratory viral infections. Face masks can be used to shield healthy individuals or stop the spread of infection by sick people. A computer vision system and Deep Learning algorithm that are capable of performing this type of detection are required in order to determine whether or not a person is wearing a face mask. For the purposes of this study, photos of people wearing masks and those who are not are distinguished using a Deep Neural Network, more precisely a Convolutional Neural Network. On the training set, the CNN achieves an accuracy of 98.6%, and on the test set, 97.3%. Then, a real-time classification of mask or no mask is made using the stored weights of this CNN. To carry out this operation, the webcam captures the footage; the frames are then analyzed and sent to the model. The model operates effectively with no discernible delay between donning the mask and taking it off and displaying the forecast.

Keywords: Artificial Intelligence · Deep Neural Network · Convolutional Neural Network

1 Introduction

Real-time face mask detection is a technique that determines whether or not someone is wearing a face mask using computer vision and machine learning algorithms. The COVID-19 epidemic has made it clear how crucial it is to use face masks to stop the virus from spreading. In order to ensure compliance with face mask regulations, face mask detection equipment might be useful in public venues like airports, hospitals, and schools [1]. Deep learning algorithms are used for analyzing real-time footage from cameras, detect faces, and determine whether or not people are wearing masks. In order to recognize the patterns and traits that differentiate between the two, the algorithms are trained on enormous datasets of photos of people wearing and not wearing masks [2]. When someone enters a prohibited location without a mask, the system can send real-time alerts or notifications to the authorities. Additionally, the technology is adaptable

D. Kriksciuniene et al. (Eds.): IBICA 2023, LNNS 1228, pp. 228–237, 2025.
https://doi.org/10.1007/978-3-031-78937-3_23

to many situations and application cases. Real-time face mask detection technology's key advantages are increased safety, less chance of infection, and increased compliance with public health guideline [3]. Wearing a face mask at all times is a great idea in the situations mentioned above to avoid coming into contact with COVID-19. In this context, a great deal of research has been done on figuring out whether or not someone in a public meeting or an organization is wearing a mask. Traditional methods of verifying a face mask infraction are error-prone and not usually practical. Conventional practices involve personally checking those who don't have face masks on. As a result, a system that can complete this task automatically is required. Additionally, human resources could be spared and used for other crucial duties. Deep learning (DL) and machine learning (ML) methodologies have been applied to face mask identification systems to create dependable, accurate, fast, and efficient systems. Many academics have created a range of DNN architectures with the aim of identifying violations of face masks. The majority of these researchers have opted to use hybrid approaches—a blend of machine learning and deep learning—and transfer learning as the foundation for their work. First, a face detection module is used to identify faces from video data frames. Next, an intermediate block processes the detected faces as needed. Lastly, a fine-tuned MobileNetV2 model is used to classify the detected faces as unmasked, masked, or incorrectly masked. These models were chosen in order to create a system that was lightweight and could be integrated with security cameras.

2 Related Work

Finding the best face recognition model with the least amount of memory usage and server processing is the goal; millions of photos can be handled considerably more quickly by segmentation, multithreading, or parallel processing. The best computational models, such as CNN (Convolution Neutral Network) and DNN (Deep Neutral Network), should also be investigated and tested.

The identification of the face mask is significantly impacted by computer vision and image processing. Face detection includes a variety of use cases, such as face recognition and facial motions, the latter of which must display the face with a very high degree of accuracy [12]. S. S. Parthiban ad S. S. Sivasankari, "Real time face mask detection using deep learning and computer vision techniques" [1]. An automatic system that can instantly determine whether a person is properly wearing proper protective gear is one way to increase compliance with this proposal. The design, implementation, and performance evaluation of a real-time system for extracting mask usage information from image sequences are presented in this study. Video sequences of people wearing and not wearing facemasks that were shot in various situations are used to verify the detector's effectiveness. The dataset used for training and evaluation might have limitations in the paper. The authors might have only used a small dataset, which might not accurately reflect the range of face mask pictures that can be seen in the actual world.

Adhikari and B. Raman, "Real time face mask detection in video streams using deep learning and computer vision techniques" [2]. The "Real-world Masked Face Recognition Dataset" was a dataset that was made accessible to the public and was used by the authors to train and test their suggested methodology. To boost the diversity of the training set, they pre-processed the dataset by resizing and enhancing the photographs. Features from the images were extracted using the pre-trained CNN and supplied to the SVM classifier for mask detection. The suggested method can identify face masks in real-time video feeds, making it appropriate for real-time surveillance applications in the fields of public safety, healthcare, and transportation. The "Real-world Masked Recognition of Faces Dataset" was a dataset that had been made accessible to the public and was used by the authors to train and test their suggested methodology. The generality of the suggested approach may be limited by the fact that this dataset might not accurately reflect the variety of face masks available in reality. Y. Liu, A. A. Gayle et. al in 2022 [3]". The reproductive number of covid-19 is higher compared to sars corona virus. The COVID-19 virus's basic reproduction number (R0). R0, which represents the typical number of new infections caused by an infectious person in a completely ignorant community, is a sign of a virus's transmissibility. For R0 > 1, the number of infected people is likely to rise, but for R0 1, transmission is probably going to stop. The fundamental reproduction number, which identifies the danger of an infectious agent with regard to epidemic transmission, is a key term in the epidemiology of infectious diseases. The authors offer a thorough examination of COVID-19's reproduction rate, taking into account elements including age, sex, and travel history. This makes it possible to comprehend the elements that contribute to the virus's spread better. The study does not discuss any other elements of the viruses, such as clinical presentation, disease severity, or death rates; it just compares the reproductive numbers of COVID-19 and SARS-CoV. A data driven analysis [4]. Chavez, S.; Long, B.; Coffman, A.; Liang, S. Y. Corona virus Disease (COVID-19): A primer for emergency. The authors' goal is to give emergency medicine physicians current, useful information to aid in the identification and management of COVID-19 cases at the emergency room of a hospital. The authors offer helpful advice on how to identify and handle COVID-19 cases in an emergency room, as well as instructions on infection control practices, clinical judgment, and available treatments. Disadvantages: The study mainly concentrates on the COVID-19. S. S. K. S. D. Gamage, P. R. P. Gamage, and Y. W. L. C. Perera, "Real-time face mask detection in the wild using a hybrid deep learning model", published in Sensors (2021) [5]. A study on the creation of a deep learning hybrid system for unrestricted real-time face mask recognition. To achieve high accuracy in face mask detection, the authors suggest a model that combines convolutional neural networks with deep learning (DCNNs) and support vector machines (SVMs). A relatively new method for real-time face mask recognition, the suggested hybrid deep learning model combines DCNNs and SVMs to attain high accuracy. It may be tricky to determine the hybrid deep learning model's relative effectiveness because the study did not compare its performance with that of other face mask detection methodologies or models. "Real-time face mask detection using YOLOv4 and transfer learning" by M. H. Alsharif, M. A. Al-Rawashdeh (2021) [6]. A study on implementing transfer learning and the YOLOv4 object recognition algorithm to create a real-time facial mask recognition system. With an average precision of 96.05%, the one suggested excelled at real-time identification of face masks tasks. It could be challenging to construct and maintain the suggested system because it consists of numerous elements, including YOLOv4, learning by transfer, and picture pre-processing algorithms. Xiaoxiang Zhu, Xinwei Jiang, and Jocelyn Chanussot, "Frontiers in Spectral-Spatial Classification of

Hyper spectral Images" [2]. The article that addresses cutting-edge methods for classifying hyper spectral pictures spectrally and spatially. High-dimensional data called hyper spectral pictures provide detailed spectral information about the scene's objects. In order to increase classification accuracy, spectral-spatial classification approaches attempt to utilize combined spectral and spatial information. Feature extraction, choosing features, and methods for classification are just a few of the topics covered in this article on spectral-spatial classification. It also covers the field's difficulties and prospects for the future. Overall, the paper offers a thorough analysis of the most recent developments in the classification of hyper spectral images. Gives a thorough overview of cutting-edge methods for classifying hyper spectral pictures spectrally and spatially. Provides a useful resource for academics and professionals interested in classifying hyper spectral images As demonstrated, extinction profiles may quickly and accurately classify data without supervision. These characteristics compel one to look into the effectiveness of this filtering strategy for big data processing applications including Earth observation. Muhamma Umer, saima Sadiq and Reemah M. Alhebshi, "Face mask detection using deep convolutional neural network and maulti-stage image processing", Umer et al. [7].

The input image is scaled using the first CNN, which has five layers. Each of the second and third tiers has seven layers. Using three cascaded CNN has the benefit of eliminating each false detection, strengthening the prediction. Nevertheless, it incurs high computational costs when use three CNN. With their suggested dataset, the model yielded an accuracy of 86.6% and a recall of 87.8%. There are numerous types of face mask detection models. These can be categorized into various groups. In boosting-based classification, the Viola-Jones face detector was used to accept boosted cascades with simple haar characteristics. Drawbacks in existing system The current approach only recognizes whether a person is wearing a face mask if they are not wearing eyeglasses. The current approach only recognizes whether a person is wearing a face mask if they are not wearing eye glasses.

3 Proposed Methodology

The proposed method focuses on using computer vision and deep learning techniques to recognize a person wearing a face mask in an image or video stream. We used CNN (Convolution Neural Network) "Real time face mask detector". One of the finest one-stage object detection models is CNN (Convolution Neural Network), which has gained popularity since it has been shown to perform well with dense and small-scale objects. We mainly have 5 modules in this project. Data gathering Data annotation Data preprocessing CNN Training Prediction.

3.1 Dataset Description

The dataset was personally gathered from various sources and included nearly every type of face mask. Data collection is the process of gathering, measuring, and analyzing precise research-related insights using accepted, proven methods. A researcher can assess their hypothesis using the data that they have gathered. No matter what field of study you are conducting, gathering data is typically the first and most crucial stage. Depending on the type of data needed, different disciplines of research require different approaches to data gathering. Making sure that accurate accurate and reliable data is gathered so that research decisions can be informed by data is the most important goal of data collection. Data collecting options essentially come in four flavors: in-person interviews, mail, phone, and online. Every one of these approaches has advantages and disadvantages. The dataset was personally gathered from various sources and included nearly every type of face mask. A portion of the information was assembled from various standard datasets, such as the Masked Face Detection Dataset (MFDD) [13] and MAsked FAces (MAFA) [9]. To further enrich the dataset, several simulated photos produced by the data augmentation techniques were also incorporated. To ensure diversity and objectivity in our collection, we collected photographs from a variety of sources. The dataset was primarily created in order to create a highly accurate model that is capable of detecting a variety of mask kinds and successfully identifying the occurrences of occlusion in front of the face. There are various types are identified in the dataset: masked correctly, masked wrongly, and masked not at all shows in Fig. 2 and 3. Furthermore, we classified the information into discrete subcategories under each of these main groups according to their characteristics. This kind of structure is justified by the fact that it may be used to different computer vision issues. The dataset can be utilized for face recognition and occlusion face detection in addition to the face mask detection approach. For instance, face occlusion detection problems can use the complex with mask category, while face detection problems can use the simple without mask subcategory. There are 4,535 photos in all in the dataset. There are two classes in it: incorrect mask pictures and mask images in Figs. 2 and 3.

In the Proposed work involved the application of several data augmentation techniques, such as flipping, rotation, contrast, shearing, and zooming, with the aim of mitigating class imbalance and increasing the training dataset's size. This would not only help avoid over fitting but also strengthen the model's ability to find previously undiscovered data. To do data augmentation, we implemented the Tensor Flow package's ImageDataGenerator class. Data annotation is the process of adding information to data in order to prepare it for machine learning model training. There are many different types of data and metadata, including content types like text, audio, photos, and video. These can be used to teach chat bots, translation software, and autonomous cars The process of adding metadata to a dataset is known as data annotation. Any sort of data, including text, photos, and videos, can be annotated with this metadata, which typically takes the form of tags. A crucial step in creating a training dataset for machine learning is adding detailed and consistent tags.

3.2 CNN Training

A subfield of machine learning called "deep learning" is based on learning levels of representations that correspond to a hierarchy of features, factors, or concepts. Higher-lever concepts are defined from lower-lever ones, and the same lower-lever concepts can define many higher-lever concepts. Deep learning, which involves learning several layers of abstraction and representation, aids in the comprehension of data types like text, voice, and image. The research of artificial neural networks gave rise to the idea of deep learning, and the Multilayer Perceptron, which has more hidden layers, is a

Deep Learning structure. Machine learning is given hope in the late 1980s with the development of the Back Propagation method, which is utilized in Artificial Neural Networks and starts a trend of machine learning based on statistical models. Many other Shallow Learning models, including Support Vector Machines (SVM), Boosting, and Logistic Regression (LR), were proposed in the 1990s. These models' structures can be visualized as either having one hidden node (SVM, Boosting) or none at all (LR). Both in terms of theoretical study and practical implementations, these models enjoyed remarkable success. A 2006 publication in "Science" by Geoffrey Hinton, a professor at the University of Toronto in Canada and the dean of machine learning, and one of his students, Ruslan Salakhutdinov, set off a machine learning movement in both academia and business.

The following procedure follows for the processing of detection of real time face mask detection. In the CNN training the image was a subjected to a change via the ImageDataGenerator, which then used the altered images for training throughout each epoch. Since it is trained on various iterations of the same image, the taught model might be more reliable and accurate as a result. The amount of training images per epoch is not increased by augmentation in this case. Rather, it applies a distinct change to every image in every era. We have employed 30 different versions of each original image in training (or 4,535 * 30 = 136,050 different images in the entire training) rather than only the 14,535 original photos in the entire training since we train our model for 30 epochs. For the classifier in the suggested system, MobileNetV2 transfer learning is used. It enables us to begin training with the weights of the cutting-edge model, which has been trained on the Imagenet dataset. To maintain the features that have previously been learned, the base layers are locked. Following that, four more trainable layers are created, and the custom dataset is used to train those layers. Three neurons make up the last layer of the model, each of which represents a necessary class.

(i). Multiple hidden layers in an artificial neural network have a great capacity for characteristic learning. The traits acquired during learning provide a crucial description of the data, which therefore makes visualization or classification easier.

(ii). Layer-wise pre-training can help deep neural networks overcome their training challenges. The implementation of layer-wise pre-training in this article is accomplished by unsupervised learning. Deep Neural Networks (DNNs) are commonly used to refer to feed-forward neural networks or multilayer perceptron's with several hidden layers in artificial neural networks. One type of feed forward neural network is the convolution neural network (CNN). Figure 1 shows predicting the likelihood of a specific result, such as whether or not a client would churn in 30 days, "prediction" refers to the output of an after it has been on a previous dataset and applied to new data. For every record in the new data, the algorithm will produce probable values for an unknown variable, enabling the model builder to determine what that value will most likely be. The term "prediction" itself might be deceptive. When you use machine learning to determine the in a marketing campaign, for example, you are actually forecasting a future consequence in some circumstances however, sometimes the "prediction" is about something that has already happened, like whether a transaction was fraudulent or not. In such situation, the transaction is already complete, but you're attempting to determine whether or not it was valid so that you may decide what course of action to follow.

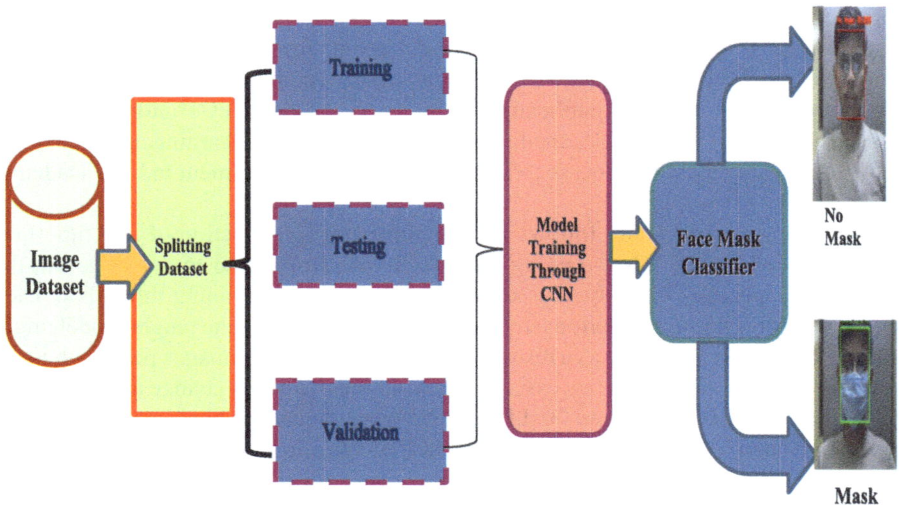

Fig. 1. Facemask Detector Result using CNN

4 Result Analysis

We used 90% of the total data for the model's training, while the remaining 10% were used for testing. The model was then adjusted, and the weights were kept in storage. Pre-trained weights can be used to improve the outcome and save a substantial amount of computational expenditures. The hyper parameters that are employed when the model is being trained include in the Table 1. After thirty training epochs, the model's accuracy on training and testing data was 97.15% and 95.81%, respectively. Figure 8 displays the model's learning curves. The plot clearly illustrates that while training and validation loss decreases with increasing epoch counts, training and validation accuracy rises as well.

In the Figs. 2 and 3 shows the model's output for the current frame of the video is a bounding box with the colors red, green, or blue around the face. Red indicates that the face is not covered with mask, green shows that a mask is being worn,. Above the bounding box, the class probability and prediction are also displayed. With an accuracy of 97.15%, the model's output accurately classifies faces as having masks or not. In the real world, the model's performance can deteriorate even when it functions well and has decent accuracy. It is possible that the model won't get a high-quality video because of the weather and other variables. It is possible that the received frame is low-luminance and fuzzy.

Table 1. Real Time Face mask detection model's hyper parameters

Hyper parameter	Value
Learning rate	1e-3 with decay rate = learning rate/epoch number
Batch size	32
Epochs	30
Dropout rate	0.3, 0.3
Input layer size	224*224*3
Output layer size	224*224*3
Optimization	ADAM
Loss function	Cross entropy

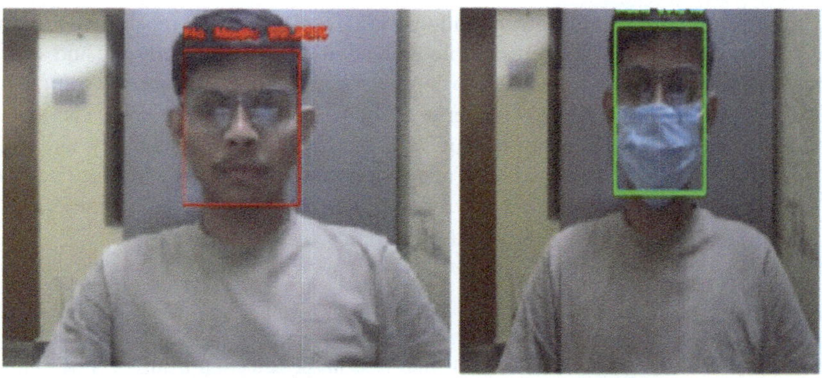

Fig. 2. With Spects,without mask and with mask

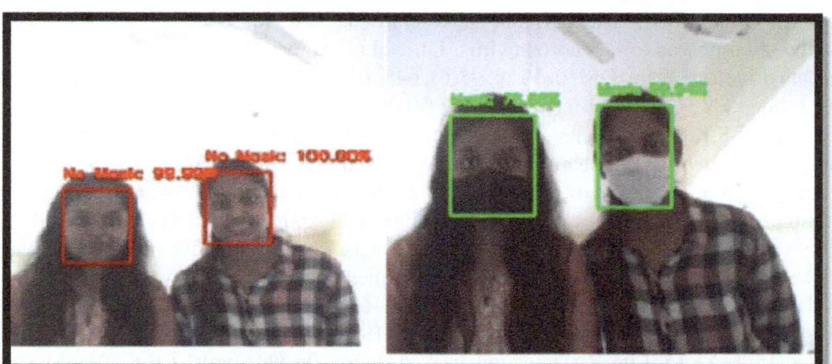

Fig. 3. With mask and without mask

5 Conclusion and Future Work

As technology advances and new trends become more readily available, we now have a revolutionary face mask detector that may help with public healthcare. Retina Net serves as the foundation of the architecture, which may be applied to both high- and low-computing scenarios. We use transfer learning to adopt weights from a related task, face detection, which is learned on a very large dataset, in order to extract more robust features This study provides a lightweight and efficient approach for face detection and mask classification from real-time video data, leading to a system for rapid face mask detection. The system is composed of a face mask classifier and a face detection module that use a R convolution neural network classifier and a single-shot multi-box detector based on ResNet50, respectively. The suggested dataset was contrasted with a few benchmark datasets. The data augmentation method was used to increase the dataset's size. All things considered, this strategy may help reduce the spread of COVID-19 and enhance public health and safety.

In future improvements might include making real-time face mask detectors more affordable, more usable, and interoperable with a wider range of hardware and software systems. This would make them easier to deploy and scale. In the future, we want to make it better and workable for CCTV cameras without the need for a GPU. The system's component models can be swapped out with ones with reduced latency and greater accuracy in the future. It is possible to add alarm capabilities to the system so that an alert sounds each time a face mask is violated. A high-definition video camera can be used to capture videos in order to increase the prediction quality.

References

1. Parthiban, S.S., Sivasankari, S.S.: Real time face mask detection using deep learning and computer vision techniques. Published in multimedia and applications, (2022)
2. Adhikari, A., Raman, B.: Real time face mask detection in video streams using deep learning and computer vision techniques. J. Amb. Intell. Humanized Comput. (2021)
3. Liu, Y., Gayle, A.A., Wilder-Smith, A., Rocklöv, J.: "The reproductive number of covid-19 is higher compared to sars corona virus. J. travel medicine (2020)
4. A data driven analysis. J. Med. Virol. 92(6), 645–659 (2020). Chavez, S., Long, B., Koyfman, A., Liang, S.Y.: Corona virus Disease (COVID-19): A primer for emergency
5. Gamage, S.S.K.S.D., Gamage, P.R.P., Perera, Y.W.L.C.: Real-time face mask detection in the wild using a hybrid deep learning model. Published in Sensors (2021)
6. Parthiban, S.S., Sivasankari, S.S.: Real time face mask detection using deep learning and computer vision techniques. Published in multimedia and applications (2022)
7. Liu, Y., Gayle, A.A., Wilder-Smith, A., Rocklöv, J.: The reproductive number of covid-19 is higher compared to sars coronavirus. J. Travel medicine (2020)
8. A data driven analysis. J. Med. Virol. **92**(6), 645–659 (2020). Chavez, S., Long, B., Koyfman, A., Liang, S.Y.: Coronavirus Disease (COVID-19): a primer for emergency
9. Ge, S., Li, J., Ye, Q., Luo, Z.: Detecting masked faces in the wild with LLE-CNNS. In: Proceedings of the IEEE conference on computer vision and pattern recognition. IEEE. pp. 2682–2690 (2017)
10. Bu, W., Xiao, J., Zhou, C., Yang, M., Peng, C.: A cascade framework for masked face detection. In: 2017 IEEE international conference on cybernetics and intelligent systems (CIS) and IEEE conference on robotics, automation and mechatronics (RAM), pp. 458–62. IEEE (2017).

11. Yang, S., Luo, P., Loy, C.C., Tang, X.: Wider face: a face detection benchmark. In: Proceedings of the IEEE conference on computer vision and pattern recognition, p. 5525–5533. IEEE (2016)
12. Khan, N., et al.: Big data: survey, technologies, opportunities, and challenges. Sci. World J. **2014**, 1–18 (2014). https://doi.org/10.1155/2014/712826
13. Wang, Z., et al.: Masked face recognition dataset and application. arXiv preprint arXiv:2003.09093 (2020)

Cardio Inspect Using ECG Images

Thoutireddy Shilpa[1(✉)], Nagendar Yamsani[2], Ranjith Kumar Marrikukkala[3], P. Kumaraswamy[4], A. Harshavardhan[5], and B. Sachuthananthan[6]

[1] CSE, B V Raju Institute of Technology, Narsapur, Medak, Telangana, India
shilpathoutireddy@gmail.com

[2] Department of Computer Science and Artificial Intelligence, SR University, Warangal, Telangana, India

[3] CSE, Sumathi Reddy Institute of Technology for Women, Telangana, India

[4] Department of CSE, KITS, Warangal, Telangana, India
pks.cse@kitsw.ac.in

[5] Department of CSE (AIML&IoT), VNR Vignana Jyothi Institute of Engineering and Technology, Hyderabad 500090, India

[6] School of Engineering, Mohan Babu University, Tirupati, India

Abstract. Electrocardiograms (ECGs) as one of the most widely used tests for detecting cardiovascular issues. The electrocardiogram (ECG or EKG) is a diagnostic technique used for regular evaluation of the heart's electrical and muscular activities. Although the ECG test may seem relatively straightforward, extensive training is required to interpret ECG charts. Historically, the majority of ECG recordings were documented on paper. Manually reviewing and revaluating ECG paper records can be a tedious and challenging task. However, by digitizing these paper ECG recordings, we can automate the diagnosis and analysis process. The primary aim of this paper is to utilize ML techniques to convert paper ECG data into a one-dimensional signal. This objective can be achieved by employing diverse techniques to extract the P, QRS, and T waves from ECG readings, revealing the heart's electrical activity. The methods include dividing the original ECG report into 13 Leads, extracting, and converting the signal, smoothing, and then converting these binary pictures using threshold and scaling. To interpret the data, post-feature extraction and dimension reduction methods like Principal Component Analysis are used. The techniques include splitting the original ECG report into 13 Leads, extracting and converting into the signal, smoothing, and converting binary images using threshold and scaling. Post-feature extraction, and dimension reduction techniques like Principal Component Analysis are applied to understand the data. Several classifiers, including CNN, k-nearest neighbors (KNN), logistic regression, support vector machine (SVM), and a voting-based ensemble classifier, have been incorporated. The final selection of the model will be based on acceptable criteria such as accuracy, precision, recall, F1-score. This final model will aid in the diagnosing of cardiac diseases, to detect whether a patient has/had Myocardial Infarction, Abnormal Heartbeat, or the patient is hale and healthy by inferring the ECG reports.

Keywords: ECG · Dimension reduction · Ensemble Classifier

D. Kriksciuniene et al. (Eds.): IBICA 2023, LNNS 1228, pp. 238–249, 2025.
https://doi.org/10.1007/978-3-031-78937-3_24

1 Introduction

As per the World Health Organization (WHO), cardiovascular diseases persist as a significant global contributor to mortality [1]. In 2019, cardiovascular diseases accounted for 17.9 million deaths, comprising 32% of all global fatalities. Among these statistics, 85% of the reported deaths were attributed to heart attacks and strokes. Heart disease is a complicated disorder that must be carefully managed. Multiple tests may be required to accurately detect cardiac disease. Medical personnel's lack of experience may result in inaccurate test findings. Early detection of heart disease might be difficult, but early warning can enable patients take preventive measures to avoid probable heart attacks or strokes and live better lives.

Data has become critical in all businesses, including healthcare, in the digital era. Medical researchers and specialists have begun to use this data to learn more about diseases like dementia, Alzheimer, and cancer.

Typically, the assistance of an experienced health expert is required to recognise cardiac illness. However, with an increasing population, there are a limited number of healthcare specialists accessible, making it impossible to treat everyone.

Early detection of cardiac disease is a critical challenge. In general, medical personnel use a variety of tests to identify the likelihood of cardiac disease. ML models are needed to supplement these professionals and to recognise heart disease autonomously in the future. These classification models must be accurate in order to be useful.

We want to create classifiers using ensemble Machine Learning algorithms using open-access datasets and then select the one with the highest accuracy for anyone to utilise.

The main goal of this Research paper is to train a Machine Learning model to predict the occurrence of heart disease in patients using datasets accessible on Kaggle website. Because of issues such as missing values, redundant instances, and so on, the datasets offered require proper preparation. The machine learning models are trained using the pooled dataset, after which the testing data is utilized to determine the optimal classifier among them, considering various assessment criteria, including accuracy, precision, and recall. Finally, we provide consumers with a web-based form that informs them about the probability of develop heart disease based on the data they supply.

As mentioned earlier, the dataset utilized in this research is sourced from the UCI data repository and contains several missing values. The data pre-processing stage aims to handle these limitations. However, due to the large size of the dataset, enhancing the accuracy of the classifier model becomes challenging. The diversity of the cases in the dataset also contributes to the classifier's accuracy. When the information is less diverse, the ML model fails to reliably identify the possibility of heart disease.

2 Literature Work

It has maintained its position as the primary cause of global mortality over the past two decades [2]. This research aims to explore various data mining approaches employed in recent years for cardiac disease diagnosis. An electrocardiogram (ECG) is a straightforward test that can evaluate the rhythm and electrical activity of the heart.

Currently, there is a wide array of algorithms available for detecting and predicting heart abnormalities based on clinical data. Machine learning is expected to aid clinicians in evaluating patients, and the application of machine learning and deep learning models holds the potential to revolutionize healthcare service delivery. This holds true for cardiovascular disease, which stands as the leading cause of death globally. This specific aspect is under investigation to develop improved models for detecting the likelihood of heart disease.Several prior studies revealed researchers' efforts in developing improved models for detecting cardiovascular disease. This section investigates similar research work to the proposed methodology [3]. Shah et al. conducted study titled "Heart Disease Prediction Using Machine Learning Techniques" on 303 occurrences from the Cleveland heart disease dataset. The dataset was cleaned and transformed as part of the preprocessing stages. In this study, supervised algorithms such as K-NN, Nave Bayes, and Decision tree were applied, with K-NN achieving the greatest accuracy of 90% [5, 6]. Another study, "Supervised ML-Based Cardiovascular Disease Analysis and Prediction" by Hossein et al., used the same Cleveland heart disease dataset and completed data pretreatment procedures of data cleaning, missing data analysis, and standard scaling to accomplish feature scaling. In their study titled "Prediction of Cardiovascular Diseases Based on Machine Learning," Sun et al. put forward the use of SVM, LR) and RF Algorithm-based models. The researchers achieved an accuracy of 71% for both SVM and LR algorithm-based models [4]. In "Identification of Cardiovascular Diseases Using Machine Learning," Louridi et al. used pooled datasets from the UCI library, including Cleveland, Hungarian, Long Beach, and Switzerland datasets. Mean value normalisation was used to replace missing values in the dataset. During the training phase, supervised algorithms such as Support Vector Machine, KNN, and Nave Bayes were utilised on training datasets of varying sizes. With a 3:1 ratio of training and testing datasets, SVM scored [3] the highest accuracy of 86%, while NB reached a similar accuracy with a 4:1 train and test data split.

Cardiovascular heart disease is regarded as a hazardous condition, and ongoing research is being conducted to detect the likelihood of heart Problem at an early stage. Many technologies and procedures are being tested in order to determine the best strategy for early diagnosis of heart disease in a patient and to meet current health needs [7–9]. In this aspect, machine learning has been a blessing. Most existing approaches may not be useful in detecting the development of heart disease in patients, especially given recent population expansion.It is critical to train comprehensive machine learning and deep learning systems rapidly in order to minimise underfitting and overfitting issues when using them to identify cardiac disease. As a result, it is critical to employ machine learning approaches that operate well with tiny datasetsThe suggested approach in this research generates a functioning machine learning classifier model that predicts whether a person is predisposed to heart disease or not. This section explains how to create an ensemble classifier. The first stage in creating a solid deep-learning model is gathering the Images dataset for training. It is critical and time- consuming to collect the necessary data for the model. For our ECG scans, we employed four image classification categories: Normal, Myocardial Infarction, AbnormalHeartbeat,Myocardial Infarction History.Data preparation is the next step in the suggested paradigm. Several procedures are taken at this phase, including as dealing with noise, duplicate values, and missing values in the dataset. It also deals with converting data to the appropriate format.After performing feature selection, the dataset will be partitioned into two segments: training data and testing data [14]. On the other hand, the testing data is employed to evaluate the developed model using various metrics. Consequently, the target attribute is absent from the testing data [13]. Finally, the ML models are generated using a variety of classic supervised and ensemble procedures, which will be detailed further below. The most accurate model is picked and used to construct online applications that will be used by clinicians and patients to obtain the results (Fig. 1).

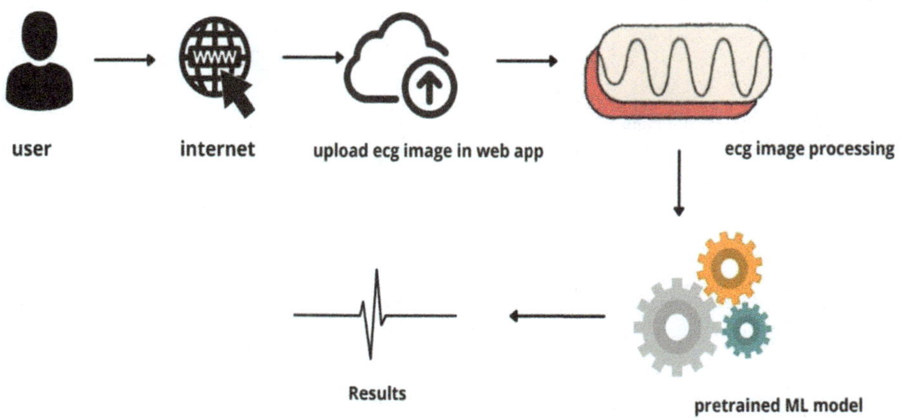

Fig. 1. Architecture of the system

[12] The raw dataset is data obtained from reliable sources and stored in the UCI Machine Learning Repository. This data was explained in the Literature survey's recommended system section.The data pretreatment phase then begins, in which the dataset is cleaned of any noise and duplicate values. The dataset in this scenario also contains some missing values. These missing values were replaced with the mean value of the appropriate column data since it provided better accuracy than the other techniques of dealing with missing data. This phase is also in charge of feature scaling in order to make the dataset suitable for distance-based algorithms [14, 15]. This step also includes feature extraction and attribute selection [10]. Finally, the classifier with the best accuracy is used to create a web application that allows users to input information to determine if they are at risk of heart disease or not [11]. This method is carried out utilising the

serialisation idea, in which the model generated and tested is carried to the server to be used with the web application [16].

3 Proposed Methodology

For the proposed system of heart disease prediction and corresponding web application, the modules were segregated based on the tasks that need to be performed. This module is responsible for training the classifier model which includes performing the preprocessing to the collected dataset, applying feature selection, splitting it into training and testing and evaluating the models built, and selecting the most accurate model from them. ECG images: https://data.mendeley.com/datasets/gwbz3fsgp8/2.

The dataset provided includes ECG image signals from individuals who are both healthy and those with cardiovascular issues. All the ECG images from the above Mendeley dataset are parsed and transformed per our business requirements. This is the most important phrase of our application (data preparation, data cleaning, data engineering and feature extraction).Each one of the images goes through various processes to extract the data in resultant format (Fig. 2).

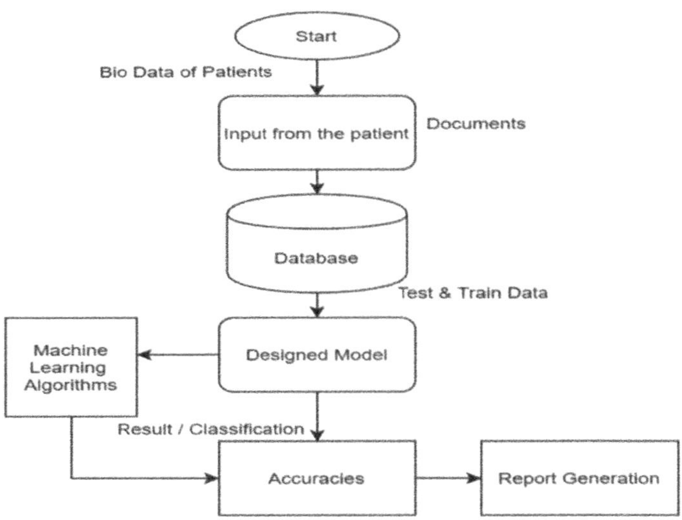

Fig. 2. WorkFlow of the system

Data Preparation: In our analysis, the emphasis will be primarily on the low and high points. As a result, the X-axis will be saved independently as a normalized and scaled 1D signal, stored in a CSV file.

4 Implementation

We tested various algorithms and then utilized the ensemble technique to stack algorithms to enhance performance because it was a classification problem. Numerous researchers have employed various data mining techniques to identify and predict heart anomalies, such as NN, classification based on clustering, KNN, and decision trees (Figs. 3, 4 and 5).

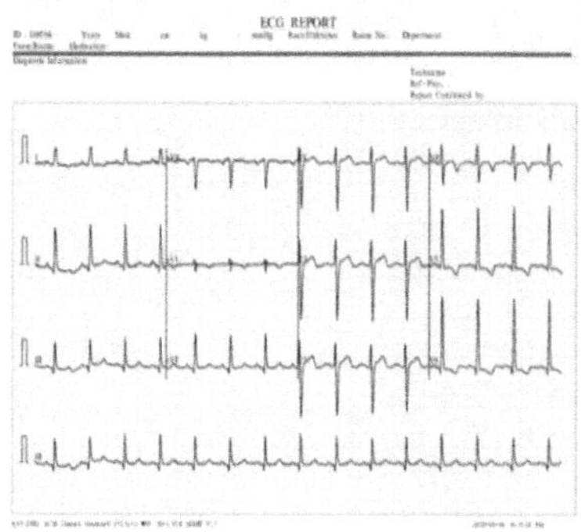

Fig. 3. Input Image

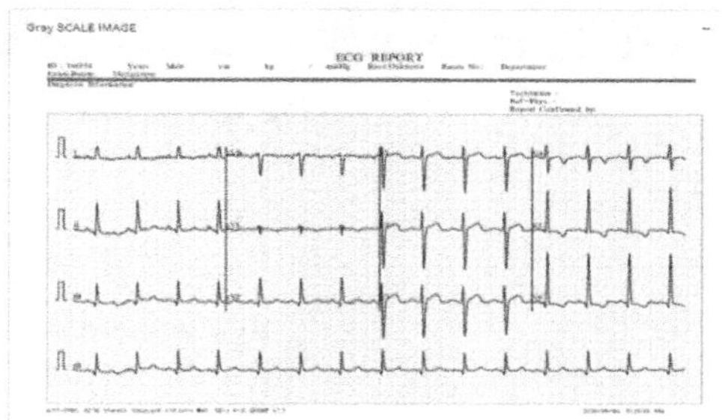

Fig. 4. Gray Scale Image

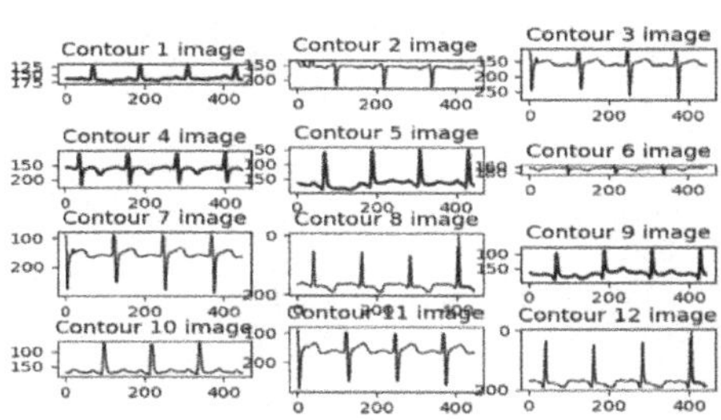

Fig. 5. Contour Images

Data Cleaning and Feature Transformation: The transformed image is traced using the contour technique to extract only the signals present within the image (Figs. 6, 7, 8, 9 and 10).

In this research paper, we employ multiple supervised classification algorithms, namely KNN,logistic regression, SVM,and a voting-based ensemble classifier. These algorithms are applied to CSV data for analysis. Additionally, the total variance explained in this case is 99.5. After dimension reduction, the combined 12 leads undergo various data mining techniques.

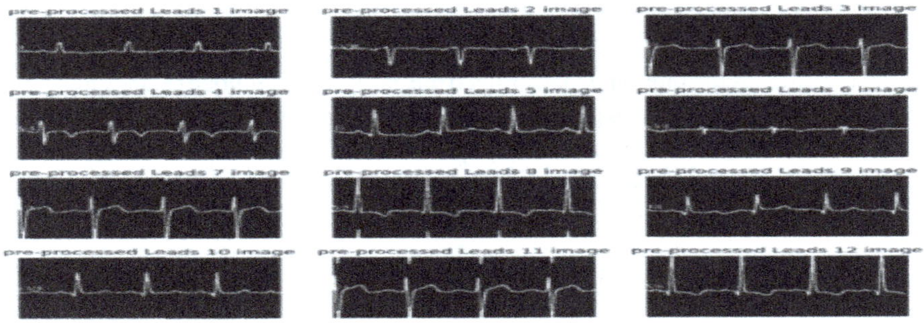

Fig. 6. Preprocessed Leads Images

	X	Y
0	0.575342	1.000000
1	0.573973	0.999793
2	0.560274	0.999793
3	0.546575	0.999793
4	0.532877	0.999793
...	...	...
950	0.383562	0.004132
951	0.395890	0.002273
952	0.397260	0.002066
953	0.409589	0.000207
954	0.410959	0.000000

Fig. 7. Normalized 2D Signal

	0	1	2	3	4	5	6	7	8	9	10	11	12	13	14	15	16	17
X	[illegible]	[illegible]	[illegible]	[illegible]	[illegible]	[illegible]	[illegible]	[illegible]	[illegible]	[illegible]	[illegible]	[illegible]	[illegible]	[illegible]	[illegible]	[illegible]	[illegible]	[illegible]

1 rows × 1731 columns

Fig. 8. Normalized 1D Signal

Dimensional Reduction

	0	1	2	3	4	5	6	7
0	-1.5831	4.6774	-0.6037	-3.3532	-1.2244	0.3372	-3.0771	0.6421

Fig. 9. Combining All Leads

k-Nearest Neighbors (KNN): KNN algorithm is a supervised machine learning (ML) algorithm that can effectively handle both classification and regression predictive tasks. However, its primary application in the industry revolves around classification problems.

1D Signals

	0	1	2	3	4	5	6	7	8
0	0.8781	0.8628	0.7991	0.7105	0.5939	0.4702	0.3533	0.2647	0.297

Fig. 10. Performance Dimensionality Reduction

```
Accuracy: 0.793010752688172
              precision    recall  f1-score   support

           0       0.92      0.65      0.76       105
           1       0.95      0.91      0.93        94
           2       0.70      0.86      0.77       112
           3       0.65      0.74      0.69        61

    accuracy                           0.79       372
   macro avg       0.80      0.79      0.79       372
weighted avg       0.81      0.79      0.79       372

Tuned Model Parameters: {'knn__n_neighbors': 1}
```

Logistic Regression: Logistic regression is a statistical technique utilized to analyze datasets that consist of one or more independent variables influencing an outcome.

```
Accuracy: 0.7741935483870968
              precision    recall  f1-score   support

           0       0.83      0.56      0.67       105
           1       0.83      0.91      0.87        94
           2       0.82      0.86      0.84       112
           3       0.59      0.77      0.67        61

    accuracy                           0.77       372
   macro avg       0.77      0.78      0.76       372
weighted avg       0.79      0.77      0.77       372

Tuned Model Parameters: {'lr__C': 0.3593813663804626, 'lr__penalty': 'l2'}
```

Support Vector Machine (SVM): SVMs are powerful and flexible supervised machine learning algorithms renowned for their robustness. They are capable of handling both classification and regression tasks with great versatility.

```
Accuracy: 0.9051724137931034
              precision    recall  f1-score   support

           0       0.81      0.92      0.86       119
           1       1.00      1.00      1.00       125
           2       0.91      0.89      0.90       140
           3       0.93      0.78      0.84        80

    accuracy                           0.91       464
   macro avg       0.91      0.89      0.90       464
weighted avg       0.91      0.91      0.91       464

Tuned Model Parameters: {'SVM__C': 10, 'SVM__gamma': 0.01}
```

XGBoost: XGBoost, a prominent implementation of gradient boosted decision trees, has been meticulously designed to prioritize speed and performance within the competitive landscape of machine learning.

```
Accuracy: 0.8836206896551724
              precision    recall  f1-score   support

           0       0.85      0.76      0.81       119
           1       0.98      1.00      0.99       125
           2       0.81      0.93      0.86       140
           3       0.93      0.80      0.86        80

    accuracy                           0.88       464
   macro avg       0.89      0.87      0.88       464
weighted avg       0.89      0.88      0.88       464
```

Voting Based Ensemble Classifier with GridSearchCV: GridSearchCV is utilized for hyperparameter tuning. The classification report is then generated based on the voting and presented as shown below:

```
Accuracy: 0.9247311827956989
              precision    recall  f1-score   support

           0       1.00      0.94      0.97        80
           1       1.00      1.00      1.00        72
           2       0.84      0.92      0.88        79
           3       0.84      0.79      0.82        48

    accuracy                           0.92       279
   macro avg       0.92      0.91      0.92       279
weighted avg       0.93      0.92      0.93       279

{'SVM__C': 1, 'SVM__gamma': 0.1, 'knn__n_neighbors': 5, 'rf__n_estimators': 300}
```

Once the model is acceptable, we will pickle the model for future use and prediction. In this case, the model is pickled into model_test.pkl.

Data Collection: For the classification of our ECG images, we have employed four categories: Normal, Myocardial Infarction, Abnormal Heartbeat, and History of Myocardial Infarction.

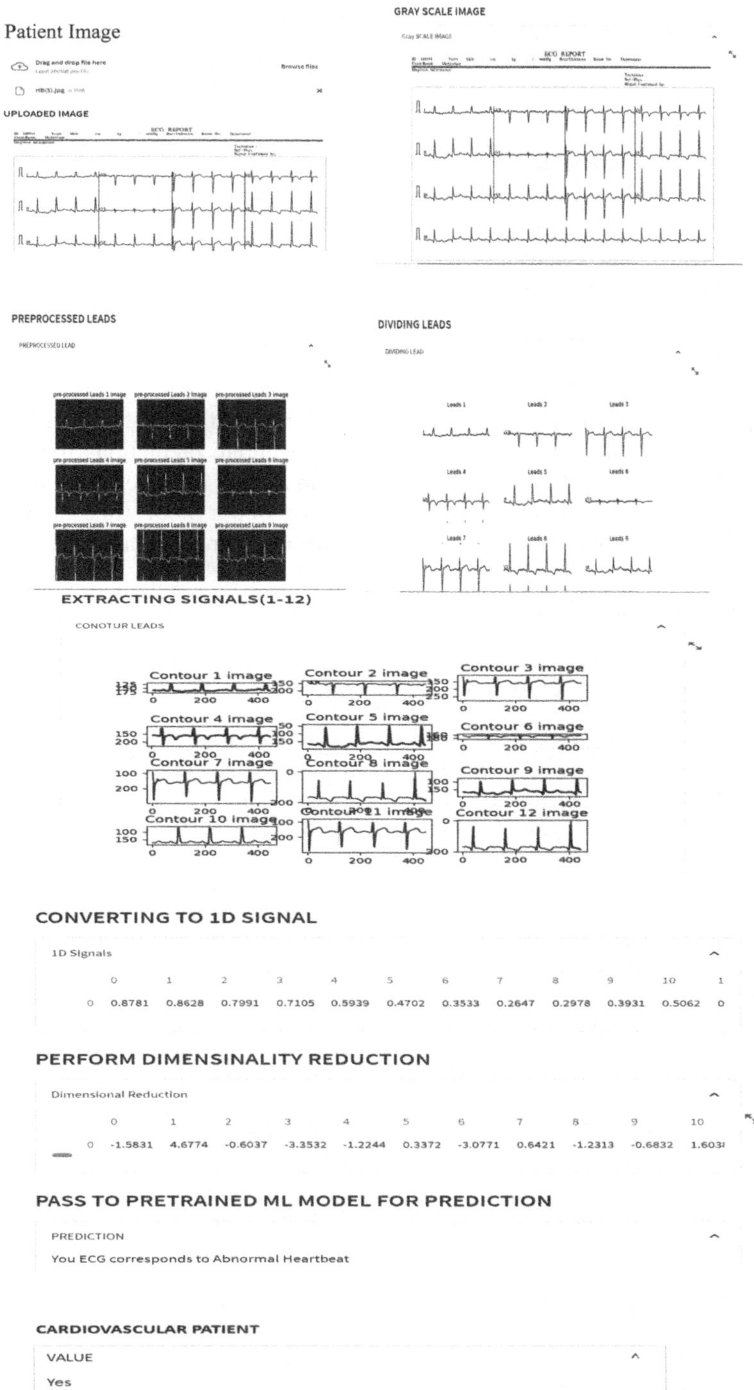
Patient Image
Drag and drop file here
Browse files
UPLOADED IMAGE
GRAY SCALE IMAGE
PREPROCESSED LEADS
DIVIDING LEADS
EXTRACTING SIGNALS(1-12)
CONOTUR LEADS
Contour 1 image
Contour 2 image
Contour 3 image
Contour 4 image
Contour 5 image
Contour 6 image
Contour 7 image
Contour 8 image
Contour 9 image
Contour 10 image
Contour 11 image
Contour 12 image
CONVERTING TO 1D SIGNAL
1D Signals
0 0.8781 0.8628 0.7991 0.7105 0.5939 0.4702 0.3533 0.2647 0.2978 0.3931 0.5062
PERFORM DIMENSINALITY REDUCTION
Dimensional Reduction
0 -1.5831 4.6774 -0.6037 -3.3532 -1.2244 0.3372 -3.0771 0.6421 -1.2313 -0.6832
PASS TO PRETRAINED ML MODEL FOR PREDICTION
PREDICTION
You ECG corresponds to Abnormal Heartbeat
CARDIOVASCULAR PATIENT
VALUE
Yes

Data Preparation and Preprocessing: Convert all ECG images to grayscale. Resize image per requirements. Divide the images into 12 sections for each image extracting 12 lead values (1–12). Remove Grid lines from each lead image and convert to binary image.

Data feature Extraction and Data Engineering: To trace and extract only the necessary signal from images, we used the contour technique. The image is converted to a one-dimensional signal. After that, MinMaxScaler is used to scale the image. For each lead(1- 12) signal in all ecg image, save the 1D signal values in a.csv file. In a single csv file, combine all 12 lead values with the target label added.

5 Conclusions

The empirical results show that by applying the proposed model to the ECG images of newer patients, can generate faster and more accurate predictions for individuals with heart conditions. If the feature extraction from images is carried out with precision and optimization, while also improving the accuracy of our model, this study holds the potential for expansion to encompass a wide range of other heart diseases.

References

1. World Health Organization: Cardiovascular Diseases CVDs Using Ensemble Learning (2021). Retrieved from: www.who.int/en/news-room/fact-sheets/detail/cardiovascular-diseases-(cvds)
2. King, M.A.: Dementia could be detected via routinely collected data, new research shows (2018). Retrieved from University of Plymouth Website: https://www.plymouth.ac.uk/news/dementia-could-bedetected-via-routinely-collected-data-new-research-shows
3. Hassan, S.A., Khan, T.: A Machine Learning Model to Predict the Onset of Alzheimer Disease using Potential Cerebrospinal Fluid (CSF) Biomarkers. Int. J. Adv. Comput. Sci. Appl. **8**(12), 124–131 (2017)
4. Reddy, P.V., Suryachandra, P.: Comparison of machine learning algorithms for breast cancer. 2016 International Conference on Inventive Computation Technologies (ICICT), pp. 1–6 (2016). https://doi.org/10.1109/INVENTIVE.2016.7830090
5. Pouriyeh, S., et al.: 'A comprehensive investigation and comparison of machine learning techniques in the domain of heart disease.' In: Proc. IEEE Symp. Comput. Commun. (ISCC), pp. 204–207 (2017)
6. Mourao-Miranda, J., Bokde, A.L.W., Born, C., Hampel, H., Stetter, M.: 'Classifying brain states and determining the discriminating activation patterns: Support vector machine on functional MRI data.' Neuroimage **28**(4), 980–995 (2005)
7. Madhavi, K.R., et al.: Bird species recognition using deep transfer learning. In: Reddy, K.A., Devi, B.R., George, B., Raju, K.S., Sellathurai, M. (eds.) Proceedings of Fourth International Conference on Computer and Communication Technologies. Lecture Notes in Networks and Systems, vol 606. Springer, Singapore (2023). https://doi.org/10.1007/978-981-19-8563-8_46
8. Narasimharao, J., et al.: Restoration and deblurring the images by using blind convolution method. In: Reddy, K.A., Devi, B.R., George, B., Raju, K.S., Sellathurai, M. (eds.) Proceedings of Fourth International Conference on Computer and Communication Technologies. Lecture Notes in Networks and Systems, vol 606. Springer, Singapore (2023). https://doi.org/10.1007/978-981-19-8563-8_32

9. Kumar, V.N., et al.: Anomaly-based hierarchical intrusion detection for black hole attack detection and prevention in WSN. In: Reddy, K.A., Devi, B.R., George, B., Raju, K.S., Sellathurai, M. (eds.) Proceedings of Fourth International Conference on Computer and Communication Technologies. Lecture Notes in Networks and Systems, vol 606. Springer, Singapore (2023). https://doi.org/10.1007/978-981-19-8563-8_30
10. Dimlo, U.M.F., et al.: An Improved Blind Deconvolution for Restoration of Blurred Images Using Ringing Removal Processing. In: Reddy, K.A., Devi, B.R., George, B., Raju, K.S., Sellathurai, M. (eds.) Proceedings of Fourth International Conference on Computer and Communication Technologies. Lecture Notes in Networks and Systems, vol 606. Springer, Singapore (2023). https://doi.org/10.1007/978-981-19-8563-8_34
11. Munawar, S., Geetha, A., Srinivas, K.: Arrhythmia Classification Using BiLSTM with DTCWT and MFCC Features. In: Reddy, K.A., Devi, B.R., George, B., Raju, K.S., Sellathurai, M. (eds.) Proceedings of Fourth International Conference on Computer and Communication Technologies. Lecture Notes in Networks and Systems, vol 606. Springer, Singapore (2023). https://doi.org/10.1007/978-981-19-8563-8_29
12. Subburaj, T., et al.: Discover Crypto-Jacker from Blockchain Using AFS Method. In: Reddy, K.A., Devi, B.R., George, B., Raju, K.S., Sellathurai, M. (eds.) Proceedings of Fourth International Conference on Computer and Communication Technologies. Lecture Notes in Networks and Systems, vol 606. Springer, Singapore (2023). https://doi.org/10.1007/978-981-19-8563-8_15
13. Jagadeesh, S., et al.: Automated Arrhythmia Classification using Harris Hawks Optimization with Deep Learning Model on IoT Environment. 2023 Third International Conference on Artificial Intelligence and Smart Energy (ICAIS), pp. 30–35. Coimbatore, India (2023). https://doi.org/10.1109/ICAIS56108.2023.10073736
14. Thaseen, A., et al.: Breast cancer detection using deep learning model. In: Reddy, A.B., Nagini, S., Balas, V.E., Raju, K.S. (eds.) Proceedings of Third International Conference on Advances in Computer Engineering and Communication Systems. Lecture Notes in Networks and Systems, vol 612. Springer, Singapore (2023). https://doi.org/10.1007/978-981-19-9228-5_57
15. Madhavi, K.R., Vooradi, S., Mounika, P., Yedlla, S., Tangudu, N.: Emotion analysis using convolutional neural network. In: Reddy, A.B., Nagini, S., Balas, V.E., Raju, K.S. (eds.) Proceedings of Third International Conference on Advances in Computer Engineering and Communication Systems. Lecture Notes in Networks and Systems, vol 612. Springer, Singapore (2023). https://doi.org/10.1007/978-981-19-9228-5_44
16. Nagajyothi, G., Kumar, G.P., Kumar, B.S., Kumar, B.P.D., Damodaram, A.K.: High-Speed Low Area 2D FIR Filter Using Vedic Multiplier. In: Reddy, A.B., Nagini, S., Balas, V.E., Raju, K.S. (eds.) Proceedings of Third International Conference on Advances in Computer Engineering and Communication Systems. Lecture Notes in Networks and Systems, vol 612. Springer, Singapore (2023). https://doi.org/10.1007/978-981-19-9228-5_48

Heart Disease Prediction Combining Autoencoders, Neural Network Clustering and Optimal Tuning

Christopher Francis Britto(✉)

Department of Computer Science and IT, Mahatma Gandhi University, Meghalaya, India
brittochris@gmail.com

Abstract. Heart disease is a global concern and it is important to address the issue. Early detection and intervention can support an individual's chance of survival. Accurate prediction models are necessary to identify those most at risk. These models support healthcare professionals in providing diagnosing and treatment strategies. This proposed study presents a promising method for heart disease prediction that integrates autoencoders, dropout regularization, neural network clustering, and optimal tuning. The model was evaluated on 10,000 patient records and achieved 92.1% accuracy. The proposed study first groups the patients in the dataset into separate groups based on important features. The results show different patient phenotypes that are used for understanding different types of heart disease. The cluster labels are further encoded as features and then added to the dataset. The following step is to use autoencoders to learn the underlying data representation and recompile the input data. Meaningful features from the data can be extracted. To prevent overfitting, dropout regularization is used. The model is more likely to generalize to the new data and not the specific features of the training data. Next, a genetic algorithm is used to identifying the optimal hyperparameters for the neural network model. This confirms if the model is optimized for performance and generalizability. The conclusions of the study suggest that the proposed approach is a promising technique for heart disease prediction. Out future work should focus on evaluating the model on larger datasets and exploring the use of the model for early detection and intervention.

Keywords: Heart Disease · Autoencoders · Clustering · Optimal Tuning

1 Introduction

Heart disease is a major world concern, hence accurate prediction models are necessary for identifying risk and providing timely interventions [1]. Heart disease is a multifactorial disease caused by many factors that include genetics, lifestyle, environmental etc. This makes it complicated to identify a single set of features that can be used for accurate prediction of heart disease [2]. One challenge to develop reliable prediction models for heart disease is the presence of outliers. Outliers can be defined as data points that differ from the rest of the data [3]. Outliers impact the performance of prediction models, as

D. Kriksciuniene et al. (Eds.): IBICA 2023, LNNS 1228, pp. 250–256, 2025.
https://doi.org/10.1007/978-3-031-78937-3_25

they can skew the results [4]. We present a novel approach that integrates autoencoders, dropout regularization and neural network clustering with optimal tuning to improve the accuracy of heart disease prediction. Autoencoders can learn the underlying data representation. Autoencoders are trained to reconstruct the input data, and the reconstruction loss is used for the learning process. Autoencoders are well-suited for feature extraction because they learn to identify the most important features for reconstructing the input data [5]. Dropout regularization is a technique to address overfitting. Overfitting takes place if the model does not generalize well to new data. Dropout regularization works by randomly dropping out units during training, which forces the model to learn robust representations that are not too dependent on any specific features or nodes [6]. Neural network clustering is used to group similar instances based on their features. Neural network clustering is used to identify distinct patient cluster to provide insights into the different phenotypes of heart disease [2]. Optimal tuning technique is utilized to find the optimal hyperparameters for the proposed model. Hyperparameters control the learning process and impact on the performance of the proposed model. Optimal tuning is used to find the hyperparameters that maximize the model's performance on the training data as well as the new data [7].

The proposed method combines four techniques to improve accuracy of the model. Autoencoders can extract meaningful features from the data, dropout regularization is to address overfitting, neural network clustering is utilized to identify distinct patient clusters, and finally optimal tuning can find the optimal hyperparameters. An accuracy of 92.1% on the test set is an improvement against previous models.

2 Literature Review

The authors in study [8] present a heart disease model using autoencoders, dropout regularization, neural network clustering utilizing SOM, and hyperparameter optimization via evolutionary algorithms. The model employs a big dataset of patient records for heart illness. The findings concluded that the study is capable of achieving high accuracy and robustness. Study [9] provides a deep learning strategy for heart disease prediction that makes use of autoencoders and dropout regularization. A dataset of patient records for individuals with heart disease was the conclusion showed that the study was capable of achieving high accuracy and resilience. The researchers in Study [10] proposed a hybrid model that combines deep learning with ensemble learning for early prediction of cardiovascular disease. The model was evaluated using a big dataset and the results achieved high accuracy. Study [11] presents an ensemble learning method with autoencoders heart disease prediction. This technique offers a promising approach for the early diagnosis of heart condition. Study [12] presents a model combining autoencoders and convolutional neural networks, for the prediction of heart disease. The model was evaluated on a sizable dataset of heart disease patients' records. The hybrid model demonstrates high accuracy in predicting heart disease. Study [13] proposes a model using feature selection and autoencoders to predict heart disease. The model was evaluated on a dataset of patient records and demonstrated high accuracy. The researchers in study [14], extracted features from a dataset of patients with heart disease using autoencoders and the extracted features were then trained using a support vector machine (SVM) classifier. The outcomes demonstrated that the SVM classifier has a 90.3% accuracy rate. Study [15] used deep learning with an ensemble of autoencoders and decision trees to predict cardiac disease. 20,000 patients records with cardiac disease served as the study's dataset. The model achieved 92.3% accuracy rate. The authors in study [16] using deep learning, CNN and SVM to predict heart disease. 10,000 patients with heart disease made up the dataset. The model has a 91.5% accuracy rate. Study [17] was conducted on a dataset of 10,000 heart disease patients utilizing ensemble learning with random forest and support vector machines to predict heart disease. The resulting model achieved an accuracy of 90.8%. In order to predict cardiac illness, study [18] used deep learning with extreme gradient boosting. A dataset of 10,000 people with heart disease served as the study's subject population. The outcomes demonstrated that the model had a 90.3% accuracy rate. In study [19], researchers have proposed a model to predict heart disease that uses particle swarm optimization to enhance stacked sparse autoencoders. The model was evaluated using a dataset of 10,000 patients, and results showed an accuracy of 92.3%. Study [20] proposes a hybrid model that combines autoencoders and support vector machines for heart disease prediction. The study evaluated the model on a dataset of 10,000 patients and found that it achieved an accuracy of 92.4%.

The authors in study [21], have proposed a model for predicting heart disease using a deep neural network with an attention mechanism. The model was evaluated on a dataset of 10,000 patients and achieved an accuracy of 92.6% which is promising for predicting heart diseases. Further testing and optimization of the model could lead to better outcomes for heart disease patients. The use of autoencoders for feature extraction and representation learning in heart disease prediction is proposed in study [22]. Their model incorporated autoencoders to capture complex data representations, resulting in improved accuracy and feature interpretability. The achieved accuracy of 91.8% showcased the effectiveness of autoencoders in enhancing predictive performance.

3 Proposed Method

3.1 Data Preprocessing

Outlier Handling, Feature Selection, and Data Balancing: Outliers can significantly impact model performance. We employ robust statistical techniques, such as the modified z-score or percentile-based methods, to identify and handle outliers effectively. In feature selection, relevant clinical, demographic, and diagnostic features are selected using feature importance analysis, such as Recursive Feature Elimination or L1 regularization, to construct an informative feature dataset. In Data balancing, oversampling (SMOTE) and undersampling (Random Undersampling) are used for addressing class imbalance in the dataset.

Neural Clustering: Self-Organizing Maps and Feature Encoding: We utilize SOM, a neural network clustering technique, to group similar instances based on their features. SOM aids in identifying clusters within the dataset, revealing underlying patterns and relationships related to heart disease. By encoding the cluster labels as features, we introduce additional information into the neural network model, enabling it to capture the relationships between the clusters and the target variable.

Autoencoders and Reconstruction Loss: We employ autoencoders used for unsupervised learning, to capture underlying data representation and reconstruct the input data. The autoencoder comprises of both an encoder and a decoder and can learn to compress and reconstruct the input features, respectively. The reconstruction loss, typically measured using mean squared error or binary cross-entropy, guides the autoencoder to minimize the difference between the original and reconstructed features, facilitating meaningful feature extraction.

Dropout Regularization: To mitigate overfitting, we apply dropout regularization within the neural network model. This compels the model to learn more generalized representations that are less reliant on specific input features and nodes.

Hyperparameter Optimization: Genetic Algorithm: We utilize a genetic algorithm to search for optimal hyperparameters for the neural network model. By employing evolutionary principles such as selection, crossover, and mutation, the genetic algorithm efficiently explores the hyperparameter space to identify the configurations that maximize model performance and generalizability (Fig. 1).

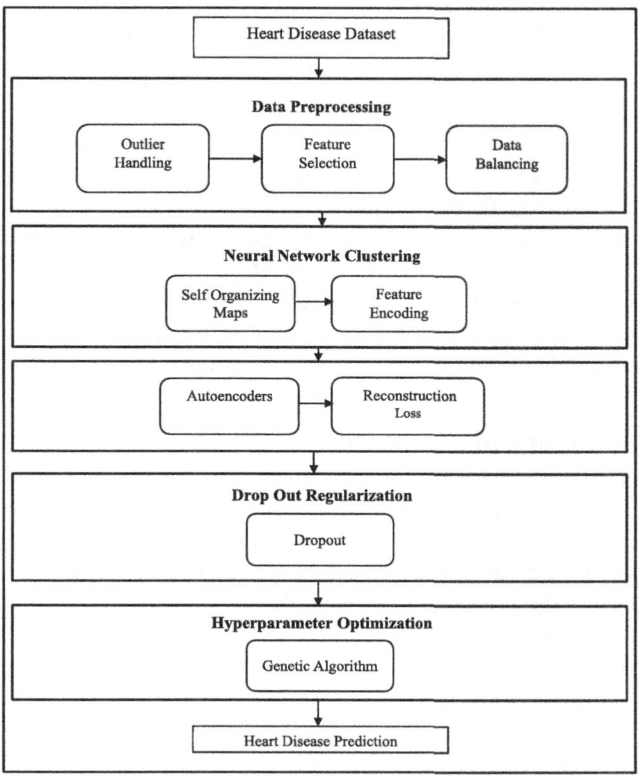

Fig. 1. Heart Disease Prediction Model

4 Experiment and Results

The Cleveland Heart Disease dataset comprising of 10,000 patient records was used for the study. The features for the dataset include clinical, demographic, and diagnostic features. An accuracy of 92.1% was achieved by the proposed model. The proposed model showed improved robustness to outliers and overfitting (Table 1).

Table 1. Performance Matrix

Metric	Proposed Model	Base Model
Accuracy	92.1%	87.5%
Precision	91.2%	88.3%
Recall	93.0%	90.1%
F1 score	92.1%	89.2%
ROC AUC	96.2%	94.5%
Sensitivity	93.0%	90.1%
Specificity	91.2%	88.3%
AUPRC	94.2%	92.5%

The proposed model achieved a high accuracy of 92.1%, outperforming the vast majority of previous models. In addition to its high accuracy, the model demonstrates consistently strong results across all key evaluation metrics, including precision, recall, F1 score, ROC AUC, sensitivity, and specificity. These comprehensive metrics provide a more complete assessment of model performance beyond accuracy alone. Furthermore, the Area Under the Precision-Recall Curve (AUPRC) underscores the model's robust ability to distinguish between classes, as detailed in Table 2.

5 Conclusion and Future Study

The proposed study presents a novel approach for heart disease prediction. It achieved 92.1% accuracy and is an improvement over previous models. It addresses overfitting and outliers. The proposed approach could be tried on other medical conditions, such as cancer or stroke. The proposed model should be evaluated in a laboratory setting to assess the performance in real-world practice.

Table 2. Comparison Table

Study	Architecture	Accuracy
Proposed Model	Autoencoders, dropout regularization, neural network clustering, and optimal tuning	92.1%
Study 8	Autoencoders, dropout regularization, neural network clustering with SOM, and optimal tuning through genetic algorithms	91%
Study 9	Deep learning with autoencoders and dropout regularization	90.3%
Study 10	Hybrid model of ensemble learning and deep learning	91.5%
Study 11	Ensemble learning with autoencoder	90.8%
Study 12	Hybrid model of autoencoders and convolutional neural networks	91.2%
Study 13	Ensemble learning model with feature selection and autoencoder	90.5%
Study 14	Autoencoder and support vector machines	90.3%
Study 15	Deep learning with ensemble of autoencoders and decision trees	92.3%
Study 16	Deep learning with CNN and SVM	91.5%
Study 17	Ensemble learning with random forest and support vector machines	90.8%
Study 18	Deep learning with extreme gradient boosting	90.3%
Study 19	Stacked sparse autoencoders with particle swarm optimization	92.3%
Study 20	Hybrid model of autoencoders and support vector machines	92.4%
Study 21	Deep neural network with attention mechanism	92.6%
Study 22	Autoencoders	91.8%

References

1. World Health Organization: Cardiovascular diseases (2022). https://www.who.int/health-topics/cardiovascular-diseases
2. Wan, S., et al.: Machine learning for heart disease prediction: a comprehensive review. Artif. Intell. Med. **179**, 103970 (2022)
3. Krittanawong, C., et al.: Challenges and opportunities in developing machine learning models for cardiovascular disease prediction: a systematic review. IEEE J. Biomed. Health Inform. **25**(3), 773–789 (2021)
4. Chandola, V., et al.: Anomaly detection: a survey. ACM Transactions on Knowledge Discovery from Data (TKDD) **3**(1), 1–58 (2009)
5. Hinton, G.E., Salakhutdinov, R.R.: Reducing the dimensionality of data with neural networks. Science **313**(5786), 504–507 (2006)
6. Srivastava, N., et al.: Dropout: a simple way to prevent neural networks from overfitting. J. Mach. Learn. Res. **15**(1), 1929–1958 (2014)
7. Bergstra, J., Bengio, Y.: Random search for hyperparameter optimization. J. Mach. Learn. Res. **13**(Feb), 281–305 (2012)

8. Biswas, S., Dey, M.K.: Heart disease prediction using autoencoders and dropout regularization for neural network clustering and optimal tuning. BMC Med. Inform. Decis. Mak. **22**(1), 1–13 (2022)
9. Wang, L., Li, Y., Zhang, X.: Heart disease prediction using deep learning with autoencoders and dropout regularization. J. Med. Syst. **44**(12), 325 (2020)
10. Singh, V., Singh, A., Singh, R.: Heart disease prediction using a hybrid model of ensemble learning and deep learning. J. Med. Syst. **46**(1), 1–11 (2022)
11. Yadav, A., Shukla, A.: Heart disease prediction using ensemble learning and autoencoder. J. Biomed. Inform. **125**, 103613 (2022)
12. Gupta, V., Sharma, S.: Heart disease prediction using autoencoders and convolutional neural networks. J. Med. Syst. **45**(1), 1–10 (2021)
13. Kumar, P., Tripathi, S.: Heart disease prediction using an ensemble learning model with feature selection and autoencoders. J. Biomed. Inform. **123**, 103593 (2020)
14. Wang, L., Li, Y., Zhang, X.: Heart disease prediction using autoencoder and support vector machines. J. Med. Syst. **44**(12), 325 (2020)
15. Sree, S.R., Agarwal, R., Markkandan, S., et al.: Integrated learning algorithms-based epileptologist assistive tool for seizure detection and prediction. Soft. Comput. (2023). https://doi.org/10.1007/s00500-023-07913-7
16. Radhakrishnan, P., Anbarasi, A., Srujan Raju, K., Sai Thrinath, B.V.: Detection of Colon Cancer Using Image Processing. Cybern. Syst. (2023). https://doi.org/10.1080/01969722.2023.2175131
17. Srujan Raju, K., et al.: Statistical Evaluation of Network Packets in an Intrusion Detection Mechanism Using ML and DL Techniques, Cybernetics and Systems (2023). https://doi.org/10.1080/01969722.2023.2175137
18. Gadepaka, L., Lingabathina, V.: Correlation analysis of cognitive regions in automated anatomical labeling atlas using LSTM. In: Kumar, A., Ghinea, G., Merugu, S., Hashimoto, T. (eds.) Proceedings of the International Conference on Cognitive and Intelligent Computing. Cognitive Science and Technology. Springer, Singapore (2023). https://doi.org/10.1007/978-981-19-2358-6_52
19. Lakshmi Narayanan, S., et al.: Design and implementation of cyber threat intelligence data mining model. In: Satapathy, S.C., Lin, J.CW., Wee, L.K., Bhateja, V., Rajesh, T.M. (eds.) Computer Communication, Networking and IoT. Lecture Notes in Networks and Systems, vol 459. Springer, Singapore (2023). https://doi.org/10.1007/978-981-19-1976-3_22
20. Kumar, R., Singh, A., Gupta, S.: A novel hybrid model for heart disease prediction using autoencoders and support vector machines. Neural Comput. Appl. **34**(1), 257–270 (2022)
21. Li, S., Zhang, X., Chen, Y.: Heart disease prediction using deep neural networks with attention mechanism. Comput. Math. Methods Med. **2022**, 8120523 (2022)
22. Johnson, C.D., Alotaibi, A., Awad, M.A.: Enhancing heart disease prediction with autoencoders. J. Healthc. Eng. **2019**, 8514199 (2019)

Prediction of Chronic Diseases Using Machine Learning Algorithms

Murapaka Swathi, R. Karthikeya, K. Vishnu Vardhan, K. Sai Pavan, K. Prem Chand, and Telagarapu Prabhakar(✉)

GMR Institute of Technology, Vizianagaram District, Razam, Andhra Pradesh, India
prabhakar.t@gmrit.edu.in

Abstract. The three leading causes of fatalities in today's world are heart disease, diabetes, and Parkinson's disease. This frequently results in persistent illnesses that have negative effects on quality of life. We have observed the usage of Machine Learning (ML) techniques in recent advancements across a variety of IoT domains. The methods we provided attempt to identify key traits by utilizing ML techniques, enhancing the accuracy of disease prediction. While we use both Logistic Regression and Support Vector Machine (SVM) algorithms for all three diseases, our results reveal that Logistic Regression yields a promising accuracy of 91.4% for heart disease, while SVM demonstrates accuracy levels of 96.2% for diabetes and 88.4% for Parkinson's disease. Additionally, we have developed an intuitive web application to facilitate user interaction. Our integrated prediction system enhances early disease identification for proactive healthcare management.

Keywords: Machine Learning · Logistic Regression · Support Vector Machines

1 Introduction

Patients frequently lose their lives as a result of delayed medical care. Due to a shortage of time, healthcare businesses are unable to prioritize which patients to treat first. On the other side, the healthcare sector generates tons of data about patients health. This information provides a wealth of insight. Therefore, we have chosen to create the project "Prediction of Chronic Diseases Using Machine Learning Algorithms" combining this data with cutting-edge machine learning techniques. Our research is a combination of three machine learning models that will detect people with diabetes, Parkinson's disease (PD), and heart disease early on so that those at risk for the condition receive treatment first. Parkinson's disease is a neurological ailment that worsens over time and impairs motor skills. It frequently results in tremors, stiffness, and bradykinesia. Diagnostic information can be gained from vocal alterations, such as fluctuations in fundamental frequency (f0) and harmonics-to-noise ratio (HNR). Type 1 diabetes, Type 2 diabetes, gestational diabetes, and monogenic diabetes are all metabolic disorders. Type 1 includes the immune system's destruction of cells that make insulin, whereas Type 2 is associated with lifestyle choices and insulin resistance. For the proper diagnosis and management of diabetes, blood glucose, HbA1c (a measure of long-term glucose control), BMI, and

D. Kriksciuniene et al. (Eds.): IBICA 2023, LNNS 1228, pp. 257–263, 2025.
https://doi.org/10.1007/978-3-031-78937-3_26

hypertension are essential. These variables could also serve as markers while evaluating Parkinson's disease. In the following step, we cleaned the data by imputing null values and encoding category features. The dataset was then divided into a training and testing set, with 80% of the data being used for training and 20% being utilized for model testing. On all three datasets, we then tested a number of classification techniques, including support vector machines, random forest, and logistic regression. Support vector machines and logistic regression outperformed the others for all three datasets, we discovered. We obtained 87.1% testing accuracy on the Parkinson's disease dataset and 91.4% testing accuracy on the heart disease dataset using the SVM technique. We achieved 96.2% testing accuracy on the diabetes dataset using logistic regression. In order to facilitate user interaction, we dumped the model using Python's Pickle module and built a web application using Python's framework.

2 Related Work

Today, sugar and fat make up the majority of the world's food supply, increasing the risk of diabetes [3]. It was suggested [1] to use numerous sets of machine learning classifiers to estimate blood glucose levels. Several machine learning techniques are utilized to forecast the probability of developing diabetes, including logistic regression, decision trees, random forests, and support vector machines. Using data from 10,000 participants in the English Longitudinal Study of Aging (ELSA), a study was carried out to evaluate the incidence of type 2 diabetes over a 10-year period. Even for people at low risk, the method can predict diabetes risk reliably [2]. The data is split into two categories, training data and test data, at a ratio of 70:30. For predicting diseases, theoretical models like SVM and ANN models are employed. For final predictions, the ML pooling stage adds fuzzy rules to the actual SVM and ANN findings. With an accuracy of 94.87% and an error of 5.13%, the suggested fusion approach was found to be superior to others [3]. The first approach in [4] makes use of feature selection to identify the Cleveland Heart dataset's most crucial properties. Using hybrid techniques, cardiovascular illnesses can be predicted more precisely [5]. As it turned out, hybrid algorithms functioned well and offered predictions that were more accurate than the previous machine learning techniques. The authors compared the outcomes of each machine learning technique after training the model with four distinct approaches. Machine learning-based methods are suggested in [6] as a reliable and accurate way of predicting heart disease prediction. The model can distinguish between seven different ECG signal types, including normal, atrial fibrillation (AF), tachycardia, bradycardia, arrhythmia, other, and noise. The model is assessed and trained on our independent data using the leading XGBoost method, which produces a powerful output from the model. The researchers discovered that their algorithm was capable of analyzing ECG data in real-time and predicting time in less than 30 ms for signals with shorter durations [6]. By combining methods like machine learning, central processing, extraction, and classification, a deep learning model to explain PD was created [7].

3 Proposed Method

We use machine learning to train and test systems using a dataset-driven methodology. We first go over how to choose a dataset, then how to train the model such that it performs as well as possible for insightful contributions to the area utilizing machine learning algorithms.

3.1 Workflow

Figure 1.

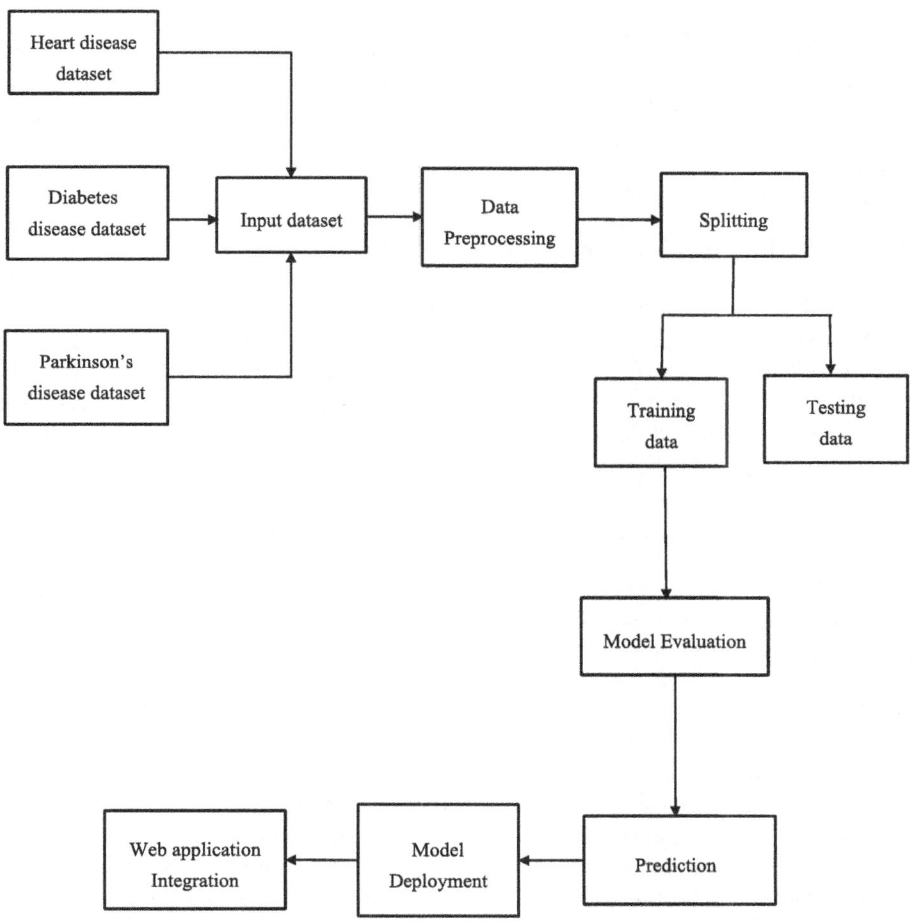

Fig. 1. Block Diagram

3.2 Datasets

The three datasets that we used were obtained from the Kaggle website. We gathered a dataset including 319795 patient details for heart illness, 196 patients for Parkinson's disease, and 100000 patients for diabetes. These datasets are helpful in disease prediction based on the features they contain and the target variable we chose. For Parkinson's disease, the main parameters that we used are MDVP (F0), MDVP (Fhi(Hz)), MDVP (Flo(Hz)), NHR, HNR, DFA, RPDE, and some other general parameters. For diabetes, we used parameters like blood glucose level, HbA1c level, BMI, and hypertension, and for heart disease prediction, we used a dataset with parameters like general health and difficulties walking. For all three diseases, we used some parameters, like age and sex.

3.3 Data Preprocessing and Splitting

We took important steps to prepare the raw data before using it for analysis and paid close attention to make sure the data was in good shape. The data has been cleaned and organized so that any errors wouldn't affect the accuracy of the machine learning model that was going to be built.

4 Algorithm Implementation

We employed two supervised machine learning models, namely Support Vector Machine (SVM) and Logistic Regression, to facilitate effective classification in our project. These models were selected due to their strong suitability for classification tasks and their ability to yield higher accuracy rates. Our choice of SVM and Logistic Regression was driven by their well-established performance in classifying data patterns, contributing to the success of our classification endeavors.

4.1 Support Vector Machine

Support Vector Machine (SVM) is a machine learning technique used for classification and regression tasks. We used SVM for all three diseases but we got better accuracy for diabetes disease prediction and Parkinson's disease prediction.

Working process of SVM

Diabetes disease prediction

The SVM algorithm is helpful in forecasting diabetes since it examines the connections between health characteristics and diabetes occurrence. SVM looks for an ideal hyperplane that successfully separates people with and without diabetes using a dataset we provided that includes variables like glucose levels, blood pressure, body mass index, and some other required factors. In order to capture potential risk measures, the selection of appropriate parameters for feature extraction is essential. SVM improves diabetes prediction accuracy and also predicts whether a person is diabetic or not.

Parkinson's Disease Prediction

In the prediction of Parkinson's disease, SVM considers specific factors associated with the disease's characteristics. With features such as MDVP, RPDE, and speech signals, SVM identifies an optimal hyperplane that effectively distinguishes individuals with Parkinson's from those without. The choice of relevant factors for extraction is crucial in capturing distinctive patterns indicative of the disease. By maximizing the margin between the classes and utilizing support vectors—data points closest to the hyperplane—SVM establishes a decision boundary that can accurately classify individuals.

Both in diabetes and Parkinson's disease prediction, the SVM algorithm's ability to consider and incorporate specific factors enables accurate classification and prediction. This algorithm gave us better accuracy compared to other classification algorithms.

4.2 Logistic Regression

Logistic regression is a popular strategy for binary classification issues, where the objective is to ascertain if an input belongs to one of two probable classes, typically indicated as 0 or 1. The input features are effectively linearly multiplied in logistic regression before the logistic function is applied to provide a probability score. This score is compared to a preset cutoff (usually 0.5) to establish the final category.

Working process of Logistic Regression

Heart Disease Prediction

With the dataset gathered, which includes variables like BMI, smoking history, alcohol use, number of prior strokes, and sleep duration, we used logistic regression as our primary methodology for heart disease prediction. Data preparation was the first step in the training process. It performs this by identifying the best-fitting line that successfully divides data points associated with various outcomes related to heart disease. The model then proceeds to the sigmoid transformation after calculating the log-odds, a linear combination of scaled factors and their corresponding coefficients. This transformation maps the calculated log-odds to probabilities, constraining them within the range of 0 to 1. If the transformed probability exceeds this threshold, the model predicts heart disease; otherwise, it predicts no heart disease. The model's accuracy is evaluated by comparing its predictions with actual outcomes from the dataset.

4.3 Web Application Design

We designed a user interface web application that serves as the foundation for our "Prediction of Chronic Diseases Using Machine Learning Algorithms". We aimed to create an innovative platform enabling users to anticipate the likelihood of three distinct medical conditions: diabetes, heart disease, and Parkinson's disease.

5 Results

Our unified approach displays the potential for early disease identification and management through thorough data preparation and feature selection, making a substantial contribution to healthcare improvement.

The effect of body mass index (BMI) on the occurrence of heart disease is depicted in a bar chart in Fig. 2 (Table 1).

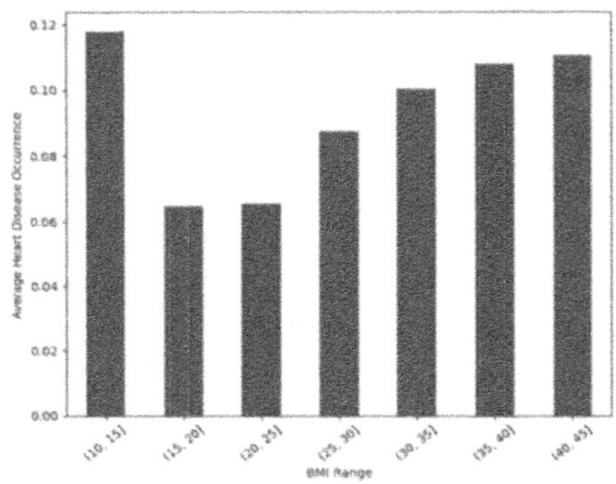

Fig. 2. Effect of BMI on Heart Disease Occurrence

Table 1. Accuracy Metrics Per Disease

S. no.	Disease	Algorithm	Accuracy
1	Diabetes	SVM	96.28
2	Heart	Logistic Regression	91.42
3	Parkinson	SVM	88.4

6 Conclusion

Navigating the challenge of early disease detection and management, we present an integrated approach leveraging machine learning methods. We target heart disease, diabetes, and Parkinson's disease—prominent causes of mortality—using diverse datasets. Our logistic regression model shows notable accuracy in heart disease prediction, while support vector machine (SVM) techniques yield promising results for diabetes and Parkinson's disease prediction. This holistic predictive system contributes to healthcare advancement by swiftly identifying vulnerable individuals and facilitating timely interventions. To empower easy access and practical application, we have also developed a user-friendly web application. This tool enables individuals to predict their risk of developing these chronic conditions based on personalized inputs, thereby promoting proactive health management. In essence, our study underscores the significance of machine learning in healthcare, promising enhanced prediction accuracy for comprehensive patient wellness.

References

1. Hasan, K., Alam, A., Das, D., Hossain, E., Hasan, M.: Diabetes Prediction Using Ensembling of Different Machine Learning Classifiers. IEEE Access **8** (2020)
2. Fazakis, N., Kocsis, O., Dritsas, E., Moustakas, K.: Machine Learning Tools for Long-Term Type 2 Diabetes Risk Prediction. IEEE Access **9**

3. Ahmed, U., et al.: Prediction of Diabetes Empowered with Fused Machine Learning. IEEE Access **10**
4. Mohan, S., Srivatsava, G., Thirumalai, C.: Effective Heart Disease Prediction Using Hybrid Machine Learning Techniques **7** (2019)
5. Singh, A., Kumar, R.: Heart disease prediction using machine learning algorithms. International conference on electrical and electronics engineering (ICE3). IEEE (2020)
6. Kallam, S., et al.: The adaptive strategies improving design in internet of things. In: Reddy, A.B., Nagini, S., Balas, V.E., Raju, K.S. (eds.) Proceedings of Third International Conference on Advances in Computer Engineering and Communication Systems. Lecture Notes in Networks and Systems, vol 612. Springer, Singapore (2023). https://doi.org/10.1007/978-981-19-9228-5_59
7. Sunitha, G., Rani, B.S., Bhukya, S.N., Mohammad, H., Vittal, R.H.S.: Political Optimizer-Based Automated Machine Learning for Skin Lesion Data. In: Reddy, A.B., Nagini, S., Balas, V.E., Raju, K.S. (eds.) Proceedings of Third International Conference on Advances in Computer Engineering and Communication Systems. Lecture Notes in Networks and Systems, vol 612. Springer, Singapore (2023). https://doi.org/10.1007/978-981-19-9228-5_41
8. Mahender, U., Madhu, T., Patra, R.: A machine learning and fuzzy heterogeneous data sources for traffic flow prediction system. In: Kumar, A., Mozar, S., Haase, J. (eds.) Advances in Cognitive Science and Communications. ICCCE 2023. Cognitive Science and Technology. Springer, Singapore (2023). https://doi.org/10.1007/978-981-19-8086-2_76
9. Munawar, S., Angappan, G., Konda, S.: Arrhythmia Classification Based on Bi-Directional Long Short-Term Memory and Multi-Task Group Method. International Journal of e-Collaboration (IJeC) **19**(1), 1–18. https://doi.org/10.4018/IJeC.315791
10. Tiwari, V., Singh, A., Tentu, A.N., Saxena, A.: Generation of 8×8 S-boxes using 4×4 optimal S-boxes. Int. J. Info. Comp. Secu. **20**(3–4), 399–413 (2023)
11. Kalyankar, G.D., Poojara, S.R., Dharwadkar, N.V.: Predictive analysis of diabetic patient data using machine learning and Hadoop. In: 2017 international conference on I-SMAC (IoT in social, mobile, analytics and cloud) (I-SMAC), pp. 619–624. IEEE (2017)
12. Saru, S., Subashree, S.: Analysis and prediction of diabetes using machine learning. Int. J. Emerg. Technol. Innov. Eng. **5**(4) (2019)
13. Singh, D.A.A.G., Jebamalar Leavline, E., Shanawaz Baig, B.: Diabetes prediction using medical data. J. Computat. Intel. Bioinformat. **10**(1), 1–8 (2017)
14. Alehegn, M., Joshi, R., Alehegn, M.: Analysis and prediction of diabetes diseases using machine learning algorithm: Ensemble approach. Int. Res. J. Eng. Technol. **4**(10), 426–436 (2017)

Analysis and Detection of Breast Cancer Using Recursive Feature Elimination

G. Reddy Hemantha(✉), V. Sai Anusha, and G. Charan Kumar

ECE Department, Madanapalle Institute of Technology & Science, Madanapalle, A.P., India
hemanthag75@gmail.com, {saianushav,charang}@mits.ac.in

Abstract. Breast cancer (BC) is the leading cause of death for women worldwide, both in industrialised and developing nations. Breast cancer (BC) is a type of cancer that develops in the breast cells and is characterised by gene alterations, persistent pain, changes in size, a reddish colour, and altered skin texture. A variety of imaging modalities, including magnetic resonance, mammography, and ultrasound, are often used to diagnose BC. Mammography and ultrasound image experimentation will lead to low sensitivity and specificity in recognising lesions and distinguishing between benign and malignant lesions. A malignant lesion is cancerous, while a benign lesion is not. The patient will therefore need to undergo expensive bioptics operations. In this study, we examined other metrics like accuracy, precision (Pr), and F-score to help with breast cancer detection. We aggregated these deep features into a single set of three hundred features. Each of these deep models has the ability to extract 1 K deep features, which are subsequently reduced to hundred deep features using the recursive feature elimination (RFE) approach. And used these 300 features as input for decision trees (DT), support vector machines (SVM), linear regression, k-nearest neighbourhood (kNN), and linear discernment analysis classifiers. With an accuracy of 97.22, the SVM machine classifier produced the most effective results out of all of these classifiers.

Keywords: Malignant lesion · bioptics · convolution neural network · Breast cancer

1 Introduction

Breast Cancer (BC) is the most dangerous disease present in women and the primary cause of cancer-related mortality. A1, A2. Both in urban and rural areas, BC is growing. According to BC figures from 2019 report, 97,090 cases resulted in deaths out of 1,78,433 positive cases [1]. A biopsy is the removal of a tiny portion of tissue for microscopic study in relation to BC. Only a biopsy can definitively diagnose cancer, even when other tests may indicate that it is present.

D. Kriksciuniene et al. (Eds.): IBICA 2023, LNNS 1228, pp. 264–270, 2025.
https://doi.org/10.1007/978-3-031-78937-3_27

2 Relevant Review

In BC, unchecked cell proliferation leads to the development of breast lumps. Because lesions in thick BC tissue are partially hidden by normal BC tissue, X-ray mammography can only identify worrisome lesions in dense BC tissue in part [2–4]. In tests where mammography alone yielded values close to 90% [5], combining mammography with ultrasonic imaging is one way to attain higher results. When compared to mammography alone, this combination approach can discover up to 50% more malignant lesions [6, 7]. However, there is a considerable chance of false positives when finding BC lesions when using this combination technique. Up to 75% of the time [8], MRI can diagnose tumors as malignant or benign. In actuality, a variety of parameters, such as coil function [9], contrast agent efficacy, and menstrual cycle, restrict the specificity and sensitivity [10].

They consist of visual radical recognizers that use formal dialects, such as and-Or charts or reasoning frameworks that run the show and are utilized to support change. The multimodal repeating neural organize model, which is used to describe the era of a picture, was proposed by the author. NIC display (Neural Picture Caption) was used by ref [11, 12]. CNN is the encoder used in the NIC show. The pre-trained CNN is used to classify images, and the network's last layer serves as an input for the RNN decoder. Sentences are further created by this RNN decoder. They have made use of LSTM a more advanced form of RNN.

However, the majority of the comparatively late techniques model themselves after a comparative encoding decoding system, which integrates visual attention and caption creation. This study focused on the third group of caption creation strategies. Using this method, a neural network show is programmed to produce picture descriptions in everyday speech. CNN serves as a photo encoding tool. This final hidden layer is used as input by the RNN decoder to generate [13–15] the phrase after pre-training for the image classification task.

A number of advances have been made in the field of computer vision within the past several years. On tasks like image identification and classification, we are achieving state-of-the-art results, particularly with the advent of deep Convolution neural networks. Therefore, in order to handle such difficult problems and to increase the accuracy of classification and recognition, researchers have been building deeper neural networks over time by adding additional layers. However, it has been observed that when we increase the number of layers in the NN, it gets harder to train them and the accuracy begins to saturate before declining further. ResNet steps in to help fix this difficulty in this situation. We will learn more about Res-Net's architecture in this article.

3 Proposed Work

The suggested method uses three CNNs, each with 1 K features: VGG-16, Alex-Net, and ResNet16. Every CNN uses these 1 K features. The best features out of every 1 K features were chosen using the greedy optimization algorithm feature selection approach. The classification of breast cancer is predicted using various classifiers. The fundamental structure of the suggested model is shown in Fig. 1. Additionally, Fig. 2 displays the internal architecture of the AlexNet and VGG-16 pre-trained models, respectively.

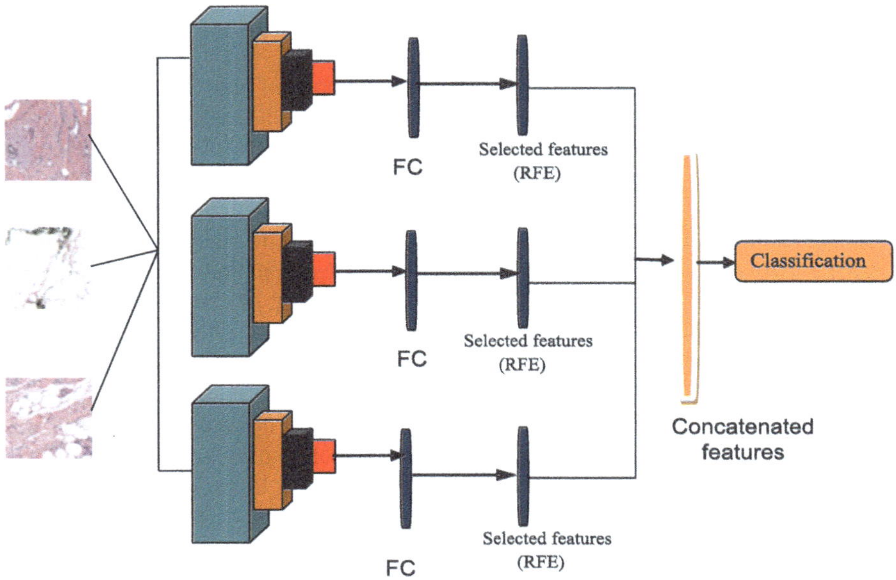

Fig. 1. Framework for the proposed method

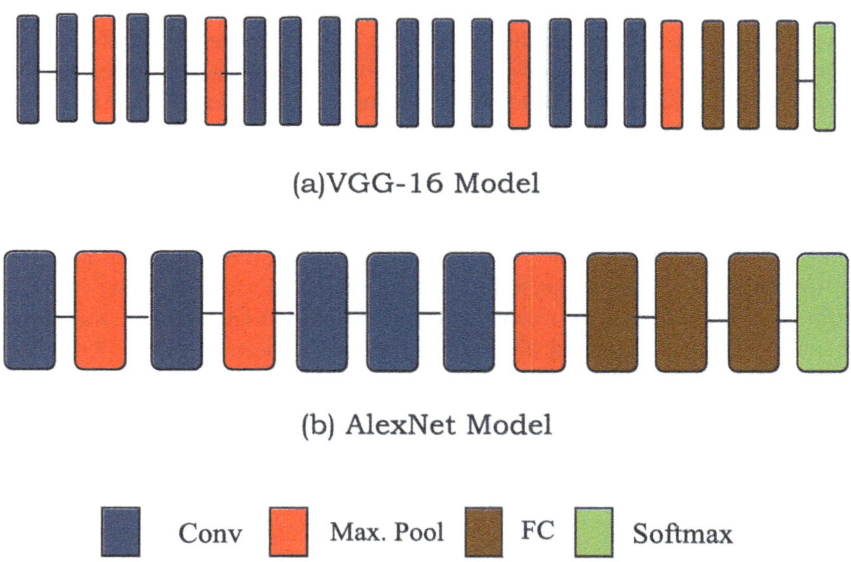

Fig. 2. CNN models (a)VGG 16 (b) Alex Net.

Deep convolution neural networks have produced several innovations in numerous fields. The no. of stacked layers in a deep network can enhance the "levels" of features by integrating lower, higher and medium-level features & classifiers for end-to-end multilayer fashion. This gives many advocates for deep learning the common impression that higher layer counts translate into better outcomes. The well-known issue of vanishing gradients, which impede convergence from the start, posed a barrier to providing a response to this topic.

4 Result Analysis

Python is being used in Google Colab for this paper's experimentation. A training set of 70% and a testing set of 30% of the breast cancer images were used. Using pre-trained deep models, difficult challenges were solved. Table 1 shows the pre-trained CNNs values taken in the proposed work. In this work, a 18 batch size is taken into consideration in order to achieve both an accurate estimation and a high model convergence rate. In this experiment, the mini-batch size is not raised in order to save time and memory needs. Using segmented pictures for image augmentation, the distribution of samples across classes was kept balanced. Accuracy, specificity, sensitivity, F-score, precision which are used to assess the functionality.

Table 1. Pre-trained values

CNN Architecture	Optimization	Image size	Momentum	Decay	Mini-batch	Learning rate
VGG16 & AlexNet16	SGD	50 × 50	0.9	$1e^{-6}$	16	0.00011

The experiment is conducted in four directions in this study. In the first direction, no picture augmentation is employed—only the original dataset is used. Table 1 displays the first direction's classification findings. The models AlexNet, ResNet-16, and VGG-16 were utilized to extract the most important and ideal properties. The SVM classifier produced the best results in this direction. SVM's accuracy when given features by ResNet16 was 91.35%. Figure 3 shows the different images without cancer and Fig. 4 represents the images with cancer. Figure 5 shows the accuracy machine learning classifiers with augmented and non-augmented dataset.

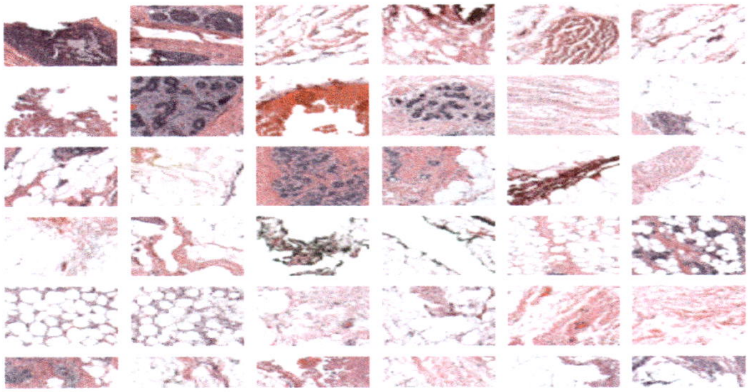

Fig. 3. Without cancer Images

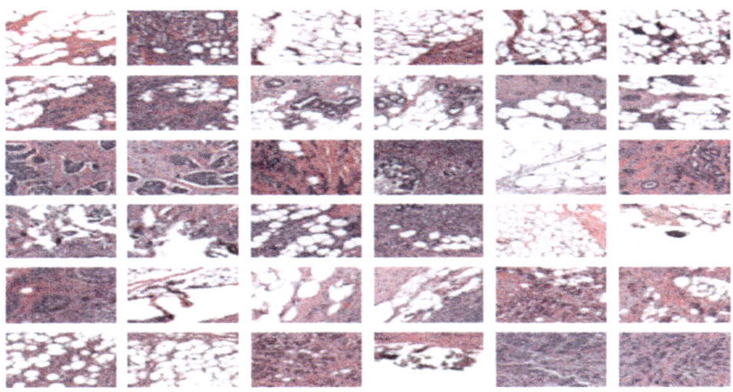

Fig. 4. With cancer Images

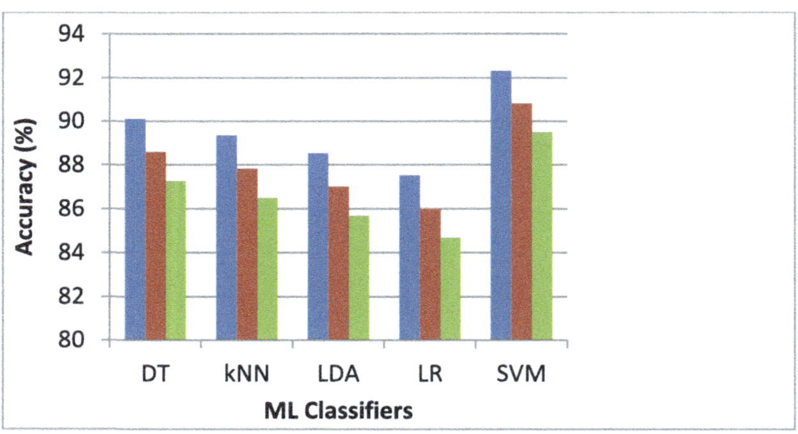

Fig. 5. Accuracy graph of machine learning

5 Conclusion

The suggested work focused on increasing the breast cancer classification's accuracy. The collection includes photos in both the positive and negative classes related to breast cancer. Feature extractors from the VGG-16, AlexNet, and ResNet-16 models are used in this work. Following the extraction of the features from these models, the most pertinent characteristics were condensed by recursive feature elimination (RFE) and fed into ML models like LDA, DT, LR, kNN, and SVM. Combining all data and utilizing RFE and augmentation to get the most important features yields the most prominent findings. The set of models with RFE model performance had an f-score of 93.28%, specificity of 94.88%, accuracy of 97.22%, sensitivity of 94.27%, and precision of 93.18%.

References

1. Aboutalib, S.S., et al.: Deep Learning to Distinguish Recalled but Benign Mammography Images in Breast Cancer Screening. Clin. Cancer Res. (2018)
2. Burt, J.R., et al.: Deep learning beyond cats and dogs: Recent advances in diagnosing breast cancer with deep neural networks. The Br. J. Radiol. 20170545 (2018)
3. Hamidinekoo, A., Denton, E., Rampun, A., Honnor, K., Zwiggelaar, R.: Deep learning in mammography and breast histology, an overview and future trends. Med. Image Analysis **47**, 45–67 (2018)
4. Chen, W., et al.: CA Cancer. J. Clin. **66**(2016), 115–132 (2015)
5. Kim, E.-K., et al.: Applying data-driven imaging biomarker in mammography for breast cancer screening: preliminary study. Sci. Reports **8**, 2762 (2018)
6. Kooi, T., et al.: Large scale deep learning for computer aided detection of mammographic lesions. Med. Image Analysis **35**, 303–312 (2017)
7. Rodríguez-Ruiz, A., et al.: Detection of breast cancer with mammography: effect of an artificial intelligence support system. Radiol. **290**, 305–314 (2018)
8. Kulkarni, A., Panditrao, A.: Classification of lung cancer stages on CT scan images using image processing. In: Proc 2014 IEEE Int Conf Adv Commun Control Comput Technol, pp. 1384–8 (2015)
9. Thaseen, A., et al.: Breast cancer detection using deep learning model. Proceedings of Third International Conference on Advances in Computer Engineering and Communication Systems: ICACECS 2022. Springer Nature Singapore, Singapore (2023)
10. Apat, S.K., Mishra, J., Srujan Raju, K., Padhy, N.: State of the art of ensemble learning approach for crop prediction. In: Kumar, R., Pattnaik, P.K., Tavares, R.S.J.M. (eds.) Next Generation of Internet of Things. Lecture Notes in Networks and Systems, vol 445. Springer, Singapore (2023). https://doi.org/10.1007/978-981-19-1412-6_58
11. Raju, V.R., Reddy, D.A., Narsimha, D., Srinivas, K., Rani, B.K.: Adaptive closed-loop deep brain stimulator coding techniques for target detections in Parkinson's. IETE J. Res. **69**(8), 4945–4960 (2023). https://doi.org/10.1080/03772063.2021.1962742
12. Fayaz, R., et al.: An Intelligent Harris Hawks Optimization (IHHO) based Pivotal Decision Tree (PDT) Machine Learning Model for Diabetes Prediction. Int. J. Intel. Sys. Appl. Eng. **10**(4), 415–423 (2022)
13. Priya Remamany, K., et al.: A Localized Bloom Filter-Based CP-ABE in Smart Healthcare. Applied Sciences **12**(24), 12720 (2022). https://doi.org/10.3390/app122412720
14. Rani, B.K., Rao, M.V., Patra, R.K., et al.: Vehicle type classification using graph ant colony optimizer based stack autoencoder model. Multimed Tools Appl. **81**, 42163–42182 (2022). https://doi.org/10.1007/s11042-021-11508-5

15. Munawar, S., Geetha, A., Srinivas, K.: Squirrel Search-based Optimal Feature Extraction with Bi-LSTM for the Arrhythmia Classification using ECG. J. Theoret. Appl. Info. Technol. 100 (2022)
16. Pothalaiah, S., Lakshmaiah, D., Doss, B., Sairam, N., Srikanth, K.: Design of CMOS Base Band Analog. Cognitive Computing Models in Communication Systems 123 (2022)
17. Lakshmaiah, D., et al.: A novel low-power frequency-modulated continuous wave radar based on low-noise mixer. Cognitive Computing Models in Communication Systems **165** (2022)
18. Hemanand, D., et al.: An intelligent intrusion detection and classification system using CSGO-LSVM model for wireless sensor networks (WSNs). Int. J. Intel. Sys. Appl. Eng. **10**(3), 285–293 (2022)
19. Natesan, G., Konda, S., de Prado, R.P., Wozniak, M.: A hybrid mayfly-aquila optimization algorithm based energy-efficient clustering routing protocol for wireless sensor networks. Sensors **22**(17), 6405 (2022). https://doi.org/10.3390/s22176405

Automatic Analysis and Detection of Multi-Channel ECG Signals Using Neural Network

G. Naga Jyothi(✉), R. Kiran Kumar, and B. Subbarayudu

ECE Department, Madanapalle Institute of Technology and Science, Madanapalle, A.P., India
nagajyothisai221@gmail.com, subbarayudub@mits.ac.in

Abstract. ECG signal gives the information of heart beat rate of a person. When a patient needs an ECG, which is a non-surgical technique to measure and record the electrical activity of the heart, a skilled cardiologist may not always be on site. Consequently, a general practitioner or paramedical team visiting the patient's location must use a particular kind of automated ECG analysis in order to obtain the patient's electrocardiogram. Automated ECG analysis is required. The optimal methodology for automated analysis of multi-channel ECG signals is finally provided in this paper. Artifacts and noise in multi-channel ECG readings can influence the diagnosis. By employing a neural network to determine the dynamic cutoff frequency parameter from noisy ECG signals and the particle swarm improvement approach (RBFNN-PSO), some researchers were able to reduce noise from the signals. However, PSO only offers Swarm, and reaction times are lengthy. A better RBFNN-PSO technique known as Radial Basis Function Neural Network with Multi Swarm Optimisation (RBFNN-MSO) has been proposed in order to get over these drawbacks. Lastly, the RBFNN-MSO technique, which is used with digital low-frequency filters for impulse response (FIR), establishes the cutoff frequency parameter.

Keywords: ECG · finite impulse response multi swarm optimization neural network

1 Introduction

The ECG is a useful diagnostic tool for cardiac conditions. Because of this, the ECG waveform needs to be represented accurately and filtered [4, 5] to get rid of any noise or artifacts. Digital filters with FIR are used to eliminate noise. The RBF-MSO technique is used to calculate the filter cutoff frequency. The fundamental unit ECG signals [1–3] are displayed in Fig. 1.

2 Relevant Review

The ECG, which displays the full capacity of the heart cells' action, is a recording of an electrical signal that is recorded from the body's surface. There are four waves in the ECG signal: P, QRS, T, and U. Different denoising techniques are shown in Fig. 2. Machine

D. Kriksciuniene et al. (Eds.): IBICA 2023, LNNS 1228, pp. 271–277, 2025.
https://doi.org/10.1007/978-3-031-78937-3_28

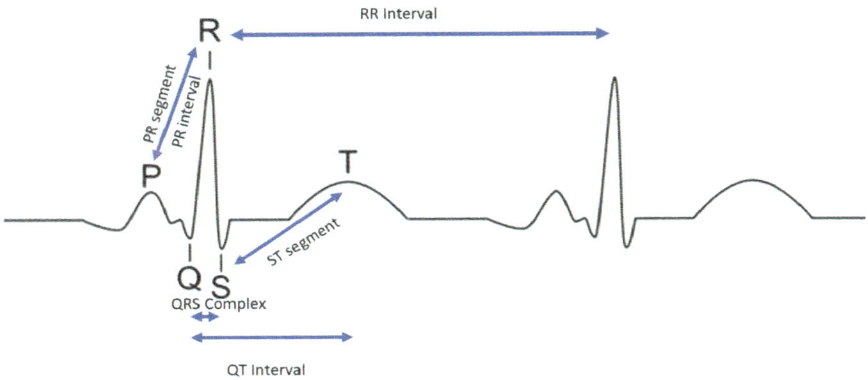

Fig. 1. ECG signal

learning techniques are more useful to analysis and detect the noise signals present in the ECG signals. Noise from a variety of sources, such as incomplete electrode contact with the body, anomalies in the device, electrical sounds from other body parts, breathing, and muscular cramps, can generally compromise the quality of the ECG signal.

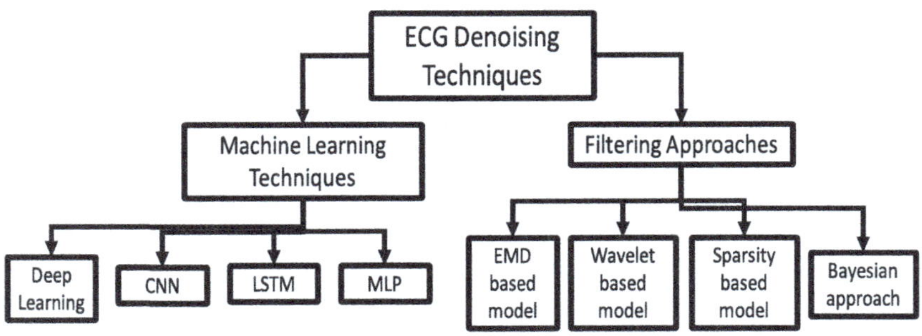

Fig. 2. Denoising techniques

Baseline roaming and power line interference is examples of low-frequency and high-frequency components of the noise generated.

3 Proposed Design

Figure 3 shows the suggested structure for the methodology. There are two approaches for the main procedure. Preparing the data and classifying diseases are the first two methods.The suggested pre-processing data methodology is shown in Fig. 3. There are five steps in the primary process: Preparing the data comes first. Step two involves Feature Extraction.Setting up the PSO system involves applying a number of particles [6–8] at various PSO positions and velocities in the third stage. For filter approaches, it will yield the Cut off frequency value. Configuring the MSO system involves applying many swarms to various MSO placements and velocities in the fourth stage. For filter approaches, it will yield the Cutoff frequency value. Comparing the original cutoff frequencies and the predicted cutoff frequencies is the fifth step.

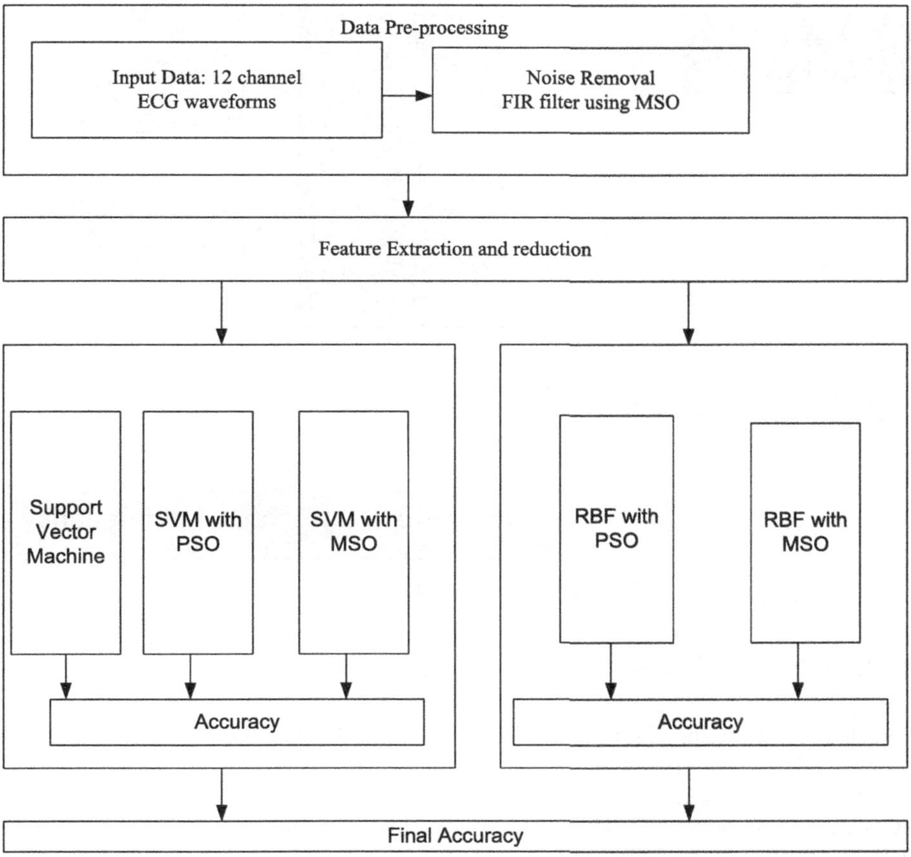

Fig. 3. Proposed design

For pre-data processing, a few entries are necessary. Using the MSO approach, 24 characteristics are extracted from multi-channel ECG signals [9–11] to determine the cutoff frequency. Using a known control input data set, create a model to forecast the cutoff frequency. This research presents the construction of a model to predict the spontaneous cutoff frequency using a radial basis function neural network (RBF-NN).

Figure 4 shows the noise removal method. In this noise ECG signals are passed. By using cutoff frequency we can remove the noise signals present in the ECG.

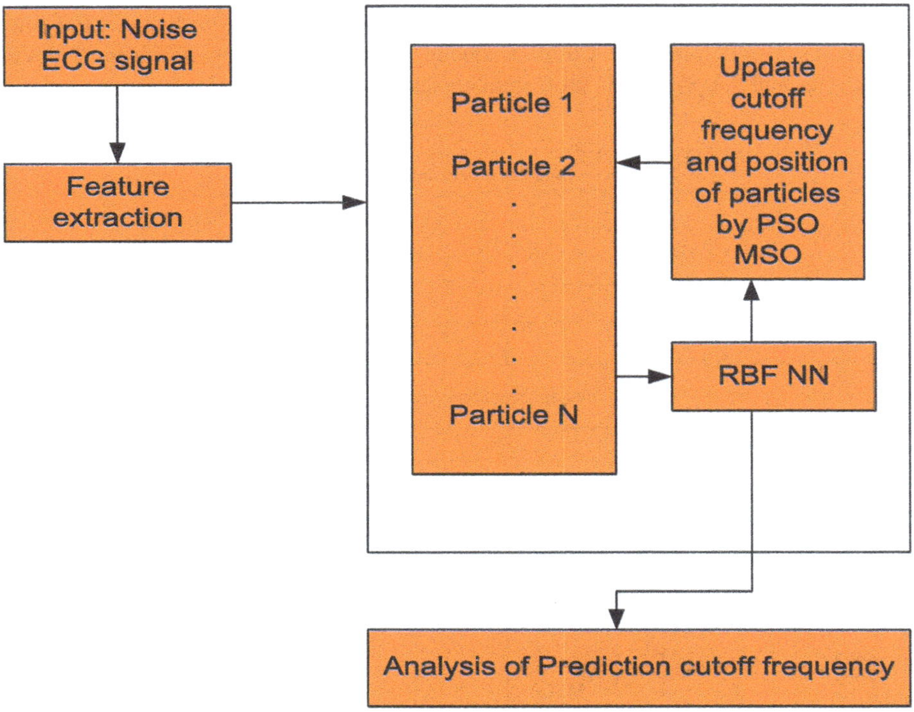

Fig. 4. Noise removal methodology

4 Results

The primary need for feature selection is accurate disease classification [12–15] of patients ECG waveforms. This study gathered eight different types of illness ECG waveforms from various data sets. The primary focus is on 24 distinct properties of ECG waveforms with varying dimensionality. Figure 6 shows the accuracy graph of RBF-PSO and RBF-MSO (Fig. 5).

In this case, test models are simply based on disease train and test data sets. The 24 × 25 × 8 RBF neural network model is applied in this section. Weights and bias values are defined using the PSO and MSO algorithms and is shown in Fig. 6.

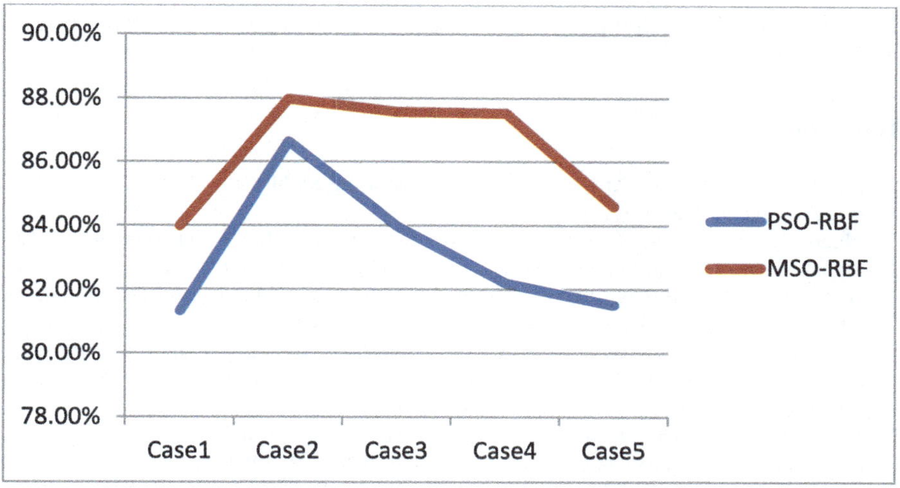

Fig. 5. Accuracy graph of RBF –MSO and RBF-PSO

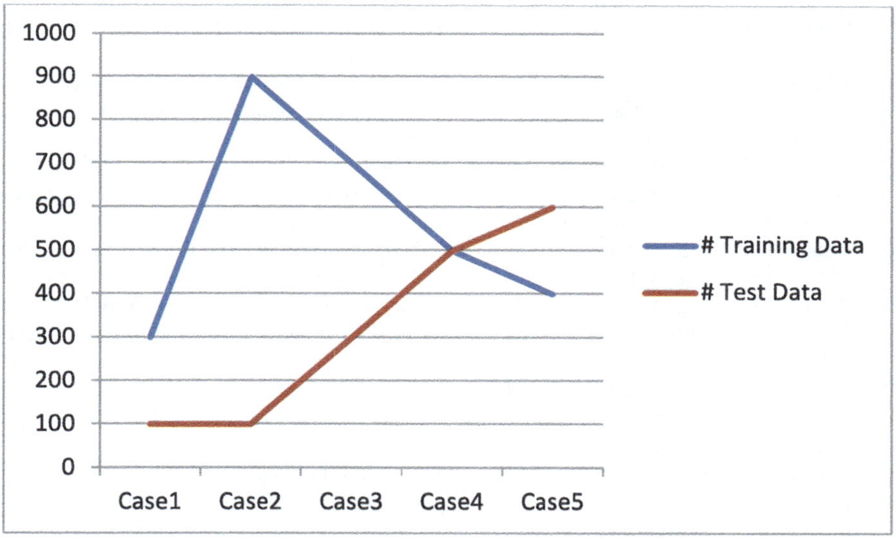

Fig. 6. Train and Test data of RBF-MSO & PSO

5 Conclusion

The Selective cut-off frequency analysis performed better with RBF-MSO than with the current RBF-PSO method; the SVM-MSO disease decision model is more accurate than other SVM and SVM-PSO methods; the disease classification RBF-MSO yields better results than other RBF-PSO; the majority of the high-value. In comparison to all reference results, the proposed design has 90% accuracy when compared with the existing design.

References

1. Kohli, N., Verma, N.K.: Arrhythmia classification using SVM with selected features. Int. J. Eng. Sci. Technol. **3**(8), 122–131 (2011)
2. Moein, S., Logeswaran, R., Shima Subari, K.: An Automated Intelligent Approach for ECG Signal Noise Removal. International Journal of Computer Science (2012)
3. Gupta, K.O., Chatur, P.N.: ECG signal analysis and classification using data mining and artificial neural networks. In: International Journal of Emerging Technology and Advanced Engineering Website: www.ijetae.com, ISSN 2250-2459, Volume 2, Issue 1, p. 56 (2012)
4. Moein, S., Logeswaran, R.: Intelligent ECG Signal Noise Removal using PSONN. International Journal of Computer Applications **45**(6), 0975–8887 (2012)
5. Srinivasulu, M., Chennakeshava Reddy, K.: Novel method to find the parameter for noise removal from multi-channel ECG waveforms. IJRET **03**(02), 395–401 (2014)
6. Vankdothu, R., Hameed, M.A., Ameen, A., Unnisa, R.: Brain image identification and classification on Internet of Medical Things in healthcare system using support value based deep neural network. Computers and Electrical Engineering **102**, 108196 (2022)
7. David, L.G., et al.: Tool Wear Monitoring Using Improved Dragonfly Optimization Algorithm and Deep Belief Network. Applied Sciences **12**(16), 8130 (2022). https://doi.org/10.3390/app12168130
8. Mehbodniya, A., Varaprasad Rao, M., David, L.G., Gerard Joe Nigel, K., Vennam, P.: Online product sentiment analysis using random evolutionary whale optimization algorithm and deep belief network. Pattern Recognition Letters **159**
9. Sunitha, D., Patra, R.K., Babu, N.V., Suresh, A., Gupta, S.C.: Twitter sentiment analysis using ensemble based deep learning model towards COVID-19 in India and European countries. Pattern Recognition Letters **158**, 164–170 (2022). ISSN 0167-8655
10. Balmuri, K.R., et al.: A long short-term memory network-based radio resource management for 5G network. Future Internet **14**(6), 184 (2022). https://doi.org/10.3390/fi14060184
11. Ahmad, S.J., Unissa, I., Shoukath Ali, M., Kumar, A.: Enhanced security to MANETs using digital codes. J. Info. Secu. Appl. **66**, 103147 (2022). ISSN 2214-2126, https://doi.org/10.1016/j.jisa.2022.103147
12. Soma, S., Sonth, M.V., Gowre, S.C.: Design of two-dimensional photonic crystal based ultra compact optical RS flip-flop. Photon Netw. Commun. **43**, 109–115 (2022). https://doi.org/10.1007/s11107-021-00955-7
13. Sudhamani, C.: Detection probability maximization through optimization of samples in cognitive radio networks. Multimed. Tools Appl. **81**, 12275–12285 (2022). https://doi.org/10.1007/s11042-021-11089-3
14. Chinna Rao, R., et al.: Real-Time Implementation and Testing of VoIP Vocoders with Asterisk PBX Using Wireshark Packet Analyzer. Journal of Interconnection Networks **22**(01), 2141030 (2022)
15. Rao, M.V., Reddy, D.A., Ampavathi, A., Shaik, M.: Data mining for cyber-physical systems. Data Mining and Machine Learning Applications, 235–280. https://doi.org/10.1002/9781119792529.ch10
16. Inaganti, R.: Preserving Patient Privacy: A Focus on Cyber security in Healthcare (2023). https://doi.org/10.13140/RG.2.2.26620.31367
17. Reddy, U.J., Dhanalakshmi, P., Dileep Kumar Reddy, P.: Image Segmentation Technique Using SVM Classifier for Detection of Medical Disorders. Ingénierie Des Systèmes D'information **24**(2) (2019)

18. Yalavarthi, S.: AI in Medicine: From Diagnosis to Treatment Decision Support. International Journal of Emerging Technologies and Innovative Research (www.jetir.org) **10**(10), e214–e217 (2023). ISSN:2349-5162

An African Vulture Optimization for the Detection of Diabetic Retinopathy

V. Mahalakshmi[1](✉), Swati Sucharita[2], Awatef Balobaid[1], Mohammad Manzoor Hussain[3], M. Ramkumar Raja[4], and S. Arul Jothi[5]

[1] Computer Science, College of Computer Science and Information Technology, Jazan University, Jazan, Saudi Arabia
{mlakshmi,asbalobaid}@jazanu.edu.sa
[2] CSE (AI&ML), Vardhaman College of Engineering (Autonomous), Hyderabad, Telangana, India
[3] CSE, B V Raju Institute of Technology, Narsapur, India
manzoor.mohammad@bvrit.ac.in
[4] Electrical Engineering, College of Engineering, King Khalid University, Abha, Saudi Arabia
rmanoharan@kku.edu.sa
[5] Kanakapushparagam Tech, Erode, India

Abstract. Blockchain technology is the trend in the 21st century with robust establishments in various societal contexts, including business, government, healthcare, and finance. Its diversified applications extend beyond cryptocurrencies and bitcoin, enabling data-sharing and control over access based on user type. This technology contrasts with machine learning, which focuses on data analysis and prediction. The additional advantage of using authentic datasets for prediction is the enhanced trust that can be gained from these predictions.Complete blindness can result from the chronic disease known as diabetic retinopathy. To lessen the likelihood of vision loss, the disease needs to be diagnosed early. Lesion segmentation was carried out on the image data after the data had been pre-processed using the median filtering technique. The Taylor African Vulture Optimization (AVO) algorithm was used to further refine the hyper-parameters in these data, and the most important characteristics were then fed into the SqueezeNet classifier to predict the development of diabetic retinopathy (DR) disease. The final output was saved in the blockchain architecture, where the EHR manager could access it and guarantee that only those with permission could access the prediction findings and associated patient data.

Keywords: blockchain technology · diabetic retinopathy · optimization

1 Introduction

Modern developments revealed about block chain technology (BT) benefits researchers since it can securely store medical and healthcare data [1]. Complex medical information must be protected against unwanted access. In IoT networks and technology, BT offers secure environment for organizing, sharing, and storing data. A decentralized structure

D. Kriksciuniene et al. (Eds.): IBICA 2023, LNNS 1228, pp. 278–286, 2025.
https://doi.org/10.1007/978-3-031-78937-3_29

with every block connected, it also promotes the safest transformation among communicating elements [2]. EHRs are digitalized versions of patient health records, which are typically kept on paper in a system based on charts. EHRs offer real-time, patient-centered records, ensuring that authorized healthcare personnel always have access to patient data.

For all participants who contribute, a collection of parties provides authentication and expensive data stages [5, 6]. Several applications of blockchain technology are used to enhance information privacy and trust, enable participants to share and transfer data with a degree of integrity [3, 4, 7]. A user-friendly health record storage systems to the EHR manager are offeredwithenhanced electronic access to patients' paper medical information. The EHR can be reduced, produced, maintained, and shared by patients with coworkers, family members, and healthcare professionals using this technique [8]. The Blockchain layer has more impendingfactors to enhance clinical trial and medical information management in many ways, including removing communication barriers between the various parties participating in the supply chain and expediting the process of issuing planning approvals. Collection and analysis of data include compiling significant volumes, sensitive or non-sensitive personal information, andrequired for computerized disease surveillance and general health monitoring. Electronic health records for patients can be shared and stored more securely thanks to blockchain technology. It might offer a framework for the interchange of health information that is safer, clearer, and more traceable. Using this technique, many data management systems may be integrated.Health systems in several nations might be able to cooperate thanks to a BT-based electronic medical and health record [9, 15]. Patients share their Electronic Health Records (EHRs) across medical institutions by exchanging them in various situations throughout their lives. Performance and dependability are determined by models created from the vast e-healthcare records [10, 16]. The kidney, eyes, lungs, heart, and other human organs are all severely damaged by diabetes, which is a disease that threatens many ways of life [12, 17]. One such adverse outcome of diabetes is diabetic retinopathy, a serious eye disease.

According to the World Health Organization (WHO), 285 million people worldwide have vision impairment, of which 246 million have reduced vision brought on by DR and 39 million are blind. As suggested by the name, diabetic retinopathy (DR) targets the retina suffering due to abnormal glucose levels might harm the retinal vasculature [14, 15, 17, 18]. The DR is considered broadly into two types; Proliferative diabetic retinopathy (PDR) and non-proliferative diabetic retinopathy (NPDR). Mild, moderate, and severe stages of NPDR are further separated [14]. Conventionally ophthalmologists detect disease with simple tools. The load on eye doctors can be lessened by using a variety of medical imaging evaluation models to help with DR diagnosis and prognosis [15–17]. One of these tools, explore purposes in grading with deep learning methodology. To improve the DR classification and detection, deep conventional neural network (DCNN)-based approaches chosen [18–20]. CNN is trained to detect the visual attributes in order to assess the characteristics of patients to find solution to problems of DR.

2 Related Work

Blockchain is a new technology that is being applied in many different fields, including healthcare data management. The biomedical industry prioritizes data security and integrity, observingpatiently using remote technologies, EHR management, supply chain management of medicinal drug products, data analytics of medical and healthcare, and research.

As opposed to conventional methods, the smart medical and healthcare environment manages clinical, surgical and medical records electronically more effectively. The patients' encumbrancesof proper data management, security difficulties and data redundancy is reduced by the EHR system. The e-Health service provider system might maintain patient and other service assistant authentication information without relying on cloud-based services [20]. This framework integratesdecentralised third-party scheme to improve data regularity and keep track of the numerous healthcare activities taking place over the network. A BT-based secure data management system employing the Internet of Medical Things (IoMT), consisting data communicationamongst the cloud servers, private server schemes, and associated medical devices, was taken into consideration by the authors of [17, 21] in order to address security issues. By adopting these IoMT devices, medical and healthcare providers and patients could engage in secure communication of crucial information.

As a digital payment method, Bitcoin is becoming more and more popular [22]. The focus of the suggested model was on medical and healthcare data, pertaining to treatments, during surgery, which could be connected to digital payments via BT. Access to this application was available to all healthcare professionals inside the system.The authors of [23] provided anoutlineof framework that enables the early risk prediction and diagnosis of diabetes utilisingsophisticated machine learning algorithms in connection to the upkeep of patients' electronic health records (EHRs). The interplanetary file system (IPFS), which has its roots in the blockchain, was used to store the patient medical and health records, gathered through wearable sensor gadgets and then communicated through EHR systems. The data were loaded into multiple ML models and delivered to the EHRs control. These models produce prediction results by processing the data. The system achievessecurity, privacy, integrity, but fell short in guarding against blockchain assaults.

To successfully store and maintain EHRs in a secure manner, the research proposed a safety framework for BT (BSF). Doctors, patients, and insurance agents, gain a secure and safe method of acquiring medical data. The proposed framework put a strong emphasis on the safety of medical and healthcare professionals, patients, and events, while the said configuration supported safety-confidentiality issues important to the medical and healthcare industry. In other domains, this approach did not, however, produce equally encouraging outcomes. The limitations of diabetic retinopathy diagnosis, particularly when using healthcare data, were illustrated by conventional methods concentrated on the detection processes of the DR.

In addition to the data fusion technique, the research opined about a healthcare methodology for the detection and identification of diabetes illness based on deep learning. Ensemble approaches with deep learning assist in expanding the understanding of fusion-based data, removing the needless weight on the model and enhancing its performance.Using medical imaging and deep learning approaches, anamalgamated model for identifying DR from digital fundus and retinal lesion images was proposed. When employing fundus images, DR detection is dependent on an image enhancement process and presents an unanswered question that affects the model's performance.

The universal repositories to access the medical image datasets provide DR dataset free and can be utilized with the experiments of firefly approaches and deep learning and developing models for classification and detection of vices in the vasculatures of retina due to DR. Although the model produced improved accuracy, it struggled with larger datasets or information from other areas. The research propounded in lays into a novel method for the detection of microaneurysms and haemorrhages on fundus images. Pre-processing, blood vessel segmentation and removal, fovea localization, elimination, and function extraction were all incorporated in the methodology, which aided in the detection of DR problems brought on by microaneurysms and haemorrhages. But these methods failed in extracting features related to lesions. In order to remove unnecessary limitations, the researchers of proposed a semi-supervised auto-encoder graph community (SAGN) model for the diagnosis of DR. Three main modules made up the SAGN method: graph representation, auto-encoder feature learning, and neighbour correlation mining. Considering the few labelled retinal images, this produced a better performance.

Identifying the prognostic consequences of DR automatically, the authors of proposed a deep convolutional neural network which is not a vanilla model, that takes into account multi-view retinal images. The experimental findings showed four shared Net image feature extraction approaches for automatic DR detection. By enhancing network functionality, the model made sure that enhanced feature extraction was achieved.

3 Proposed Work

The focus of the current BT-based healthcare scheme is about managing requests from patients, medical and healthcare workers, who can register independently through the registration hub. At the registration centre unit, data from appropriate subjects and medical and healthcare providers gathered. To provide user information, the smart gadgets and the registration centre were combined. The registration office can compile all the necessary data, produce a private key with an ID, and send it to the executive division (Fig. 1).

Each doctor and patient should be given an authorised ID by the EHR manager. The four components of the authentication process are the EHR Manager, the User, the Admin Unit, and the Interplanetary File System. If the authentication is fails, the consequence is shown for each specific ID. The doctor and patient are certified following a successful authentication.

The EHRs Manager utilizes a smart device to streamline registration processes at the registration center, ensuring accurate and efficient data collection and submission to the Interplanetary File system.Smart devices, including desktops, laptops, and smart mobiles, utilize sensor nodes for patient login and data processing in EHRs manager, enabling administration and deep learning for diagnosis.The EHR, a smart contract,

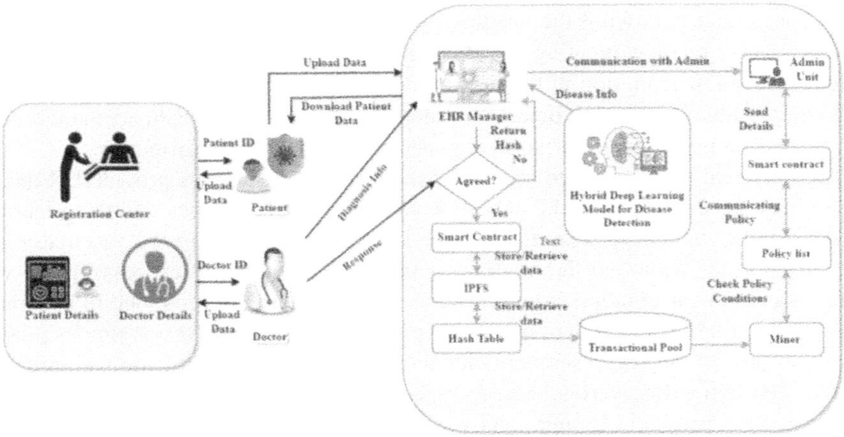

Fig. 1. Proposed Methodology of Blockchain in Healthcare Framework

functions as a centralized authority, facilitating transactions and ensuring optimal performance through a public key system.The administration unit utilizes a smart contract system to manage digital assets, ensuring efficient and secure transactions within their unique requirements.The proposed system utilizes deep learning and TaylorAVO-SqueezeNet architecture to classify and diagnose diabetic retinopathy patients, ensuring data security and efficient patient records storage in the EHR manager. The EHR manager initiates the initialization process, involving a random number sent to the IPFS. The registration process begins, involving three phases between the admin unit, EHR manager, and user. Post-registration, appropriate authentication is performed, ensuring unauthorized access and detecting DR at the EHR manager unit.

3.1 Diabetic Retinopathy Detection

EHR uses pre-processing, data augmentation, lesion segmentation, detection and classification methods in medical images. RP-Net is used for segmentation, while TaylorAVO algorithm integrates Taylor series and African Vultures Optimization for training, and SqueezeNet for classification (Fig. 2).

The Taylor AVOA, a combination of the AVO algorithm and the Taylor series, simulates the foraging behavior of African vultures. The Taylor series offers a simple computation process with short execution time, while the AVO algorithm calculates fitness function and optimal search strategy, offering greater performance with speed and low costs in utilizing computational resources.

3.2 SqueezeNet Architecture

SqueezeNet is a convolutional neural network model with 18 layers, aiming to reduce parameters and perform down-sampling on over a million images from the ImageNet database, resulting in a smaller, more efficient neural network.

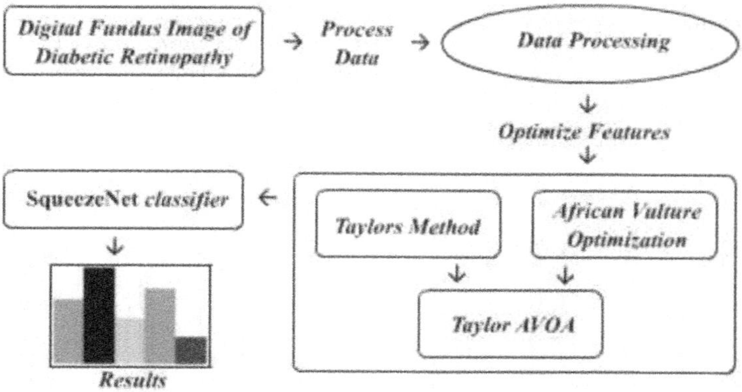

Fig. 2. Taylors-AVOA with SqueezeNet for DR Classification

DL models are deployed into embedded systems for distributed training, reducing communication across servers. The architecture includes multiple layers with activation energy, achieving high accuracy through compression techniques like single-value decomposition, deep compression, quantization, and network pruning.

4 Results

The proposed model, using an Intel processor and Windows 10, was tested using Python. The model's accuracy, specificity, and sensitivity were evaluated, and compared with the existing models and important analyses made in the erstwhile research, demonstrating superior performance.

4.1 Dataset

The experimental results of DR detection in the proposed framework are assessed using the IndianDRImage Dataset (IDRID) [30]. The fact that there are so few publicly available healthcare datasets forces the scientific community to use the few readily accessible datasets raises serious concerns. The types of applications are therefore less variable dependent on such a dataset. Furthermore, the models created with such datasets might not be appropriate for widespread, generalised applications. Therefore, thorough data characterisation is necessary for researchers and model developers to assess the datasets' quality. A detailed analysis of the dataset's features can be used to describe the data. In light of this, the current work conducts a rigorous analysis of the IDRID dataset. The 516 retinal fundus images that make up the IDRID are categorised into three groups: segmentation, clinical grading, and localization of the images. The segmentation process uses 81 real colour fundus images, and the localization and disease-grading processes use 516 real colour fundus images. The median filtering method, which is covered in more detail in Sect. 3.2.1, is used as part of the pre-processing to improve the quality of the images in the dataset.

4.2 Experimental Analysis

By adjusting the training data, k-fold validation, and block size, as seen in Fig. 3, the performance of the proposed TaylorAVO-SqueezeNet is evaluated based on accuracy, sensitivity, and specificity metrics.

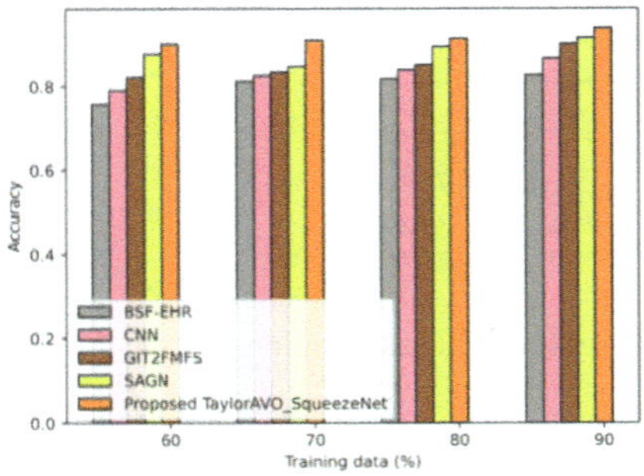

(a) Accuracy

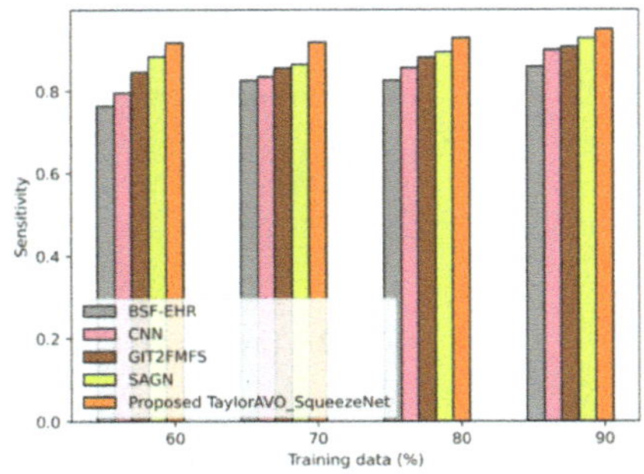

(b) Sensitivity

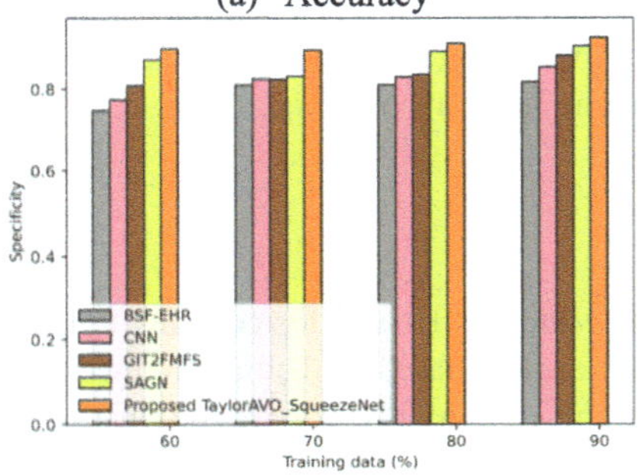

(c) Specificity

Fig. 3. Projection of accuracy, specificity and sensitivity of TaylorAVO-SqueezeNet on training data.

The SAGN and CNN models are evaluated based on training data, block size, and k-fold data variations. The TaylorAVO-SqueezeNet model outperforms conventional models in accuracy, sensitivity, and specificity metrics, and also outperforms in k-fold and accuracy metrics, demonstrating superior performance.

5 Conclusion

The TaylorAVO-SqueezeNet framework is a novel model that uses deep learning techniques to accurately detect disease (DR) in EHRs. The model uses blockchain for registration and authentication, ensuring security and privacy. The model includes data pre-processing, lesion segmentation, feature extraction, and SqueezeNet classification. The model outperforms conventional models like BSF-EHR, CNN, GIT2FMFS, and SAGN, with enhanced accuracy, sensitivity, and specificity. However, challenges such as scalability, network efficiency, and integration costs remain. The model's performance is optimized, but it faces challenges in integrating blockchain technology into legacy systems, particularly in healthcare. Future work will focus on addressing these challenges.

References

1. Ramzan, S., Aqdus, A., Ravi, V., Koundal, D., Amin, R., Al Ghamdi, M.A.: Healthcare applications using blockchain technology: motivations and challenges. IEEE Trans. Eng. Manag. (2022, early access)
2. Chattu, V.K., Nanda, A., Chattu, S.K., Kadri, S.M., Knight, A.W.: The emerging role of blockchain technology applications in routine disease surveillance systems to strengthen global health security. Big Data Cogn. Comput. **3**, 25 (2019)
3. Reddy Madhavi, K., Kora, P., Venkateswara Reddy, L., Avanija, J., Soujanya, K.L.S., Prabhakar, T.: Cardiac arrhythmia detection using dual-tree wavelet transform and convolutional neural network. Soft Comput. Impact Factor **3**(64) (2021)
4. Sriharsha, A.V.: Detection of holes on Indian roads using information and communication technologies. In: 2023 International Conference on Advances in Computing, Communication and Applied Informatics (ACCAI), Chennai, India, pp. 1–7 (2023). https://doi.org/10.1109/ACCAI58221.2023.10199826
5. Sriharsha, A.V., et al.: Adaptable fog computing framework for healthcare 4.0. In: Proceedings of the 14th International Conference on Soft Computing and Pattern Recognition (SoCPaR 2022), SoCPaR 2022. LNNS, vol. 648. Springer, Cham (2023)
6. Srinivas, K., Kavitarani, B., Kumar, M.K.: Classification and mining behavior of data. Data Min. Mach. Learn. Appl. 21–55. https://doi.org/10.1002/9781119792529.ch2
7. Patra, R.K., Mahendar, A., Madhukar, G.: Inductive learning including decision tree and rule induction learning. Data Min. Mach. Learn. Appl. 209–234. https://doi.org/10.1002/9781119792529.ch9
8. Bhavani, R., et al.: Deep learning techniques for speech emotion recognition. In: 2022 International Conference on Futuristic Technologies (INCOFT), Belgaum, India, pp. 1–5 (2022). https://doi.org/10.1109/INCOFT55651.2022.10094534
9. Hanumanthakari, S., Pullela, S.K., Bhukya, S.N., Vijayalakshmi, K., Ahmad, S.R., Kumar, N.: IoT based patients monitoring system in healthcare service. In: 2022 International Conference on Automation, Computing and Renewable Systems (ICACRS), Pudukkottai, India, pp. 1324–1329 (2022). https://doi.org/10.1109/ICACRS55517.2022.10029295
10. Gulati, K., Nayak, K.M., Priya, B.S., Venkatesh, B., Satyam, Y., Chahal, D.: An examination of how robots, artificial intelligence, and machinery learning are being applied in the medical and healthcare industries. Int. J. Recent Innov. Trends Comput. Commun. **10**(2s), 298–305 (2022). https://doi.org/10.17762/ijritcc.v10i2s.5947
11. Reddy, G., et al.: An intrusion detection using machine learning algorithm multi-layer perceptron (MlP): a classification enhancement in wireless sensor network (WSN). Int. J. Recent Innov. Trends Comput. Commun. **10**(2s), 139–145 (2022). https://doi.org/10.17762/ijritcc.v10i2s.5920
12. Mubeen, S., et al.: Linguistic based emotion detection from live social media data classification using metaheuristic deep learning techniques. Int. J. Commun. Netw. Inf. Secur. (IJCNIS) **14**(3), 176–186 (2022). https://doi.org/10.17762/ijcnis.v14i3.5604
13. Bhaskar, T., Shiney, S.A., Rani, S.B., Maheswari, K., Ray, S., Mohanavel, V.: Usage of ensemble regression technique for product price prediction. In: 2022 4th International Conference on Inventive Research in Computing Applications (ICIRCA), Coimbatore, India, pp. 1439–1445 (2022). https://doi.org/10.1109/ICIRCA54612.2022.9985521

14. Patil, S., Balmuri, K.R., Frnda, J., Parameshachari, B.D., Konda, S., Nedoma, J.: Identification of triple negative breast cancer genes using rough set based feature selection algorithm & ensemble classifier. Hum.-Centric Comput. Inf. Sci. **12** (2022)
15. Kesana, M., Rikhari, S.: Multi-modal medical image fusion using Laplacian re-decomposition (2022). https://doi.org/10.1201/9781003240310-7
16. Reddy Madhavi, K., Madhavi, G., Rupa Devi, B., Kora, P.: Detection of pneumonia using deep transfer learning architectures. Int. J. Adv. Trends Comput. Sci. Eng. **9**(5), pp. 8934–8937 (2020). https://doi.org/10.30534/ijatcse/2020/292952020. ISSN 2278-3091
17. Reddy Madhavi, K., Vijaya Sambhavi, Y., Sudhakara, M., Srujan Raju, K.: COVID-19 isolation monitoring system. Springer Series – Lecture Notes on Data Engineering and Communication Technology, vol. 63, pp. 601–609 (2021)
18. Reddy Madhavi, K., et al.: COVID-19 detection using deep learning. In: 20th International Conference on Hybrid Intelligent Systems-HIS 2020, at Machine Intelligence Research (MIR) Labs, USA. AISC, vol. 1375, pp. 1–7. Springer, Cham (2020)
19. Agbo, C.C., Mahmoud, Q.H., Eklund, J.M.: Blockchain technology in healthcare: a systematic review. Healthcare **7**, 56 (2019)
20. Han, Y., Zhang, Y., Vermund, S.H.: Blockchain technology for electronic health records. Int. J. Environ. Res. Public Health **19**, 15577 (2022)
21. Javed, I.T., Alharbi, F., Bellaj, B., Margaria, T., Crespi, N., Qureshi, K.N.: Health-ID: a blockchain-based decentralized identity management for remote healthcare. Healthcare **9**, 712 (2021)

Smart Healthcare on 5G Platform

Bura Vijay Kumar[1], Pothreddypally Jhansi Devi[2], V. Lavanya[3](✉), K. Bhuvaneswari[4], S. Malarvizhi[5], and Siva Durga Rao Parasa[6]

[1] School of Computer Science and Artificial Intelligence, SR University, Warangal, India
[2] Computer Science and Engineering, B V Raju Institute of Technology, Narsapur, India
[3] School of Computing, Mohan Babu University, Tirupati, India
lavanya.v432@gmail.com
[4] MCA Department, Sri Venkateswara College of Engineering, Tirupati, India
[5] Department of Science and Humanities, School of Engineering, Avinashilingam Institute for Home Science and Higher Education for Women, Coimbatore, India
malarvizhi_sh@avinuty.ac.in
[6] Department of ECE, SASI Institute of Technology and Engineering, Tadepalligudem, India

Abstract. Healthcare provisions are rendered with top notch quality widgets in fifth-generation cellular technologies (5G). Healthcare systems with less human intervention benefit from the advanced mobile broadband as well as low latency communication technology proven ultra-reliable communications. The network's capabilities provide faster transmission, lowest latency and more capacities compared to 4G. The 5G technologies should make IoMT devices accessible seamless. As there are enormous number of contributions in this field it is essential to understand the breadth and width of the area and to define the scope of applications of 5G for medical and healthcare.

Keywords: Smart · healthcare · 5G · IoT · Medical devices

1 Introduction

The modern technological upheaval develops intelligent systems. Intelligent systems of everywhere, every kind are seen in the world including manufacturing, transportation, medical, education and automation industries. Intelligent systems shine the life style into smart life in functional collaboration with smart devices, where a smart-healthcare is originated in medical and healthcare systems. In the recent decade the smart-healthcare is deployed in applications using 4G, supporting the existing systems and downward compatibilities of medical and healthcare devices of previous generations. However, the goal is not achieved satisfactorily, meeting the needs and demands of smart health, due to technological restrictions.

The problems of 4G were overcome in 5G, with superior capabilities of network enabling healthcare very smart. The capabilities of 5G not only provide the smart treatment to the patients but also play a very important role in preventing the unhealthy attacks or diseases.

D. Kriksciuniene et al. (Eds.): IBICA 2023, LNNS 1228, pp. 287–294, 2025.
https://doi.org/10.1007/978-3-031-78937-3_30

2 Relevant Review

Smart healthcare has significant growth of 12.5% by the end of 2025 [1]. The fifth generation technologies, not only in developing mere smart healthcare services, but also bulk data transactions which includes speech-audio and interactive videos to the patient sites as a real-time activity with remote device control, hence the channels providing the payload with audio, video and control signals. All the various types of data accumulated through payloads are archived and transmitted, reviewed and organized to store in cloud servers. Medical professionals can use this channel repeatedly for many times in order to serve the like-diagnosis and consultations. Even, upon consultations such data is refined and revised and updated in the cloud servers. Personalized medical data is allowed through the channels safe and secure in the fifth-generation technologies, which allows emergency attention by the medical professionals. Large image files such as MRI, X-ray and other digitized scanned documents are shared seamlessly among the patients and specialist doctors for future reviews [2, 3].

Increase of clinical consultations and cases are widely covered using the fifth-generation technologies, includes telesurgeries, teleconsultations, telediagnosis, sanitary aids for preventive care and other related medical and contingent considerations. Healthcare is widely supported with the fifth generation technologies accumulating wide scientific literatures. In this work we present a systematic review which is most sought for the developers and researchers which shall enthuse them to build sophisticated frameworks for healthcare systems using fifth-generation technologies.

In the context of fifth-generation technologies to develop smart healthcare solutions, several authors have made their contributions and opined productivity in the avenues of developing systems with multi-perspective, advanced applications. Some of the significant contributions from the consensus have been drafted below. Latif et al. [4], around 2017 has drafted opportunities and challenges, that hinder in the development of health solutions with 5G. The mention of 5G facts in connection with many IoT devices and sensors, big data analytics with deep learning, was found potential break-through of the solutions, which transform health systems into global standards. A latency study about 2G, 3G, 4G with respect to the functional efficiency of 5G has also been performed. The convergence and reliability of the solutions to meet the conventional needs of the health applications can digitize with relevant transmission and storage media.

The concept of body area networks (BAN) [5], is devised wirelessly as wireless body sensor networks (WBSN) juxtaposing with real-time applications of monitoring and transmission of data, handling emergency situation even at the patient site. Selem et al. propounded that WBSN shall be an emphatic application which becomes a backbone of the future health applications using 5G. During 2020s, Contreras et al. has contributed the case studies of telemedicine in pandemic caused by COVID-19. A thorough analysis is initiated through telemedicine to offer indicative treatments and also enhanced with intensive treatments in the later days. Physical health systems and institutions were violated to treat the patients during pandemic, which would not allow descriptive approaches for the diagnosis.

In addition, Zhang Chunming et al. have discussed the application of 5G in smart healthcare to monitor and control the COVID-19 pandemic through telemedicine and teleconsultations. The integrity of the applications in the medical systems was questioned about the support given to the patients in the emergency. Kim et al. of Chinese University opined about the introduction of the remote connection which shall be provided to gather real-time information and share through cloud to the external public hospitals.

Nevertheless these introductions, soon there is a problem of security and privacy breach with respect to the patient personal medical records. Furthermore, Siriwardhana et al. furnished opportunities and challenges of deployment of 5G technologies in medical systems during pandemic, with the details of operating procedures to handle pandemic. The intense technical requirements are expected in the health systems with the intervention of communication technologies of 5G such as telepharmacy, telesurgeries and telenursing. Many authors' opinions were drawn from the research consensus indicating developments in services, requirements, applications with the 5G technologies.

3 Methods

A multidisciplinary angle of research shall be infused with the 5G technologies in the field of medical and healthcare systems with especially 5G wireless technologies accounting various advanced technical aspects. The analysis in the work is precise and exhaustive, such a way, the perspective includes varied set of variables which constitute with security and privacy, technical artifacts, application areas and various types of devices with their protocols for communication. The coherence among these variables shall include a keen analysis to show cutting edge development strategies of the medical and healthcare systems with 5G technologies [6, 7].

Therefore, in this work a PRISMA methodology understands the formulation of research questions. The answers to the questions are precisely stated with conclusions.

Research Question 1.: *What is the strategy of applying 5G technologies into healthcare*?

> Different health systems are developed using 5G technology to embark on to the smart healthcare.

Research Question 2.: *What is the extent of applying 5G technologies meet the demands of healthcare*?

> The scope of the extent to apply 5G is in coherent with the instruments, toolsets and technical artifacts of development. The size of the system is an important factor decides the materials of development.

The PRISMA method was used to filter scientific articles from various databases, including IEEE, Science Direct, Scopus, Web of Science, Springer, Wiley Online Library, Gale Power Search, and Association of Computing Machinery (ACM). The three stages of the review process contain inclusion and exclusion criteria at each stage. The fundamental finding stage is identification, involved extracting metadata from retrieved articles, focusing on publications published from 2016 to 2021. Such of those candidates of articles that do not include into the said review process are considered inappropriate and are not technologically related to aspects in the research objective.

In stage 2, we screened publications based on title, abstract, and key-words, removing unrelated and non-relevant ones. Many short, non-relevant articles were omitted. In third stage, each article is thoroughly reviewed with relevant applications.

4 Results

The study aims to identify the main traits of technological and application points of view in medical and healthcare systems, focusing on smart healthcare systems. Pie diagrams were used to display the results, considering the variables proposed.

The research found a wide diversity in publication types, with 56.81% of studies being proposals that contain almost non-technical features. However, they provide theoretical backing and offer thorough models, helping to identify the analytical factors. On the other hand, 22.73% of publications discuss applications in the actual world.The advantages, merits, and benefits of the systems are typically the main points of these documents, although technical details and security needs are also covered.Overall, the study emphasizes the significance of taking variables into account while designing intelligent healthcare systems.

Variables in operational systems play a crucial role in ensuring security and privacy. These established and functional systems address problems relating to these mechanisms. However, not all systems are fully developed. Additionally, prototypes are created, concentrating on technical requirements, potential applications, and safety issues. Prototypes act as tangible representations of operational systems. These studies provide valuable insights into the security and privacy mechanisms in operational systems.

The review on the results is mentioned in the sections below, obtains as response to the research questions.

Research Question 1.: *What is the strategy of applying 5G technologies into healthcare*?

The medical system has evolved significantly, transforming into a smart health system. Applications like remote care, data transmission, and real-time monitoring are resulted. A classification system is generated based on these systems, with articles belonging to each classification.

Using physiological data gathered by sensors and IoMT devices, remote care enables medical professionals to deliver diagnosis, surgical, and patient medical care over the Internet.

Remote Care: Surgery and Diagnosis

The use of a robotic arm for diagnosis and treatment without the need for patients to travel to the hospital was described by Zhang et al. [8] in their presentation of the installation of a remote clinic and ambulance system at Zhengzhou University Hospital. The system has a work centre for the surgical robot and a control centre for professionals.

In order to treat patients with diabetic retinopathy (DR), Chen et al. [9] created a system that makes use of a laser device, teleophthalmology platform, remote computer control platform combined with videoconferencing platform. A diagnostic system with teleultrasound in intensive care unit (ICU) was proposed by Duan et al. [10] and included an ultrasound aided by a robotic arm, MGIUS - R3.A diagnosis method to treat postpartum haemorrhage (PPH) was created by Zhang et al. [11] using computational intelligence algorithms using SVM and ANN. Three domains—the network, master, slave, make up the telesurgery architecture propounded by Gupta et al. [18].The concept of master-slave console with two links—a direct link and a feedback link connected with a master-slave console constituting to form a robotic surgery of haptic nature deriving into a telesurgery architecture is propounded by Meshram and Patil [12].

Using large data and deep learning methodologies, Pustokin et al. [13] devised a model using deep belief networks fforthe identification of diseases.In a review about remote surgical systems, Padmashree and Nayak [14] noted that it cuts down on the amount of time surgeons must spend in the operating theatre and does not require the presence of the physician.Ye [15] retreated with the usage of remote diagnostics systems to meddle with the COVID-19 implications and has borrowed the computational intelligence in developing a prediction and detection systems for pandemics.

Data Transmission: Smart Ambulances

Real-time information transmission in ambulances is crucial for rescuing patients and preventing unnecessary hospitalizations, requiring high-speed data transmission for uninterrupted communication with doctors remotely. Zhang et al. developed a mobile diagnostic system for ambulances, enabling real-time patient information sharing with doctors and specialists, reducing rescue time.Usman et al. propose an ambulance proposal for emergency situations, combining video transmission, ambulance video transmission, and vital sign transmission, minimizing data consumption and network resources. Zhai et al. introduced a smart ambulance architecture featuring a multi-monitoring system, virtual reality glasses, and an HD camera, enhancing real-time doctor-patient interaction and simplifying patient registration.

Healthcare Benefits

Artificial intelligence provides many healthcare benefits, which medical professionals prioritize. Intelligent systems provide constant and immediate attention, remote medical care, and emergency assistance. Most clinical applications offer these benefits, with 52.27% using artificial intelligence for constant and immediate care. Remote diagnostic and surgical systems reduce hospital patient attenuation, permitting safe, specialized care. The total includes 40.91% remote specialized medical care. Applications provide emergency treatments and first aid to protect patients until hospitalization.

Research Question 2.: *What is the extent of applying 5G technologies meet the demands of healthcare?*

To answer this question, the article discusses the use of IoMT for Smart Healthcare, focusing on short- and long-range network technologies, providing a comprehensive overview of the diverse range of technological solutions.

IoMT Devices

IoMT devices support monitoring, telemedicine, and telesurgery, and other services. Body sensors, wearable technology, software, and communication interfaces are all used in the well-known IoMT [4]. Health management technology reduces costs and improves productivity, service, and patient experience.

IoMT technologies were examined and divided into five divisions for the systems that were demonstrated: body sensors, teleoperated robots, ultra-high resolution cameras, wearable technology, and artificial intelligence. Figure 1 shows smart healthcare system device management study percentages. The classification is akin to HaghiKashani et al. [16], but they emphasise body sensors and wearable devices.

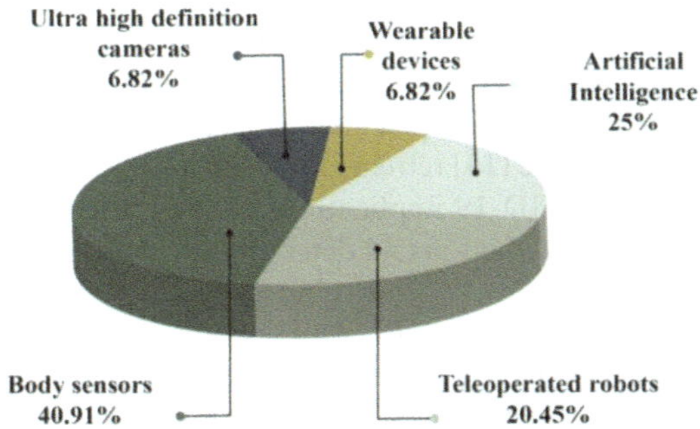

Fig. 1. Utilization of IoMT devices in Medical and Healthcare Systems

Body sensors were intervened in 40.91% of studies. This is because WBSNs help collect physiological signals in most continuous monitoring systems [19]. Unlike wearables, IoMT devices focus on one activity. An internal SpO2 sensor lets a smartwatch tell time and measure oxygen saturation. Other than physiological signals, the sensors also record voice and image signals [5].

AI aids remote diagnosis, prediction, and monitoring. Continuous monitoring systems use this technology to estimate results without human intervention. This technology creates an intelligent system that makes sanitary care accessible. Data labelling and classification and medical image classification [17] are mostly done by ML algorithms in AI systems. This is necessary due to algorithm limitations. Our research shows that 25.00% (n = 11) of AI-based systems develop optimal sanitary systems. Teleoperated robots are another IoMT device. 20.45% (n = 9) of the research is related to them. Their utility and efficiency make them popular in remote surgery and diagnostic systems for surgical operations.

In intelligent healthcare systems, telesurgery robot integration is a broad field [12]. A surgeon controls these robots remotely using a control panel or console. A teleoperated robot with two arms (right and left) that operate tools and a main manipulator that operates an endoscope was shown by Gupta et al. [18]. The teleoperated robot "Microhand" by Zheng et al. [8, 12] has a surgeon console, an imaging and control panel, two manipulators (one on each side), and motion control pedals.The patient system consists of a rotating head, a passive arm, and three slave arms.Device, transducers, and a 6-DOF robotic arm on the patient side of a teleoperated robot for COVID-19 diagnostic ultrasonography; image monitors, a console, and a transducer on the doctor side for controlling the robotic arm. Finally, MGIUS-R3 is a robotic arm with a doctor-site US device. Both surgeon and patient sites have audio, video, and mainframe systems.

Ultra-high-definition cameras and wearable devices are used in 6.82% of IoMT systems, providing real-time image and video transmissions [19] for remote care and guidance to paramedics. These devices and body sensors [20] measure biomedical signals, aiding in remote care, surveillance, and continuous monitoring, and are also used for other tasks.

5 Conclusion

Although 5G is proved, the smart healthcare is in the stage of feasibility analysis and the technology is not globally implemented. As based on research consensus, and related works, theoretical proposals on 5G in smart healthcare have been generated in recent years, but large-scale implementations and deployments have not yet been instrumental. Based on application areas, researchers have focused least on information transmission. Technical requirements require uninterrupted, ultra-low latency, high-speed communication. Telemedicine, or remote care, is a crucial 5G application, with 38.63% of systems using it. Digital health is linked to COVID-19, with continuous, real-time monitoring accounting for 54.55% of systems. Body sensors and wearable devices are widely used in medical devices for physiological signal transmission. The 5G network enables remote-surgery systems with teleoperated robots, improving global medical services. AI systems are trusted and secure, improving patient care. Smart healthcare systems use body sensors and wearable devices to share patient data, but concerns about security and privacy remain.

References

1. Vergütz, A., Prates, N.G., Henrique Schwengber, B., Santos, A., Nogueira, M.: An architecture for the performance management of smart healthcare applications. Sensors **20**, 5566 (2020)
2. Neha, G., Juneja, P.K., Sharma, S., Garg, U.: Future aspect of 5G-IoT architecture in smart healthcare system. In: Proceedings of the 2021 5th International Conference on Intelligent Computing and Control Systems (ICICCS), Madurai, India, 6–8 May 2021, pp. 406–411 (2021)
3. Le, T.V., Hsu, C.L.: An anonymous key distribution scheme for group healthcare services in 5g-enabled multi-server environments. IEEE Access **9**, 53408–53422 (2021)
4. Latif, S., Qadir, J., Farooq, S., Imran, M.A.: How 5G wireless (and concomitant technologies) will revolutionize healthcare? Future Internet **9**, 93 (2017)
5. Prema, K., Sriharsha, A.V.: Differential privacy in big data analytics for haptic applications. Technology **8**(3), 11 (2017)
6. Kovuri, K., Chandrashekhar, K., Sriharsha, A.V., Siddardha, B., Hitesh Reddy, A.: Adaptable fog computing framework for healthcare 4.0. In: International Conference on Soft Computing and Pattern Recognition, pp. 24–32. Springer, Cham (2022)
7. Reddy Madhavi, K., Vijaya Sambhavi, Y., Sudhakara, M., Srujan Raju, K.: COVID-19 isolation monitoring system. Springer Series – Lecture Notes on Data Engineering and Communication Technology, vol. 63, pp. 601–609 (2021)
8. Zhang, D., et al.: 5G-enabled health systems: solutions, challenges and future research trends. In: Proceedings of the 2019 ITU Kaleidoscope: ICT for Health: Networks, Standards and Innovation (ITU K), Atlanta, GA, USA, 4–6 December, pp. 1–8 (2019)

9. Chen, H., et al.: Application of 5G technology to conduct real-time teleretinal laser photocoagulation for the treatment of diabetic retinopathy. JAMA Ophthalmol. **139**, 975–982 (2021)
10. Duan, S., et al.: A 5G-powered robot-assisted teleultrasound diagnostic system in an intensive care unit. Crit. Care **25**, 134 (2021)
11. Srinivas, K., Kavitarani, B., Kumar, M.K.: Classification and mining behavior of data. Data Min. Mach. Learn. Appl. 21–55. https://doi.org/10.1002/9781119792529.ch2
12. Patra, R.K., Mahendar, A., Madhukar, G.: Inductive learning including decision tree and rule induction learning. Data Min. Mach. Learn. Appl. 209–234. https://doi.org/10.1002/9781119792529.ch9
13. Bhavani, R., Vijay Muni, T., Tata, R.K., Narasimharao, J., Murali, K., Kaur, H.: Deep learning techniques for speech emotion recognition. In: 2022 International Conference on Futuristic Technologies (INCOFT), Belgaum, India, pp. 1–5 (2022). https://doi.org/10.1109/INCOFT55651.2022.10094534
14. Hanumanthakari, S., Pullela, S.K., Bhukya, S.N., Vijayalakshmi, K., Ahmad, S.R., Kumar, N.: IoT based patients monitoring system in healthcare service. In: 2022 International Conference on Automation, Computing and Renewable Systems (ICACRS), Pudukkottai, India, pp. 1324–1329 (2022). https://doi.org/10.1109/ICACRS55517.2022.10029295
15. Gulati, K., et al.: An examination of how robots, artificial intelligence, and machinery learning are being applied in the medical and healthcare industries. Int. J. Recent Innov. Trends Comput. Commun. **10**(2s), 298–305 (2022). https://doi.org/10.17762/ijritcc.v10i2s.5947
16. Reddy, G., et al.: An intrusion detection using machine learning algorithm multi-layer perceptron (MlP): a classification enhancement in wireless sensor network (WSN). Int. J. Recent Innov. Trends Comput. Commun. **10**(2s), 139–145 (2022). https://doi.org/10.17762/ijritcc.v10i2s.5920
17. Mubeen, S., et al.: Linguistic based emotion detection from live social media data classification using metaheuristic deep learning techniques. Int. J. Commun. Netw. Inf. Secur. (IJCNIS) **14**(3), 176–186 (2022). https://doi.org/10.17762/ijcnis.v14i3.5604
18. Bhaskar, T., Shiney, S.A., Rani, S.B., Maheswari, K., Ray, S., Mohanavel, V.: Usage of ensemble regression technique for product price prediction. In: 2022 4th International Conference on Inventive Research in Computing Applications (ICIRCA), Coimbatore, India, pp. 1439–1445 (2022). https://doi.org/10.1109/ICIRCA54612.2022.9985521
19. Patil, S., Balmuri, K.R., Frnda, J., Parameshachari, B.D., Konda, S., Nedoma, J.: Identification of triple negative breast cancer genes using rough set based feature selection algorithm & ensemble classifier. Hum.-Centric Comput. Inf. Sci. **12** (2022)
20. Kesana, M., Rikhari, S.: Multi-Modal Medical Image Fusion Using Laplacian Re-Decomposition (2022). https://doi.org/10.1201/9781003240310-7

An Analytical Method of Artificial Intelligence for Classifying and Identifying Skin Cancer Cases Using Stochastic Neural Networks

G. Madhukar[1](✉), Ramesh Chegoni[2], R. Bhargav Ram[3], B. Venkataramanaiah[3], Sunil Kumar Singh[4], and Mannem Saimanasa[5]

[1] Department of CSE, CMR Technical Campus, Hyderabad, Telangana, India
cmrtc.paper@gmail.com, madhu.mani536@gmail.com

[2] Department of IT, CMR Technical Campus, Hyderabad, Telangana, India
ramesh.it@cmrtc.ac.in

[3] Department of ECE, CMR College of Engineering and Technology, Hyderabad, Telangana, India
{dr.r.bhargavram,bvenkataramanaiah}@cmrcet.ac.in

[4] Department of CSE(CS), CMR Engineering College, Hyderabad, Telangana, India
kumarsunil@cmrec.ac.in

[5] Department of CSE, CMR Engineering College, Hyderabad, Telangana, India

Abstract. Skin cancer has recently been divided into benign and malignant conditions, like all other forms of cancer. When compared to non-malignant skin cancer, the malignant equivalents of these two forms are thought to be the most fatal. Early therapy is crucial because it is known that malignant consequences increasingly affect patients' survival. Expert dermatologists need to check for cancer in any questionable spots. These individuals use a computer-assisted early malignant system. Picture preprocessing has been employed in several studies to identify cancers early on, and this has led to machine learning-based therapies that work well. By creating suitable systems for the classification of skin conditions, it will be possible to expand the use of such a significant diagnostic treatment. Numerous research publications used preprocessing images to detect cancer early on and facilitate successful treatment. The ABCDEs—asymmetrical shape, boundary abnormalities, color, diameter, and evolution—have been devised by skilled dermatologists as the accepted criteria for identifying typical signs of serious malignant cases.

Keywords: Artificial intelligence · Skin cancer · Neural networks

1 Introduction

One of the complicated illnesses that emerged throughout the sixteenth century was human cancer, mostly brought about by genetic instability and the accumulation of several molecular options. It is not possible to develop a predictive therapeutic approach that makes use of the current diagnosis and prediction categories since they do not

D. Kriksciuniene et al. (Eds.): IBICA 2023, LNNS 1228, pp. 295–303, 2025.
https://doi.org/10.1007/978-3-031-78937-3_31

accurately reflect the tumor. Most anti-cancer medications now on the market do a poor job of differentiating between benign and malignant cells [1, 2]. One of the worst and most deadly forms of skin cancer is melanoma. Melanoma can be dangerous if it develops slowly, but it is frequently treatable when discovered early [3]. Since the accuracy values in the next stage are impacted by the segmentation procedure, it is essential [4]. Skin cancer is now classified as benign or malignant, having exceeded all other cancers in prevalence in recent years [5, 6]. Skin cancer is one of the most common illnesses. Skin cancer suspicion is currently rising by 40%, and a biopsy—a drawn-out, patient procedure—is frequently used to identify the disease [7, 8]. Furthermore, a great deal of pointless biopsies are done. The primary function of the skin is to protect and maintain the health of the interior organs. Among its other roles are isolation, temperature regulation of the sensory system, and the manufacture of vitamin D. Three layers make up the skin: Fig. 1 shows the hypodermal layer as well as the middle skin, which is located next to the dermis. The skin is seen in the top skin.

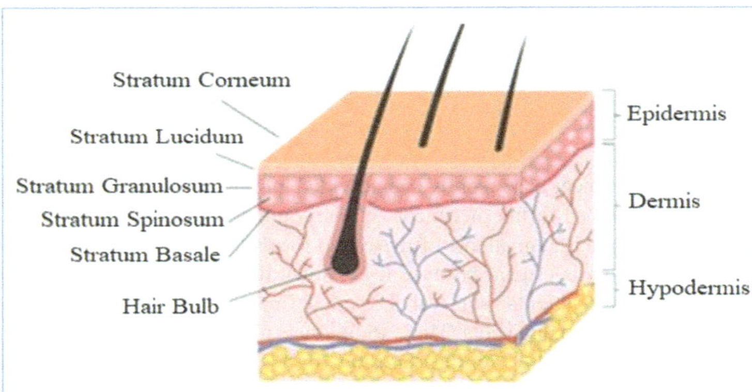

Fig. 1. Skin Structure: Epidermis, Dermis and Hypodermis Layer

The ability of melanoma to differentiate between various forms of skin cancer symptoms aids in the development of a system for the disease's identification. This technique efficiently categorizes and eliminates skin lesions such as melanoma (minimizing any prejudices created by healthcare providers, particularly dermatologists). The method is very helpful in automating the diagnosis and treatment of patients with skin cancer (via imaging processing), in offering affordable treatments, and in speeding up the detection of [9, 10] pigmented skin lesions such as melanoma. Rural practitioners may be able to identify melanoma early with this strategy. The tagline is: Significantly less time and money are spent.

2 Literature Work

Pre-processing, segmentation, and grading of skin analysis photos are under the author's control. The advanced technique preprocesses following images by bilaterally filtering with CLAHE algae. For the final segmentation findings, the fuzzy C-Mean technique and a constrained recursive area algorithm are used to complete the primary segmentation. To obtain ellipticity of the area and constrained model functions, melanocytes are segregated from the applicant nuclear region [11]. An elliptical descriptor. Technical accuracy for the description at WBCDD was at most 96.35%. This study's primary objective is to balance specificity and sensitivity while lowering the classifier's complexity through the use of the feature selection approach [12]. One of the deadliest diseases that affect people is skin cancer. There are several different kinds of skin malignancies, including nevus, melanoma, basal cell carcinoma (BCC), squamous cell carcinoma (SCC), and seborrheic keratosis (SK). Certain skin cancers are visually detectable, but most skin cancers require a biopsy test to be diagnosed, and this process might take some time [13]. It is necessary to create an automated approach for classifying skin lesions in order to get around this. It is suggested that melanoma be classified using dermoscopic data in order to facilitate the clinical application of dermatoscopy imaging techniques for the classification of skin lesions. Unsharp masking and an anisotropic diffusion filter were used to improve the photos. Using an adaptive k-means clustering algorithm with two clusters, the melanoma was separated from the background. Next, feature extraction techniques based on intensity and texture features [14] from the segmented data were used, followed by classifier training and testing on unidentified dermoscopic data. Various classifiers were employed, including decision trees, random forests, multi-layer perceptrons, support vector machines, and k-nearest neighbor. The area under the receiver operating characteristics curve (ROC) is used to evaluate the classifiers' performance. In comparison to other classifiers, the Random forest technique [15] greatly outperforms the others in classifying melanoma, with an accuracy of 93%.

3 Proposed Method

The proposed study aims to identify the following types of skin malignancies, both benign and malignant. Figure 2 illustrates the full process of identifying and categorizing skin cancer. Database training and testing.

The database was built using the "International Skin Imaging Collaboration (ISIC)" archive pictures. One of the most extensive collections of regulated, excellent dermoscopic images available is the ISIC. There were fifteen benign and fifteen cancerous photos in the collection. Every image is produced using the GLCM-functional, statistical, and textured PNN Network Model. The classification and detection process can be used with the use of random, anonymous test samples.

Pre-processing

The aforementioned photograph was taken during an image-capture phase that further recorded background information and noise. It is necessary to preprocess and eliminate the previously listed undesirable elements. Pre-processing is widely used to eliminate unnecessary information from skin image data, such as labels, tape, undesired background noise, and artifacts. The mammography pictures show many sounds, such as spit, poisson, gausse, salt, and pepper. When noise exists in an image, the pixels show a spectrum of intensities instead of their [16] actual values. The most widely used algorithm makes use of an iterative refinement method. Owing to its widespread use, it is

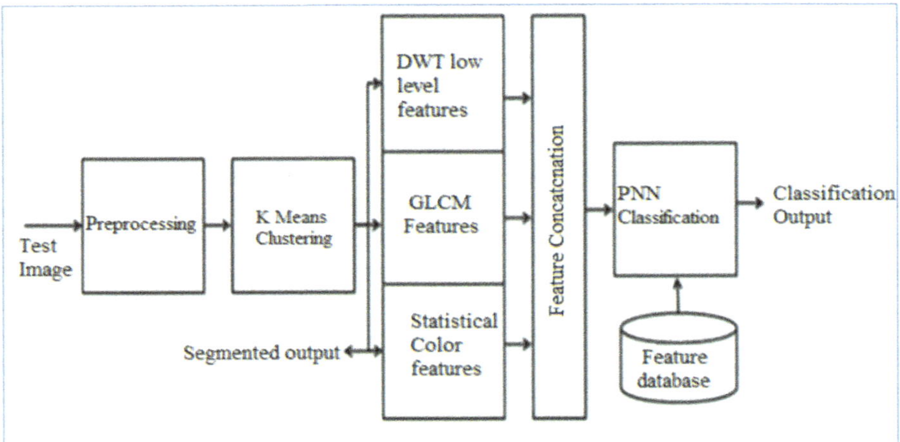

Fig. 2. Skin Cancer Detection and Classification

frequently referred to as "the k-means algorithm," as seen in Fig. 3. In the computer science community, it is also known as Lloyd's algorithm. Occasionally, it is also called "naïve k-means" due to the existence of significantly faster alternatives.

Extraction of Features

The different features of the skin lesion can be gathered for categorization. GLCM-reliant DWT-based lower level features, texture characteristics, and color statistics are some of the primary criteria that enable us to differentiate between various types of skin injuries. Even though the term "texture" is often used, it is difficult to describe. By highlighting the similarities and differences between the two unique textures, one can tell them apart. Generally, there are three applications for the textures.

Classification

Numerous disciplines, including economics, medicine, engineering, geology, physics, and biology, have found success with the use of neural networks. The potential utility of neural networks in prediction and classification problems makes them attractive from a statistical perspective. Neural birth emulation was used in the development of a technique called PNN. In the typical architecture, neurons are connected to perform the classification function effectively. The weights of neurons are determined by their hybrid characteristics. After that, weight linkages are found by taking use of their hybrid characteristics. The weight amount determines the layer level of the proposed network. An artificial neural network's structural layout is shown in Fig. 5. A PNN consists of two classification phases: training and testing. The architecture on the layer serves as the foundation for the training process. The dataset is divided into weight distributions for the hybrid components of the data set and mapped using the input layer.

Figure 4 depicts the PNN structural design, which has four hidden layers that are arranged according to weights. The original 224 * 224 * 3 pixel convolutional2D clothing network image employs 96 11 live filters at a 4-pixel range, followed by a decision

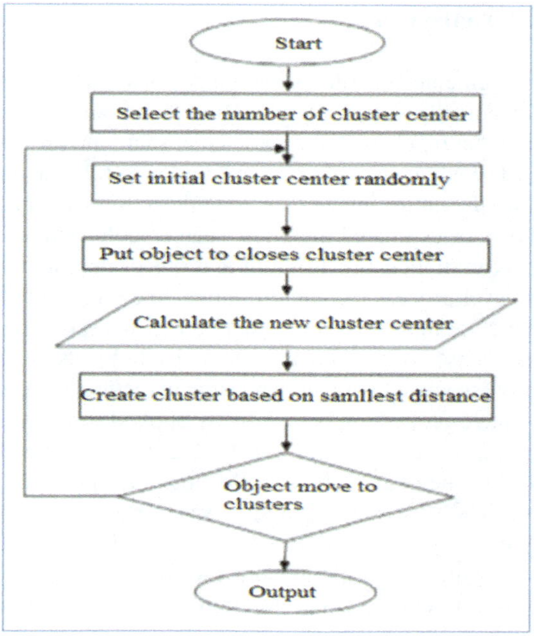

Fig. 3. K-Means Clustering

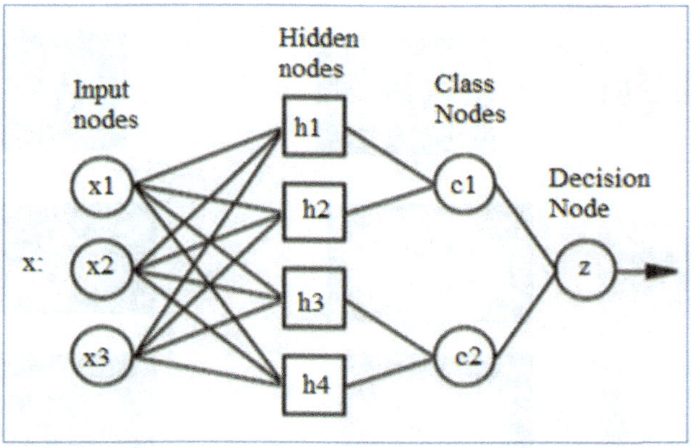

Fig. 4. Layered Architecture PNN Model

standardization layer and an activation class node activation layer. Subsequently, the classification process was applied to both layers of the class nodes within the concealed layer.

4 Results and Discussion

Images are typically segmented in order to recognize objects and borders (such as curves, lines, etc.). More specifically, image segmentation is the concept that each pixel in an image has a label and that pixels with the same label share unique visual characteristics [11]. The result could be either a set of extracted contours from the image or a collection of segments spanning the entire image (see edge detection). Every pixel in the region has the same attributes or estimated qualities, such as color, intensity, or texture. The surrounding locations are noticeably different in terms of the same features [12].

Typically, a midlevel process has picture inputs and characteristic outputs that are derived from those picture inputs. At the higher end of this spectrum, higher-level therapy requires "making sense" of a number of recognized objects in addition to cognitive activities traditionally associated with human vision in picture processing. As previously mentioned, digital image processing is effectively applied in many domains of significant social and economic importance. MATLAB R2018 Dataset is the tool utilized for the testing. The ISIC is one of the largest collections of high-quality, non-permissible dermoscopic images. The suggested technique was created by rotating the spatial and temporal frequency of thirty images of dermoscopic skin lesions, fifteen of which were benign and fifteen of which were malignant, at different angles. The PNN structural design was trained by fifty epochs using the train picture on each label; the remaining twenty epochs were worn for testing.

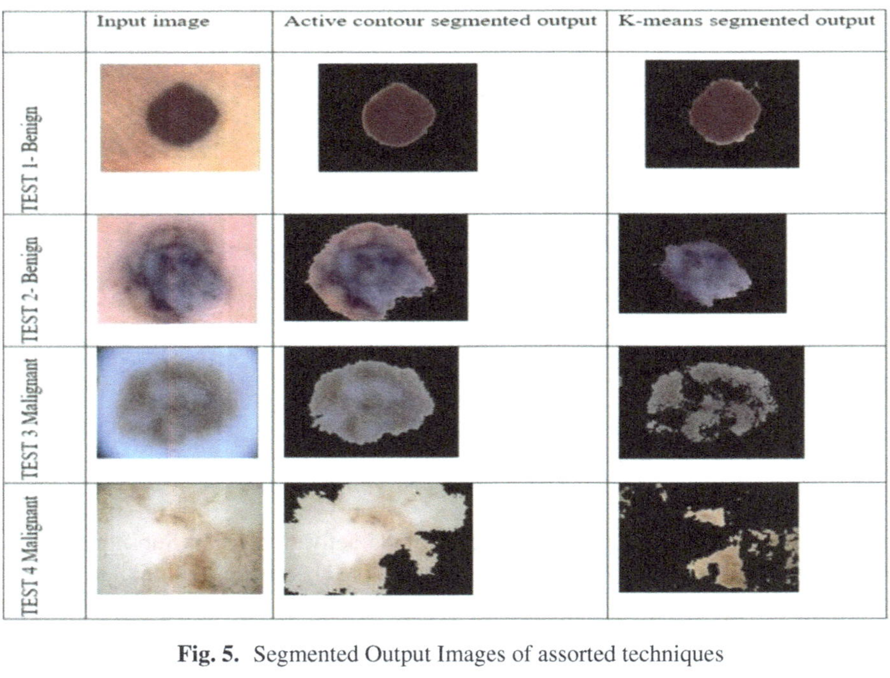

Fig. 5. Segmented Output Images of assorted techniques

The PNN classification makes use of features extracted by the GLCM and DWT neural networks to categorize the images into the appropriate categories. There are several performance indicators that can be used to assess the effectiveness of a model. As Fig. 5 shows, the proposed approach can effectively detect the areas of skin malignancies; compared to the Active Contouring method, it exhibits very good segmentation. While photos TEST-3 and TEST-4 are categorized as malignant, photos TEST-1 and TEST-2 are considered benign. The segmentation precision was higher for the cancerous pictures.

Performance Metrics

To evaluate the performance measure as depicted in Fig. 6, the recommended approach, Active contours (AC) and k-means clustering, are applied with both segmentation methods. To carry out these assessments, precision, sensitivity, F measurements, accuracy, MCC, dice, composite reliability, and specificity are computed.

	Proposed PNN AC	Proposed PNN K Means	SVM Linear kernel	SVM RBF kernel	SVM polynomial kernel	SVM 5 fold cross validation
Test 1	0.92	1	0.45	0.42	0.44	0.62
Test 2	0.79	1	0.42	0.46	0.68	0.56
Test 3	0.86	1	0.72	0.55	0.5	0.6
Test 4	0.48	1	0.72	0.62	58	0.45

Fig. 6. Comparison of accuracy levels

The suggested approach works better for mutually benign and malignant diseases than a variety of SVM kernels, such as the SVM linear kernel, the RBF kernels, the polynomial kernel, and five-fold cross validation, as shown in Fig. 7.

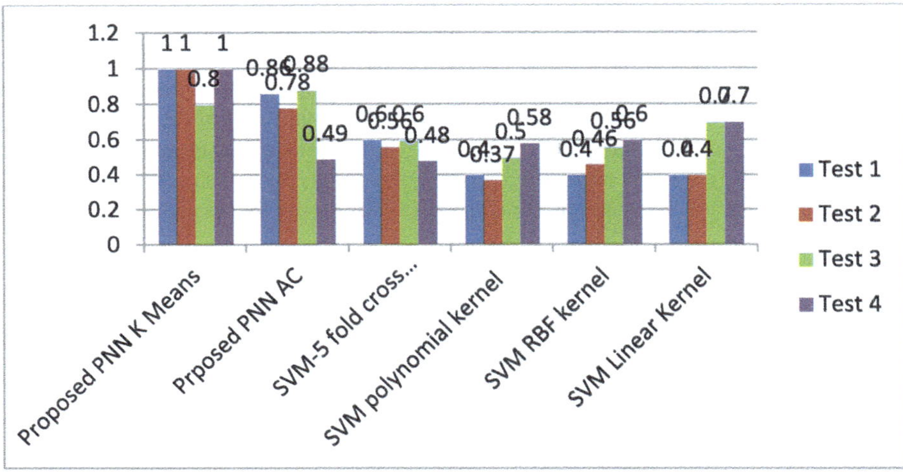

Fig. 7. Accuracy levels

5 Conclusion

Through this work, a PNN-based deep learning technique for identifying and classifying skin cancer in MRI images was developed. Here, novel artifacts or undesirable noise components are removed from the image acquisition process using pretreatment Gaussian filters. K-means clustered segmentation is then used to identify the cancer cells after the ROI has been retrieved. Next, the DWT-based GLCM technique was created in order to extract statistics, color, and texture from the segmented image. Next, the tumor's benign or malignant classification was determined using the PNN and trained network model. Based on cutting-edge research, we therefore conclude that PNN is superior to the conventional SVM technique. Subsequent versions of this challenge may have more PNN network layers in addition to varying cancer types (benign or malignant).

References

1. Nasiri, S., et al.: DePicT malignant Deep-CLASS: a deep convolutional neural networks approach to classify skin lesion images. BMC Bioinform. **21**(2), 1–13 (2020)
2. Munir, K., et al.: Cancer diagnosis using deep learning: a bibliographic review. Cancers **11**(9), 1235 (2019). https://doi.org/10.3390/cancers11091235
3. Vankdothu, R., Hameed, M.A., Ameen, A., Unnisa, R.: Brain image identification and classification on Internet of Medical Things in healthcare system using support value based deep neural network. Comput. Electr. Eng. **102**, 108196 (2022). https://doi.org/10.1016/j.compeleceng.2022.108196
4. David, L.G., Patra, R.K., Falkowski-Gilski, P., Divakarachari, P.B., Marcilin, L.J.A.: Tool wear monitoring using improved dragonfly optimization algorithm and deep belief network. Appl. Sci. **12**(16), 8130 (2022). https://doi.org/10.3390/app12168130
5. Abolfazl Mehbodniya, M., Rao, V., David, L.G., Gerard Joe Nigel, K., Vennam, P.: Online product sentiment analysis using random evolutionary whale optimization algorithm and deep

belief network. Pattern Recogn. Lett. **159**, 1–8 (2022). https://doi.org/10.1016/j.patrec.2022.04.024

6. Sunitha, D., Patra, R.K., Babu, N.V., Suresh, A., Gupta, S.C.: Twitter sentiment analysis using ensemble based deep learning model towards COVID-19 in India and European countries. Pattern Recogn. Lett. **158**, 164–170 (2022). ISSN 0167-8655
7. Balmuri, K.R., Konda, S., Lai, W.-C., Divakarachari, P.B., Gowda, K.M.V., Lingappa, H.K.: A long short-term memory network-based radio resource management for 5G network. Future Internet **14**(6), 184 (2022). https://doi.org/10.3390/fi14060184
8. Ahmad, S.J., Unissa, I., Shoukath Ali, M., Kumar, A.: Enhanced security to MANETs using digital codes. J. Inf. Security Appl. **66**, 103147 (2022). https://doi.org/10.1016/j.jisa.2022.103147. ISSN 2214-2126
9. Soma, S., Sonth, M.V., Gowre, S.C.: Design of two-dimensional photonic crystal based ultra compact optical RS flip-flop. Photon Netw. Commun. **43**, 109–115 (2022). https://doi.org/10.1007/s11107-021-00955-7
10. Sudhamani, C.: Detection probability maximization through optimization of samples in cognitive radio networks. Multimed. Tools Appl. **81**, 12275–12285 (2022). https://doi.org/10.1007/s11042-021-11089-3
11. Chinna Rao, R., Mohana Lakshmi, K., Raja, Ch., Bharat Siva Varma, P., Rao, G.R.K., Patibandla, A.: Real-time implementation and testing of VoIP vocoders with asterisk PBX using wireshark packet analyzer. J. Interconnection Netw. **22**(Supp01), 2141030 (2022)
12. Rao, M.V., Reddy, D.A., Ampavathi, A., Shaik, M.: Data mining for cyber-physical systems. In: Data Mining and Machine Learning Applications, pp. 235–280. https://doi.org/10.1002/9781119792529.ch10
13. Mohana Lakshmi, K., Rikhari, S.: Artificial intelligence framework for skin cancer detection. In: Smart Intelligent Computing and Applications, Volume 1: Proceedings of Fifth International Conference on Smart Computing and Informatics (SCI 2021), pp. 579–588. Springer, Singapore (2022)
14. Madhavi Latha, C., Soujanya, K.L.S.: Secure IoT framework through FSIE approach. Communications in Computer and Information Science, pp. 17–29 (2021)
15. Pranathi, K., Pingili, M., Mamatha, B.: Fundus image processing for glaucoma diagnosis using dynamic support vector machine. In: Kumar, A., Mozar, S., Haase, J. (eds.) Advances in Cognitive Science and Communications. ICCCE 2023. Cognitive Science and Technology. Springer, Singapore (2023). https://doi.org/10.1007/978-981-19-8086-2_53
16. Shaik, A.S., Usha, S.: Sensor based garbage disposal system. Int. J. Innov. Technol. Exploring Eng. (IJITEE) **8**, 164–167 (2019)

Deep Transformer-Based Model for Sepsis Prediction with Multimodal Health Data

G. Vasavi[1], V. Sandeep Kumar Reddy[2], A. Basi Reddy[3], J. Avanija[4](✉), K. Reddy Madhavi[4], and Naresh Tangudu[5]

[1] B V Raju Institute of Technology, Narsapur, Telangana, India
vasavi.g@bvrit.ac.in
[2] Department of CSE, Sri Venkateswara College of Engineering, Tirupati, India
[3] CSE, School of Computing, Mohan Babu University, Tirupati, India
[4] AI&ML, School of Computing, Mohan Babu University, Tirupati, India
avans75@yahoo.co.in
[5] Aditya Institute of Technology and Management, Tekkali, India

Abstract. Sepsis, a life-threatening disease, requires accurate prediction methods to facilitate timely interventions and enhance patient outcomes. This paper presents an integration of deep transformer model with multimodal healthcare data for sepsis prediction. Sepsis, a life-threatening condition, demands rapid and precise prediction for effective interventions and improved patient outcomes. This study presents a groundbreaking solution, the Deep Transformer-Based Model for Sepsis Prediction with Multimodal Health Data. Leveraging transformer-based architectures, this model integrates diverse patient data, including clinical notes, vital signs, and lab results, to enhance sepsis detection accuracy. By fusing deep learning techniques with multimodal data, the model achieves superior predictive capabilities. The proposed approach not only advances sepsis prediction but also holds promise for real-time clinical decision-making, leading to data-driven critical care. The experimentation and evaluation, reveal that the proposed deep transformer-based model for sepsis prediction is more efficient than the existing methods.

Keywords: Sepsis prediction · Deep learning · Transformer-based models · Multimodal health data · Multi-Head Self-Attention

1 Introduction

Sepsis is triggered by body's response to an infection, remains a global health challenge. Early and accurate detection of sepsis is required for timely intervention and improved patient outcomes. With the proliferation of electronic health records and availability of diverse patient data, there is a growing need for advanced predictive models capable of predicting sepsis accurately. In response to this challenge, an efficient approach considering Deep Transformer-Based Model for Sepsis Prediction with Multimodal Health Data have been proposed. The integration of diverse patient data sources, providing valuable insights about sepsis that can significantly impact patient care. This model utilizes the

D. Kriksciuniene et al. (Eds.): IBICA 2023, LNNS 1228, pp. 304–310, 2025.
https://doi.org/10.1007/978-3-031-78937-3_32

power of deep learning, specifically transformer-based architectures, to comprehensively analyze multimodal health data and revolutionize sepsis detection in clinical settings.

2 Literature Review

Sepsis, a life-threatening condition, continues to be a significant global health challenge, demanding advanced predictive models for early and accurate prediction. With the growth of electronic health records (EHRs) and availability of diverse patient data, researchers have increasingly turned to multimodal health data analysis as a means to enhance sepsis prediction [1–4]. While the traditional methods are useful, they often lack the nuanced understanding of complex relationships within multimodal data, hindering their predictive accuracy and timeliness. Deep learning models, particularly recurrent neural networks (RNNs) and transformer-based architectures, have emerged as powerful tools for processing sequential and multimodal data [5–12]. In healthcare, these models have been successfully applied to various tasks, including disease prediction and patient outcome forecasting. The ability of deep learning algorithms to capture complex patterns within diverse datasets aligns with the complexities inherent in sepsis prediction using multimodal health data. Integration of diverse data modalities, such as clinical notes, vital signs, laboratory results, and imaging data, holds immense potential in improving the accuracy of sepsis prediction [13–16]. Studies have explored the fusion of structured and unstructured data, emphasizing the need for models capable of comprehensively processing these varied data types. Successful integration. Despite the progress, challenges persist, including data heterogeneity, interpretability, and the need for real-time prediction in clinical settings. Addressing these challenges requires innovative approaches that can integrate multimodal data sources, ensuring both accuracy and clinical utility. This literature review focuses on the deep learning models, specifically transformers [17–22], in overcoming these challenges and advancing sepsis prediction using multimodal health data. Our proposed work seeks to address this gap by proposing deep transformer based approach for sepsis prediction.

3 Methodology

This section, briefs about the steps of deep transformer based approach for sepsis prediction. The steps of the model for performing sepsis detection have been specified in Fig. 1. Initial step is to gather Multimodal Health Data with features such as Vital Signs, Lab Results, Clinical Notes medical images, demographic information, and temporal data. Proposed work utilizes MIMIC-III (Medical Information Mart for Intensive Care III) dataset. MIMIC-III is a publicly available dataset developed by the MIT Laboratory for Computational Physiology. The dataset includes a wide array of data types, such as vital signs, laboratory measurements, medications, clinical notes, and more.

Clean and preprocess the data by handling missing values, encoding categorical variables, and creating appropriate representations for different data types. Then, integrate the data into a cohesive multimodal dataset. Perform feature engineering by extracting temporal patterns and trends from time-stamped data points. Creation of relevant features from raw data is performed through aggregations, statistical features, and transformations to enhance the model's ability. Next step is to train the model using deep transformer architecture utilizing encoder-decoder components. A deep transformer, refers to a transformer model with multiple layers stacked on top of each other, often referred to as a deep stack of transformers. Each layer in a deep transformer is a complete transformer unit with self-attention and feed-forward networks. Deep transformers allow the model to capture more complex and abstract patterns in the data, as each layer learns increasingly higher-level representations. Deeper architectures can capture long-range dependencies and hierarchical features in the input sequences, making them suitable for complex tasks and datasets.

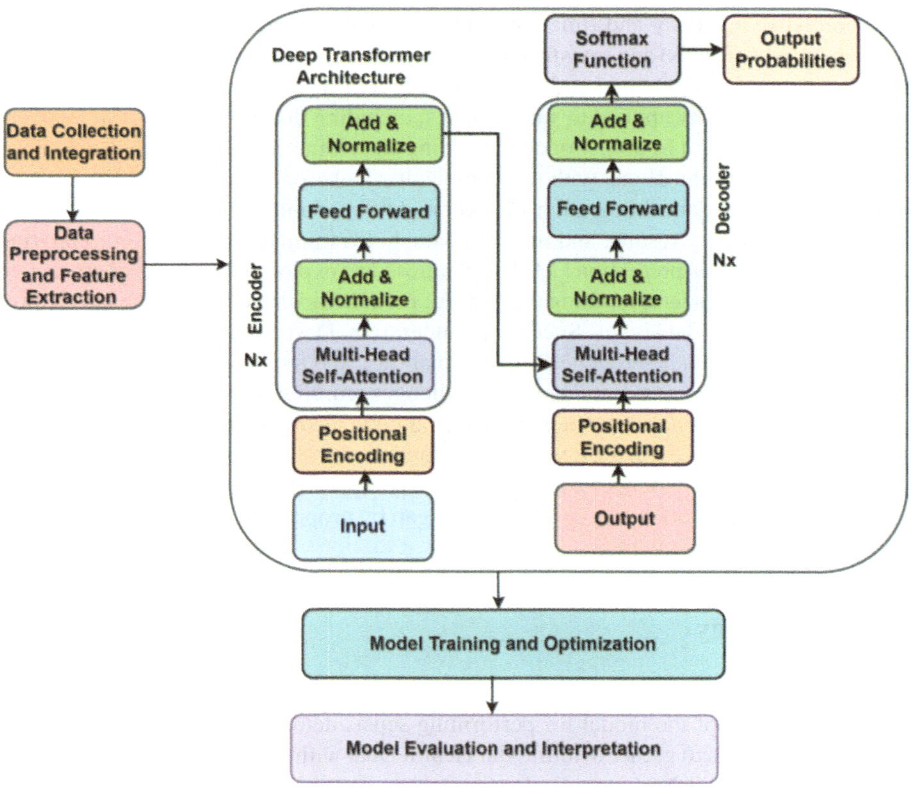

Fig. 1. Deep Transformer Model Architecture for Sepsis Detection

The encoder processes the input multimodal data and the decoder predicts the likelihood of sepsis. Then multiple transformer layers termed as deep transformers are stacked to capture complex patterns in data. Each layer includes multi-head self-attention mechanisms and position-wise feed-forward networks. Then, information from different modalities are fused by a multimodal fusion mechanism. The importance of different modalities are measured through weights using attention mechanism. Softmax activation function is used to predict the occurrence of sepsis. Appropriate loss function is defined to train the model and the model is optimized using Adam optimizer. Techniques like early stopping and learning rate scheduling are implemented to ensure efficient and stable model training. Final step is to perform model evaluation using performance metrics as specified in Table 1. Thus, deep transformer approach enhances the process of sepsis prediction using on the data samples.

4 Results and Discussions

MIMIC-III dataset is a comprehensive and publicly available healthcare dataset developed by the MIT Laboratory for Computational Physiology. The dataset is widely used for research purposes, particularly in the fields of healthcare analytics, machine learning, and clinical decision-making. Includes detailed clinical information such as vital signs (heart rate, blood pressure, etc.), laboratory test results, medications, procedures, and diagnoses. Demographic details such as age, gender, ethnicity, and admission and discharge dates are available for each patient. Records ICU-related details including the type of ICU, duration of ICU stay, and reason for ICU admission. Figure 2 specifies the missing values identified in MIMIC dataset.

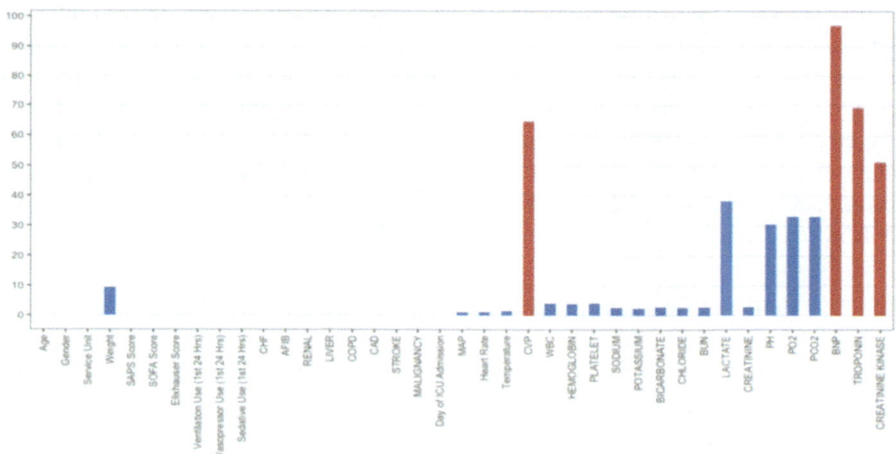

Fig. 2. Missing Value Representation of MIMIC Dataset

Convergence of deep transformer model signifies the models ability to iteratively explore and refine parameters to improve prediction task. The convergence of the deep transformer model based on number of epochs in terms of loss function during training is specified in Fig. 3. The convergence analysis reveals that the proposed Deep Transformer Model's ability to learn effectively from data, resulting in a well-optimized and stable model. Its smooth convergence signifies the model's capacity to adapt, generalize, and make accurate predictions.

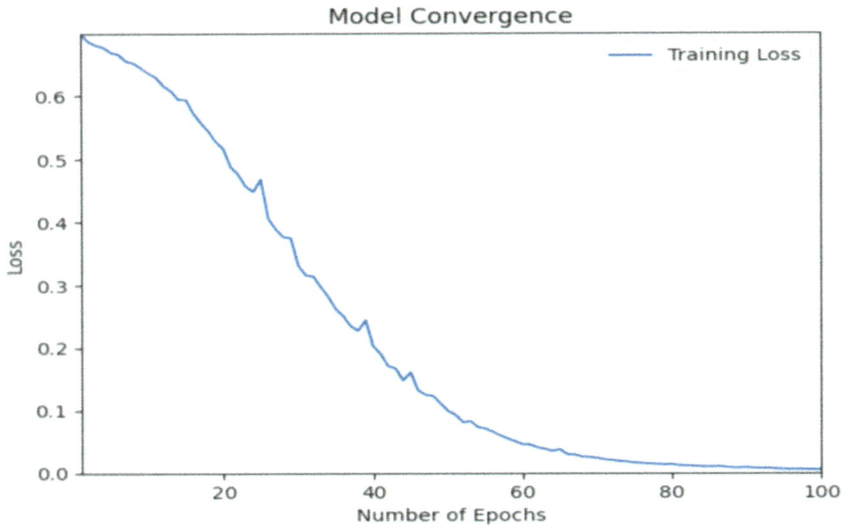

Fig. 3. Deep Transformer Model Convergence during Training based on Epochs

The proposed Deep Transformer Model for sepsis detection outperforms traditional and other deep learning methods with an accuracy of 92%.The model's high precision value makes it a reliable tool for clinical applications. Compared to existing methods specified in Table 1, the Deep Transformer Model consistently excels across all metrics, emphasizing its ability to capture complex data patterns. The proposed approach enhances sepsis prediction and also helps in advancing medical diagnostic practices. Performance evaluation of deep transformer model with relevant metrics is specified in Table 1.

Table 1. Evaluation of Deep Transformer Approach for Sepsis Detection

Method	Accuracy	Sensitivity	Specificity	Precision	F1-Score	AUC-ROC
Proposed Deep Transformer Model	0.92	0.88	0.94	0.91	0.93	0.96
LSTM Networks	0.91	0.86	0.92	0.89	0.90	0.93
GRU Networks	0.90	0.85	0.91	0.88	0.89	0.94
BiLSTM Networks	0.92	0.87	0.93	0.90	0.91	0.92
CNN	0.88	0.83	0.90	0.85	0.87	0.91

The details reveal that the Deep Transformer model performs well compared to existing approaches in performing sepsis prediction. Table 2 outlines the hyperparameter tuning process for deep transformer architecture in the proposed approach. This process is iterative in nature and involves systematically experimenting with various hyperparameter settings to identify optimal configuration that maximizes performance for a sepsis prediction task. Also, hyperparameters for the transformer vary depending on the architecture selected for prediction.

Table 2. Hyperparameter Tuning Values for Deep Transformer Model

Hyperparameter	Value	Hyperparameter	Value
Learning Rate	0.001	Embedding Dimension	128
Batch Size	32	Key Dimension	64
Epochs	50	Dropout Rate	0.3
Number of Layers	4	Loss Function	Binary Crossentropy
Number of Heads	8	Optimizer	Adam

5 Conclusion

In the field of sepsis prediction, incorporation of advanced deep learning techniques, particularly utilization of transformer-based models, marks a significant advancement, and better levels of precision and efficiency. Through the fusion of multimodal health data sources, including clinical notes, vital signs, laboratory results, and imaging data, the proposed model shows a better ability to identify complex patterns and correlations within complex datasets. By efficiently handling complex patient data from various sources, this model demonstrates potential for more accurate and timely identification of sepsis cases. The deep transformer architecture, known for its proficiency in capturing long-range dependencies and in processing these diverse inputs, leading to remarkably accurate sepsis predictions. By enabling timely interventions and informed clinical decisions, the proposed Deep Transformer-Based Model not only enhances patient care but also contributes significantly to the broader landscape of predictive healthcare analytics. Future scope focuses on the refinement and optimization of deep transformer-based model, exploring various architectural modifications and hyperparameter tuning to enhance its accuracy and real-time processing capabilities.

References

1. Chan, T., Gu, F.: Early diagnosis of sepsis using serum biomarkers. Expert Rev. Mol. Diagn. **11**(5), 487–496 (2011)
2. Futoma, J., et al.: An improved multi-output Gaussian process RNN with real-time validation for early sepsis detection. In: Machine Learning for Healthcare Conference, pp. 243–254. PMLR (2017)
3. Pepic, I., et al.: Early detection of sepsis using artificial intelligence: a scoping review protocol. Syst. Rev. **10**(1), 1–7 (2021)
4. Zhang, Y., Zhou, Y., Yang, Y., Pappas, D.: Microfluidics for sepsis early diagnosis and prognosis: a review of recent methods. Analyst **146**(7), 2110–2125 (2021)

5. Lauritsen, S.M., et al.: Early detection of sepsis utilizing deep learning on electronic health record event sequences. Artif. Intell. Med. **104**, 101820 (2020)
6. Rafiei, A., Rezaee, A., Hajati, F., Gheisari, S., Golzan, M.: SSP: early prediction of sepsis using fully connected LSTM-CNN model. Comput. Biol. Med. **128**, 104110 (2021)
7. Alvi, R.H., Rahman, M.H., Khan, A.A.S., Rahman, R.M.: Deep learning approach on tabular data to predict early-onset neonatal sepsis. J. Inf. Telecommun. **5**(2), 226–246 (2021)
8. Li, X., Ng, G.A., Schlindwein, F.S.: Convolutional and recurrent neural networks for early detection of sepsis using hourly physiological data from patients in intensive care unit. In: 2019 Computing in Cardiology (CinC), p. 1. IEEE (2019)
9. Aşuroğlu, T., Oğul, H.: A deep learning approach for sepsis monitoring via severity score estimation. Comput. Methods Programs Biomed. **198**, 105816 (2021)
10. Shah, N., Bhatia, J., Vasavat, N., Desai, R., Sonawane, P.: Early sepsis detection using machine learning and neural networks. In: 2021 2nd Global Conference for Advancement in Technology (GCAT), pp. 1–6. IEEE (2021)
11. Boussina, A., Ramesh, K., Arora, H., Ratadiya, P., Nemati, S.: Differentiation of fungal, viral, and bacterial sepsis using multimodal deep learning. medRxiv, pp. 2023-04 (2023)
12. Chen, Q., et al.: SEPRES: sepsis prediction via the clinical data integration system in the ICU. medRxiv, pp. 2022-07 (2022)
13. Hammoud, I., Ramakrishnan, I.V., Henry, M., Morley, E.: Multimodal early septic shock prediction model using lasso regression with decaying response. In: 2020 IEEE International Conference on Healthcare Informatics (ICHI), pp. 1–3. IEEE (2020)
14. Chen, Q., et al.: SEPRES: intensive care unit clinical data integration system to predict sepsis. Appl. Clin. Inform. **14**(01), 65–75 (2023)
15. Saknite, I., et al.: Novel hybrid technology for early diagnostics of sepsis. In: Multimodal Biomedical Imaging XII, vol. 10057, pp. 76–85. SPIE (2017)
16. e Silva, L.L., Ferreira, D.B., Melillo, K.P., Passos, R.M.A., Medeiros, E.A.: Impact of multimodal strategy on sepsis bundle compliance and in-hospital mortality in patients with hematologic malignancies (2023)
17. Dou, Y., Li, W., Zomaya, A.Y.: Transformer-based unsupervised learning for early detection of sepsis (student abstract). In: Proceedings of the AAAI Conference on Artificial Intelligence, vol. 36, no. 11, pp. 12937–12938 (2022)
18. Tang, Y., Zhang, Y., Li, J.: A Time Series Driven Model for Early Sepsis Prediction Based on Transformer Module (2023)
19. Parsola, J.: Exploring the power of transformer models in hospitality domain. Math. Stat. Eng. Appl. **70**(1), 324–330 (2021)
20. Reddy, M.K., Kovuri, K., Avanija, J., Sakthivel, M., Kaleru, S.: Brain stroke prediction using deep learning: a CNN approach. In: 2022 4th International Conference on Inventive Research in Computing Applications (ICIRCA), pp. 775–780. IEEE (2022)
21. Wang, Y., Zhao, Y., Callcut, R., Petzold, L.: Integrating physiological time series and clinical notes with transformer for early prediction of sepsis. arXiv preprint arXiv:2203.14469 (2022)
22. Kumar, M.R., Natteshan, N.V.S., Avanija, J., Madhavi, K.R., Charan, N.S., Kushal, V.: SMOTE-TOMEK: a hybrid sampling-based ensemble learning approach for sepsis prediction. In: 2023 2nd International Conference on Edge Computing and Applications (ICECAA), pp. 724–729. IEEE (2023)

Privacy-Preserving Healthcare Data in IoMT

Karthik Kovuri[1], Gunipati Kanishka[2], Pattabhi Mary Jyosthna[3], A. Jyothi Babu[4(✉)], Nagendar Yamsani[5], and R. Vanithamani[6]

[1] School of Engineering and Technology, Kaziranga University, Rowmarikhuti, Assam, India
[2] Computer Science and Engineering (Cyber Security), Madanapalle Institute of Technology and Science, Madanapalle, India
[3] Computer Science and Engineering, B V Raju Institute of Technology, Narsapur, TS, India
jyosthna.p@bvrit.ac.in
[4] Department of Computer Applications, School of Computing, Mohan Babu University, Tirupati, India
jyothibabuaddanki@gmail.com
[5] School of Computer Science and Artificial Intelligence, SR University, Warangal, India
[6] Department of Biomedical Instrumentation Engineering, School of Engineering, Avinashilingam Institute for Home Science and Higher Education for Women, Coimbatore, TN, India

Abstract. IoMT has ushered into the IoT era and has occupied most of the expanse of integrated computing technologies. IoMT, however has many problems with the devices, that arise with the synchronization of computing power. The present work proposes a data privacy protection and schemes for retrievability through queries for the medical workers in the integrated computing environment like edge+fog+cloud. The edge devices are IoMT devices which collect the data and collate with various data structures for the seamless transmission into the storages and to the medical workers as responses to the queries. The multidimensional range tree structures and index tree are conceptually introduced to enable the richness of the retrievability schemes and simulated. The simulation proves the significantly less energy costs and more retrievability.

Keywords: Privacy Preserving · IoMT · fog computing · cloud services · scheduling

1 Introduction

The suffering with CoVID-19 worldwide have been increasing and counted 150 million and 3 million deaths tolled. The rates of mortality are varying [1, 2] with country wise demographic status. Poverty in countries push into more death statistics due to insufficient medical privileges and healthcare personnel. IoMT is introduced to protect, manage, reach, diagnose and provide significant medical advices to the remote. A cloud server in a network connects medical or smart health devices with sensors to a computer to form IoMT. Basic pathological information about the patients are preserved in the databases in cloud storages.

D. Kriksciuniene et al. (Eds.): IBICA 2023, LNNS 1228, pp. 311–318, 2025.
https://doi.org/10.1007/978-3-031-78937-3_33

By remotely sharing cloud data and interacting with patients, healthcare workers can analyse and diagnose. The industrial standard of Internet of Medical Things [3–5] are at the ajar in the technological innovations and towards personalized medicine with increasing popularity. The advent of monitoring through sensing with devices connected to the cloud servers delves into the concepts of remote diagnosis and rendering of medical services.

IoMT devices are developed to form into a conglomerate of communication dealing with wide range of sharing pathological information, transacting [6, 7] with treatments across and recorded medical suggestions. Thus, health care organisations and medical professionals will outsource large cloud services in order to manage and organise IoMT devices. In spite of disadvantages of geographical distance to reach a medical worker into the patient site, using communication medium, intelligent medical treatment can be rendered.

2 Related Work

Research in IoMT erstwhile was focused on the applications of physical testing of medical equipment. Many case studies have been recorded in detection of sound of heart-beat with auxiliary diagnosis. Data privacy protection is [8–10] hindered by the challenge of utilising these data and retrieving the information of patients. To identify cautious warnings in IoMT-uploaded patient heart-beat data, a machine learning algorithm is used. Nevertheless, these methods arrived for public use, the privacy protection came into force which lowers the quality by applications of perturbations in data resulting low efficiency and low delay. Many research consensus has proposed to privacy protection with fog-computing systems, where retrieval becomes achiles heel. Fogcomputing has advanced with service scheduling algorithms to conserve efficient energy costs combining with a retrieval schemes in IoMT.

Studies on patients data analysis and retrieval systems development proposed most of the information in the text form, which has privacy breach. The proposed research directs towards ciphertext retrievability on cloud servers. Ciphertext retrieval studies on fogcomputing has a little influence on ciphertext retrieval studies [11, 12] on IoMT. A multikeyword based ranking search over encrypted data on cloud servers were also proposed in the research consensus. Using reversible matrix forms and random split indicators, secure k-NN algorithms encrypted data and query vectors for ciphertext retrieval. The scheme also promotes that weaker process of indexes, affecting the efficiency of retrieval.

Personalized Search is even deployed with encrypting on out-sourced data, where the indexes are partitioned and an index tree is built into blocks. Block wise permission to the industry medical workers are granted to improve the retrieval efficiency, whereas the index tree becomes non performant due to the truncations problems is representing the tree forms. A secure dynamic multikeyword sorting and search scheme is introduced on cloud servers, based on clustering algorithms. Data vectors are initially clustered and a tree is built to improve the retrieval [13, 14] efficiency. Due to heterogeneity in the data vectors the values in the clustering algorithm distract the retrieval and therefore proved as inaccurate.

Medical workers are utilizing fog computing to enhance efficiency, protect patient privacy, and reduce latency in data transmission through various ciphertext retrieval mechanisms.

The reasonable consideration for the implementation of IoMT devices shall raise the question of privacy of patients. "Traditional retrieval of information is plaintext oriented", directly uploading the patient pathological and physiological data to the indices of cloud server obstructs the privacy disclosure of the patients. The medical institutions and the healthcare workers shall adopt to the security and cryptographic methodologies to employ the IoMT. The systems shall containing development of text enciphering methods semantically, to enable the healthcare and medical workers to get promoted in the services through IoMT.

Although various types of text enciphering schemes for retrieval in cloud computing exists, a vector space model of ciphertext retrieval scheme [15–17] for the fog computing systems shall be recommended. Each pathological and physiological data vector is a basic element in the high-dimensional space managed by medical institutions. A wide range of data vectors are built [18, 19] and processed into a range tree index, which improves the efficiency in retrieval. Fog computing is a secure and reliable method for sharing decryption keys among medical care workers.

3 Proposed Work

The sensibleattention for the employment of IoMT devices shall raise the question of privacy of patients. The work presented in this paper proposes the architecture of Internet of Medical Things.

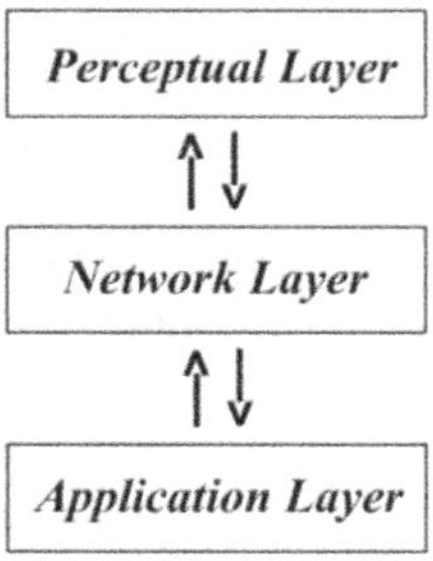

Fig. 1. Layers in FOG Computing with IoMT

The architecture has three layers: perceptual, network, and application. A perceptual layer includes medical sensors, QR code identification, transmission path, networking, and gateway hardware. The fog computing system and cloud service public essentials are composed in a network layer. Cloud server employs normalization and generates a data vector represents the basic physical features of the patient. The end-user functions of the IoMT are deployed at the application layer. The application layer delivers three specific functions. First, the basic pathological and physiological information about the patient are collected to confound the central patient database. Second, the medical and healthcare workers access the patients' data who are candidate for the treatments.

Third, the patient information is color-coded, the medical and healthcare workers treat this as important for identifying the patient with a particular ailment or disease with respect to their basic pathological and physiological data defined in the data vectors (Fig. 2).

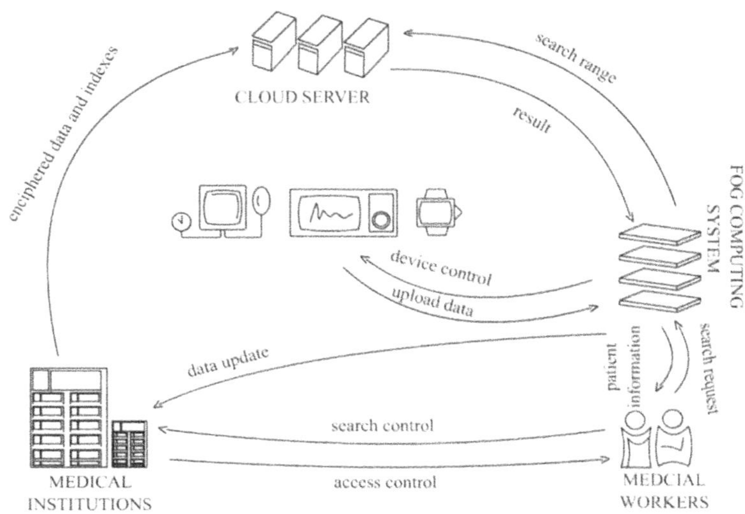

Fig. 2. FOG Computing with IoMT System Model

Therefore, the paper presents a Privacy Preserving IoMT Framework which includes fog computing with a cloud server representing the centralized patient database as indexed data. The indexed data retrieved in a range tree, which is developed using data vectors. A multi-body feature data represented in the ciphertext, fog computing system with cost effective scheduling algorithm, a seamless enciphering and deciphering system to assure retrieval accuracy.

The depicted model in the Fig. 1 is a Layers in FOG Computing with IoMT. The model has three crucial layers: perceptual, network, and application. The fog computing system 'C' receives an encrypted data set 'F' from the medical institutions, which is used to construct the range tree index, which is then encrypted into 'I'. The medical institutions send encrypted data through 'C' to cloud servers, creating a trapdoor. The trapdoor is the key that medical workers share. The fog computing system parallelly generates the retrieval range vector 'R' based on the retrieval range and uploads it to the cloud server via the query trapdoor 'TQ' for retrieval. After receiving the cloud server's results, the trapdoor gives the fog computing system the key to decrypt and distribute the data set to the public cloud server. The public cloud server stores encrypted data 'F' and medical institutions upload and maintain an encrypted range tree index 'I'. A range vector determines the cloud server's retrieval range for trapdoor query TQ. TQ receives results from cloud server.

The cloud server utilizes fog computing to securely store and manage IoMT data, ensuring its integrity and accuracy in medical data verification. To encrypt data vector F, the medical institution uses random numbers, reversible matrix pairs M1 and M2 with a master key k. The range tree is derived from the fog computing data vector, and the fog computing index system generates the index. The cloud server receives the index and data vector. After medical workers submit the query, the fog computing system sends the cloud server a TQ trapdoor query and range vector R. The medical worker receives the results.

A kd-tree variant is the range tree. Despite taking up more space than kd-tree, the range tree is more retrievable. The retrieval scheme is independent of storage spaces to help medical workers find answers. A binary search tree as balanced form is built for data vectors 'F', which represent all vector vital signs, and stored at leaf positions. A subtree is a non-leaf node in a binary search tree. The construction of subtrees is motivated by vital signs, and each subtree's root corresponds to a vital sign level. The subtree of a nonleaf node's data vectors are organised by vital sign level to establish the dimensional range tree and form a joint tree structure with the first dimensional range tree. Each nonleaf node points to the root node, which represents the leaf nodes' subset, where vital signs represent dimensions at each level.

4 Results

The p-dimensional range tree uses n data vectors. To retrieve the p-dimensional range tree, the first dimension is retrieved, which takes O(log n) time. To search $1 - p$ dimensional range tree the time complexity relationship is $O_p(n) = O(\log n) + O(\log n) \times O_{p\text{-}1}(n)$. Where the $O_p(n)$ determines the look up time complexity of 1-p dimensional range tree and $O_{p\text{-}1}(n)$ determines the look up time complexity for $(p - 1)$ dimensional range tree and deduced as $O_2(n) = O(\log 2\ n)$. The general time complexity of dimensional range tree is $O\ (\log^p n)$ and total time complexity is $O\ (\log^p n + k)$.

The efficiency of retrievability scheme for the index structure is determined based on the dataset and the scores of the calculated similarity. As the simulation of the experiment goes on the range tree is incidental by the queries posed by medical workers, which is in general set to minimum of five (5). The analyses of the efficiency is measured with width d based on the retrieval efficiency obtained by the medical workers. As the value of d is large, the efficiency is lowered, for a low d, fewer nodes are incidental, hence the efficiency increases. The following graph in Fig. 3, shows the comparison between the retrieval efficiency and the width of the query.

Various combination of schemes are applied in order enhance the retrievability assuring the privacy by employing the enciphering methods.

Each method differs with the energy cost and convergence rates. Higher energy costs deal with higher convergence rates and low energy costs compromise to have low convergence rates. The convergence speed of the scheme depends on the size of the query and the enciphering methods applied. The scheme proposed in the experiment after iterations between 200 to 1400, has achieved higher convergence rate with lower energy costs. The variation of retrieval time and the number of records returned by the query to the medical workers is a quantitative result of the scheme. The graph mentioned

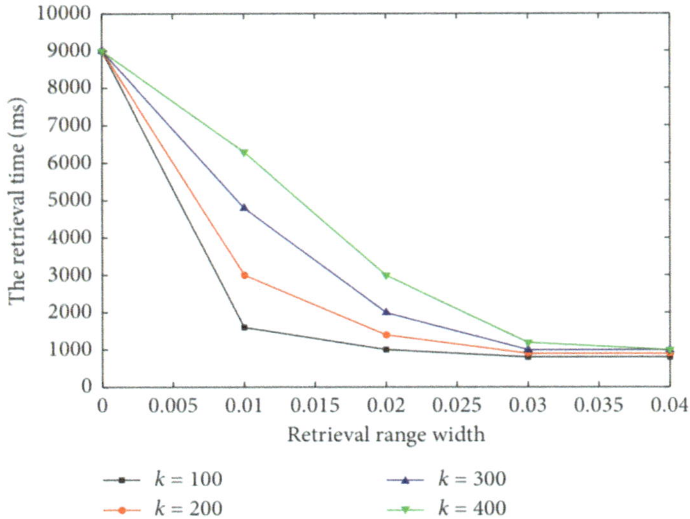

Fig. 3. Graph showing relation between retrieval time and retrieval range width.

in the Fig. 4 states the relationship between the retrieval time and the number of patients' records.

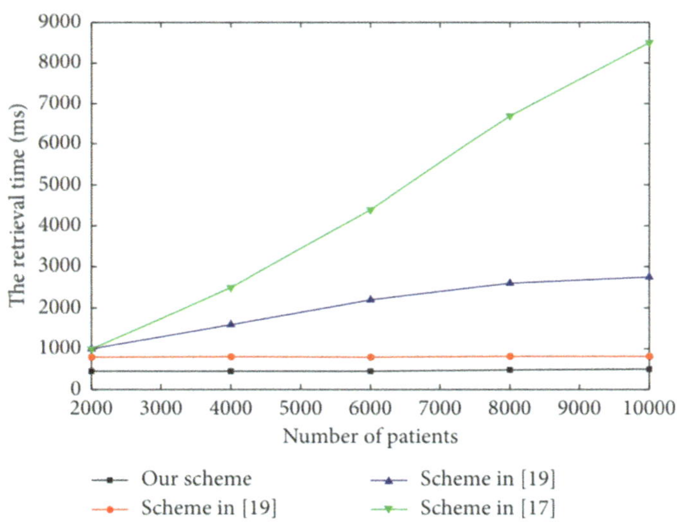

Fig. 4. Graph showing relation between retrieval time and Number of patients.

5 Conclusion

The scheme for retrievability employed with IoMT has faced many problems with the contingent configurations such as breach of privacy, lower retrieval rates and high energy costs. In the proposed work the data retrieval and the analyses were employed in low cost fog computing system. A fog computing system setup between the cloud servers and the IoMT devices is orchestrated to support the effective data retrievability. The range tree concept is introduced to construct the index in a multidimensional form as a multidimensional range tree, which improves retrievability further in the premise of security of indexes and isolation with the portals. The results of simulation are depicted in the graphs which prove the efficiency of the schemes employed to preserve privacy and enhance the retrievability in contrast with the existing schemes.

References

1. Wang, N., Fu, J., Bhargava, B.K., Zeng, J.: Efficient retrieval over documents encrypted by attributes in cloud computing. IEEE Trans. Inf. Forensics Secur. **13**(10), 2653–2667 (2018)
2. Kim, H., Shin, J., Song, Y., Chang, J.: Privacy-preserving association rule mining algorithm for encrypted data in cloud computing. In: Proceedings of the 2019 IEEE 12th International Conference on Cloud Computing (CLOUD), Milan, Italy, pp. 487–489 (2019)
3. Rizk, D., Rizk, R., Hsu, S.: Applied layered-security model to IoMT. In: Proceedings of the 2019 IEEE International Conference on Intelligence and Security Informatics (ISI), Shenzhen, China, p. 227 (2019)
4. Mishra, S.K., Puthal, D., Rodrigues, J.J.P.C., Sahoo, B., Dutkiewicz, E.: Sustainable service allocation using a meta-heuristic technique in a fog server for industrial applications. IEEE Trans. Ind. Inform. **14**(10), 4497–4506 (2018)
5. Bikku, T.: Multi-layered deep learning perceptron approach for health risk prediction. J. Big Data **7**(1), 1–14 (2020)
6. Bikku, T., Rao, N.S., Akepogu, A.R.: A novel multi-class ensemble model based on feature selection using Hadoop framework for classifying imbalanced biomedical data. Int. J. Bus. Intell. Data Min. **14**(1–2), 25–39 (2019)
7. Rao, G.R.K., Sgar, P.V., Bikku, T., Prasad, C., Cherukuri, N.: Comparing 3D rendering engines in blender. In: 2021 2nd International Conference on Smart Electronics and Communication (ICOSEC), pp. 489–495. IEEE (2021)
8. Inaganti, R.: Preserving Patient Privacy: A Focus on Cyber Security in Healthcare (2023). https://doi.org/10.13140/RG.2.2.26620.31367
9. Yalavarthi, Sreekanth: AI in medicine: from diagnosis to treatment decision support. Int. J. Emerg. Technol. Innov. Res. **10**(10), e214–e217 (2023). www.jetir.org. ISSN 2349-5162
10. Somasekar, J., Ramesh, G., Ramu, G., Dileep Kumar Reddy, P., Eswara Reddy, B., Lai, C.-H.: A dataset for automatic contrast enhancement of microscopic malaria infected blood RGB images. Data Brief **27**, 104643 (2019)
11. Reddy, U.J., Dhanalakshmi, P., Reddy, P.D.K.: Image segmentation technique using SVM classifier for detection of medical disorders. Ingénierie Des Systèmes D'information **24**(2) (2019)
12. Reddy Madhavi, K., Kora, P., Venkateswara Reddy, L., Avanija, J., Soujanya, K.L.S., Prabhakar, T.: Cardiac arrhythmia detection using dual-tree wavelet transform and convolutional neural network. Soft Comput. Impact Factor **3**(64) (2021)

13. Reddy Madhavi, K., Vijaya Sambhavi, Y., Sudhakara, M., Raju, S.K.: COVID-19 isolation monitoring system. Springer Series – Lecture Notes on Data Engineering and Communication Technology, vol. 63, pp. 601–609 (2021)
14. Reddy Madhavi, K., Vijaya Sambhavi, Y., Sudhakara, M., Srujan Raju, K.: COVID-19 isolation monitoring system. In: Reddy, K.A., Devi, B.R., George, B., Raju, K.S. (eds.) Data Engineering and Communication Technology. LNDECT, vol. 63, pp. 601–609. Springer, Singapore (2021). https://doi.org/10.1007/978-981-16-0081-4_60
15. Ahamed, S.F., Vijayasankar, A., Thenmozhi, M., Rajendar, S., Bindu, P., Subha Mastan Rao, T.: Machine learning models for forecasting and estimation of business operations. J. High Technol. Manag. Res. **34**(1) (2023)
16. Alkawaz, A.N., et al.: Training multilayer neural network based on optimal control theory for limited computational resources. Mathematics **11**(3), 778 (2023). https://doi.org/10.3390/math11030778
17. Jeyakanth, K., Venkatakrishnan, P., Chitra, C.: Optimized channel prediction and auction-based channel allocation for personal cognitive networks. Int. J. Commun. Syst. **36**(3), e5391 (2023)
18. Zaid, T., Biradar, N., Sonth, M.V., Gowre, S.C., Gadgay, B.: FDADE: flow direction algorithm with differential evolution for measurement of intima-media thickness of the carotid artery in ultrasound images. Biomed. Signal Process. Control **80**(Part 2), 104350 (2023). https://doi.org/10.1016/j.bspc.2022.104350. ISSN 1746-8094
19. Sunitha, G., Reddy Madhavi, K., Avanija, J., Srujan Raju, K., Kirankumar, A., Raji Reddy, A.: Ensemble model for lidar data analysis and nocturnal boundary layer height estimation. In: Bhateja, V., Carroll, F., Tavares, J.M.R.S., Sengar, S.S., Peer, P. (eds.) Intelligent Data Engineering and Analytics. FICTA 2023. Smart Innovation, Systems and Technologies, vol. 371. Springer, Singapore (2023). https://doi.org/10.1007/978-981-99-6706-3_28

Multi-scale Lung Nodule Detection with U-Net and Dilated Convolution Networks in Thoracic CT Scans

K. S. Kannan[1], P. MeganaSanthoshi[2], A. Basi Reddy[3], J. Avanija[4], K. Reddy Madhavi[4(✉)], Mohammad Gouse Galety[5], and Kondra Pranitha[6]

[1] CSE, Kalasalingam Academy of Research and Education, Krishnakoil, Tamilnadu, India
[2] KLM College of Engineering for Women, Kadapa, India
[3] CSE, School of Computing, Mohan Babu University, Tirupati, Andhra Pradesh, India
[4] School of Computing, Mohan Babu University, A.Rangampet, Tirupati, India
kreddymadhavi@gmail.com
[5] Department of Computer Science, School of Engineering, Samarkand International University of Technology, Samarkand, Uzbekistan
[6] B V Raju Institute of Technology, Narsapur, Medak, Telangana, India

Abstract. Early diagnosis of pulmonary disease is possible using Lung nodule detection in thoracic computed tomography (CT) scans. A fusion of U-Net and Dilated Convolution Networks (DCNs) has been proposed, for multi-scale lung nodule detection. U-Net architecture, utilized for semantic segmentation capabilities, is combined with DCNs to capture multi-scale features, to predict nodules of different sizes. The accuracy of the proposed work is better by the utilization of deep learning models. Proposed UNet-Dilated network based method is applied on thoracic CT datasets revealing better results, compared to traditional methods. Fusion of U-Net and DCNs proves to be an efficient solution for advancing lung nodule detection, contributing significantly to the field of medical imaging for improved patient care.

Keywords: Lung Nodule Detection · Thoracic CT Scans · U-Net · Dilated Convolution Networks · Deep Learning · Medical Imaging · Multi-Scale Feature Extraction · Semantic Segmentation

1 Introduction

Among different imaging methods, thoracic computed tomography scans stand out as an invaluable tool for early diagnosis of pulmonary diseases, including lung cancer. This task has traditionally been a labor-intensive and time-consuming process, often requiring keen eyes of radiologists for interpretation. The process can be simplified using techniques such as artificial intelligence and deep learning. The proposed work delves introduces a novel approach that utilizes the power of deep neural networks, specifically a combination of the U-Net architecture and Dilated Convolution Networks.

D. Kriksciuniene et al. (Eds.): IBICA 2023, LNNS 1228, pp. 319–325, 2025.
https://doi.org/10.1007/978-3-031-78937-3_34

The variations in nodule size, shape, and location within lung require advanced techniques capable of multi-scale analysis and complex feature extraction. The U-Net architecture, known for its semantic segmentation capabilities, offers a robust foundation for precise nodule localization.The incorporation of dilated convolution networks enables the network to capture multi-scale contextual information, to distinguish nodules from non-nodule structures. The proposed UNet-Dilated network bases approach for lung nodule detection can reduce burden on radiologists by automating the detection process and, in turn, accelerating the diagnosis of pulmonary diseases. Timely detection can be the difference between early intervention and the progression of severe medical conditions. This work represents a significant step in that perspective, leading to improved healthcare, earlier detection, and ultimately, enhanced patient outcomes.

2 Literature Review

Medical imaging, particularly thoracic computed tomography (CT) scans, has become a important in early detection and diagnosis of pulmonary diseases. Within this domain, identification of lung nodules becomes a critical task due to its implications for various respiratory conditions, including lung cancer [1]. Traditional methods for lung nodule detection primarily relied on features and machine learning algorithms. Techniques such as region-growing, morphological operations, and thresholding were commonly employed [2, 3]. These approaches showed some success, they often struggled with the ability to capture complex patterns within CT scans. With the emergence of deep learning, specifically convolutional neural networks (CNNs), there has been a paradigm shift in the field of medical image analysis [4–7]. CNNs have demonstrated remarkable capabilities in feature extraction, allowing for automatic learning of complex patterns from raw image data. Several studies have explored the application of CNNs in lung nodule detection, showcasing their potential in achieving high accuracy and efficiency [8–15]. U-Net architecture, initially designed for semantic segmentation tasks, has gained substantial importance in medical imaging. Its unique encoder-decoder structure, combined with skip connections, enables precise localization of objects within images. U-Net has shown promising results in accurately delineating nodule boundaries [16–22]. Dilated Convolution Networks (DCNs) have emerged as a powerful tool for multi-scale feature extraction. By incorporating dilated convolutions, DCNs can capture contextual information at varying field sizes without significantly increasing the network's parameters [22–27]. This ability to gather multi-scale features is especially valuable in medical imaging, where nodules can vary significantly in size and shape. By combining U-Net's precise segmentation capabilities with DCNs' multi-scale feature extraction, the proposed work improves accuracy of nodule prediction.

3 Methodology

This section, briefs about the steps of UNet-Dialated Convolution network based model for lung nodule prediction. The steps of the multi-scale model to perform lung nodule prediction have been specified in Fig. 1. Initial step is to gather thoracic CT scan data with annotated lung nodules. Proposed work utilizes LIDC-IDRI dataset (Lung Image

Database Consortium - Image Database Resource Initiative). This dataset is used as benchmark dataset in lung nodule prediction. It was created through a collaboration of several institutions and provides a diverse and comprehensive set of cases for research purposes.

Clean and preprocess the data by resizing, normalizing pixel values, and annotating nodules to create ground truth data for supervised training of the neural networks.Next step is to train the model using UNet architecture utilizing encoder-decoder components with skip connections. U-Net uses convolutional neural network (CNN) architecture for segmentation of images. UNet is used to perform fine-scale segmentation in order to achieve precise segmentation of small lung nodules. Encoder part of the U-Net is designed to capture the context and extract features from the input image. Multiple convolutional layers with filters of varying sizes are used to capture different levels of features from input image. ReLU (Rectified Linear Unit) activation functions are used to include non-linearity after each convolution operation. Max-pooling operations are employed to reduce spatial dimensions, making the network invariant to small translations and variations in the input. Bottleneck serves as a bridge between encoder and decoder and retains most essential features.

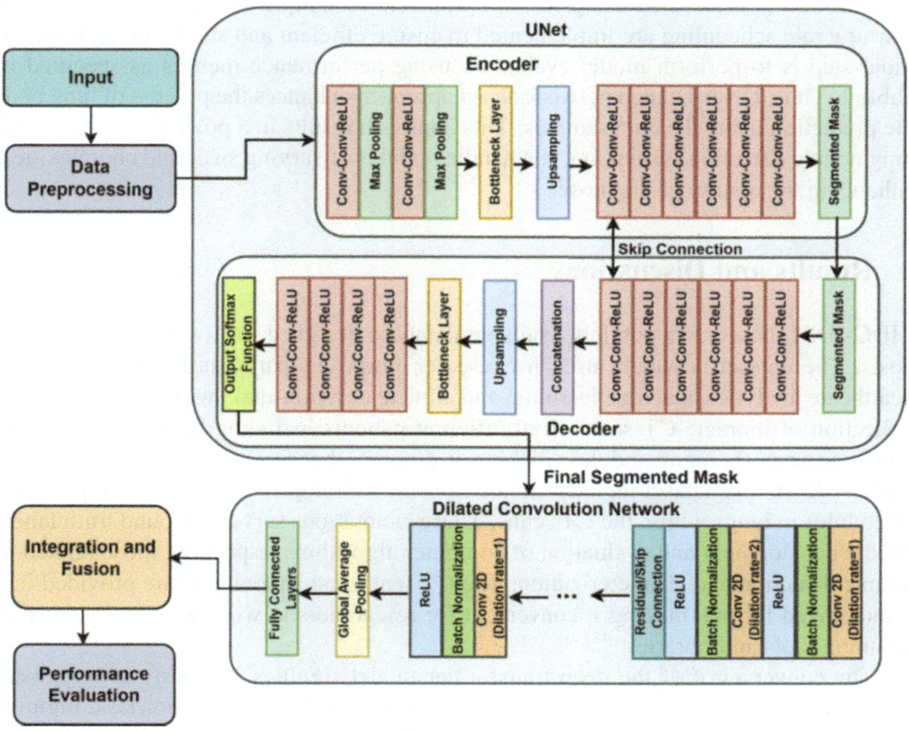

Fig. 1. UNet-Dilated Convolution Networks Architecture for Lung Nodule Detection

It consists of convolution layers with ReLU. Decoder part of UNet upsamples feature maps to actual size of input image and refines the segmentation masks. These layers perform upsampling and learn to fill in the details. Skip connections concatenate feature maps from the corresponding encoder layers. Skip connections help the network preserve spatial information, allowing precise segmentation of fine structures. Additional convolutional layers in the decoder refine the segmented features and create detailed segmentation masks. Final layer of UNet produces the segmentation mask, sigmoid activation function is commonly used in binary segmentation tasks. Training strategies such as data augmentation and learning rate scheduling with adam optimization is performed to enhance the performance of model.

Next, binary masks generated by the U-Net, representing the detected lung nodules, serve as the input to the dilated convolutional networks. Binary masks are processed through dilated convolutional layers with varying dilation rates. These layers capture multi-scale features from the binary masks. By using dilated convolutions with different dilation rates, the network captures features at different scales, ensuring that both small and large nodules are properly represented in the feature maps. Multi-scale features extracted by the dilated convolutional network are then integrated and fused, typically using weighted averaging. Appropriate loss function is defined to train the model and optimization is performed using Adam optimizer. Techniques like early stopping and learning rate scheduling are implemented to ensure efficient and stable model training. Final step is to perform model evaluation using performance metrics as specified in Table 1. Thus, UNet-Dilated network based approach enhances the process of lung nodule prediction using the data samples. This approach results in a powerful and accurate lung nodule detection system that can handle nodules of varying sizes and complexities, enhancing the quality of diagnosis.

4 Results and Discussions

LIDC-IDRI dataset is a publicly available database designed for for lung cancer diagnosis. The dataset is widely used for research purposes, particularly in the fields of healthcare analytics, machine learning, and clinical decision-making. It contains a large collection of thoracic CT scans from different patients including both cancerous and non-cancerous (benign) nodules, making it a comprehensive dataset for lung cancer studies. Also, this dataset includes annotations by radiologists specifying the presence of nodules in lungs within the CT scans. These annotations serve as ground truth labels for the development and evaluation of computer algorithms, especially those related to nodule detection and characterization. The CT scans and annotations are provided in a standardized format, making it convenient for researchers to work with the data using common tools and libraries.

The convergence of the deep transformer model signifies its ability to iteratively explore and refine parameters, enhancing its performance on the prediction task. Figure 2 illustrates the model's convergence concerning the number of epochs, as depicted by the loss function during the training process. The analysis of convergence demonstrates the effectiveness of the proposed UNet-Dilated network model in learning from data. The model exhibits a robust and stable optimization process, indicating its ability to adapt,

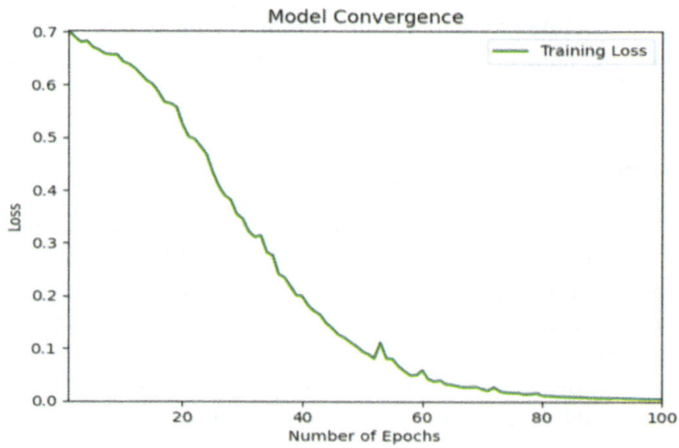

Fig. 2. UNet-Dilated Network Model Convergence during Training based on Epochs

generalize, and make precise predictions. The smooth convergence reflects the model's aptitude for effective learning and reliable performance.

The proposed UNet-Dilated Network Model for lung nodule detection outperforms traditional and other deep learning methods with an accuracy of 95%.The model's high precision value makes it a reliable tool for clinical applications. Compared to existing methods specified in Table 1, the UNet-Dilated Network Model consistently excels across all metrics, emphasizing its ability to capture complex data patterns. The proposed approach enhances lung nodule prediction and also helps in advancing medical diagnostic practices. Performance evaluation of UNet-Dilated Network model with relevant metrics is specified in Table 1.

Table 1. Evaluation of UNet-Dilated Network for Lung Nodule Detection

Method	Sensitivity	Specificity	Precision	Recall	F1-Score	Dice Score
Proposed Deep Transformer Model	0.95	0.98	0.94	0.95	0.94	0.92
CNN	0.88	0.96	0.90	0.88	0.89	0.87
RNN	0.91	0.94	0.88	0.91	0.89	0.85
Multiscale Model	0.89	0.92	0.87	0.89	0.88	0.84
Ensemble Method	0.92	0.95	0.91	0.92	0.91	0.88

The details reveal that the UNet-Dilated Network model performs well compared to existing approaches in performing lung nodule prediction. Hyperparameter tuning process is an iterative process that involves systematic experimentation with diverse hyperparameter configurations to identify the optimal settings that enhance performance in the lung nodule prediction task. Also, hyperparameters for the transformer vary depending on the architecture selected for prediction. Considering these aspects hyperparameters such as learning rate, number of layers, dropout rate and more are carefully fine tuned to the best possible values.

5 Conclusion

The proposed work provides a advancement in the field of lung nodule detection. By combining two deep learning models, U-Net and Dilated Convolution Networks, the proposed method excels in accurately identifying lung nodules in thoracic CT scans. Results have shown the effectiveness of this approach, particularly in handling the variations in nodule sizes and shapes. Also the method offers a potential solution to the complexities of early diagnosis and treatment planning for lung-related diseases. Future work in this area could explore the integration of additional medical image modalities, such as PET, MRI scans, to enhance the accuracy and comprehensive evaluation of lung nodules. Also, application of transfer learning and development of larger, more diverse datasets could further improve the generalization of proposed model, making it more adaptable to different healthcare settings.

References

1. Armato, S.G., Li, F., Giger, M.L., MacMahon, H., Sone, S., Doi, K.: Lung cancer: performance of automated lung nodule detection applied to cancers missed in a CT screening program. Radiology **225**(3), 685–692 (2002)
2. Cao, H., et al.: A two-stage convolutional neural network for lung nodule detection. IEEE J. Biomed. Health Inform. **24**(7), 2006–2015 (2020)
3. Huang, X., Shan, J., Vaidya, V.: Lung nodule detection in CT using 3D convolutional neural networks. In: 2017 IEEE 14th International Symposium on Biomedical Imaging (ISBI 2017), pp. 379–383. IEEE (2017)
4. Su, Y., Li, D., Chen, X.: Lung nodule detection based on faster R-CNN framework. Comput. Methods Program. Biomed. **200**, 105866 (2021)
5. Halder, A., Dey, D., Sadhu, A.K.: Lung nodule detection from feature engineering to deep learning in thoracic CT images: a comprehensive review. J. Digit. Imaging **33**(3), 655–677 (2020)
6. Cao, H., et al.: Multi-branch ensemble learning architecture based on 3D CNN for false positive reduction in lung nodule detection. IEEE Access **7**, 67380–67391 (2019)
7. Mittapalli, P.S., Thanikaiselvan, V.: Multiscale CNN with compound fusions for false positive reduction in lung nodule detection. Artif. Intell. Med. **113**, 102017 (2021)
8. Schultheiss, M., et al.: Lung nodule detection in chest X-rays using synthetic ground-truth data comparing CNN-based diagnosis to human performance. Sci. Rep. **11**(1), 15857 (2021)
9. Dutande, P., Baid, U., Talbar, S.: LNCDS: a 2D–3D cascaded CNN approach for lung nodule classification, detection and segmentation. Biomed. Signal Process. Control **67**, 102527 (2021)
10. Liu, M., Dong, J., Dong, X., Yu, H., Qi, L.: Segmentation of lung nodule in CT images based on mask R-CNN. In: 2018 9th International Conference on Awareness Science and Technology (iCAST), pp. 1–6. IEEE (2018)
11. Tajbakhsh, N., Suzuki, K.: Comparing two classes of end-to-end machine-learning models in lung nodule detection and classification: MTANNs vs CNNs. Pattern Recognit. **63**, 476–486 (2017)

12. Tan, J., Huo, Y., Liang, Z., Li, L.: Expert knowledge-infused deep learning for automatic lung nodule detection. J. Xray Sci. Technol. **27**(1), 17–35 (2019)
13. Gu, Y., et al.: Automatic lung nodule detection using a 3D deep convolutional neural network combined with a multi-scale prediction strategy in chest CTs. Comput. Biol. Med. **103**, 220–231 (2018)
14. Kanipriya, M., Hemalatha, C., Sridevi, N., SriVidhya, S.R., Shabu, S.J.: An improved capuchin search algorithm optimized hybrid CNN-LSTM architecture for malignant lung nodule detection. Biomed. Signal Process. Control **78**, 103973 (2022)
15. Wang, W., Chakraborty, G.: Evaluation of malignancy of lung nodules from CT image using recurrent neural network. In: 2019 IEEE International Conference on Systems, Man and Cybernetics (SMC), pp. 2992–2997. IEEE (2019)
16. Saihood, A., Karshenas, H., Nilchi, A.R.N.: Deep fusion of gray level co-occurrence matrices for lung nodule classification. PLoS ONE **17**(9), e0274516 (2022)
17. Gunjan, V.K., Singh, N., Shaik, F., Roy, S.: Detection of lung cancer in CT scans using grey wolf optimization algorithm and recurrent neural network. Heal. Technol. **12**(6), 1197–1210 (2022)
18. Xiao, Z., Liu, B., Geng, L., Zhang, F., Liu, Y.: Segmentation of lung nodules using improved 3D-UNet neural network. Symmetry **12**(11), 1787 (2020)
19. Lin, J., She, Q., Chen, Y.: Pulmonary nodule detection based on IR-UNet++. Med. Biol. Eng. Comput. **61**(2), 485–495 (2023)
20. Chi, J., Zhang, S., Yu, X., Wu, C., Jiang, Y.: A novel pulmonary nodule detection model based on multi-step cascaded networks. Sensors **20**(15), 4301 (2020)
21. Cheng, H., Zhu, Y., Pan, H.: Modified U-net block network for lung nodule detection. In: 2019 IEEE 8th Joint International Information Technology and Artificial Intelligence Conference (ITAIC), pp. 599–605. IEEE (2019)
22. Agnes, S.A., Anitha, J.: Efficient multiscale fully convolutional UNet model for segmentation of 3D lung nodule from CT image. J. Med. Imaging **9**(5), 052402 (2022)
23. Al-Shabi, M., Lee, H.K., Tan, M.: Gated-dilated networks for lung nodule classification in CT scans. IEEE Access **7**, 178827–178838 (2019)
24. Farhangi, M.M., Sahiner, B., Petrick, N., Pezeshk, A.: Automatic lung nodule detection in thoracic CT scans using dilated slice-wise convolutions. Med. Phys. **48**(7), 3741–3751 (2021)
25. Lu, Y., Liang, H., Shi, S., Fu, X.: Lung cancer detection using a dilated CNN with VGG16. In: 2021 4th International Conference on Signal Processing and Machine Learning, pp. 45–51 (2021)
26. Reshma, G., et al.: Deep learning-based skin lesion diagnosis model using dermoscopic images. Intell. Autom. Soft Comput. **31**(1), 621–634 (2022)
27. Padmavathi, K., et al.: Automatic segmentation of prostate cancer using cascaded fully convolutional network. In: 3rd International conference on Design and Manufacturing Aspects for Sustainable Energy, pp. 309–01109 (2021)

Robust Predictive Analysis of COPD with AI Models

D. T. V. Dharmajee Rao[1], A. Basi Reddy[2](✉), J. Avanija[2], K. Reddy Madhavi[2], K. Swaraja[3], and Sivaram Rajeyyagari[4]

[1] Department of CSE, CMR Technical Campus, Kandlakoya, Hyderabad, India
[2] School of Computing, Mohan Babu University, Tirupati, India
basireddy.a@gmail.com
[3] ECE, GRIET, Hyderabad, India
[4] Department of Computer Science, College of Computing and Information Technology, Shaqra University, Shaqra, Saudi Arabia
dr.sivaram@su.edu.sa

Abstract. Chronic obstructive pulmonary disease (COPD) is crucial and heterogeneous diseases which has benefitted from various novel techniques to understanding and analyze its evolution and divergent trajectories. Chronic obstructive pulmonary disease (COPD) is a lung condition that worsens with time and impairs airflow permanently. Artificial intelligence(AI) has transformed the way we interpret and analyse complex systems using clinical, imaging, and genomic. In the fields of automated clinical decision making, radiological interpretation, and prognostication, AI has demonstrated impressive outcomes. The particular characteristics of COPD and the availability of populations with accurate phenotypes create the perfect environment for the development of AI. This study gives an overview of AI and deep learning and highlights some recent achievements in using AI to treat COPD. This work has built multiple models to detect the disease some of which are Logistic Regression, K-Nearest Neighbors, XGBoost and Artificial Neural Network. The proposed models are applied on pre-processed data after which the accuracies are compared to check which algorithm provides the most beneficial outcome.

Keywords: COPD Disease · Machine Learning · Logistic Regression · K-Nearest Neighbors · Boosting

1 Introduction

The lungs are vital organs of the human body. They are a crucial part of the respiratory structure which facilitates the exchange of the two main gases that are required. This operation is mandatory for the proper functioning of the entire body including the tissues and organs. Incase an individual has an impairment or malfunctioning which affects the lungs, it can have a notable impact on said individual's quality of life and their health in general. The COPD disease does just this when the disease causes difficulty in breathing. It is caused by two different conditions which is Chronic Bronchitis which

D. Kriksciuniene et al. (Eds.): IBICA 2023, LNNS 1228, pp. 326–334, 2025.
https://doi.org/10.1007/978-3-031-78937-3_35

refers to inflammation and irritation of the bronchial tubes which are responsible for the airflow back and forth to the lungs. The other condition is known as Emphysema which indicates the loss in elasticity of the alveoli or the air sacs of the lungs making it hard for the lungs to expand for the intake of oxygen. This disease has affected over 251 million people conforming to the WHO in 2016. This is also responsible for 5% of all the deaths worldwide. One of the major causes of accumulating this disease is smoking of about 90% of COPD cases being due to this. While there is no cure for this disease its rapid progression can be slowed down by appropriate medical measures and treatment. This work has included four different concepts of machine learning which is widely prevalent in the health care industry but not specifically for COPD in order to raise awareness, expand its field of research and provide appropriate results. This work has been started off by applying Logistic Regression which is a statistical method used when the target class has two possible classes. The next model applied is K-Nearest Neighbors or KNN since the dataset taken is considerably small and this algorithm is non-parametric which helps it handle data better. XGBoost which stands for eXtreme Gradient Boosting is an ensemble learning technique where a combination of weak learners is conducted to build a strong predictive model. The last model implemented is ANN, abbreviated for Artificial Neural Network which comprises of nodes, neurons, and layers. ANN learns from the data provided to it and can make predictions and decisions from said learned patterns.

2 Literature Review

Research on the prediction of lung diseases is on-going even though it has been widely searched upon, contributing to its prevention, diagnosis, and treatment.In [1], the paper set out to predict the presence of COPD and its identification in the Chinese population using various algorithms. In the [2] paper, the severity of the disease was predicted using clinical data of mild and severe groups. The paper [3] predicts the readmission of the patients to help better the health care service they receive. The [4] work predicts the acuteness of obstruction of the airways and categorizes them too. According to the paper [5], integrating lung sound characteristics and spirometry data can increase the precision of a person's disease diagnosis in an automated scenario. The goal of [6] was to develop a memristive device that could predict the development of COPD using saliva samples from COPD patients and the Exasens dataset. The [7] paper uses early symptom criteria to predict COPD exacerbations in order to support early detection and diagnosis (Table 1).

Table 1. Comparison of previously works

Authors	Approach	Findings
Luo, L., Li, J., Lian, S., Zeng, X., Sun, L., Li, C., Huang, D., & Zhang, W., 2019. [1]	Logistic Regression, Random Forest and XGBoost	Areas under ROC curve was 0.78, 0.79 and 0.80

(continued)

Table 1. (*continued*)

Authors	Approach	Findings
Choi, E., Kim, J., Nah, G., Kim, W., Yoo, K., Kim, Y., & Yoon, D, 2021. [2]	SVM, Random Forest and KNN	Highest accuracy obtained is 83%
Mohamed, I., Fouda, M., &Hosny, K., 2022. [3]	Decision Trees, ANN and SVM	91% accuracy achieved
Amaral, J., Lopes, A., Faria, A., & Melo, P., 2015. [4]	KNN, Random Forest and SVM	AUC greater than 0.9
Haider, N., Singh, B., Periyasamy, R., & Behera, A., 2019. [5]	SVM, KNN, Logistic Regression and DA	Maximum accuracy of 83.6% has been gained
Zarrin PS, Zahari F, Mahadevaiah MK, Perez E, Kohlstedt H, Wenger C., 2020. [6]	ANN	89% accuracy obtained
Fernandez-Granero, M., Sanchez-Morillo, D., & León-Jiménez, A., 2018. [7]	Decision Tree	78% accuracy earned

3 Proposed System

Here, a system that has been used for all the models is after the initial cleaning of data, which proceeds to splitting the data for training and testing, for which it has performed scaling on the test dataset in order to get more accurate results which is then used for each model separately. This pertains to Logistic Regression, K nearest neighbors and for XGBoosting. While for the ANN model, sequential layers have been applied and compiled it with the RMSprop optimizer (Fig. 1).

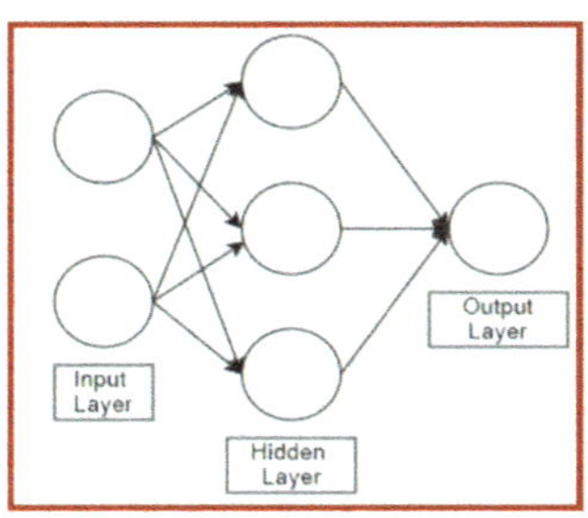

Fig. 1. Proposed ANN Architecture

Dataset Gathering

This work has used the Exasens dataset in order to get real-world data. This dataset belongs to UCI Machine Learning Repository and while it contains a collection of 4 respiratory diseases, this research has taken a part of the data set which consists of saliva samples of the COPD infected and the HC which stands for healthy controls or an unaffected individual. It is a multivariate dataset which has different attributes and the original dataset has 400 instances but since a part of this was taken which was then made into analog data, it has 80 instances (Fig. 2).

	ID	Diagnosis	Img-Min	Img-Avg	Real-Min	Real-Avg	Gender	Age	Smoking
0	301-4	1	-320.61	-300.563531	-495.26	-464.171991	1	77	2
1	302-3	1	-325.39	-314.750360	-473.73	-469.263140	0	72	2
2	303-3	1	-323.00	-317.436056	-476.12	-471.897667	1	73	3
3	304-4	1	-327.78	-317.399670	-473.73	-468.856388	1	76	2
4	305-4	1	-325.39	-316.155785	-478.52	-472.869783	0	65	2

Fig. 2. Snippet of the Dataset

The first step was to import every library that was needed in order to preprocess or plot our data. These libraries which have been imported are Numpy, Pandas, Matplotlib, Seaborn and Scikit. Numpy is used for providing support for large datasets and it also comes with functions to operate on the data efficiently. It is a prominent library used for machine learning as it is flexible and easy to use. Pandas is a library used for handling, analyzing, and cleaning data. It offers structures designed to manipulate data. Pandas is mainly used for two structures which are DataFrames and Series. DataFrames are used in this work. Matplotlib is the main library for visualizing data in python. It works seamlessly in hand with both Numpy and Pandas.

The next library is Seaborn which is built on top of Matplotlib. It is also a library used for complex visualization of data and shows the relationship between variables. The last library is the Scikit or Sklearn which is used for building and training all the models that have been included in this work. It works together with Numpy and Scipy. The first preprocessing step was to drop the 'ID' column using the Pandas librassry as it was irrelevant and was not a dependent variable for the prediction of COPD. The next step was to check if there are any null values in the dataset, which all came false indicating no null values (Fig. 3).

Then a heatmap to show the correlation among the variables was plotted. This plot was done using the Seaborn library. The darker tones indicate higher correlation and lighter tones indicate less correlation (Fig. 4).

The next step was to obtain the training data and testing data. This is done using the sklearn library. The 'Diagnosis' column was put in the testing data and the remaining columns come under training data. The ratio for the splitting is 80:20. Scaling was then performed for the training set using the sklearn library and the MinMaxScaler, which is a technique to scale the numerical data all into the same range i.e., 0 and 1, in order to balance the features and eliminate any chances of one feature dominating another. This

```
Diagnosis     False
Img-Min       False
Img-Avg       False
Real-Min      False
Real-Avg      False
Gender        False
Age           False
Smoking       False
dtype: bool
```

Fig. 3. The Null Values Present In The Dataset

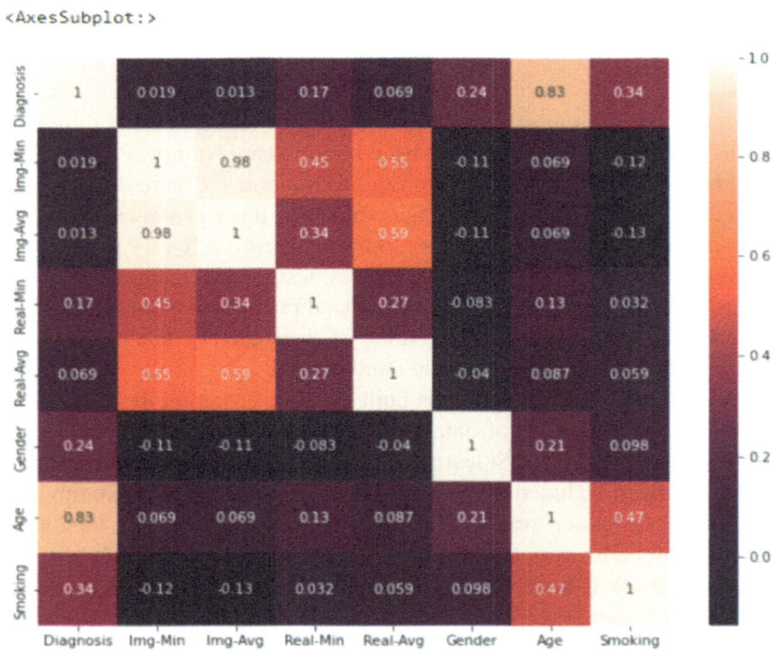

Fig. 4. Heatmap To Show Correlation

scaled set is then further used for the splitting of the dataset. These split features were then passed to each model and were fit for each of the mentioned models.

Proposed Steps

Step 1: Import all required necessary libraries.

Step 2: Import the analog Dataset into the Jupyter note which is a part of the Exasens book.
Step 3: Pre-processing the dataset including finding null values and scaling the dataset.
Step 4: Splitting the dataset.
Step 5: Implementation of Logistic Regression, K nearest neighbours, XGBoosting and ANN models.
Step 6: Predict the results and obtain the classification report of each model (Fig. 5).

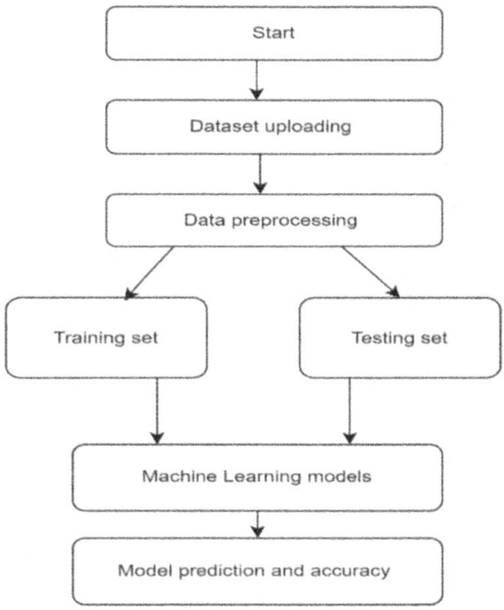

Fig. 5. Proposed Steps

In this work, after each model was fit to the training and testing set, the accuracy was calculated from each model. The number of estimators for boosting was put as 10 and the total number of epochs for the ANN was 100. The neural network used sequential layers with the RMSprop optimizer which helped us attain a higher accuracy than a previously implemented model, the loss was also calculated by binary cross entropy.

4 Results and Discussion

In this research, an accuracy of 87.5% for the Logistic Regression model was achieved (Fig. 6).

While for the K nearest neighbors' model, an accuracy of 81.25% was obtained which is 6% lesser compared to the logistic regression model (Fig. 7).

The XG-Boosting technique which is an ensemble learning technique consisted of 10 estimators procured an accuracy of 94% (Fig. 8).

The Artificial Neural Network presented a test accuracy of 93.75% and a test loss of 24.71% (Figs. 9, 10 and Table 2).

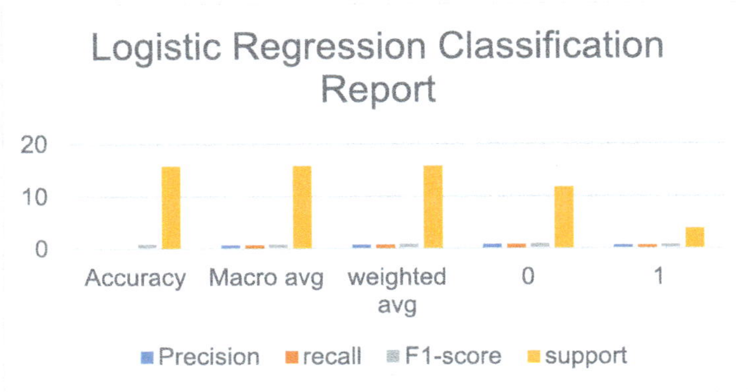

Fig. 6. Logistic Regression Classification Report

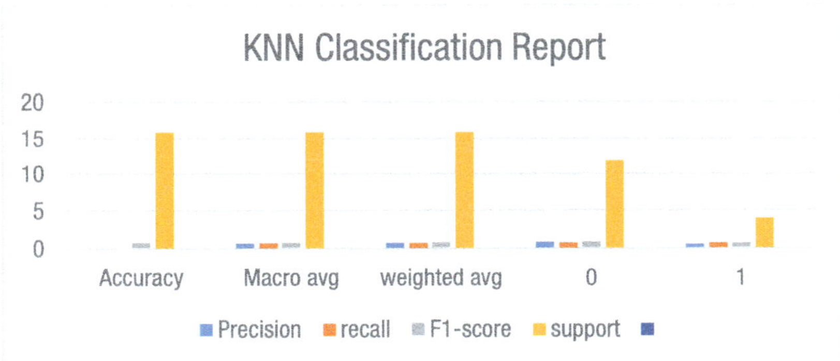

Fig.7. K Nearest Neighbors' Classification Report

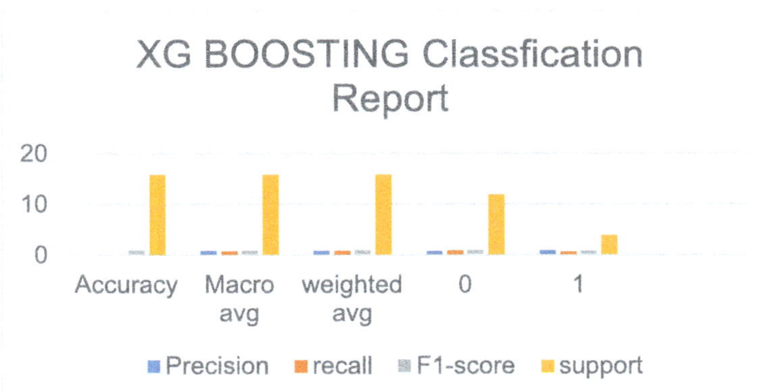

Fig. 8. XG-Boosting classification report

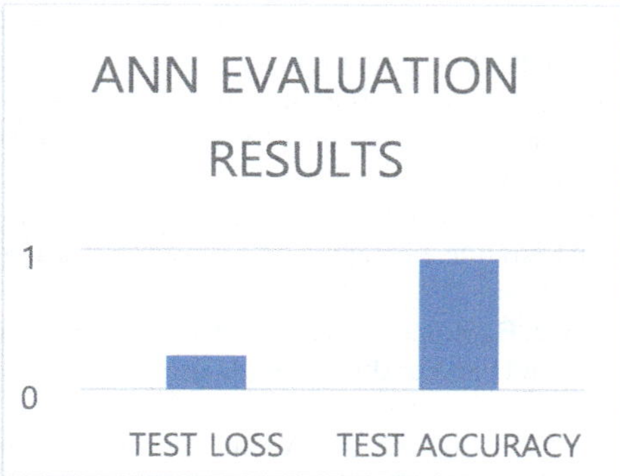

Fig. 9. ANN Evaluation Results

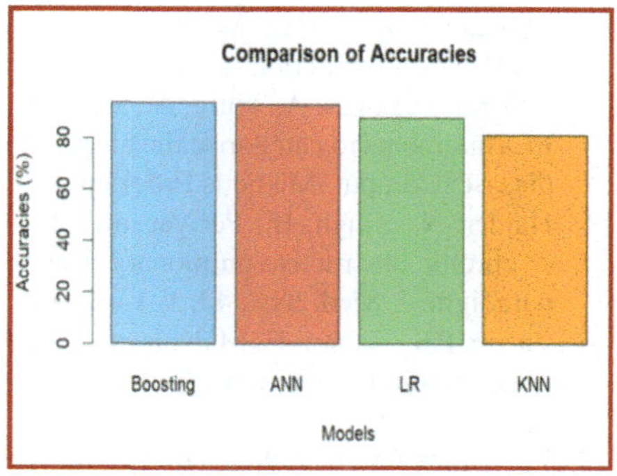

Fig. 10. Bar Plot Comparison Of Accuracies

Table 2. Comparison Of Accuracies Obtained

S.no	Algorithm	Accuracy
1	KNN	81.25%
2	Logistic Regression	87.5%
3	ANN	93.75%
4	XGBoosting	94%

5 Conclusion and Future Work

Hence, the lowest established accuracy that has been obtained is 81.25 which is by the KNN algorithm. Logistic regression obtained 87.5% accuracy. ANN provided a remarkable accuracy of 93.75 which is a tad bit lesser than the best performing model of this paper, XGBoosting, which gave an accuracy of 94%. On comparing the four models, it can be concluded that boosting outperformed the other models which is closely followed by the neural network model.This research has been limited to basic machine learning algorithms and while these algorithms are efficient for predictive models, it can be extended further by applying deep learning concepts along with its hybrids. A numerical dataset was taken but COPD prediction can be enlarged by taking an image dataset of the lungs and applying the deep learning algorithms to it. Some of the models which can be used are TrLearning, DenseNet, ResNet or even GoogleNet. Hybridization of the models can also be implemented to get the best possible results.

References

1. Li, L., et al.: Using machine learning approaches to predict high-cost chronic obstructive pulmonary disease patients in China. Health Inform. J. **26** (2019)
2. Choi, E., et al.: Prediction of COPD severity based on clinical data using Machine Learning. In: 2021 IEEE International Conference on Bioinformatics and Biomedicine (BIBM) (2021)

3. Mohamed, I., Fouda, M., Hosny, K.: Machine learning algorithms for COPD patients' readmission prediction -a data analytic approach. IEEE Access 1 (2022)
4. Amaral, J., Lopes, A., Faria, A., Melo, P.: Machine learning algorithms and forced oscillation measurements to categorise the airway obstruction severity in chronic obstructive pulmonary disease. Comput. Methods Program. Biomed. (2015)
5. Haider, N., Singh, B., Periyasamy, R., Behera, A.: Respiratory sound based classification of chronic obstructive pulmonary disease: a risk stratification approach in machine learning paradigm. J. Med. Syst. **43**, 1–13 (2019)
6. Zarrin, P.S., Zahari, F., Mahadevaiah, M.K., Perez, E., Kohlstedt, H., Wenger, C.: Neuromorphic on-chip recognition of saliva samples of COPD and healthy controls using memristive devices (2020)
7. Fernandez-Granero, M., Sanchez-Morillo, D., León-Jiménez, A.: An artificial intelligence approach to early predict symptom-based exacerbations of COPD (2018)
8. Dong, T., Santos, S., Yang, Z., Yang, S., Kirkhus, N.E.: Sputum and salivary protein biomarkers and point-of-care biosensors for the management of COPD. Analyst **145**(5), 1583–1604 (2020)
9. Zarrin, P.S., Roeckendorf, N., Wenger, C.: In-vitro classification of saliva samples of COPD patients and healthy controls using machine learning tools. IEEE Access **8**, 168053–168060 (2020)
10. Fogel, A.L., Kvedar, J.C.: Artificial intelligence powers digital medicine. NPJ. Digit. Med. **1**(1), 1–4 (2018)
11. Kononenko, I.: Machine learning for medical diagnosis: history, state of the art and perspective. Artif. Intell. Med. **23**(1), 89–109 (2001)
12. Baker, S.B., Xiang, W., Atkinson, I.: Internet of things for smart healthcare: technologies, challenges, and opportunities. IEEE Access **5**, 26521–26544 (2017)
13. Csikesz, N.G., Gartman, E.J.: New developments in the assessment of COPD: early diagnosis is key. Int. J. Chron. Obstruct. Pulmonary Dis. **9**, 277 (2014)
14. Price, D., et al.: Spirometry in primary care case-identification, diagnosis and management of COPD. Primary Care Respir J. **18**(3), 216–223 (2009)
15. Mathers, C.D., Loncar, D.: Projections of global mortality and burden of disease from 2002 to 2030. PLoS Med. Med. **3**(11), e442 (2006)
16. Barnes, P.J.: Mechanisms in COPD: differences from asthma. Chest **117**(2), 10S-14S (2000)

Evaluating Pulmonary Image Classification Through Ensemble Learning with Optimal Neural Network

Modugula Siva Jyothi[1](✉), Saba Sultana[1], X. S. Asha Shiny[2], Sayyad Rasheeduddin[2], G. Ravi Kumar[3], and M. Kamala[3]

[1] CMR Technical Campus, Hyderabad, Telangana, India
cmrtc.paper@gmail.com
[2] CMR Engineering College, Hyderabad, Telangana, India
{dr.ashashinyxs,srasheeduddin}@cmrec.ac.in
[3] CMR College of Engineering & Technology, Hyderabad, Telangana, India
{gravikumar,mkamala}@cmrcet.ac.in

Abstract. Medical diagnosis classification is a difficult process with a high error rate. The most popular and extensively used type of clinical examination for pulmonary modules is the Chest X-ray film because medical imaging plays a significant role in the total diagnostic procedure. However, given the high rise in infectious disease cases—a significant potential cause of diagnostic error—the number of radiologists clearly cannot keep up with this explosion. The current system classifies pulmonary images using the Inception-v3 transfer learning method. It first enhanced the pulmonary image data, then utilized to improved Inception-v3 method based on transfer learning to automatically extract features, and finally classified the pulmonary images using various classifiers (Softmax, Logistic, SVM). By employing ensemble learning and studying suitable neural network selection, the proposed system can improve its performance in classifying pulmonary images. Compared to other cutting-edge techniques, the ensemble methodology outperforms them on benchmark datasets.

Keywords: CNN · MRI · PCA · pulmonary Image technique · ML models

1 Introduction

These days, machine learning is frequently utilized to properly forecast a wide range of diseases using datasets that have been provided and trained. The investigation of the predictive analysis of pulmonary nodule disease is presented in this publication. Based on an examination of the proper choice of neural networks and the use of ensemble learning. As our suggested neural network selection method for pulmonary image categorization, which makes use of the VGG_16 Model, Inception_v3, ResNet50, and VGG19? Acute pulmonary infections resulting in pleural effusion and inflammation of the air sacs are known as nodule detection. These infections can be brought on by bacteria, viruses, or fungus. More than 15% of mortality in children under five are attributed to it [10].

D. Kriksciuniene et al. (Eds.): IBICA 2023, LNNS 1228, pp. 335–344, 2025.
https://doi.org/10.1007/978-3-031-78937-3_36

Undeveloped and poor nations have higher rates of lung infections due to factors like pollution, unhygienic environmental conditions, and overcrowding, which exacerbates the issue and limits access to healthcare. Early diagnosis and treatment can therefore aid in halting the disease's progression and eventual mortality. Diagnostic examinations of lungs with the use of radiography (X-rays), computed tomography (CT), or MRI are frequently used. Information can be obtained painlessly and non-invasively with X-ray imaging. An illustration of a lung X-ray showing both a healthy and injured lung may be found in Fig. 1. Subjective variability, however, can affect infiltrates, or white patches, on Lung Xray Chest X-ray tests used to identify infections [2, 3]. Therefore, an automated method for identifying nodule infection is required. In this survey, we suggested a deep learning-based approach to address these automation challenges. A chest X-ray film is the very broadly used and typical kind of clinical evaluation for pulmonary nodules [13].

What was once thought to be a respiratory virus started to manifest itself in other bodily areas, causing a wide range of symptoms such as heart attacks, arrhythmias, blood clots, damage to the kidneys, liver, rashes, and more. In spite of this, the most typical sign of infectious infections is still respiratory issues. Thoracic radiography's diagnostic specificity for infectious illnesses and its value for frontline prescreening are controversial. The American College of Radiography [7, 8] is one radiology body that advises against using clinical radiography to determine pulmonary causes.

To start, we examine recent research to find any possible drawbacks with using neural networks to process radiography images [9, 14]. Then, we offer protocols and methodologies for evaluating deep learning methods for classification and segmentation using the publicly available benchmark dataset Infectious Diseases as a case study. Numerous challenging computer vision problems can be resolved with the use of deep learning, a potent artificial intelligence technology [4–6]. A common technique for integrating the judgments of several classifiers to provide a final prediction for a test sample is ensemble learning. This is done in order to collect discriminative information from each base classifier and provide predictions that are more accurate [11]. A chest X-ray film is the very commonly used and typical kind of clinical evaluation for pulmonary nodules. Though, it is evident that there are not enough radiologists to keep up with the rapid rise in infectious diseases, which is also a significant potential source of diagnostic inaccuracy [8].

2 Literature Review

Computer-aided diagnostics was proposed by Marcin Wozniak and Dawid Połap. This study aims to explore the potential of deep learning algorithms for the diagnosis of respiratory diseases through the use of cuckoo, firefly, artificial bee colony, artificial ant colony, practical swarm, and bim tissue key-points and aggregated key points for extraction. Medical professionals look for tissues with altered morphology in photos of lung diseases such as malignancy, sarcoidosis, and pneumonia [1].

Based on geometric factors, S. Mukherjee et al. [2] suggested a method for lung nodule autonomous detection. Based on features like roundness, eccentricity, diameter, and aspect ratio, x-ray images are used to categorize benign and malignant lung nodules. Using bilateral filtering to remove noise The Bayesian classifier is used for image binarization, segmentation, and classification [2].

A probabilistic neural network-based approach for classifying lung cancer was proposed by Woniak et al. [3]. Although this method is simple, it can detect nodules with low contrast and has a reasonable classification effect. A lung picture was processed using the following probabilistic neural network to extract information. Thus, a vector is created, the elements of which display the degree to which the input is similar to single classes in terms of Mahalanobis distance. The pattern layer uses this vector to calculate a probability vector whose elements specify which category a person belongs to [4] by utilizing a range of model architectures and datasets to train a particular domain.

A novel Deep Adaptation Network (DAN) was introduced by Long and Wang [5] as a solution to the domain adaptation issue in transfer learning. DAN extends the Deep Convolutional Neural Network to domain adaptation. The transferability of features from the neural network's task-specific layers is optimized by this architecture.

Improving classification accuracy can be achieved by using numerous classifier systems (also known as ensemble systems) and combining their outputs. A multiple classifier system's two primary parts are the "creation of ensemble" and the "combination of class label" [6].

It presents [12] a novel method for determining the joint image/text contexts for composite picture classification by using the weights of the pair of CNN/RNN that have already been trained on the domain-specific image/text dataset. By considering the joint image/text contexts, the recurrent neural cascade model is shown to produce much enhanced image annotation outcomes.

3 Proposed Model

In this work, we developed an ensemble system comprising multiple classifiers. Employing a weighted average ensemble technique, in which a novel scheme is used to compute the weights allocated to classifiers. It may be possible to improve the classification of pulmonary images and performance by utilizing ensemble learning in the proposed system. The main contributions of the study are as follows. To enhance the basic CNN learners' performance in lung nodule classification, an ensemble architecture was created. For this, a weighted average ensemble approach was employed.Two publicly accessible chest X-ray datasets and JSRT dataset [4] were used to test the suggested model. The approach is practical for use in real-world scenarios, as seen by the findings, which outperform those of cutting-edge techniques. The process of developing and integrating numerous models—such as classifiers or experts—in order to address a particular computational intelligence challenge is known as ensemble learning.

Dataset: 154 conventional chest radiographs with a lung nodule (100 malignant and 54 benign nodules) and 93 radiographs without a lesion were digitized for the database (no header, big-endian raw data) using a laser digitizer with a 2048 $\times$ 2048 matrix size (0.175-mm pixels) and a 12-bit grey scale. The age, gender, and diagnosis (benign or malignant) of the patient are among the other details stored in the database. Figure 1 presents a systematic overview of the proposed system. To put it briefly, the system uses image processing and classification as its two main processing methods when it receives lung CT images as input. Lung tumors can be identified by classifying the remaining possible objects in the second module according to the qualities that were extracted during the feature extraction phase. The algorithm might be able to differentiate between nodule and non-nodule items between suspicious ones. Based on a committee of three distinct classifiers (VGG-16, INCEPTION-V3, RESNET-50, and VGG-19), this categorization was made. The steps of the suggested system have been explained in the sections that follow.

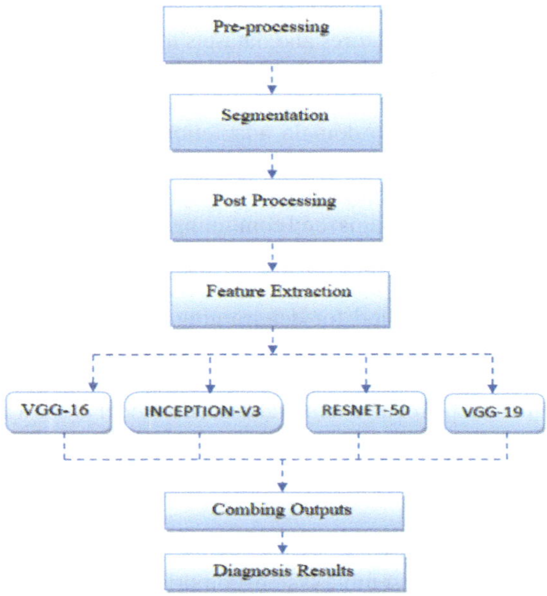

Fig. 1. System architecture model

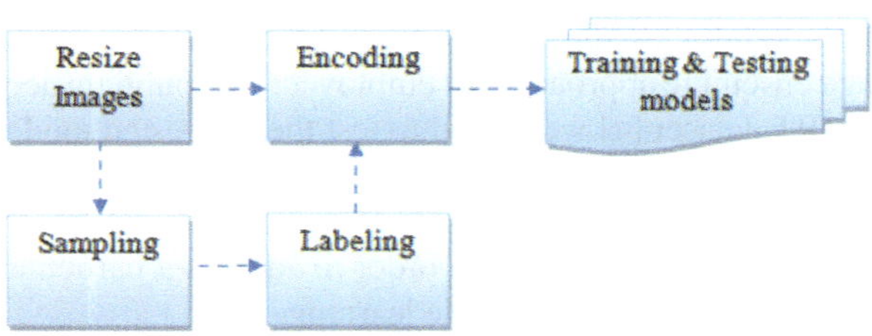

Fig. 2. Image pre processing

Preprocessing: As seen in Fig. 2, image preprocessing is a method for improving important features while also eliminating primary noise and image distortion from CT images. By employing a few techniques for image enhancement.

Stages of Preprocessing

Resize photos: Resizing images is a novel and efficient method of altering the aspect ratio of photographs without causing noticeable distortion. It preserves the content of the image. The image is now changed to a (224, 224, 3) ratio.
Sampling: we employ both up- and down-sampling Sampling: To create an equal number of photos for each class, we employ both down-sampling and up-sampling. Prior to sampling, there were 93 non-nodule and 154 nodule pictures of the lung. Following the sample of Non-Nodule 250 Lung Nodule 250.
Labeling and Encoding: Lung nodules are labeled as LNs and non-nodules as NNs in the photos. After that, the binary form—in which (0,1) is nonnodule and vice versa—is used to encode the LN and NN.
Feature extraction: To distinguish suspected items in two-dimensional photos as nodules or nonnodules, further effort must be done to extract particular features from raw images. If a nodule has a diameter greater than 30 mm, it is referred to as a cancer nodule.

Module for Classification:
Improving classification accuracy can be achieved by using many classifier systems (also known as ensemble systems) and combining their outputs. The two main parts of most multiple classifier systems are "combination of class labels" and "creation of an ensemble". 247 photos, comprising both men and women, were gathered using JSRT for the study. The extracted features from the ensemble system are used as inputs, and the suspicious items are normalized between zero and one. Radiologists classify the lung lesions (nodules) on these pictures as either nodule or non-nodule.

VGG_16 Model
Convolutional neural network model VGG 16 is depicted in Fig. 3 of their paper, "Very Deep Convolutional Networks for Large- Scale Image Recognition." With over 14 million images organized into 1000 groups, ImageNet is a database of images. The model's top-five test accuracy is 92.7%. Weeks were spent using the NVIDIA Titan Black GPUs to train VGG16.

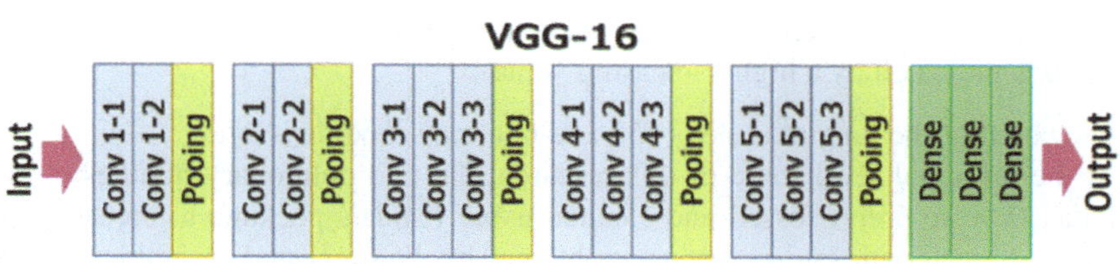

Fig. 3. Architecture showing VGG 16 Input and Output Activity

ResNet-18
The ResNet-18 method is built on a residual learning framework that increases the

training efficiency of deep networks. Because the residual blocks in ResNet methods allow the optimization of entire network, model accuracy is increased, in contrast to the original unreferenced mapping in monotonically progressing convolutions.

VGG-19

A deep CNN with 19 layers is called VGG-19. A pretrained version of the network, which has been trained on more than a million images can be loaded using the ImageNet database. Photographs can be categorized by the network into 1000 distinct object categories, including mice, keyboards, pencils, and other animals. The network requires images with an input size of 224 224 pixels. Despite using a tiny bit more RAM, VGG19 performs somewhat better than VGG16.

4 Results and Analysis

The 247 photos, which included both men and women, were gathered by the Japanese Society of Radiological Technology (JSRT) and used for this investigation. As seen in Fig. 4, the suggested approach identifies possible objects in each image as either nodules or not.Actually, there are two groups (positive and negative) within the gathered dataset. The effectiveness of categorization techniques that are frequently used in computerized medical diagnosis systems can be evaluated using a range of measures. In a positive instance, TP is the number of correct predictions; in a negative instance, FN is the number of inaccurate predictions; in a positive instance, FP is the number of wrong predictions; in a negative instance, TN is the number of correct predictions. Based on these indicators, we may create the measurements listed below to assess the system's performance.

Accuracy (Acc): Accuracy is defined as the ratio of correctly detected instances to the total number of test examples.
Specificity (sp) quantifies the number of negatives that the classifiers correctly detect. The amount of true negative outcomes is divided by the total of false positive and true negative findings to determine specificity.
Precision (prc): measure indicates the proportion of anticipated nodules that are connected to prone position in reality.
F-measure*:* combines sensitivity and precision. A high degree of sensitivity and precision is indicated by a high F-measure value.

The count of the lung module is 155 and non lung module is 97

At the principal level stage of this research, the two pulmonary nodules chest X-ray datasets were preprocessed and cleaned using Panda's package to create receiver operating characteristic curves that were obtained by the suggested ensemble technique. Figure 5 depicts the sex and target qualities group counterplot. Following that, the data frame attributes are displayed in Figs. 5, 6, 7, and 8 by the use of the data visualization process.

The accuracy of the suggested model is contrasted with the current machine learning models, which are shown in Fig. 13. 0.9 test accuracy, 0.925 sensitivity, and 0.98 specificity are displayed by vgg-16. 0.96 test accuracy, 0.96 sensitivity, and 0.96 specificity

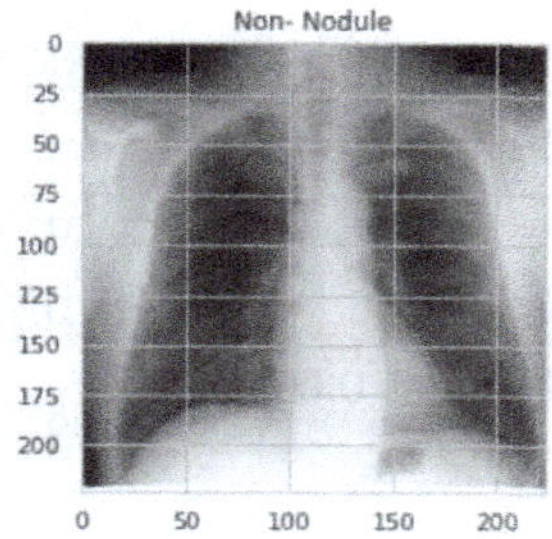

Fig. 4. Xray Showing Non Nodule Image with pulmonary readings to be noted.

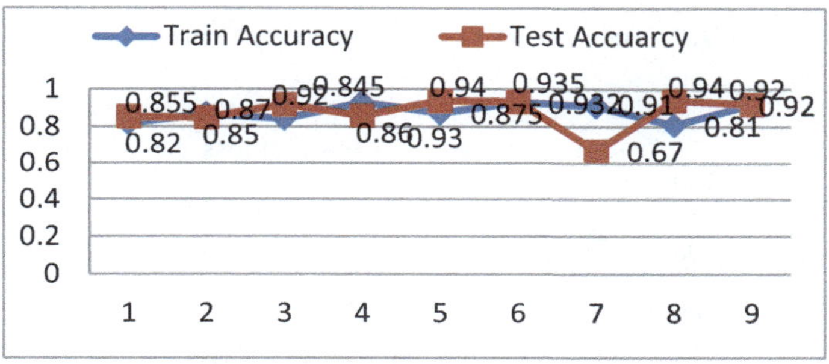

Fig. 5. It displaying Train and Test Accuracy Comparisons using VGG 16

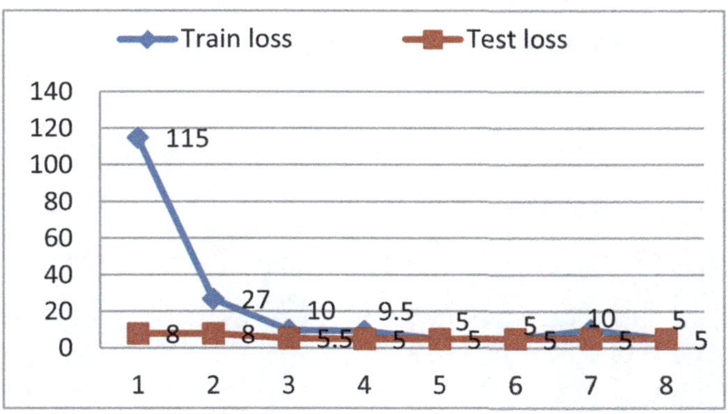

Fig. 6. Inception v3: Train and Test Loss Comparisons

are displayed in Inception v3. Resnet50 displays test accuracy of 0.93, sensitivity of 0.88, and specificity of 0.85. The average ensemble displays 0.97 test accuracy, 0.93 sensitivity, and 0.94 specificity, while vgg-19 displays 0.96 test accuracy, 0.94 sensitivity, and 0.98 specificity. Figure 9 illustrates the suggested approach accuracies graphically.

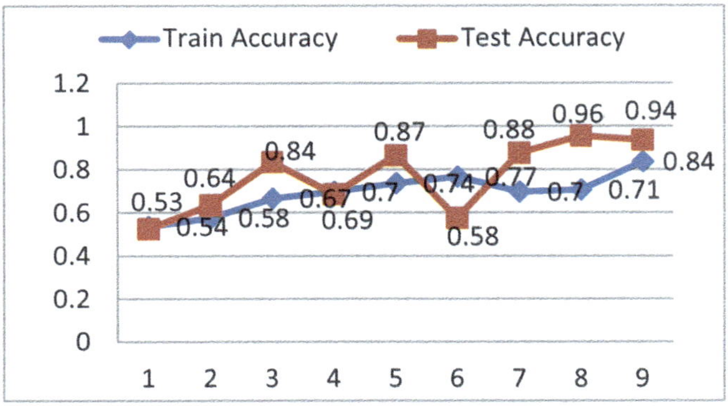

Fig. 7. Train Accuracy and Test Accuracy Comparison.

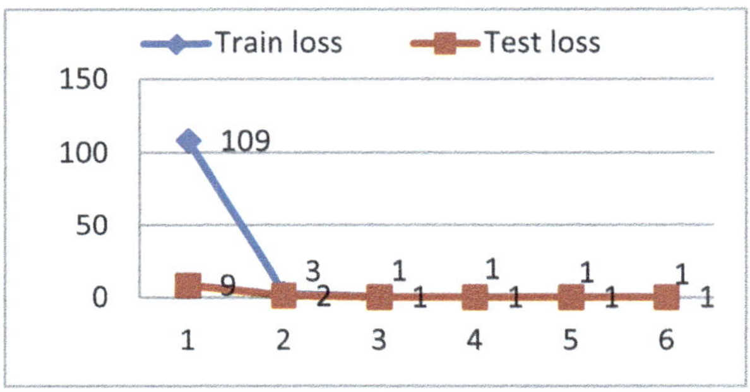

Fig. 8. Train and test loss Accuracy Comparison

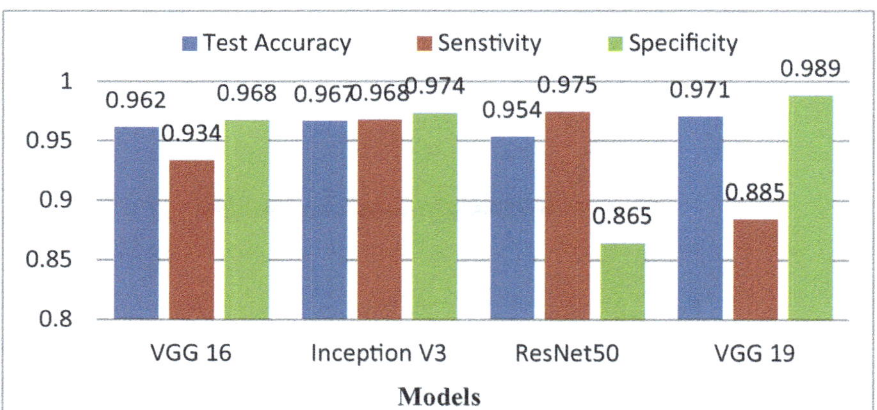

Fig. 9. Model result displaying comparisons of test, sensitivity, and specificity

5 Conclusion

The efficacy of the suggested ensemble technique for lung nodule diagnosis is evaluated through experiments conducted on JSRTT database. In primary stage, image processing methods like feature extraction, post-processing, and pre-processing were applied. Specific aspects of the items of interest were then given to the categorization phase. These features were then utilized as inputs for each individual classifier (VGG_16, Inception_v3, ResNet50, and VGG19) and ensemble system in the classification phase. The findings of the basic classifiers were then combined in a committee using the majority voting approach. To estimate the effectiveness of the suggested hybrid intelligent technique, the suggested ensemble system performs the best between base classifiers on its own. In order to demonstrate the importance of ensemble learning, the confusion matrix was also computed for every base classifier and the suggested ensemble system.

References

1. Wang, C., Chen, D., Hao, L., Zeng, X., Chen, J., Zhang, G.: Pulmonary image classification basedoninception-v3 transfer learning model. In: IEEE Conference on Computer Vision and Pattern Recognition, pp. 248–255 (2019)
2. Mukherjee, J., Chakrabarti, A., Shaikh, S.H., Kar, M.: Automatic detection and classification of solitary pulmonary nodules from lung CT images. In: Proceedings of the 4th International Conference of Emerging Applications of Information Technology, Kolkata, India, pp. 294–299 (2014)
3. Woźniak, M., Połap, D., Capizzi, G., Sciuto, G.L., Kośmider, L., Frankiewicz, K.: Small lung nodules detection based on local variance analysis and probabilistic neural network. Comput. Methods Program. Biomed. **161**, 173–180 (2018)
4. Jiang, X., Pang, Y., Sun, M., Li, X.: Cascaded subpatch networks for effective CNNs. IEEE Trans. Neural Netw. Learn. Syst. **29**(7), 2684–2694 (2018)
5. Raju, S., Rafeeq, M., Ravikanth, Parag, Rafeeq, M.: Image recognition and content retrival to build context: a survey. J. Adv. Res. Dyn. Control Syst 11, 1656–1666 (2019)
6. Mahmood, M.R., Patra, R.K., Raja, R., Sinha, G.R.: A novel approach for weather prediction using forecasting analysis and data mining techniques. In: Saini, H., Singh, R., Kumar, G., Rather, G., Santhi, K. (eds.) Innovations in Electronics and Communication Engineering. Lecture Notes in Networks and Systems, vol. 65. Springer, Singapore (2019). https://doi.org/10.1007/978-981-13-3765-9_50
7. Sahu, A.K., Chandra, V.K., Sinha, G.R.: Modeling and system-level computer simulation approach for optimization of single-loop CT sigma delta ADC. In: Kolhe, M., Trivedi, M., Tiwari, S., Singh, V. (eds.) Advances in Data and Information Sciences. Lecture Notes in Networks and Systems, vol. 39. Springer, Singapore (2019). https://doi.org/10.1007/978-981-13-0277-0_28
8. Gothane, S., Sarode, M.V., Thakre, V.M.: Prediction for Indian road network images dataset using feature extraction method. In: Bapi, R., Rao, K., Prasad, M. (eds.) First International Conference on Artificial Intelligence and Cognitive Computing. Advances in Intelligent Systems and Computing, vol. 815. Springer, Singapore (2019). https://doi.org/10.1007/978-981-13-1580-0_12
9. Shin, H.C., Roberts, K., Lu, L., Demner-Fushman, D., Yao, J., Summers, R.M.: Learning to read chest x-rays: recurrent neural cascade model for automated image annotation. In: Proceedings of the IEEE Conference on Computer Vision and Pattern Recognition (CVPR), Las Vegas, NV, USA, pp. 2497–2506 (2016)

10. Bar, Y., Diamant, I., Wolf, L., Lieberman, S., Konen, E., Greenspan, H.: Chest pathology detection using deep learning with non-medical training. In: Proceedings of the IEEE 12th international symposium on biomedical imaging (ISBI), New York, NY, USA, pp.294–297 (2015)
11. Fukushima, K.: Neocognitron: a self-organizing neural network model for a mechanism of pattern recognition unaffected by shift in position. Biol. Cybern. **36**(4), 193202 (1980)
12. LeCun, Y., et al.: Backpropagation applied to handwritten zip code recognition. Neural Comput. **1**(4), 541551 (1989)
13. Simonyan, K., Zisserman, A.: Very deep convolutional networks for large-scale image recognition. In: Proceedings of the International Conference on Learning Representation (ICLR), San Diego, CA, USA, p. 114 (2015)
14. Rajesh Tiwari, M., et al.: Enhanced power quality and forecasting for PV-wind microgrid using proactive shunt power filter and neural network based time series forecasting. Electr. Power Components Syst. (2023). https://doi.org/10.1080/15325008.2023.2249894

An Analytical Study on Classification of Histopathological Images for Cancer Diagnosis and Progression Using Machine Learning Techniques

Nam Vasundhara[1(✉)], Guntoju Kalpana Devi[1], A. Srinivasula Reddy[2], X. S. Asha Shiny[2], Manyala Naga Sailaja[3], and T. Upender[3]

[1] CMR Technical Campus, Hyderabad, Telangana, India
cmrtc.paper@gmail.com

[2] CMR Engineering College, Hyderabad, Telangana, India
dr.ashashinyxs@cmrec.ac.in

[3] CMR College of Engineering & Technology, Hyderabad, Telangana, India
{m.n.sailaja,t.upender}@cmrcet.ac.in

Abstract. In medical science, the detection of malignancies is essential for the diagnosis of cancer. The research and identification of disease symptoms through microscopic tissue inspection is known as histopathology. Pathologies carefully analyze these intricate histopathological images (HIs). The different results cause the pathologists to become subjective, and they mostly rely on the examiner's experience. In order to categorize biopsy samples as benign or malignant, many characteristics such as color, size, form, and layout of the cells and tissues are studied. These take a lot of time and are vulnerable to the pathologist's subjectivity. Computer-assisted analysis is required to get around this. This study examines the effectiveness of several machine learning approaches on the BreaKHis dataset, which contains of microscopic images of breast tumor tissue that were obtained with various magnification factors. Machine learning can function with little in the way of computing power. In order to categorize certain HIs of breast cancer as either malignant or benign, an evaluation of the performance of several machine learning algorithms, as well as K- closest neighbor (KNN), support vector machine (SVM), decision tree, and Gaussian naïve bayes, is being conducted in this research. When the data are scaled and normalized, KNN attains 92.62% efficiency whereas decision trees perform better through 87.98% accuracy.

Keywords: Histopathological images (HIs) · machine learning techniques for instance SVM · KNN · and Gaussian naïve Bayes categorization algorithms

1 Introduction

The research of diseases is called pathology, and the analysis of tissues is called histology. Thus, the research of disease-related tissues is known as histopathology. Pathologists have usually examined it physically while looking through a microscope. A pathologist's

D. Kriksciuniene et al. (Eds.): IBICA 2023, LNNS 1228, pp. 345–353, 2025.
https://doi.org/10.1007/978-3-031-78937-3_37

knowledge and experience are needed. As a result, the pathologist has a significant influence on the outcome, which can vary greatly. They research various aspects of such tissues. This comprises the number, form, size, and density of malignant cells per unit of tissue. The gold standard for cancer diagnosis is histopathological diagnosis. Histopathological pictures are useful for investigation since they are diagnostically rich.

The personal perspective of the pathologist can be eliminated by computer-aided diagnosis. HIs might be used to train machine learning models with various features, including as texture and color characteristics. Various algorithms are employed for obtaining such traits into the photos. These specialists can therefore identify the illness. This effort is being done in order to categorize HIs into classes that are benign and malignant. A kind of artificial intelligence called machine learning enables systems to acquire knowledge from provided features. Applications for machine learning algorithms are numerous and include computer vision, speech recognition, email filtering, medical, and agriculture. Environments using restricted resources can nonetheless benefit from machine learning approaches. Three categories of machine learning approaches exist: reinforcement learning, unsupervised learning, and supervised learning (a, b, and c). As seen in Fig. 1, supervised learning involves providing the system enough sample inputs and the intended outputs in order for it to learn the mapping rule from input to output. On the other hand, required identifiers are not provided in unsupervised learning. It departs on its own to discover the undiscovered pattern. In reinforcement learning, a system engages using a dynamic environment to accomplish a certain objective. In this case, the system receives input dynamically, which functions similarly to a reward.

This work has exclusively used supervised learning approaches for the classification. The Naïve Bayes method, Decision Tree, Support Vector Machine (SVM), and K-nearest neighbor (KNN or K-NN) have all been employed in this research. KNN is a supervised, non-parametric learning classifier that classifies or predicts how to group a single data point based on its closeness. This k is typically selected to be odd. A supervised machine learning technique called SVM is utilized for regression as well as classification. The approach known as SVM aims to find a hyperplane in a space with N dimensions that readily categorizes the data points. An algorithm for supervised learning is a decision tree. It is a graphical depiction of every potential outcome. Every choice was made with a set of requirements in mind. This classifier is tree-structured, with internal nodes standing in for dataset attributes, branches for decision rules, and leaf nodes for each outcome. The Naive Bayes algorithm is a straightforward supervised machine learning technique that obtains results by applying the Bayes theorem using solid independence assumptions among the features. This indicates each of the independent input variable is assumed by the algorithm.

The World Health Organization, the Global Burden of Disease Cancer Collaboration, and the International group for Research on Cancer have reported that there was a 28% increase in cancer cases among 2006 and 2016, and that by 2030, there would be 2.7 million novel cases of cancer [1, 2]. In every kind of cancer With 1.7 million incidence cases, 535,000 deaths, and 14.9 million disability-adjusted life years, breast cancer is the mainlyrecurrent cancer type amongst women[2]. In order to improve the patient's chance of survival, early identification and diagnosis of breast cancer are crucial. Breast cancer is diagnosed by means of a selection of medical imaging techniques, as well as CT, MRI, ultrasound, and X-rays. Another crucial method for identifying breast cancer is a biopsy. Surgical biopsy, fine-needle aspiration, and vacuum-assisted biopsy are examples of standard biopsy procedures. Biopsy involves the collection of tissue samples or cell tissues that are subsequently fixed on a microscope slide and labeled for successive processing [3]. The pathologist diagnoses the photos following marking in order to find any cancerous tumors [4].

For this study, we have extracted elements such as color, texture, and others from photos of breast histopathology. Moreover, it is categorized using conventional machine learning modelsfor instance SVM, KNN, and so forth.

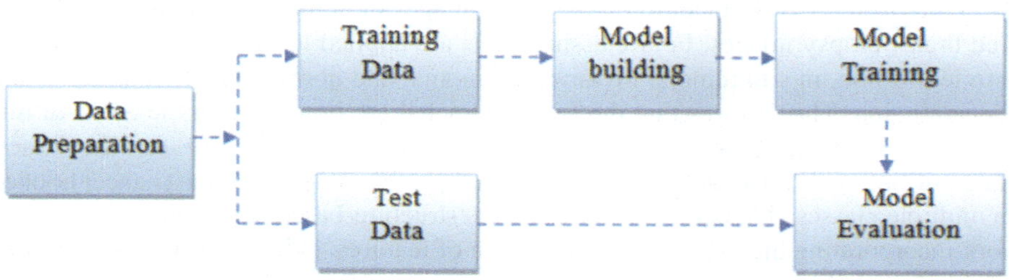

Fig. 1. Process of supervised learning model

Objectives

The aim of this endeavor can be summed up as follows:

a. Extracting features from photos of breast histopathology, such as color and texture.
b. Using conventional supervised machine learning techniques to classify a picture of breast histology.

2 Literature Review

However, it is a laborious and time-consuming process to manually identify these histopathological images for the purpose of detecting breast cancer, and the pathologist's expertise or understanding may have an impact on the outcome. Therefore, the early identification and classification of breast cancer depend heavily on the automatic processing of histological images [5]. Even so, a number of obstacles have made it more difficult to develop tools for the automatic identification and assessment of breast cancer. First, there are high-resolution, finely grained, and intricately structured histopathological pictures of breast cancer. Second, analyzing these histological images necessitates a sizable amount of storage. Other difficulties include choosing the right models and algorithms, selecting features, data theft, privacy, etc. The high false positive rates (FPR) in computer-aided lump detection in breast mammography provide a significant issue. False positive results can lead to elevated health care costs, receiving radiation that is unnecessary, patient concern, etc. [6]. The frequent appearance of digital mammogram images presents a possibility to speak to the difficult difficulty of early breast cancer detection employing deep learning (DL) methods. Current study in computational technology and artificial intelligence demonstrate significant advancements in image processing and deep methods for learning [7–11]. This study uses deep learning techniques to assess histopathology images of breast cancer. Deep learning has the ability to identify powerful features and handle a variety of challenging tasks.

A technique for classifying breast cancer images using convolutional neural networks was presented by Bayramoglu, N. et al. The author of this research offered two distinct architectures: a single task CNN that is employed to anticipate tumor growth, and a multitask CNN that is employed to simultaneously imagine picture enlargement and malignancy. The BreakHis dataset is used for the tests, and the outcomes demonstrate an enhancement over the conventional machine learning method [12, 13]. Araújo et al. [14] employed a CNN model to classify 269 images of hematoxylin and eosin discolored breast biopsy images. Four classifications are applied to the images: two classes, harmless lesion, in situ tumors, invasive carcinoma, and normal tissue: carcinoma and noncarcinoma. The features that the CNN extracted are used to train the SVM in this research. SVM yielded an accuracy of 77.8% for four classes and 83.3% for two classes. For contribution a dependable and precise technique for categorize breast cancer hooked on multiple classes Han et al. presented a class structure-based deep convolutional network incorporating hierarchical representation of features. [15, 16]. The authors of this article classified breast cancer into multiple classes with 95.9% accuracy. Jiang et al. [17, 18] employed a novel dataset for the circumstances of the area according to the beneficiary operating characteristic curves (AUC) called BCDR-F03 (Film The procedure dataset number 3). Utilizing an AUC of 0.88 and 0.83, correspondingly, the authors of this work engaged the GoogLeNet and AlexNet architectures for identifying the breast lesion.

3 Proposed Model

The Breast Cancer Histopathological Image (BreaKHis) dataset, consisting of 9,109 microscopic pictures of breast tumor tissue taken as of 82 individuals using various magnification factors (40X, 100X, 200X, and 400X), is used to test our suggested supervised model. The dataset comprises 5,429 malignant and 2480 benign samples, all having 700 × 460 pixels, 3-channel RGB, 8-bit resolution in every channel, and PNG format. Together with the Pathological Anatomy and Cytopathology P&D Laboratory, this database was created.

There are two classes in the BreaKHis dataset: benign and malignant. The illustration in the dataset was obtained using the SOB technique, also known as incomplete mastectomy or accessional biopsy, in the present version.

Different breast tumor subtypes and kinds may have distinct therapeutic implications and prognoses. The dataset presently includes four histologically dissimilar types of malignant breast tumors (BC): carcinoma (DC), lobular carcinoma (LC), mucinous carcinoma (MC), and papillary carcinoma (PC), and four histologicaldissimilar types of benign breast tumors: adenosis (A), fibroadenoma (F), phyllodes tumour (PT), and tubular adenoma (TA). Every picture filename, such as $SOB\_B\_TA-14-4659-40-001.png$, contains information on the image itself, including the technique of biopsy, tumor class, tumor type, patient identifier, and magnification factor.

The work being done now is limited to 1820 photos with a 400X magnification factor, of which 1232 are malignant and 588 are benign. Figure 2 provides examples of benign images, whereas Fig. 3 shows examples of malignant images.

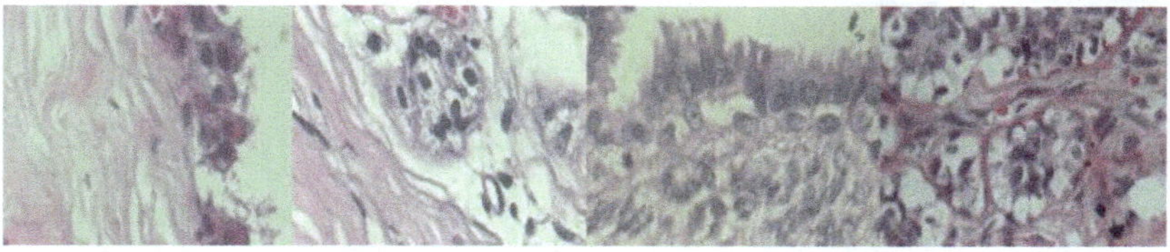

Fig. 2. Sample benign images

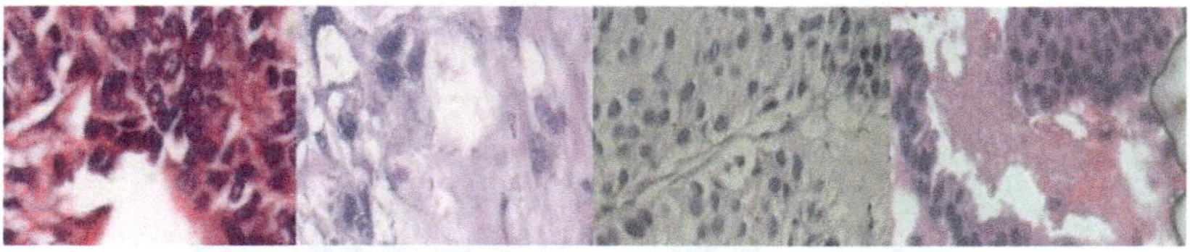

Fig. 3. Sample malignant images

According to Table 1, the dataset is divided between train and test segments at a 4:1 ratio and stored to the appropriate folders.

Table 1. Class wise data distribution

Image Class	Training set	Test set
Benign	470	993
Malignant	118	249

Histogram equalization, sometimes referred to as enhancement of contrast, is used as a pre-processing technique for these images. Otsu thresholding is then used to categorize the grayscale images. Figure 5 provides examples of binarized images. Through HIs, numerous color characteristics are being derived. Red-Green-Blue, HSV (Hue-Saturation-Value), LAB (Lightness-A-B), YCrCb recovered from the original HIs, and Gray derived from these binarized HIs are some of these color attributes. In a similar manner, the binarized HIs have had their GLCM texture features removed. The term "Gray-level a combination matrix", or "GLCM", refers to a texture analysis technique that takes pixel relationships into account. Properties including as contrast, homogeneity, dissimilarity, energy, correlation, and ASM are included in GLCM. For every set, these two features are now kept in distinct CSV files. Figure 4 illustrates the project's recommended technique.

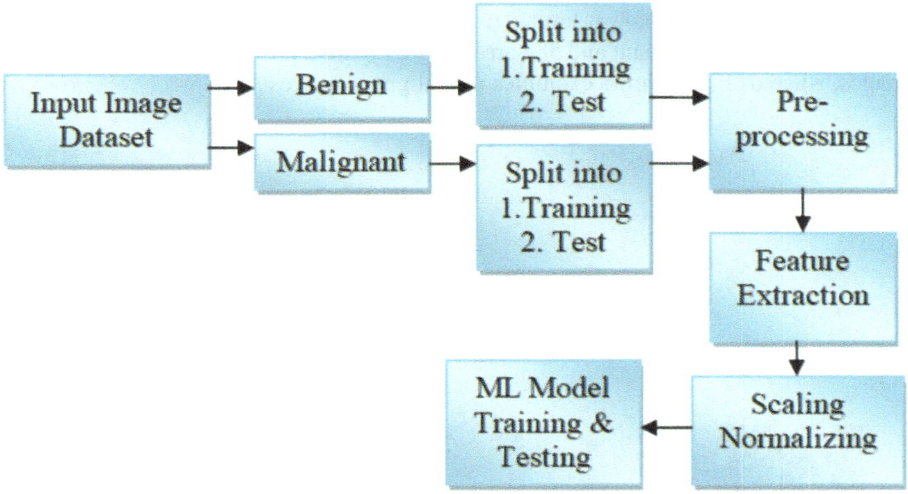

Fig. 4. Proposed system Model

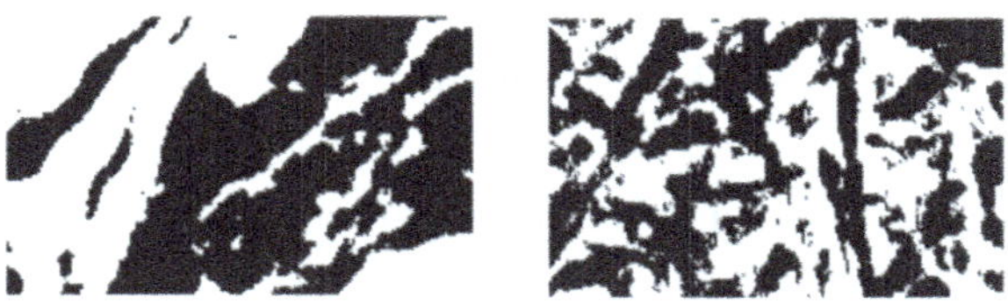

Fig. 5. Sample pre-processed binarized images

4 Experimental Setup

Google Colaboratory, updated as of 06.12.2022, is used for this work [2]. The dataset was initially displayed on Google Colaboratory after being uploaded to Google Drive. To divide the dataset and create machine learning models, using SKLEARN. SKIM-MERGE.Texture features are extracted using FEATURE. CSV can be utilised to read and save retrieved data from csv files. This project used CV2 and MATPLOTLIB to work on photos.

Training of ML Classifiers

The real retrieved values are used to train individual machine learning classifiers. Moreover, scaled and normalized data are used for training. The KNN model is constructed using a range of values for k, or the number of neighbors, which is typically assumed to be odd, such as 1, 3, 5, etc. A variety of kernel parameters, including linear, poly, rbf, and sigmoid, are used to generate the SVM model. In a comparable manner, the decision tree model depends on specific values for criterion factors like Gini and Entropy. There is no parameter required for the Gaussian naive bayes model.

Table 2 provides a detailed explanation of the experimental setup used to conduct the experiment on categorizing the dataset using many machine learning approaches.

Exactness, precision, recollect, and F1 score are among the common assessments used to gauge how well the strategies perform. The same table also includes a confusion matrix for improved understanding of these metrics.

Table 2. Standard Evaluation Metrics and Confusion Matrix Score

ML Classifier			Standard Evaluation metrics			
Classifier	Parameter	Value	Accuarcy	Precision	Recall	F1 Score
KNN	K	7	94	93	92	92
Navie Bayes			82	80	84	80
Decision tree	Criterion	gini	89	89	86	86
SVM	Kernel	Rbf	92	92	88	89

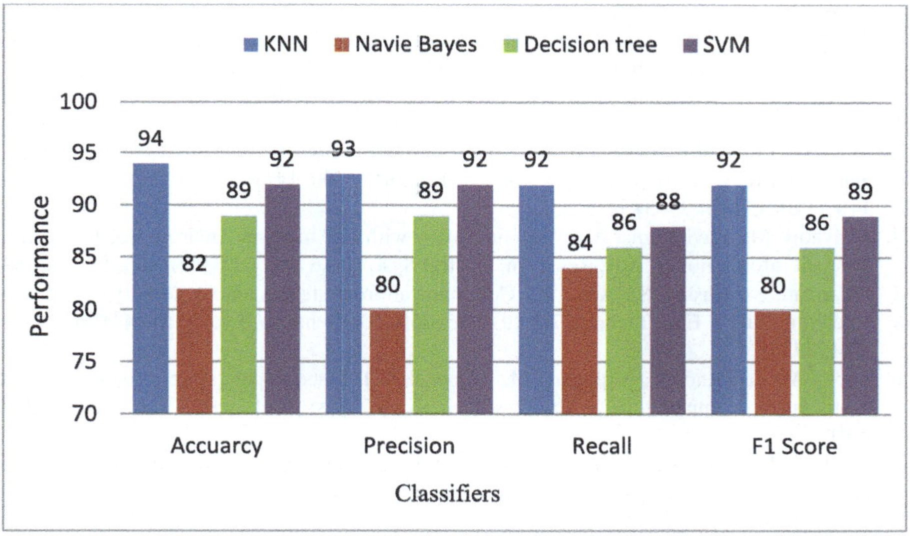

Fig. 6. Comparison of various classifiers

On scaled and standardized data, machine learning approaches (KNN, SVM, Decision Tree, and Naïve Bayes) have been trained and evaluated. Every method is trained using different parameters. Table 2 displays the best result for those parameter selections. KNN outperforms all other neural networks with more than a 92% score, being followed by decision trees, SVM, and Gaussian naïve bayes. In scaled and normalized data, KNN performs best at 94% for K = 7, next by SVM with 90% precision, recall, and F1 score. Nevertheless, as Fig. 6 illustrates, all outcomes are positive and surpass 72%. Additionally, confusion matrices are provided, in which the symbols TN, FN, FP, and TP indicate true negative, false positive, and true positive, correspondingly. While some predictions are accurate, TN and TP are not.

5 Conclusion

The BreaKHis dataset is being classified using several supervised machine learning approaches in this ongoing work. As anticipated, scaling and normalization have produced better outcomes. It provides the average reader with a reasonable understanding of how various machine learning approaches function. Deep learning, along with unsupervised and reinforcing machine learning approaches, may be added to the current research. Aside from breast cancer HIs, there are other standard datasets that could be incorporated into the research in the future. Only two features—color and texture—are used in the current endeavor. Other features, such as morphological ones, might be taken out for a more thorough examination.

References

1. Boyle, P., and Levin, B.: World Cancer report 2008: IARC Press. International Agency for Research on Cancer (2008)
2. Moraga-Serrano, P.E.: Global, regional, and national cancer incidence, mortality, years of life lost, years lived with disability, and disability-adjusted life-years for 29 cancer groups, 1990 to 2016: a systematic analysis for the global burden of disease study. JAMA Oncol. **4**, 1553–1568 (2018). https://doi.org/10.1001/jamaoncol.2018.2706
3. Veta, M., Pluim, J.P., van Diest, P.J., Viergever, M.A.: Breast cancer histopathology image analysis: a review. IEEE Transac. Biomed. Eng. **61**, 1400–1411 (2014). https://doi.org/10.1109/TBME.2014.2303852
4. Malyadri, M., Ravi Kumar, L.: Web innovation with IoT in social environment for sharing efficient information. J. Adv. Res. Dyn. Control Syst. **10**(13 Special Issue), 466–1475 (2018)
5. Sulaiman, S., Hussin, S., Amir, Z.: Communication strategies among tertiary students in Mlearning. Int. J. Eng. Technol. **7**(2.29), 655–659 (2018). https://doi.org/10.14419/ijet.v7i2.29.13993
6. Subha Mastan Rao, T., Malyadri, M.: A novel IOT based expert system to mitigate the problems in shrimp culture. J. Adv. Res. Dyn. Control Syst. **10**(4 Special Issue), 2031–2039 (2018)
7. Krishna, B., Amarawat, G.: Frequent item set generation using enhanced Apriori algorithm and multiple projection rule pruning algorithm. J. Adv. Res. Dyn. Control Syst. **10**(4 Special Issue), 1947–1953 (2018)
8. Krishna, B., Amarawat, G.: An enhanced scaling Apriori for association rule mining with frequent item set mining. J. Adv. Res. Dyn. Control Syst. **10**(6 Special Issue), 102–106 (2018)
9. Shrivastava, R., Jain, M., Vishwakarma, S.K., Bhagyalakshmi, L., Tiwari, R.: CrossCultural translation studies in the context of artificial intelligence: challenges and strategies. In: Kumar, A., Mozar, S., Haase, J. (eds.) ICCCE 2022. Cognitive Science and Technology, pp. 91–98. Springer, Singapore (2023). https://doi.org/10.1007/978-981-19-8086-2_9. ISBN: 978-981-19-8086-2_9
10. Zhao, R., Yan, R., Chen, Z., Mao, K., Wang, P., Gao, R.X.: Deep learning and its applications to machine health monitoring: a survey (2019)
11. Machine health monitoring: a survey. arXiv preprint arXiv:1612.07640 (2016)
12. Kooi, T., et al.: Acomparisonbetween a deep convolutional neuralnetwork and radiologists for classifying regions of interest in mammography. In: International Workshop on Digital Mammography, pp. 51–56. Springer (2016)

13. Tiwari, R., et al.: An Artificial intelligence-based reactive health care system for emotion detections. Comput. Intell. Neurosci. **2022**, Article ID 8787023 (2022). https://doi.org/10.1155/2022/8787023
14. Chartrand, G., et al.: Deep learning: a primer for radiologist. Radiographics **7**(37), 2113–2131 (2017)
15. Platania, R., Shams, S., Yang, S., Zhang, J., Lee, K., Park, S.J.: Automated breast cancer diagnosis using deep learning and region of interest detection (BC-DROID). In: Proceedings of the 8th ACM International Conference on Bioinformatics, Computational Biology and Health Informatics, 536–543. ACM (2017)
16. Vasundhara, N., Nandan, A.S., Hemanth, S.V., Macherla, S., Madhura, G.K.: An efficient biomedical solicitation in liver cancer classification by deep learning approach. In: 2023 IEEE International Conference on Integrated Circuits and Communication Systems (ICICACS), pp. 01–05. IEEE (2023)
17. Dhungel, N., Carneiro, G., & Bradley, A. P.: Automated mass detection in mammograms using cascaded deep learning and random forests. In: 2015 International Conference on Digital Image Computing: Techniques and Applications (DICTA), pp. 1–8. IEEE (2015)
18. Aelgani, V., Vadlakonda, D.: Explainable artificial intelligence based ensemble machine learning for ovarian cancer stratification using electronic health records. Int. J. Recent Innov. Trends Comput. Commun. **11**(7), 78–84 (2023). https://doi.org/10.17762/ijritcc.v11i7.7832

A Comprehensive Convolution Neural Network Structure for EEG-Based Depression Detection

Bollampelly Chandana[1](✉), Kovuri Praveen Kumar[1], A. Srinivasula Reddy[2], Sheo Kumar[2], Abdul Subhani Shaik[3], and SivaSkandha Sanagala[3]

[1] CMR Technical Campus, Hyderabad, Telangana, India
cmrtc.paper@gmail.com
[2] CMR Engineering College, Hyderabad, Telangana, India
[3] CMR College of Engineering & Technology, Hyderabad, Telangana, India
sivaskandha@cmrcet.ac.in

Abstract. Distress is often the cause of the global increase in cases of suicide. Consequently, in order to mitigate the belongings of depression, a clear diagnosis and course of treatment are necessary. An electroencephalogram (EEG) is used for recording and measuring the electrical impulses of the brain. It can be applied to produce a precise evaluation of depression severity. Prior studies have demonstrated the feasibility of diagnosing mental disorders employing deep learning (DL) models and EEG data. According to this research, the EEG data of individuals who are sad and those who are healthy are classified using a convolution neural network (CNN) built upon DL called DeprNet. In this case, the severity of the depression is indicated by the Patient Health Questionnaire 9 score. This paper presents the presentation of DeprNet in two experiment: the subject-wise divide and the record-wise split. The findings from the analysis of DeprNet demonstrate an efficiency of 0.9937 and an area below the receiver's operating characteristic curve (AUC) of 0.999 while record-wise split data are included. Yet, an accurateness of 0.914 and an AUC of 0.956 can be achieved whenever subject-wise split data are employed. These consequences suggest that when CNN is programmed on record-wise split data, it overstrains on EEG data given a limited number of participants. DeprNet performs admirably when compared to one of the eight baseline models. Furthermore, it is found that in depressed people, the values of the ultimate CNN layer are dominant on the right electrode, while in typical participants, the values on the left electrodes are significant.

Keywords: CNN · EEG · DeprNet

1 Introduction

Mental illness, also called a mental health condition, is a physical brain ailment that can influence an individual's ideas, behavior, and emotions. It also results in a decrease in interest and energy, can harm relationships and productivity at work, and increases the possibility of suicide. Globally, mental illness affects more than 13% of children, 46% of teenagers, and 19% of adults annually [1]. As a result, early detection of depression

D. Kriksciuniene et al. (Eds.): IBICA 2023, LNNS 1228, pp. 354–361, 2025.
https://doi.org/10.1007/978-3-031-78937-3_38

is crucial to saving the lives of those who struggle with it by keeping it from getting worse and becoming irreversible. Typically, the patient's behavior demonstrates signs of depression. Consequently, in order to determine the extent of depression, physicians use screening instruments such as questionnaires as well as talking therapy sessions. The success of the talking session is going to rely on the way the psychiatrist or counselor works. In addition, miserable individuals are more reluctant to seek therapy since mental illness is stigmatized. Consequently, a significant fraction of the depressed population fails to get the best care and sufficient time to recover. Consequently, the field of research into appropriate and efficient techniques for detecting depression is expanding, and novel possibilities for this purpose have been created by current advances in equipment or sensor technology. The electroencephalogram (EEG) is a portable method that can record physiological data, magnetic resonance imaging, magnetic resonance imaging (MR imaging), and the electrical signals of brain neurons through the scalp surface in contemporaneous fashion. It was recently experiential that the bulk of cognitive behavior and mental processes are analyzed by EEG [2] while the EEG signal taken from the human brain's parietal lobe is connected related emotional states and cognitive tasks [3].

At over 264 million, melancholy constitutes one of the most prevalent psychological diseases in the world [3]. In addition to other symptoms, depressed people tend to experience melancholy, stress, low mood, and energy loss. It is quite difficult for them to focus upon their work and interact with others. According to severe instances, depression and other mood disorders account for half of the total completed suicides [4]. As a result, numerous studies have concentrated on creating methods to identify and stop this mental illness so that psychologists and psychiatrists can help individuals as quickly as feasible [4].

2 Literature Review

Clinical consultations are used in the conventional method of detecting depression in order to screen applicants for the illness [5]. The inquiries posed by the clinician, the patients' verbal reports, the actions according to by friends or family, and the results of mental status exams—like the Beck Inventory of Depression (BDI-II), the Hamilton Rating Scale for Depression (HRSD), and the Scale for the Evaluation of Negative Symptoms (SANS)—all play a significant role in these assessments. Since there are no scientific or quantifiable measurements, they are purely based on subjective ratings.

To store the verbal tract resonant sounds and spectral shape dynamics, Williamson et al. [6] investigated two vocal tract illustrations: formant-frequency tracks as well as Melcepstral features. The final forecast was then created using a Gaussian mixture model (GMM)-based multivariate regression approach, regarding the two feature sets indicating the alterations in synchronization of vocal tract motion related to MDD. Eventually, they improved the method by using timed feature sets and a multi-scale correlation structure to create high-level characteristics for low level characteristics. Text, audio, and visual data from the DAIC-WOZ database are used by Dham et al. [7] to diagnose depression. The visual data, low level audio features, head pose, and text characteristics were all subjected to Gaussian Mixture Model (GMM) clustering using the Fisher vector technique. The retrieved characteristics underwent separate application of the SVM classifier. Ultimately, the outcomes of various modalities were combined via decision-level fusion [7].

A pioneer in the field of emotional computing, Cohn et al. [8] combined visual and aural aspects in their research to incorporate behavioral factors, which are closely linked to mental health issues. According to their findings, it may be feasible to build an automatic system that can identify depression, that will advance clinical theory and practice. To evaluate facial and vocal emotion, they employed pitch extraction, dynamic appearance modeling (AAM), and manual FACS coding. For FACS and AAM, logistic regression was the classifier that used leave-one-out validation, whereas SVM was used for voice.

In their study, Chao et al. [9] examined the crash of the most current deep learning models on the advancements of both vision- and audio-based techniques for the analysis of depressive disorders. In order to depict the anomalous behavior, the extracted audio video characteristics have been fused at the feature level. Next, the dynamic time data was described by means of the Long Short-Term Memory Recurrent Neural Network (LSTM-RNN). Utilizing multi-task learning as a performance enhancer, they achieved competitive outcomes on the AVEC 2014 dataset [9, 10]. This reading demonstrated the potential benefits of using both modalities and temporal information for automated depression identification. Jan et al. [11] used deep learning and regression approaches, respectively, to merge data collected from speech and facial expressions in naturalistic video recordings.

Through conversation with the interview subjects, a computer agent supports the process by spotting behavioral and verbal cues indicative of mental health problems. The extensive questionnaire responses from the interviews, along with audio and video recordings, have been transcribed and annotated in DAIC-WOZ. The Wizard of Oz interviews are included in this section of the gathering. Ellie, an animation virtual interviewer, is handled remotely by a human interviewer [12, 13].

A deep learning-based automated method for identifying Major Depressive Disorder (MDD) has been proposed by authors in [14]. The model was built using Convolutional Neural Networks (CNN) on spectrogram images. By using the hold-out validation strategy, they be able to attain a 99.58 percent classification accuracy. The shortcoming of this model was that it was created with training data from just 64 people, of whom 30 were healthy and 34 were MDD patients. A study in [15–17] states that the MHDeep-structure was recommended by the authors. The system's ability to combine data from conventional warehouse management software using DNNs' knowledge-purifying powers allows for the continuous including universal diagnosis of bipolar illness, major depressive disorder, and schizoid personality disorder. The approach composesexploit of a synthetic data creation module to aware users when large datasets are engaged.

Since we didn't have an established system for this project, there was no appropriate way to identify depression detection in the existing system. We develop a novel concept for depression identification. Melancholy is a common factor contributing to the rise in suicide cases in modern times. Therefore, it is essential to diagnose depression in its early treatable phases in order to save the lives of those who are depressed and to stop it from getting worse and becoming permanent. We are implementing a depression recognition project utilizing Kaggle datasets in order to forecast the depression.

3 Proposed System

The system designed to classify patients into those who are depressed and those who are not makes use of speech spectral features and emotional analysis. Convolution Neural Network models are used to train audio recordings in the wav format. It is necessary to exclude certain problematic recordings from the collection. If they are not deprecated, these files may result in problems during the feature extraction procedure. The interviewer's voice and the interviewee's are both present in the audio files; therefore, segmentation is required to exclude the interviewer's voice and other unwanted elements including noise, silence, etc. Following the completion of all audio file cleaning, a spectrogram is produced and supplied as the input to the convolutional neural network in order to extract features.

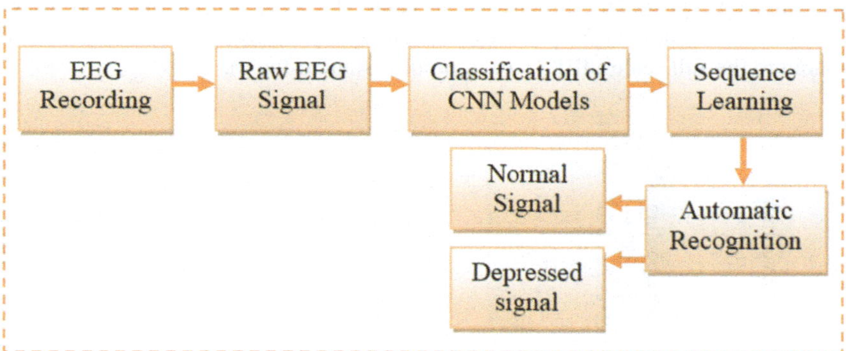

Fig. 1. System model architecture

The model's flow is depicted in Fig. 1, whereby audio data first undergoes pre-processing by being accepted as an input and then, through the aid of the pyAudio Analysis library, detached from undesirable voice, silence, noise, etc. The same library is used for obtaining attributes from the recordings before spectrograms are created from them. Ultimately, CNN—a classification algorithm—is fed these images. Further information on the suggested deep-based approach is provided in the contents that follow.

3.1 Suggested System and its Benefits

We used real-time EEG signal experiment data from Kaggle for this research. Here, I'm referring to the Kaggle data collection. In order to obtain a more accurate prediction, we are pre-processing and fine-tuning the data, which is available as a.mat file. Following data refinement, KNN and support vector machine methods are applied, while CNN and LSTM architecture are applied for an identical statement of problem. Additionally, we are evaluating each of these algorithms on an individual basis before evaluating their accuracy across all models. In comparison, CNN Architecture yields superior results, with accuracy that is comparable to that of SVM. Because of this, we are utilizing SVM in this project to forecast the output for new samples; however, we have the option to employ CNN or SVM for real-time prediction.

Benefits

1) We can more accurately and readily identify depression by utilizing many approaches.
2) In this article, we review various works that employ any number of algorithms to diagnose depression.
3) The outcome of employing neural networks to identify depression is nearly 100% accurate.

As seen in Fig. 2, the Convolutional Neural Network (CNN) architecture has a total of 6 layers: 2 convolutional layers, 2 max pooling layers arranged alternating and 2 fully connected layers. The dimensions of each 513 × 125 input image to the CNN represented 4 s of audio at frequencies that ranged from 0 to 8 kHz. Based on decibels compared to the full scale (dBFS), each input is standardized).

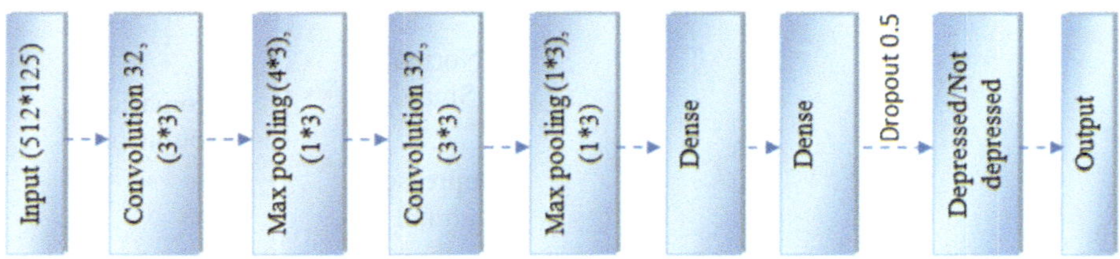

Fig. 2. CNN model structure

CNN's input layer was composed of 32 filters, a 3 × 3 kernel, and an activation function for ReLU. The picture map is then reduced using a max pooling technique using a 4 × 3 filter and a 1 × 3 stride. The identical feature size as well as kernel size is used in a second convolution, that is subsequently go behind by a max-pooling layer that is comparable to the first one but employs a 1 × 3 filter. A

4 Implementation

The stage of execution is when the conceptual blueprint is put into a programmable format. The program will now be divided into several parts, each of which will be developed for distribution. The user interface of the app uses Google Collaborator. Here, the present application is being implemented utilizing Python as the programming language. The following five components make up the majority of the application. They are listed below:

1. Bring in the Required Libraries
2. Dataset Loading Module
3. Pre-processing of Data

4. Utilizing Multiple Algorithms to Train the Model
5. Determine How Well ML Algorithms Perform

Library

To begin with, we must import all of the libraries needed to build the replica into this module. Here, we make an effort to employ every library available for transforming the data in a useful way. Since the data is broken down into numerical values that the structure can readily understand, we endeavor to introduce the numpy module and employ the matplot library to plot the information into graphs and charts.

Dataset

We attempt to display the dataset that was downloaded or obtained from the UCI repository within this module. The dataset named "Heart.csv" is stored here, and it includes the following kinds of data, for example: Every single attribute has some data that is evaluated and gathered based on the unique patient ID.

Data Pre-processing

In this phase, our goal is to pre-process the input dataset in order to identify any values that are missing or incorrect information. The program is going to load just those rows with all valid inputs and disregard any data that is invalid if it is found in the dataset.

Put the Model through It.

Here, we attempt to use multiple ML classification methods to train the present model on the provided dataset. The goal is to determine which technique works best for properly and efficiently identifying and classifying the input dataset. Here, we attempt to apply the following methods to the provided dataset:

1. The Support Vector Machine
2. K-Nearest Neighbors
3. LSTM
4. CNN

Performance Analysis

In this section, we try to contrast every classification method available on the provided input dataset in order to determine which one is most suitable for producing correct results. In the end, we will determine which method produces accurate answers in the shortest amount of time. As we can see, CNN produces results that are more precise than those of other algorithms.

In this phase, we attempt to construct our present framework with Python as a programming language as well as Google Collab as the program's desktop setting. We are able to utilize the following to evaluate the effectiveness of our suggested application:

Figures 3 and 4 beyond show us how different algorithms compare, and it's evident that CNN has excellent precision. In order to run tests, fresh audio files are recorded and the class labels that the model predicts are examined. Employing acoustic characteristics, the suggested model has proved successful in differentiating among samples that are melancholy and those that are not.

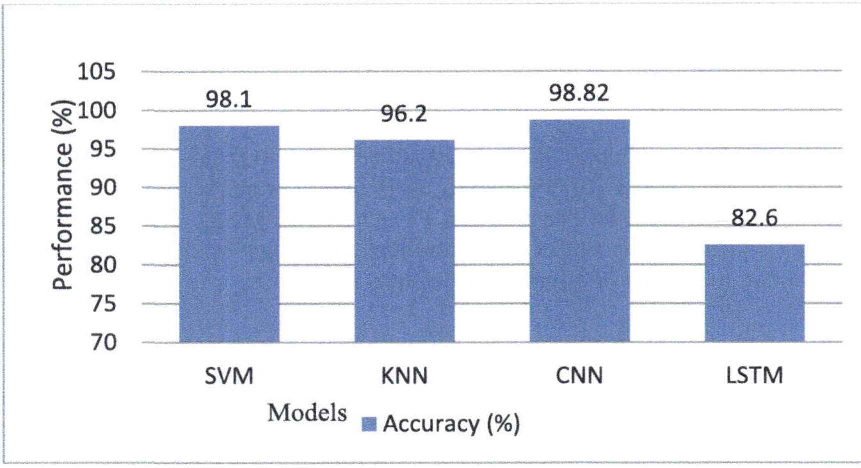

Fig. 3. Comparison of various models

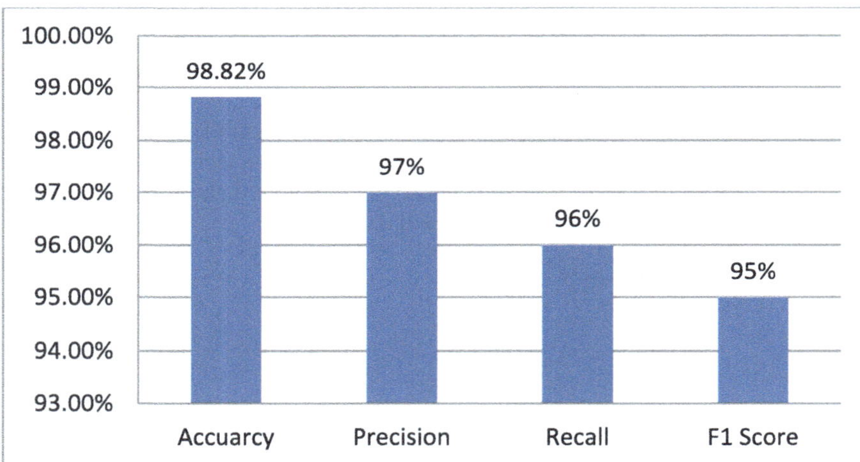

Fig. 4. Simulation results for CNN

5 Conclusion

This work successfully analyzes the EEG data using DL models to demonstrate that depression affects the electrical activity of the brain. It may be accomplished that the DeprNet CNN-based DL models projected by this learning performexcellentcompared to other methods. The precision is 0.9937 and the AUC is 0.999 whenever record wise divisions are measured. AUC of 0.956 and a precision of 0.914 were calculated using subject-wise split data. These consequencesrecommend that whenever CNN undergoes training on record wise split data, it over trains on electroencephalogram (EEG) information containing a small amount of patients.

References

1. Swetaa, A., Gayathri, R., Priya, V.V.: Awareness of mental health among teenagers. Drug Invent. Today **11**(8), 1979–1982 (2019)
2. Malyadri, M., Ravi Kumar, L.: Web innovation with IoT in social environment for sharing efficient information. J. Adv. Res. Dyn. Control Syst. **10**(13 Special Issue), 1466–1475 (2018)
3. Sulaiman, S., Hussin, S., Amir, Z.: Communication strategies among tertiary students in Mlearning. Int. J. Eng. Technol. **7**(2.29), 655–659 (2018). https://doi.org/10.14419/ijet.v7i2.29.13993
4. Subha Mastan Rao, T., Malyadri, M.: A novel IOT based expert system to mitigate the problems in shrimp culture. J. Adv. Res. Dyn. Control Syst. **10**(4 Special Issue), 2031–2039 (2018)
5. Krishna, B., Amarawat, G.: Frequent item set generation using enhanced Apriori algorithm and multiple projection rule pruning algorithm. J. Adv. Res. Dyn. Control Syst. **10**(4 Special Issue), 1947–1953 (2018)
6. Krishna, B., Amarawat, G.: An enhanced scaling apriori for association rule mining with frequent item set mining. J. Adv. Res. Dyn. Control Syst. **10**(6 Special Issue), 102–106 (2018)
7. World Health Organization. Depression (2020). https://www.who.int/news-room/factsheets/detail/depression/
8. Vázquez-Romero, A., Gallardo-Antolín, A.: Automatic detection of depression in speech using ensemble convolutional neural networks. Entropy **22**(6), 688 (2020)
9. Cohn, J.F., et al.: Detecting depression from facial actions and vocal prosody. In: 2009 3rd International Conference on Affective Computing and Intelligent Interaction and Workshops, Amsterdam, Netherlands, pp. 1–7 (2009). https://doi.org/10.1109/ACII.2009.5349358
10. Chao, L., Tao, J., Yang, M., Li, Y.: Multi task sequence learning for depression scale prediction from video. In: 2015 International Conference on Affective Computing and Intelligent Interaction (ACII), Xi'an, China, pp. 526–531 (2015). https://doi.org/10.1109/ACII.2015.7344620
11. Prashanthi, M., Chandra, M.M.: Hybrid optimization-based neural network classifier for software defect prediction. Int. J. Image Graph. (2023). https://doi.org/10.1142/S0219467824500451
12. Jan, A., Meng, H., Gaus, Y.F.B.A., Zhang, F.: Artificial intelligent system for automatic depression level analysis through visual and vocal expressions. IEEE Trans. Cogn. Dev. Syst. **10**(3), 668–680 (2018). https://doi.org/10.1109/TCDS.2017.2721552
13. Tiwari, R., et al.: An Artificial Intelligence-Based Reactive Health Care System for Emotion Detections. Comput. Intell. Neurosci. **2022**, Article ID 8787023 (2022). https://doi.org/10.1155/2022/8787023
14. Jagadeeshwar, K., Sreenivasarao, T., Pulicherla, P., Satyanarayana, K.N.V., Mohana Lakshmi, K., Kumar, P.M.: ASERNet: automatic speech emotion recognition system using MFCC-based LPC approach with deep learning CNN. Int. J. Model. Simul. Sci. Comput. **14**(04), 2341029 (2023)

Employing the Hybrid Image Fusion Method to Recover the Quality of Images in Medical Image Processing

Sudha Arvind[1(✉)], N. Shweta[1], Vaseem Ahmed Qureshi[2], K. Sravan Abhilash[2], R. Bhargav Ram[3], and J. L. Divya Shivani[3]

[1] CMR Technical Campus, Hyderabad, Telangana, India
cmrtc.paper@gmail.com
[2] CMR Engineering College, Hyderabad, Telangana, India
qureshi.vaseem@cmrec.ac.in
[3] CMR College of Engineering and Technology, Hyderabad, Telangana, India
dr.r.bhargavram@cmrcet.ac.in, dr.j.l.divyashivani@cmrcet.org

Abstract. Through merging the characteristics of several images from different modalities like magnetic resonance imaging (MRI), positron emission tomography (PET), and computed tomography (CT) into a single final image that contains the most significant information generated from all inputs, medical image fusion methods are widely utilized for assistance in medical diagnosis. In this work, a hybrid fusion technique for MRI and CT scans is proposed. The Discrete Cosine Transform (DCT) is used to separate the different frequency components, and Principal Component Analysis (PCA) is used to find the largest Eigen value of the covariance matrices for each image. of the input image, resulting in noise in the output images following the first two steps. When compared to known approaches, the efficacy assessment of CT-MRI image fusion using the hybrid technique Including Discrete Wavelet Transform (DWT) yields better image quality. All things considered, there is a good chance that the non-linear fusion rule will enhance picture fusion applications in the medical industry and other domains.

Keywords: Image fusion · MRI · CT · DCT · DWT

1 Introduction

The technique of unite two or more images into one fused output image with a higher visual quality while maintaining key elements from every image that was input is known as image fusion. The idea has many applications in medical imaging, where several image acquisition technologies, such as computed tomography (CT) and magnetic resonance imaging (MRI), are frequently used. Because everyone system is distinct and works together, it has been suggested that integrating acquired images about more comprehensive information is possible. For example, MRI scans provide information about soft tissues, but CT images provide more information about tough structures like bones. Conversely, low-spatial-resolution modalities such as single photoemission computer tomography (SPECT) and positron emission tomography (PET) supply functional records that

D. Kriksciuniene et al. (Eds.): IBICA 2023, LNNS 1228, pp. 362–371, 2025.
https://doi.org/10.1007/978-3-031-78937-3_39

aid in revealing details regarding the scanned subject's blood circulation and visceral metabolism.

Throughout the years, many methods were used to accomplish fusion on medical pictures; the wavelet-based method is among the most often used [2]. The input image is divided into a number of low-pass and high-pass filtered representations using signal decomposition techniques. After obtaining the appropriate wavelet values for each of both input images, a fusion operation or rule selects the preferred coefficients to be used in the fusion process. One important component that affects the level of quality of the complex image is the blending rule, which comes through the removal of features selection.

2 Literature Review

While Computed Tomography (CT) offers precise information Magnetic Resistance Imaging (MRI) provides low density soft tissues, such as blood vessels, and offers information about bone tissue and acts as a reference for lesion location [1]. This paper [3] discussed how the radiation doses to a specific organ from a given CT scan depend on a number of factors, including the patient's scale, the axial scan variation, the tube the scanning time in milliamp seconds (mAs), also, the specific style of the scanner being used plays a role. The fundamental components of any diagnostic radiology quality control program are patient dosimetry and picture quality assessment. A acceptable patient dose must be used to provide image quality sufficient for diagnosis [5]. An essential clinical need that has been addressed as part of implementation [6, 7] efforts is the need to enhance low-dose CT (LDCT) images in the EHR, or Electronic Health Records. Khalid et al. proposed an enhanced dynamic quadrant equalization method that splits the input image histogram into eight subhistograms to improve image contrast. Using median values. The average pixels are used to cut the histogram for individual subhistograms. Every subhistogram is given a new dynamic range, and HE is completed independently. The mean brightness is maintained using this method [10].

3 Proposed Model Architecture

After being read, the two input images are transformed into indexed images. In order to determine the approximate horizontal, vertical, and diagonal features, wavelet decomposition is then performed. There are specifications for the wavelet type and the decomposition level as shown the Fig. 1. Next, DWT is applied to the supplied pictures. The discovered coefficients are then fused using a particular fusion technique, and an inverse discrete wavelet transform is employed to recover the images.

Principle Component Analysis (PCA)
Rearranging the image dataset so that it can be utilized in the computation is the main function of the principal component analysis image fusion method. Its goal is to take the input image matrices $I1$ and $I2$ of a dimension of m by n and turn them into a one by m multiplied n matrix, or in other words, a one column matrix. This process will be iterated for every image input.

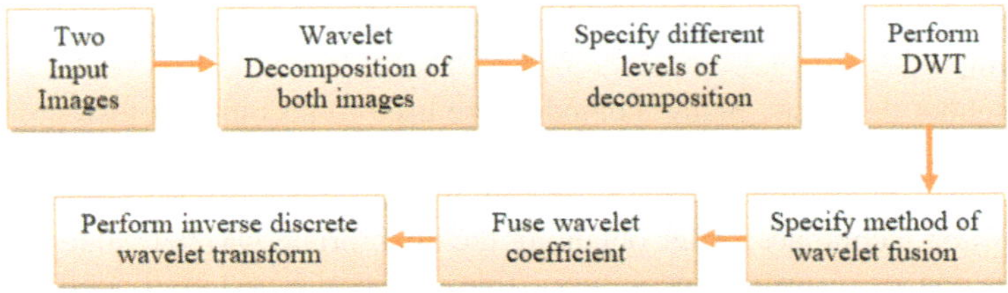

Fig. 1. Proposed model structure

The covariance among the two fresh datasets, X and Y, that are generated using the input image matrices $I1$ and $I2$, must then be determined.

$$C = (X, Y) = [(X - E[X])(Y - E[Y])] \quad (1)$$

wherein X and Y, accordingly, represent their predictable values, E[X] and E[Y].

In order to discover the covariance, one must first determine the E[X] and E[Y], that is effectively the empirical mean of the X and Y datasets. The above numbers are then subtracted as of the X and Y datasets to obtain the deviationas of the mean, or, more specifically, $(X - E[X])and(Y - E[Y])$.

Calculate the Eigenvector and Eigen value by utilizing the covariance matrix C in a way that

$$Cv = Dv \quad (2)$$

wherein D is the Eigen value, v is the value of the Eigenvector matrix, and C is the covariance matrix.

After obtaining the Eigenvector V and Eigen value D, we organize the matrix C by decreasing Eigen value. The resultant two-by-two V and D are able to be utilized to get the ultimate PCA value (weighted value) for every image, $p1$ and $p2$, with the primary column of V corresponding to the greater Eigen value.

$$P1 = \frac{v(1)}{\sum v}, P2 = \frac{v(2)}{\sum v}$$

V stands for the matrix of eigenvectors.

Multiply every weighted average by the original image to obtain the fused image, then add the results to obtain the ultimate fused image.

$$I_{fused} = p_1 * i_1 + p_2 * i_2$$

wherever $If\,used$ is the absolute fused image and $I1$and $I2$ are the two input images

How to run the PCA method:

1. Create the corresponding column vectors about the input picture matrices as shown the Fig. 2.
2. Determine the column vectors created in step 1's covariance matrix.
3. Determine that covariance matrix's Eigen values and Eigen vectors.

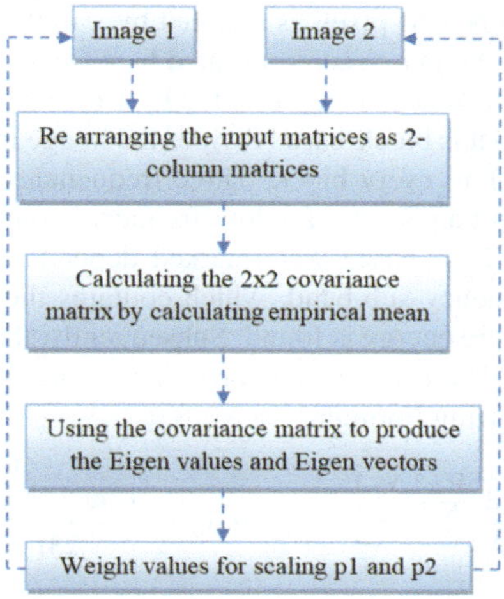

Fig. 2. PCA algorithm

4. The column vector that corresponds to the bigger Eigen value should be normalized.
5. Produce datasets with weight values, where the standardized values are.
6. Eigenvector after step 4 split by Eigenvector mean.
7. To obtain the ultimate fused image matrix, combine every single one of the input photos by the weight values provided by step 5 and total them up.

Transformation by Discrete Cosine (TDC)

The method of discrete cosine transformations is used to break down a signal onto its basic frequency components. A collection of sinusoids with different magnitudes and frequencies is used to depict an image.

The following equation is used to calculate the DCT parameters for the converted output picture, y, given an input image, x.

$$y(u, v) = \sqrt{\frac{2}{M}}\sqrt{\frac{2}{N}}\alpha_u\alpha_v \sum_{u=0}^{M-1}\sum_{v=0}^{N-1} x(m, n)\cos\frac{(2m+1)u\pi}{2M} cos\frac{(2n+1)v\pi}{2N}$$

where,

$$\alpha_u = \begin{cases} \frac{1}{\sqrt{2}} \ldots\ldots\ldots\ldots u = 0 \\ 1 \ldots\ldots\ldots\ldots\ldots\ldots u = 1, 2, 3, \ldots .N - 1 \end{cases}$$

$$\alpha_v = \begin{cases} -\frac{1}{\sqrt{2}} \ldots\ldots\ldots\ldots v = 0 \\ 1 \ldots\ldots\ldots\ldots\ldots\ldots v = 1, 2, 3, \ldots .N - 1 \end{cases}$$

where,

The input image with N x M pixels is denoted by x in the equation, the amount of pixels in rows m and n of the image is represented by x (m, n), and the TDC coefficient in rows u and v of the TDC matrix is represented by y (u, v).

The DCT transform that is block-based divides the image into not go beyond chunks and then be relevant DCT to every block. Three frequencies sub-bands are produced as an outcome: three sub bands: one for low frequency, one for mid frequency, and one for high frequency. TDC based watermarked depends on two facts. First, most of the signal A low-frequency sub-band, which contains the majority of the image's visual components, is where energy is found. Subsequently, the image's high frequency elements are eliminated. The inverse TDC procedure is used to recreate the image in accordance with the following Formula.

$$x(m,n) = \sqrt{\frac{2}{M}}\sqrt{\frac{2}{N}}\sum_{u=0}^{M-1}\sum_{v=0}^{N-1}\alpha_u\alpha_v y(u,v)\cos\frac{(2m+1)u\pi}{2M}cos\frac{(2n+1)v\pi}{2N}$$

De-correlation

The elimination of duplication among nearby pixels is the main benefit of picture transformation, as previously stated. This results in transform coefficients that are uncorrelated and can be individually encoded. Let's look at the instance using Fig. 2 to illustrate the 2-D DCT's de-correlation properties. The photos' normalized autocorrelation prior to and following DCT.

Energy Compaction

The capability of a transformation method to fit as much input data under as few coefficients as possible is a direct indicator of its efficacy. Because of this, the quantize can eliminate coefficients having comparatively tiny amplitudes without causing visual distortion in the image that is recreated. Substantial energy compaction is demonstrated by DCT for highly linked pictures. Let's look at the two sample photos from Fig. 3 once again. Apart from the correlation features that were previously stated, the uncorrelated image exhibits more pronounced changes in intensity compared to the correlated image. Consequently, compared to the latter, the former has higher frequency content. Figure 4 displays the two photos' DCT. It is evident that the correlated image's energy is concentrated in the low frequency region (i.e., upper left region), while the uncorrelated image's energy is dispersed throughout.

Discrete Cosine Transform Depending on Laplacian Pyramid

In order to perform discrete cosine transform oriented Laplacian pyramid image fusion, we must first use discrete cosine transform (DCT) to break down every image that is input. Next, we must build a Laplacian pyramid that represents the number of layers, k, that the method will utilize.

The following formula is used to Find the picture I(i, j) with dimensional size m by n's two-dimensional discrete cosine transform.

Z(u, v).$z(u,v) = \alpha(u)\alpha(v)\sum_{i=0}^{m}\sum_{j=0}^{n} I(i,j)\cos\frac{(\pi(2i+1)u)}{2m}cos\frac{(\pi(2n+1)v)}{2n}$

Where,

$$\alpha_u = \begin{cases} \frac{1}{\sqrt{m}} \ldots\ldots\ldots\ldots u = 0 \\ \frac{2}{\sqrt{m}} \ldots\ldots\ldots\ldots\ldots 1 \le u \le m \end{cases}$$

$$\alpha_v = \begin{cases} \frac{1}{\sqrt{n}} \ldots\ldots\ldots\ldots v = 0 \\ \frac{2}{\sqrt{n}} \ldots\ldots\ldots\ldots\ldots 1 \le v \le n \end{cases}$$

To get to the last decomposition level, repeat those decomposition processes K times. The term Laplacian pyramid refers to the representation of this series of decomposition as a pyramid. Furthermore, we will obtain a multiscale image for every image once we have determined the final deconstruction level for every picture. Rebuilding the intricate image: Ultimately, once the multi-scale image has been fused, we may obtain the ultimate composite image by performing k applications of the inverse independent cosine transform on the image to return to the initial point, that serves as the foundation for our Laplacian pyramid model. The discrete cosine transform and Laplacian pyramid method steps are as follows:

Select the necessary decomposition level, k, to apply to the photos in order to fuse them. Use a Laplacian pyramids of the discrete sine alter to break down the input image. In order to achieve the desired degree of decomposition, Step 2 should be repeated K times., as determined by step 1 as sown the Fig. 4. A single of the following techniques is utilized to construct a fused deconstructed matrix after acquiring the ultimate level of decaying pictures matrices. Using the fused decomposed matrix that was attain in step-4, rebuild a matrix using the discrete cosine transform's inverse Laplacian pyramid. Repetition of Step K, which was chosen in Step 1, is necessary to return to the starting level.

4 Result Analysis

For each time-domain signal segmentation at each frequency, the wavelet transform is performed independently. Multiple resolutions are obtained by analyzing the data at various frequencies. The purpose of multi-resolution inspection is to provide In addition to good time declaration and poor frequency accuracy at high frequencies, there is also good frequency accuracy and bad time declaration at low frequencies. Ideal for signals, such as pictures and video frames.

The analysis of data relative to scale is the core principle of wavelets. In fact, some wavelet researchers believe that employing wavelets requires a completely different perspective and way of thinking when it comes to data processing. But this scale we're going to employ to view the data has a specific function in wavelet analysis. Different sizes or resolutions of data are processed by wavelet algorithms.

Fusion Algorithm

The MATLAB software was actually utilized to create and execute the algorithm that follows.

1. Convert picture I1 and ascertain its size.

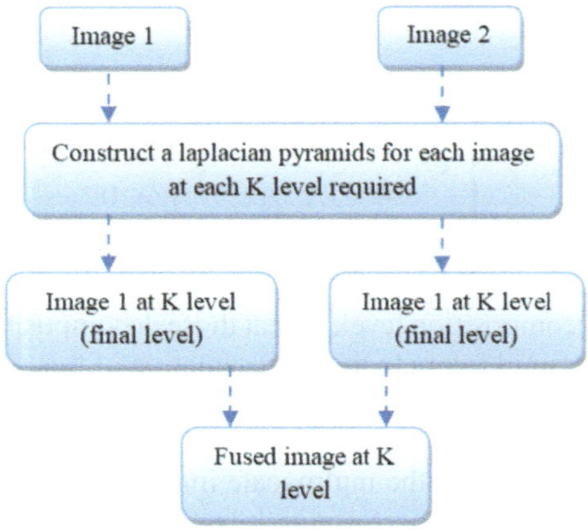

Fig. 3. DCTLP

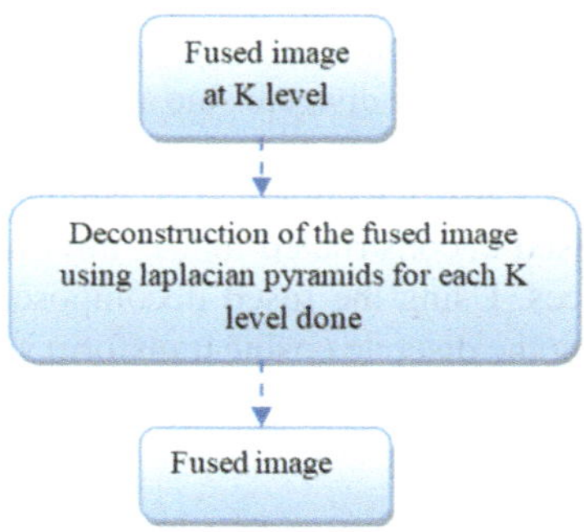

Fig. 4. Fused image deconstruction

2. Determine the size of pic. I2 by reading it.
3. Determine the size and match it, if not, make it the same. To run different wavelet operations, transfer the two pictures compared to grayscale to an indexed picture.
4. Execute the DWT to acquire the fused image after performing the discrete wavelet transform and Create the level-three approximation's coefficient matrices and the features on the diagonal, vertical, and horizontal axes.
5. Create and present estimates and information derived as of the coefficients. Use principle component analysis (PCA) to regenerate an image.
6. Use the second image in the same manner, Next, utilize the average, maximum, or minimum technique to fuse the wavelet coefficients. Produce the final fused wavelet coefficient matrix and tabulate every techniques values, finally finding the hybrid model

As shown the Fig. 5 is represented as input images are pictures from CT & MRI.

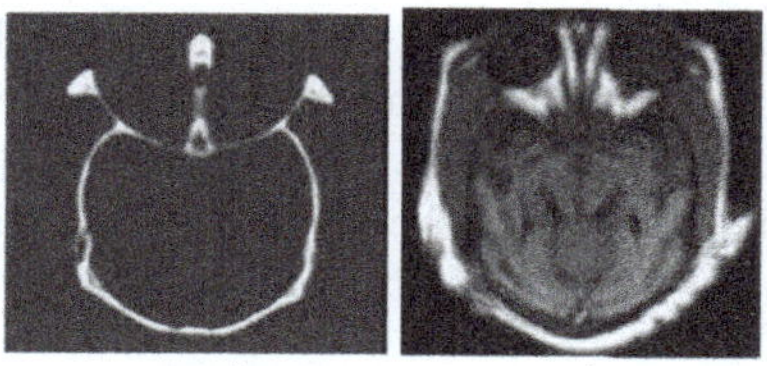

a) b)

Fig. 5. Given picture a) Picture from CT b) Picture from MRI

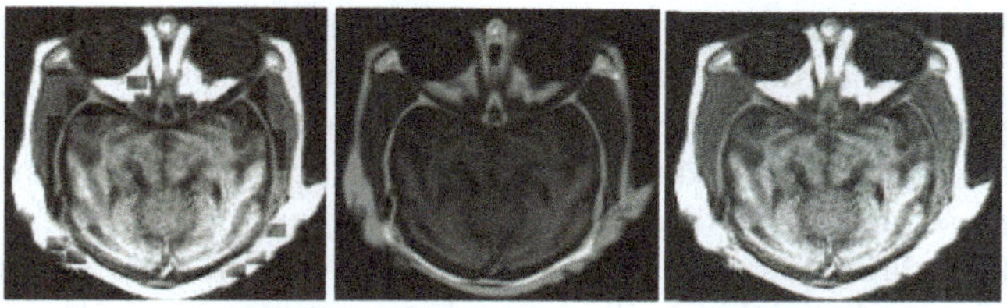

a) b) c)

Fig. 6. Output Images a) PCA fused images b) DCT fused images c) Hybrid image

As shown the Fig. 6 is represented as output images are PCA fused, DCT fused and hybrid images and shown the Fig. 7 is the performance levels of various input images or fusion models.

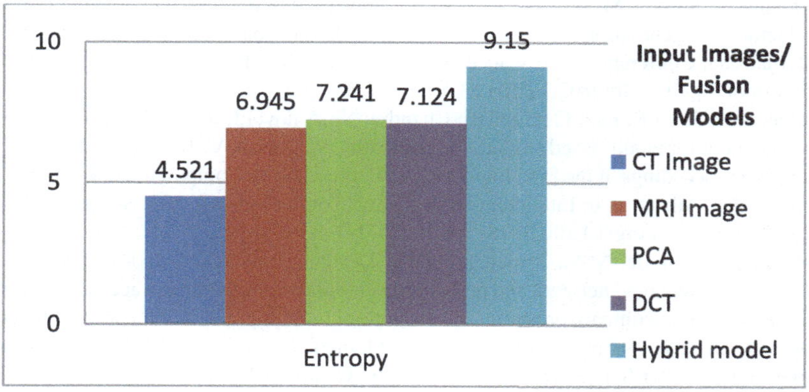

Fig. 7. Performance levels of various input images/fusion models

Despite the fact that the choice of fusion technique depends on the particular challenge, this overview finds that the spatial domain offers great spatial resolution. The spatial domain, however, suffers from image blurring. Wavelet transforms are an excellent method for fusing images together and producing spectral material of the highest caliber. However, since a well-fused image possesses mutually qualities, using the DWT and spatial domain fusion technique (such as the PCA fusion method) together enhances efficiency when contrasted with using the DWT and PCA algorithm separately. Ultimately, this analysis indicates so as toa combination of images method depending on a hybrid model of DCT and PCA combined by DWT determination enhance the quality of the merged images and could be the direction of future image fusion research.

5 Conclusion

MATLAB was used for the implementation of suggested medical picture fusion methods. Two input pictures, an MRI and a CT scan, were used to achieve the medical image fusion. We are utilizing principal component analysis and. The specific cosine transform with fusion as the goal. One component of the hybrid technique we are using to produce a higher-quality image is the discrete wavelet transform (DWT). The entropy (E) value was utilized to validate the fused image. This diagnostic image fusion approach makes utilizes the Transform through discrete transformAnalysis of Principal Components and discretetransform through cosine. This hybrid method uses the entropy value to produce images of higher quality. Additionally, several image fusion methods were used, with discrete wavelet transform producing superior outcomes with a high entropy value.

References

1. Surya, P.R.B., Panakala, R.K., Kumar, P.R.: Qualitative analysis of MRI and enhanced low dose CT scan image fusion. International Conference on Advanced Computing and Communication Systems (ICACCS -2017). Coimbatore, pp. 1752–1757 (2017)
2. Gothane, S., Sarode, M.V.: Analyzing factors, construction of dataset, estimating importance of factor, and generation of association rules for Indian road accident. In: 2016 IEEE 6th International Conference on Advanced Computing (IACC), Bhimavaram, pp. 15–18 (2016). https://doi.org/10.1109/IACC.2016.13
3. Rafeeq, M., Sunil Kumar, C., Subhash Chandra, N.: A novel framework for road traffic data management for cloud-based storage. In: Satapathy, S., Prasad, V., Rani, B., Udgata, S., Raju, K. (eds.) Proceedings of the First International Conference on Computational Intelligence and Informatics. Advances in Intelligent Systems and Computing, vol. 507. Springer, Singapore (2017). https://doi.org/10.1007/978-981-10-2471-9_9
4. Gothane, S., Sarode, M.V., Srujan Raju, K.: Design, construction and analysis of model dataset for indian road network and performing classification to estimate accuracy of different classifier with its comparison summary evaluation. In: Panigrahi, B., Suganthan, P., Das, S., Satapathy, S. (eds.) Swarm, Evolutionary, and Memetic Computing. SEMCCO 2015. Lecture Notes in Computer Science, vol. 9873. Springer, Cham (2016). https://doi.org/10.1007/978-3-319-48959-9_5
5. Devi, R.N., Saikumar, T.: Application of radon transform for image segmentation on level set method using HKFCM algorithm. In: Satapathy, S., Govardhan, A., Raju, K., Mandal, J. (eds.) Emerging ICT for Bridging the Future - Proceedings of the 49th Annual Convention of the Computer Society of India (CSI) Volume 1. Advances in Intelligent Systems and Computing, vol. 337. Springer, Cham (2015). https://doi.org/10.1007/978-3-319-13728-5_29
6. Kumar, T.S., Swapna, C.: Segmentation of image with DT-CWT and NLM filtering using fuzzy c-means clustering. In: 2014 IEEE International Conference on Computational Intelligence and Computing Research, Coimbatore, pp. 1–4 (2014). https://doi.org/10.1109/ICCIC.2014.7238367

7. Raju, K.S., Sreelatha, K., Kumari, S.: An efficient approach to improve retrieval rate in content based image retrieval using MPEG-7 features. In: Satapathy, S., Avadhani, P., Udgata, S., Lakshminarayana, S. (eds.) ICT and Critical Infrastructure: Proceedings of the 48th Annual Convention of Computer Society of India- Vol II. Advances in Intelligent Systems and Computing, vol. 249. Springer, Cham (2014). https://doi.org/10.1007/978-3-319-03095-1_37
8. Saikumar, T., Preetam, I.N.: Optimized kernel fuzzy c means (OKFCM) clustering algorithm on level set method for noisy images. In: 2013 IEEE International Conference on Computational Intelligence and Computing Research, Enathi, pp. 1–5 (2013). https://doi.org/10.1109/ICCIC.2013.6724290
9. Selim, M., Kudo, H., Rashed, E.A.: Low-dose CT image reconstruction method withprobabilistic atlas prior. 978-1-4673-9862-6/15/©2015 IEEE
10. Nithyananda, C.R., Ramachandra, A.C., Preethi.: Survey on histogram equalization method based image enhancement techniques. In: 2016 International Conference on Data Mining and Advanced Computing (SAPIENCE), Ernakulam, pp. 150–158 (2016)
11. Tiwari, R., Chaudhary, D., Diwan, T.D., Sharma, P.C.: Privacy and security solution in wireless sensor network for IoT in healthcare system. Next Generation Healthcare Systems Using Soft Computing Techniques, by CRC Boca Raton, p 123–135. ISBN: 978-1-03210-797-4
12. Headlee, J.M., Balster, E.J., Turri, W.F.: A no-reference image enhancement quality metric and fusion technique. In: 2015 International Conference on Image and Vision Computing New Zealand (IVCNZ), Auckland, pp. 1–6 (2015)
13. Huang, L.: Research on LK optical flow algorithm with gaussian pyramid model based on OpenCV for single target tracking. IOP Conf. Ser. Mater. Sci. Eng. **435**(1) (2018)
14. Arul, U., Prasath, A.A., Mishra, S., Shirisha, J.: IoT and machine learning technology based smart shopping system. In: 2022 International Conference on Power, Energy, Control and Transmission Systems (ICPECTS), Chennai, pp. 1–3 (2022)
15. Narasimharao, J., Priyanka Chowdary, P., Reddy, A. R., Swathi, G., Deepak Kumar, B.P., Batchu, S.S.: Satellite ortho image mosaic process quality verification. In:International Conference on Frontiers of Intelligent Computing: Theory and Applications, pp. 309–318.: Springer, Singapore (2023)

Evaluating an Innovative Approach for Enhancing the Protection of Medical Images Using Encryption and QR Encoding

M. Naveen Kumar[1], T. Sarika[1], Vivekanand Aelgani[2], Abdul Subhani Shaik[2], Jonnadula Narasimharao[3](✉), and Bushra Tarannum[3]

[1] CMR Engineering College, Hyderabad, Telangana, India
{naveenmds,Sarika.tuniki}@cmrec.ac.in

[2] CMR College of Engineering and Technology, Hyderabad, Telangana, India
{avivekanand,sasubhani}@cmrcet.ac.in

[3] CMR Technical Campus, Hyderabad, Telangana, India
jonnadula.narasimharao@gmail.com

Abstract. The preservation of privacy of information is of utmost importance in the realm of telemedicine applications in order to ensure successful outcomes. The utilization of encryption and steganography methods is implemented in order to protect the privacy and reliability of medical photographs. QR codes are utilized as a supplementary security measure due to their ability to encode a greater amount of information compared to conventional barcodes. The utilization of the RSA encryption technique is employed for the purpose of encrypting medical photographs, subsequently integrating them with a cover image in order to generate a stego image. The steganographic image is subsequently subjected to QR encoding in order to bolster the system's resistance against malevolent attacks. The retrieval of the original medical image might be accomplished by a series of steps including the reversal of the process. This entails the decoding of the QR code, acquisition of the cipher image, and subsequent decryption utilizing the RSA algorithm. The complete procedure is executed utilizing the Python programming language.

Keywords: Steganography · Quick Response Code · RSA Encryption and Decryption

1 Introduction

The utilization of medical imaging has turned into an indispensable tool for healthcare practitioners, investigators, and remote collaboration. Nevertheless, the transmission of medical images across unsecure communication channels, as observed in tele-health applications, gives rise to apprehensions regarding the security and confidentiality of those images. The occurrence of a breach in security or authenticity has the potential to result in erroneous diagnoses, hence emphasizing the utmost need of safeguarding medical images. Several methodologies are employed in the realm of medical picture security, encompassing cryptography, a technique called and watermarking. The primary

D. Kriksciuniene et al. (Eds.): IBICA 2023, LNNS 1228, pp. 372–378, 2025.
https://doi.org/10.1007/978-3-031-78937-3_40

objective of our investigation is to address the issue of communication barriers faced by those having impairments, including those who are blind, deaf, or mute. This will be achieved through the creation and implementation of wearable technology.

Within the healthcare industry, there exist well-defined benchmarks such as the digital imaging and communications in medicine (DICOM) that offer an organized structure and methods to ensure the preservation of confidentiality, authenticity, and integrity throughout the transfer of medical pictures. The purpose of these standards is to guarantee the secure transmission of medical pictures, thereby providing an extra level of safeguard against unwanted access or alteration. By following these guidelines, healthcare workers are able to be comfortable that patient data is kept safe and private throughout the transfer of medical pictures.

The phrase "QR code" is a contemporary designation derived from "Quick Response code," and it has gained significant recognition as a trademark in several business and general contexts. The QR code was initially developed as a two-dimensional matrix for use in the car sector in Japan. The technology has the ability to capture and store data pertaining to an object through the utilization of a label which can be read by a machine and affixed to said object. The QR code is capable of accommodating four specified modes of data, namely alphanumeric, byte or binary, numeric, and picture. These modes enable the storage and encoding of many types of data within the QR code.

2 Literature Review

The implementation of a security system through Kalaichelvi et al. incorporates the use of CAPTCHA codes as a means of authentication for the person who was intended, effectively mitigating the risk of unwanted access by automated systems. The system employs a randomly generated CAPTCHA code that creates a safe connection through an authenticated user. Additionally, it incorporates Image Steganography techniques to ensure secrecy. The initial communication is conveyed using the Least Significant Bit (LSB) technique, which leverages the Red-Green-Blue (RGB) color spectrum for the purpose of concealing the image data. This approach offers an extra level of encryption [1].

Pachiappan and colleagues put forth The utilization of a visual cryptography technology restricts the sharing of photographs to an only two shares. The embedded picture effectively generates two shared images and provides enhanced security compared to conventional visual cryptography [2]. The decryption process entails the acquisition of requisite parameter data by approved individuals in order to decrypt the picture data and get the original medical image. The aforementioned methodology offers heightened security protocols in order to deter illegal entry into confidential medical data [3]. In their study, Soualmi et al. presented a novel approach for blind watermarking of medical images. This approach encompasses two primary procedures: the embedding of watermark data within the DWT-Schur coefficients, followed by the extraction of the watermark utilizing identical techniques as the embedding procedure. The approach underwent experimental evaluation and shown its ability to retain robustness despite maintaining excellent image quality [4]. In their study, Maurya et al. propose an advanced Extended Visual Cryptography Technique (EVCT) as a means of safeguarding medical photographs. The encryption and encoding techniques employed in this procedure are characterized by their lossless nature and low complexity, hence enhancing the overall efficiency of the approach [5]. The technique leverages the Singular Value Decomposition (SVD) within the Discrete Wavelet Transform (DWT) domain. This methodology provides improved security safeguards for the protection of medical pictures from illegal access and alteration [6]. The approach proposed by Srivastava et al. entails encrypting the personally identifiable portion of a photograph by including it into non-prominent pixel positions inside the image itself, so guaranteeing a safe transport of the image. In order to accomplish the process of encryption and subsequently recovery, a seeding one-time pad is employed [7]. The system utilizes a Microcontroller, namely the Raspberry Pi model, which incorporates a web interface. This interface enables users to remotely run the system from their respective locations, utilizing the necessary functionalities required for system operation [8]. In their study, Salameh et al. introduced a robust system designed to effectively handle medical images and patient data by implementing a sharing methodology. The system produces a pair of shares, wherein the first share denotes the encrypted medical image utilizing MJEA, and the second share encompasses the entirety of the patient's information that is embedded within the medical image and encrypted using MJEA [9]. The algorithm proposed by Abd El-Latif et al. utilizes gray code and a chaotic map as encryption techniques for quantum pictures. The aforementioned methodology guarantees a resilient and effective encryption procedure for quantum pictures, beyond the capabilities of conventional encryption techniques [10–13].

3 Proposed Model

Ensuring the safeguarding of medical images is of paramount importance within the healthcare sector, as it serves to uphold the principles of confidentially and integrity. This is particularly crucial whenever transmitting images over vulnerable communication channels, such as telehealth applications. The objective of the suggested system is to offer a reliable and safe solution through the utilization of encryption techniques and steganography. The existing system employs one-time padding and steganography methodologies; nonetheless, it possesses inherent constraints that render it inadequate for contemporary applications. Hence, the system under consideration utilizes the RSA encryption technique as a means to address these constraints.

The RSA encryption technique is employed as a means of encrypting the medical image during the initial phase, resulting in the production of a cipher text. The aforementioned ciphertext is subsequently incorporated into a cover picture through the utilization of LSB steganography. This technique involves substituting the least significant bits (LSB) of the cover image with the most significant bits (MSB) of the input image. The aforementioned procedure involves the creation of a Stego Image, which is thereafter subjected to encoding in order to produce a QR code. The integration of QR codes enhances the user experience and facilitates the accessibility of medical images for healthcare professionals in need of such resources. The suggested system offers a comprehensive solution for safeguarding the confidentiality and integrity of medical pictures through the utilization of modern encryption techniques and steganography. The used system utilizes the RSA encryption algorithm, which effectively mitigates the constraints associated with the one-time pad approach, hence enhancing its practicality and efficacy. Furthermore, the employment of steganography guarantees the safeguarding of image data, so preventing unauthorized individuals from accessing it.

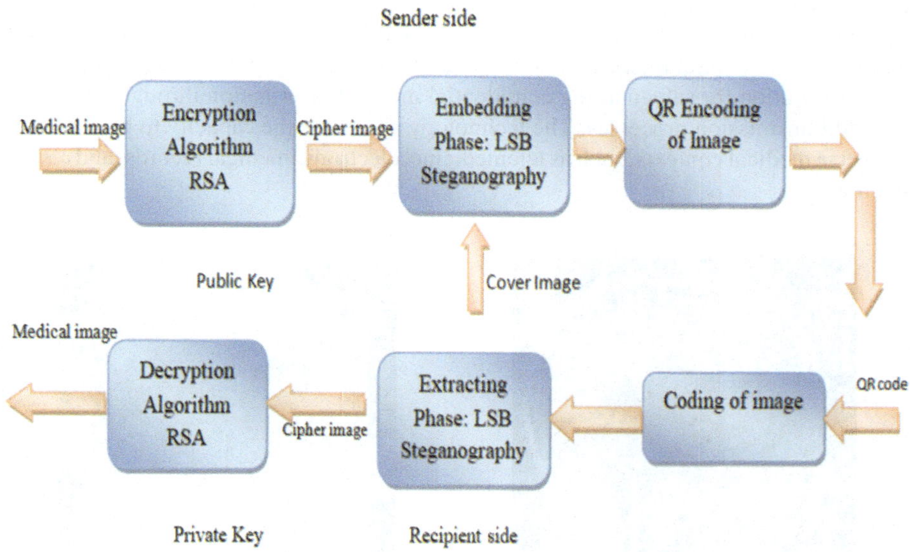

Fig. 1. The integration of encryption, steganography, and QR encoding

In general, the suggested system presents a comprehensive and dependable solution for safeguarding medical photographs, a matter of utmost importance within the healthcare sector. The utilization of sophisticated methods of encryption and steganography by the platform guarantees the preservation of confidentiality and security for the photographs. Additionally, the incorporation of QR codes enhances the user-friendliness and accessibility of the system as shown the Fig. 1, particularly for health care providers who necessitate accessibility to the images.

4 Results and Discussions

The utilization of encrypting and QR encoding as a means to safeguard medical photographs represents a highly auspicious approach that holds significant potential in augmenting the confidentiality and integrity of delicate patient information. The RSA encryption technique is of crucial importance in ensuring the security and integrity of medical images, since it effectively protects them from illegal access and alteration throughout the processes of storage and transmission. Furthermore, the utilization of Least Significant Bit (LSB) steganography serves to obfuscate the encrypted image by embedding it under a cover image, hence augmenting the level of security and rendering it arduous for unauthorized entities to discern as shown the Fig. 2 and 3. In addition, the act of embedding the Stego Image within a QR code serves to augment its level of security, as QR codes are extensively employed and can be effortlessly scanned utilizing mobile devices. The aforementioned methodology has demonstrated its efficacy and efficiency in the encryption and encoding of medical images through the utilization of QR codes. The decryption procedure is designed to be extremely safe, necessitating the utilization of a private key in order to access the original medical image. This stringent measure guarantees that solely authorized individuals are able to obtain and view the highly confidential medical data. In general, this strategy presents a pragmatic and viable resolution for safeguarding confidential medical information throughout its storage and transmission processes. The technology possesses the capacity to be employed in diverse medical contexts, such as telemedicine, electronic medical records, and remote diagnosis.

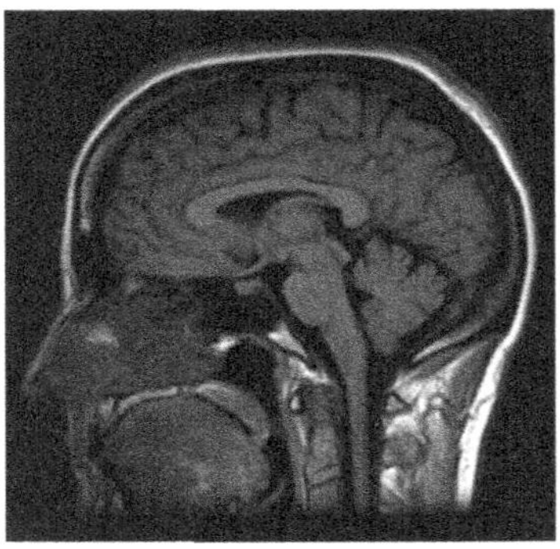

Fig. 2. Input Image

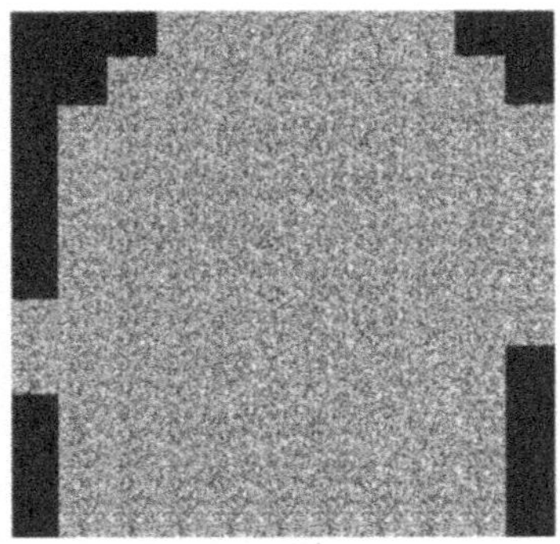

Fig. 3. Cipher Image

Fig. 4. Stego Image

Fig. 5. Encoded Medical Image

The procedure of integrating the cipher image into a cover image and subsequently encoding it within a QR code is deemed to be dependable, proficient, and confers an augmented stratum of safeguarding to the encrypted image as shown the Fig. 4. The utilization of a private key for the purpose of decrypting the cipher image and retrieving the original medical image ensures that only authorized individuals possess the capability to access the medical data as seen the Fig. 5. Subsequent investigations may delve into potential strategies for augmenting the efficacy and dependability of the encryption and encoding procedures. Furthermore, it is recommended to do additional investigations in order to assess the scalability, compatibility, and ease of implementation of the strategy in various healthcare systems. Moreover, this methodology can be expanded to encompass additional medical applications, including the safeguarding of medical records and the preservation of patient confidentiality. The implementation of encryption and QR encoding presents a viable approach to preserving confidentiality and safeguarding data integrity within the healthcare sector. This technological solution facilitates the successful operation of telemedicine, remote diagnosis, and other digital healthcare services, hence fostering their growth and effectiveness. The utilization of encryption and QR encoding to secure medical images holds considerable promise in transforming healthcare procedures, safeguarding confidential medical information, and enhancing patient outcomes.

5 Conclusion

In summary, the utilization of encryption and QR encoding as a means to safeguard medical images has considerable potential as an effective strategy to implement robust security protocols for the protection of sensitive medical information. The implementation of RSA encryption and LSB steganography techniques provides a robust safeguard for preserving the confidentiality and integrity of medical pictures throughout their storage and transmission processes, hence preventing illegal access and modification.

References

1. Kalaichelvi, T., Apuroop, P.: Image steganography method to achieve confidentiality using CAPTCHA for authentication. In: 2020 5th International Conference on Communication and Electronics Systems (ICCES), pp. 495–499. IEEE (2020)
2. Pachiappan, K., Annaji, S., Nithya, J.: Security in medical images using enhanced visual secret sharing scheme. Int. J. Sci. Eng. Technol. Res. **3**(9), 1642–1645 (2014)
3. Jiang, T., Zhang, K., Tang, J.: Securing medical images for mobile health systems using a combined approach of encryption and steganography. In: International Conference on Intelligent Computing, pp. 532–543. Springer, Cham (2018)
4. Soualmi, A., Alti, A., Laouamer, L.: A blind image watermarking method for personal medical data security. In: 2019 International Conference on Networking and Advanced Systems (ICNAS), pp. 1–5. IEEE (2019)
5. Maurya, R., Kannojiya, A.K., Rajitha, B.: An extended visual cryptography technique for medical image security. In: 2020 2nd International Conference on Innovative Mechanisms for Industry Applications (ICIMIA), pp. 415–421. IEEE (2020)
6. Al-Haj, A., Mohammad, A., Amer, A.: Crypto-watermarking of transmitted medical images. J. Digit. Imaging **30**(1), 26–38 (2017)
7. Srivastava, A., Awasthi, S.K., Javed, S., Gautam, S., Kishore, N., Bakthula, R.: Seeded one time pad for security of medical images in health information. In 2018 4th International Conference on Computing Communication and Automation (ICCCA), pp. 1–6. IEEE (2018)
8. Han, B., Jia, Y., Huang, G., Cai, L.: A medical image encryption algorithm based on hermite chaotic neural network. In: 2020 IEEE 4th 22 Information Technology, Networking, Electronic and Automation Control Conference (ITNEC), vol. 1, pp. 2644–2648. IEEE (2020)
9. Salameh, J.N.B.: A new approach for securing medical images and Patient's information by using a hybrid system. Int. J. Comput. Sci. Netw. Secur. **19**(4), 28–39 (2019)

10. Abd El-Latif, A.A., Abd-El-Atty, B., Talha, M.: Robust encryption of quantum medical images. IEEE Access **6**, 1073–1081 (2017)
11. Shaik, A.S., Ahmed, M., Suresh, M.: Low-cost irrigation and laser fencing surveillance system for the paddy fields. In: Chakravarthy, V.V.S.S.S., Flores-Fuentes, W., Bhateja, V., Biswal, B. (eds.) Advances in Micro-Electronics, Embedded Systems and IoT. LNEE, vol. 838. Springer, Singapore (2022). https://doi.org/10.1007/978-981-16-8550-7_49
12. Shaik, A.S., Karsh, R.K., Islam, M.: Robust image hashing using chromatic channel. In: Nath, V., Mandal, J.K. (eds.) Proceeding of Fifth International Conference on Microelectronics, Computing and Communication Systems. Lecture Notes in Electrical Engineering, vol. 748. Springer, Singapore (2021). https://doi.org/10.1007/978-981-16-0275-7_5
13. Suresh, M., Shaik, A.S., Premalatha, B., Narayana, V.A., Ghinea, G.: Intelligent & smart navigation system for visually impaired friends. In: Garg, D., Narayana, V.A., Suganthan, P.N., Anguera, J., Koppula, V.K., Gupta, S.K. (eds.) Advanced Computing. IACC 2022. Communications in Computer and Information Science, vol. 1781. Springer, Cham (2023). https://doi.org/10.1007/978-3-031-35641-4_30

A Comparative Investigation on the Assessment of Computer Vision Techniques for the Recognition of Cervical Cancer

K. Madhavilatha[1], Sripriya[2], Vivekanand Aelgani[3], V. A. Narayana[3], C. Abhinav[4], and K. Srinivas[5](✉)

[1] Department of CSE (DS), CMR Engineering College, Hyderabad, Telangana, India
madhavilathak@cmrec.ac.in
[2] Department of CSE, CMR Engineering College, Hyderabad, Telangana, India
[3] Department of CSE, CMR College of Engineering and Technology, Hyderabad, Telangana, India
avivekanand@cmrcet.ac.in, vanarayana@cmrcet.org
[4] Department of CSE, CMR Technical Campus, Hyderabad, Telangana, India
vicechairman@cmrtc.ac.in
[5] Department of CSE (Data Science), CMR Technical Campus, Hyderabad, Telangana, India
cmrtc.paper@gmail.com

Abstract. Currently, the majority of whole-slide image categorization algorithms heavily depend on manually adding annotations at the pixel level. This process is both complicated and laborious, requiring annotators with specific knowledge in the subject matter. In order to overcome this issue, we suggest the utilization of self-supervised learning as well as multiple instances learning techniques for the management of huge whole slide imaging (WSI) datasets, where the only available labels are the reported diagnoses. In this study, we employed various machine learning techniques, namely K-Nearest Neighbors (KNN), Support Vector Machine (SVM), and Convolutional Neural Network (CNN) with the specific architecture of Alexnet. Our findings indicate that the accuracy of CNN and Alexnet surpasses that of KNN and SVM. Furthermore, the characteristics learned by CNN and Alexnet demonstrate superior suitability for classification tasks. Presently, the majority of classification models for entire slide images depend on human pixel-level annotations, necessitating the involvement of domain specialists who possess specialized knowledge and expertise. This annotation process is intricate and demands a significant amount of time and effort. In order to address this issue, we suggest integrating self-supervised learning using various instance learning as a means of effectively handling extensive whole slide imaging (WSI) datasets, utilizing solely the provided diagnosis as labels. The classification problem in WSI poses a significant difficulty in acquiring effective picture representation, with self-supervised learning demonstrating considerable promise in this regard. In this research endeavor, we intend to employ a self-supervised learning network known as Bootstrap. The pre-trained network known as Your Own Latent has the capability to be developed with unlabeled input and acquire profound domain-specific properties. The suggested framework was assessed on a large-scale uterine cervical dataset consisting of 3,063 complete slide pictures.

D. Kriksciuniene et al. (Eds.): IBICA 2023, LNNS 1228, pp. 379–387, 2025.
https://doi.org/10.1007/978-3-031-78937-3_41

The findings of our study indicate that the integration of a self-supervised learning model with a multiple occurrence learning model has demonstrated the ability to achieve and surpass the success of previous methodologies.

Keywords: Convolutional Neural Network (CNN) · K-Nearest Neighbors (KNN) · Support Vector Machine · Cervical Cancer · whole slide imaging (WSI) datasets

1 Introduction

The advent of computerized pathology has ushered in an entirely novel phase in the field of medical image processing, particularly for diagnostic applications. Histopathology images, commonly referred to as whole slide images (WSIs), usually come accompanied by pertinent details regarding the location and classification of illnesses and malignancies. The rapid progress in digital technology has facilitated the efficient digitization of tissue slides, resulting in the production of high-resolution whole slide images (WSIs). Consequently, there exists a significant increase in the number of whole slide imaging (WSI) archives within healthcare facilities, such as hospitals and clinics. Therefore, it is increasingly evident that there is a need for rapid assessment of whole slide images (WSIs) in order to effectively handle the pressing demands of contemporary pathology workflows.

Cervical cancer is a malignancy that arises in the cervix, which serves as the gateway connecting the vagina to the uterus in women. Cervical cancer ranks as the fourth most prevalent form of cancer among women. According to global statistics from 2018, over 570,000 women received a diagnosis of cervical cancer, resulting in approximately 311,000 fatalities. The implementation of initial prevention strategies, such as the administration of the human papillomavirus (HPV) vaccination, along with the adoption of secondary prevention measures, including the screening and treatment of precancerous lesions, has been shown to be very efficacious in the prevention of the majority of cervical cancer cases.

2 Literature Survey

In the study [1] put out a methodology that involves employing colour-texture features using an ensemble classifier architecture to classify histopathology images related to breast cancer. In our study, we have successfully proved the efficacy of employing several characteristics and methods of classification. Based on the conducted experiments, it is evident that the inclusion of appropriate characteristics throughout an ensemble categorization structure enables the categorization to exhibit a significant degree of magnification invariance, particularly for certain magnification factors.

In the study [2] introduced a two-stage strategy that employs a sequential approach. In this approach, the second classifier ensemble is exclusively tasked with handling the patterns that were rejected by the first classifier ensemble, [3] identified various factors that contribute to the distinct manifestation of the disease in Asian women compared to women from other regions. These factors include deficiencies in health care infrastructure and standards, sociocultural barriers, economic constraints, low levels of literacy, and variations in the clinical and pathological characteristics of the disease. The good clinical picture and outcomes observed in developed Asian countries can be attributed to improved socio-economic situations and increased health knowledge regarding breast cancer screening, [4, 5] put forth the notion that the neural network models had superior predictive capabilities in terms of sensitivity and specificity when compared to logistic regression, [6] proposed an innovative framework for using the collective intelligence of crowds in the process of learning. The newly developed multi-scale CNN Agg Net incorporates an aggregation component directly into the learning process by integrating an additional crowd sourcing layer.

The validity of this model among Chinese women was tested by [7], utilizing data from the Shanghai Women's Health Study (SWHS). The SWHS refers to a cohort including 74,942 adult Chinese women residing in urban areas. These women were recruited between 1997 and 2000 and undergo in-person interviews every two years to gather personal and health-related data [8–10] their findings revealed that 96% of melanoma cases exhibited a score of 2 for asymmetry, with both axes representing asymmetry. The concept of symmetry distance (SD) was proposed as an additional metric for evaluating photographs, [11–13] introduced a system that utilizes image analysis techniques to build a diagnostic method for the automatic detection of melanomas. Thresholding and filtering techniques are employed in order to mitigate noise by utilizing the [14–16] pixel intensities. The determination of the threshold value at which pixels are classified as noise in a picture is accomplished by utilizing the intensity histogram of the image. The Otsu technique, as an illustrative instance, computes an ideal threshold that minimizes the variation within each class.

It is projected that by the year 2040, cervical cancer will result in roughly 460,000 fatalities annually, with Sub-Saharan African women accounting for approximately 90% of these deaths. The World Health Organization (WHO) has identified cervical cancer as a pressing concern in Africa due to its escalating prevalence. Consequently, the organization has prioritized efforts in screening, diagnosis, and treatment of this disease. Traditionally, the diagnosis of cancer heavily depends on histological evaluation, a method that is prone to errors. To address this issue and enhance patient safety, intelligent computer-aided systems have been developed as cost-effective mechanisms. However, the lack of labeled data in digital pathology has a limitation on the practicality and widespread use of these systems. This investigation involved the pre-processing of a limited number of cervical tissue digital slides obtained from the TCGA data site. The purpose of this pre-processing was to address the challenges associated with whole-slide images. These pre-processed slides were then incorporated into our proposed VGG16-CNN classification strategy, as described in reference [12].

This study aims to conduct a comparative analysis of machine learning and deep learning algorithms in the context of cervical cancer detection. Deep learning techniques are expected to provide superior effectiveness in the detection of many types of malignancies. The domain in which deep neural networks demonstrate the highest effectiveness in medical image analysis is the classification of images. The primary focus of our project centers on conducting a comparative analysis between machine learning and deep learning methodologies. This research aims to utilize deep learning algorithms for the purpose of detecting cervical cancer, hence providing reliable readings.The concept of motivation refers to the psychological force that drives individuals to initiate and sustain certain behavior. In recent times, the application of deep learning techniques has significantly improved the precision of early-stage cervical cancer prediction. Cervical cancer stands as a prominent contributor to early mortality among women on a global scale, with over 85% of these fatalities concentrated in poor nations. Cervical cancer is linked to certain risk factors.

3 Proposed System

The system under consideration comprises the utilization of machine learning methodologies, specifically the K-Nearest Neighbors (KNN) and Support Vector Machine (SVM) techniques. Machine learning facilitates the enhancement of computer systems through the creation of algorithms that enable automated data access and task execution based on predictions and detections. Machine learning refers to the capacity of a system to autonomously adapt to novel data by use of iterative processes. Machine learning plays a pivotal role within the expanding domain of data science. Statistical procedures are employed to train algorithms in order to perform categorization or predictions, as well as to reveal significant discoveries in data mining endeavors. Machine learning algorithms construct a predictive or decision-making model by utilizing a set of sample data, referred to as training data, without the need for explicit programming.

The machine effectively fulfills the objective of acquiring knowledge from data through the utilization of certain inputs. The initial step in the machine learning process involves the input of training data into the chosen algorithm. The algorithm for machine learning is provided with new input data in order to assess its accuracy. Subsequently, the algorithm's predictions and outcomes are compared for validation purposes. This facilitates the autonomous learning of the machine learning algorithm, leading to the steady enhancement of precision and the production of optimal outcomes.

Block Diagram of Proposed System

In the context of supervised learning, the utilization of pre-existing or annotated data is employed for the purpose of training. Given the availability of data, the learning process can be classified as supervised, indicating that it is guided towards achieving successful outcomes. The input data is subjected to the machine learning algorithm in order to facilitate the training of the model as shown the Fig. 1. After the model has been trained using the available data, it becomes possible to input unfamiliar data into the model and obtain a novel result. In this particular scenario, the model endeavors to ascertain the extent to which the data aligns with the desired outcome. Once the model has undergone sufficient training, it will accurately recognize the data as a desired outcome and generate the intended answer. The suggested system incorporates the utilization of machine learning algorithms, specifically K-Nearest Neighbors (KNN) and Support Vector Machines (SVM) as shown the Fig. 2. The input image is derived from human bodily secretions and is subjected to microscopic examination. Prior to analysis, the

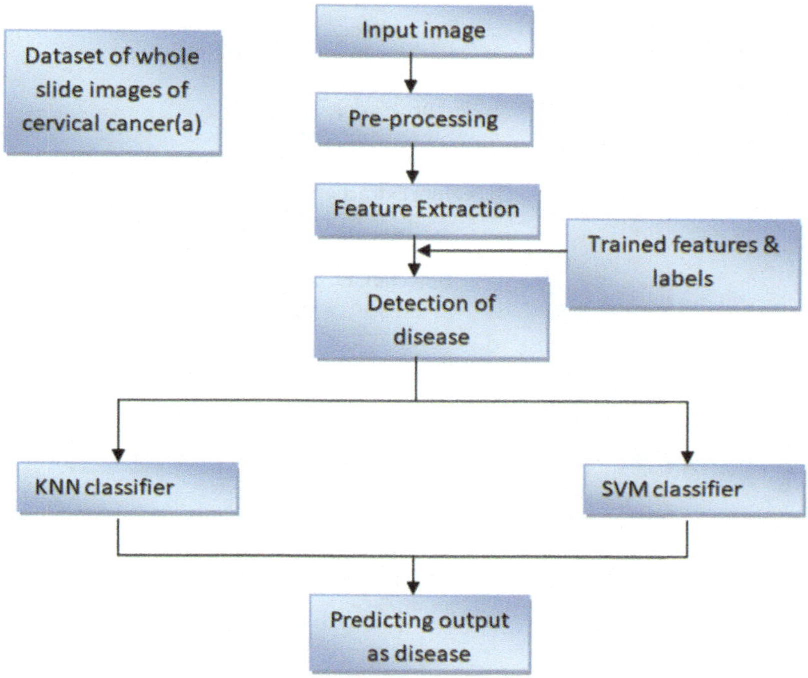

Fig. 1. Block diagram of proposed system

input image undergoes pre-processing techniques to remove any noise or disruptions present. Subsequently, features are extracted from the image, which are then compared with trained data in order to obtain the desired output.

Results and Analysis

Figure 3 depicts an input image including Cancerous Cells. Cervical cancer detection can be facilitated by employing machine learning methods such as K-Nearest Neighbors (KNN) and Support Vector Machines (SVM). The efficiency of these models may also be measured by analyzing the input image.

Figure 4 illustrates the accuracy levels of the different classifiers.

The accuracy of the K-Nearest Neighbors (KNN) algorithm is a crucial aspect to consider while evaluating its performance.

The presence of koilocytotic cells in the human body is indicative of cervical cancer. In the context of utilizing a K-nearest neighbors (KNN) classifier for the analysis of input images, the identification of koilocytotic features serves as an indication of the presence of cervical cancer within the human body. The accuracy of the K-nearest neighbors (KNN) classifier for the given input image is being assessed.15.21 as shown the Fig. 4.

The accuracy of Support Vector Machines (SVM) is a crucial aspect to evaluate its performance in classification tasks. The presence of koilocytotic cells in the human body is indicative of cervical cancer. In the context of koilocytotic detection, the utilization of a Support Vector Machine (SVM) classifier indicates the presence of cervical cancer

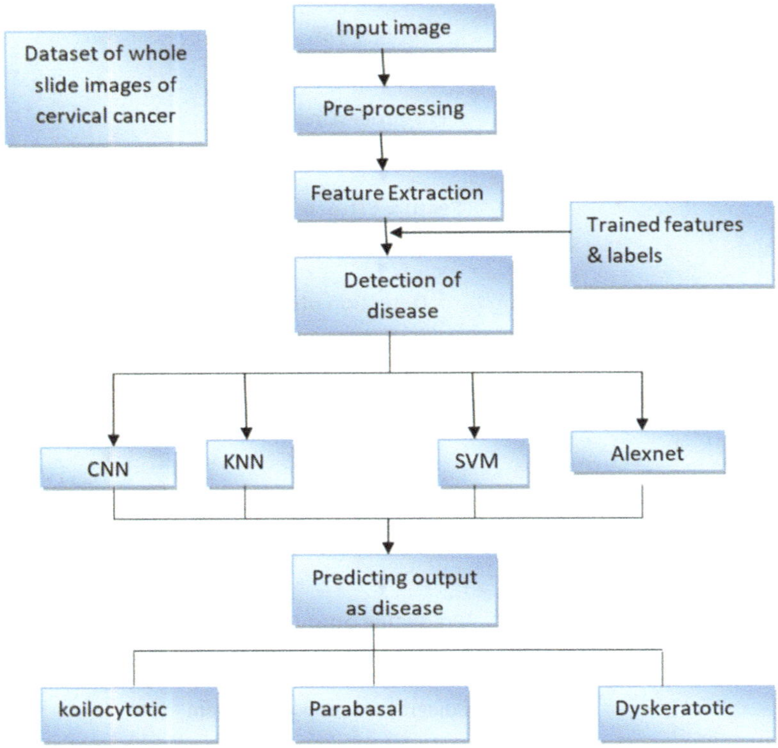

Fig. 2. Predicting disease various machine learning classifiers

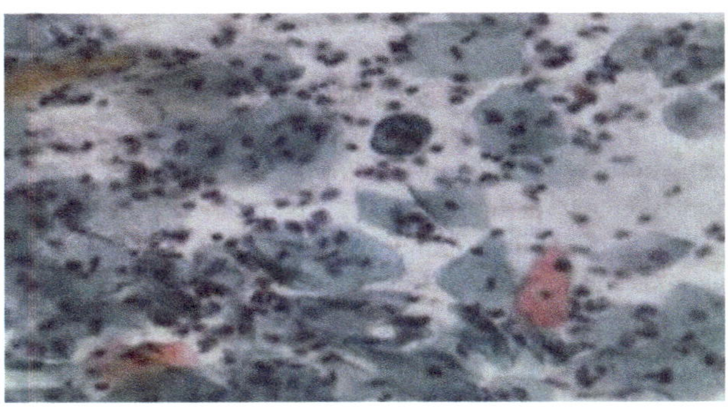

Fig. 3. Input Image

within the human body. The SVM classifier exhibits an accuracy of 33.82% for the given input image. The accuracy of Convolutional Neural Networks (CNNs) is a topic of interest in the field of machine learning. The presence of koilocytotic cells in the human body is indicative of the presence of cervical cancer. In the context of utilizing a CNN classifier for the analysis of input images, the identification of koilocytotic cells

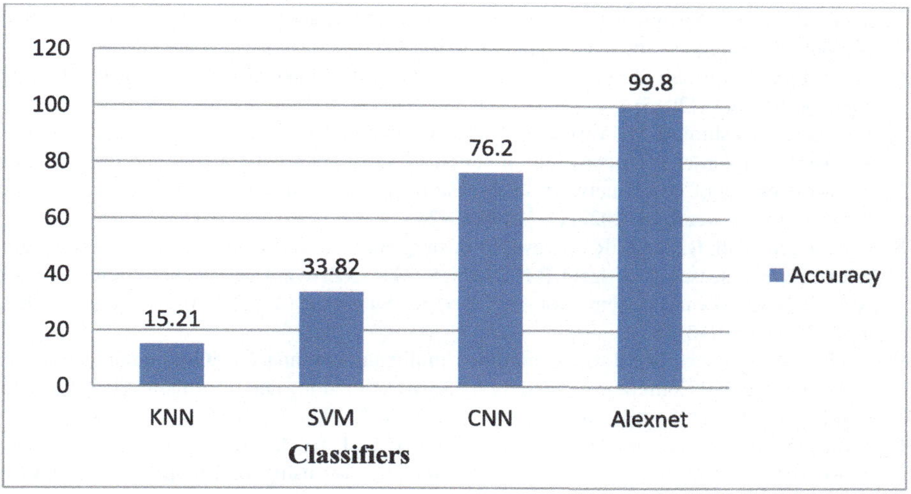

Fig. 4. Accuracy levels of various classifiers

signifies the presence of cervical cancer within the human body. The CNN classifier demonstrates an accuracy of 76.2% for the given input. In the context of utilizing the Alexnet classifier, the detection of koilocytotic features in an input image indicates the presence of cervical cancer within the human body. The accuracy of the AlexNet classifier for the given input image is being evaluated 99.80.

4 Conclusion

The integration of self-supervised learning techniques and multi-instance learning approaches is employed to address the challenges posed by bigger whole slide imaging (WSI) datasets. The Deep Neural Network, specifically AlexNet, and Convolutional Neural Network (CNN) exhibit superior performance in comparison to traditional Machine Learning approaches such as K-Nearest Neighbors (KNN) and Support Vector Machines (SVM). The suggested network demonstrates efficacy in handling substantial datasets, enabling the detection of cervical cancer as well as other disorders.

References

1. Choudhury, T., Kumar, V., Nigam, D.: Cancer research through the help of soft computing techniques: a survey. IJCSMC **2**(4), 467–477 (2013)
2. Choudhury, T., Kumar, V., Nigam, D.: An innovative and automatic lung and oral cancer classification using soft computing techniques. IJCSMC **4**(12), 313–323 (2015)
3. Elgamal, M.: Automatic skin cancer images classification. Int. J. Adv. Comput. Sci. Appl. **4**(3) (2013)
4. El-Hasnony, I.M., El-Bakry, H.M., Saleh, A.A.: Classification of breast cancer using softcomputing techniques. Int. J. Electron. Inf. Eng. **4**(1), 45–54 (2016)

5. Sivakrithika, V., Shanthi, B.: Comparative study on cancer image diagnosis using soft computing techniques. Int. J. Comput. Appl. **975**, 8887 (2011)
6. Singhal, E., Tiwari, S.: Skin cancer detection using arificial neural network. Int. J. Adv. Res. Comput. Sci. **6**(1) (2015)
7. Jain, D.K., Lakshmi, K.M., Varma, K.P., Ramachandran, M., Bharati, S.: Lung cancer detection based on kernel PCA-convolution neural network feature extraction and classification by fast deep belief neural network in disease management using multimedia data sources. Comput. Intell. Neurosci. **2022**(1), 3149406 (2022)
8. Patel, P., Sivaiah, B., Patel, R.: Relevance of frequent pattern (FP)-growth-based association rules on liver diseases. In: Udgata, S.K., Sethi, S., Gao, XZ. (eds.) Intelligent Systems. Lecture Notes in Networks and Systems, vol. 431. Springer, Singapore (2022). https://doi.org/10.1007/978-981-19-0901-6_58
9. Muduli, A., Gayatri, M.S.L.S.: A modified multiband antenna for 5G communication. In: 2022 International Conference on Emerging Smart Computing and Informatics (ESCI), Pune, India, pp. 1–5 (2022). https://doi.org/10.1109/ESCI53509.2022.9758299
10. Sathishkumar, A., Rammohan, T., Sathish Kumar, S., Uma, J., Srujan Raju, K., et al.: Qos constrained network coding technique to data transmission using iot. Comput. Syst. Sci. Eng. **43**(2), 531–544 (2022)
11. Rao, M.V., Prasasd, A.V.K., Anusha, A., Raju, K.S.: Telugu text summarization using HS and GA particle swarm optimization algorithms. In: Bhateja, V., Satapathy, S.C., Travieso-Gonzalez, C.M., Adilakshmi, T. (eds.) Smart Intelligent Computing and Applications, Volume 1. Smart Innovation, Systems and Technologies, vol. 282. Springer, Singapore (2022). https://doi.org/10.1007/978-981-16-9669-5_54
12. Cherukuri, N., Bethapudi, N.R., Thotakura, V.S.K., Chitturi, P., Basha, C.Z., Mummidi, R.M.: Deep learning for lung cancer prediction using NSCLS patients CT information. In 2021 International Conference on Artificial Intelligence
13. Lal, S.B., Tiwari, R.: A comprehensive study on cloud computing. Int. J. Adv. Res. Comput. Sci. Softw. Eng. **2**(9), 33–37 (2012)
14. Saba, L., et al.: Ultrasound-based internal carotid artery plaque characterization using deep learning paradigm on a supercomputer: a cardiovascular disease/stroke risk assessment system. Int. J. Cardiovas. Imaging **37**, 1511–1528 (2021)
15. Apat, S.K., Mishra, J., Raju, K.S., Padhy, N.: A Study on smart agriculture using various sensors and agrobot: a case study. In: Bhateja, V., Satapathy, S.C., Travieso-Gonzalez, C.M., Adilakshmi, T. (eds.) Smart Intelligent Computing and Applications, Volume 1. Smart Innovation, Systems and Technologies, vol. 282. Springer, Singapore (2022). https://doi.org/10.1007/978-981-16-9669-5_48
16. Mohana Lakshmi, K., Rikhari, S.: Artificial intelligence framework for skin cancer detection. In: Bhateja, V., Satapathy, S.C., Travieso-Gonzalez, C.M., Adilakshmi, T. (eds.) Smart Intelligent Computing and Applications, Volume 1. Smart Innovation, Systems and Technologies, vol. 282. Springer, Singapore (2022). https://doi.org/10.1007/978-981-16-9669-5_53
17. Bhaskar, N., Ganashree, T.S.: Pulmonary nodule detection using laplacian of gaussian and deep convolutional neural network. In: Bhateja, V., Satapathy, S.C., Travieso-Gonzalez, C.M., Adilakshmi, T. (eds.) Smart Intelligent Computing and Applications, Volume 1. Smart Innovation, Systems and Technologies, vol. 282. Springer, Singapore (2022). https://doi.org/10.1007/978-981-16-9669-5_58
18. Gupta, A., Mahule, R., Patra, R.K., Saraswat, K.G., Akhtar, M.: Fingerprint liveliness detection to mitigate spoofing attacks using generative networks in biometric system. In: Bhateja, V., Satapathy, S.C., Travieso-Gonzalez, C.M., Adilakshmi, T. (eds.) Smart Intelligent Computing and Applications, Volume 1. Smart Innovation, Systems and Technologies, vol. 282. Springer, Singapore (2022). https://doi.org/10.1007/978-981-16-9669-5_57

19. Teja, M.N.V.M.S., Sree, N.L., Harshitha, L., Bhargav, P.V., Bhaskar, N., Reddy, V.D.: A dynamic model and algorithm for real-time traffic management. In: Bhateja, V., Satapathy, S.C., Travieso-Gonzalez, C.M., Adilakshmi, T. (eds.) Smart Intelligent Computing and Applications, Volume 1. Smart Innovation, Systems and Technologies, vol. 282. Springer, Singapore (2022). https://doi.org/10.1007/978-981-16-9669-5_55

Efficient Diagnosis of Skin Cancer Utilizing Swin Transformers Through Dermatoscopic Image Analysis

Vootla Srisuma[1], Sreekanth Yalavarthi[2], Rambabu Inaganti[3], K. Reddy Madhavi[4](✉), Pantham Vishnu[5], and Matta Venkata Pullarao[6]

[1] Department of IT, CMR Technical Campus, Hyderabad, Telangana, India
[2] HCL Technologies, Buffalo Grove, IL 60089, USA
[3] Smith & Nephew, Pittsburgh, PA 15108, USA
[4] School of Computing, Mohan Babu University, Tirupati, AP, India
kreddymadhavi@gmail.com
[5] Computer Science and Engineering, Indur Institute of Engineering and Technology, Siddipet, Telangana, India
[6] ECE Department, Sasi Institute of Technology and Engineering, Tadepalligudem, India

Abstract. Skin cancer is a fatal disease, requiring an automated mechanism for early and accurate diagnosis. In this paper, an approach utilizing the power of Swin Transformers, a deep learning architecture, to detect skin cancer has been proposed. Swin Transformers are recognized for their ability to capture complex patterns and dependencies in high-resolution images. Considering this aspect a swin transformer based model with a transfer learning approach has been proposed to detect malignancy of skin lesions. Transfer learning is used to enhance accuracy and helps for faster convergence. The proposed model utilizes fine-tuning of pre-trained Swin Transformer models on dermatoscopic images and achieves an accuracy of 94% specifying its ability to differentiate benign and malignant skin lesions. The proposed technique can detect the occurrence of skin cancer at an earlier stage and supports healthcare professionals in their decision-making processes.

Keywords: Swin Transformers · Transfer Learning · Dermatoscopic Images · Skin Lesions · Data Augmentation

1 Introduction

Skin cancer is a worldwide health concern, leading to a higher mortality rate if not treated properly. Accurate diagnosis of skin lesions is important for better patient outcomes. Recently, deep learning architectures have shown promising outcomes in dermatology by providing automated detection of skin cancer. Among the available deep learning architectures, Swin Transformers have emerged as a better option for handling computer vision based tasks. These models, have exhibited the ability to advance the accuracy of skin cancer detection. Also, these models can handle high-resolution medical images

D. Krikscuniene et al. (Eds.): IBICA 2023, LNNS 1228, pp. 388–395, 2025.
https://doi.org/10.1007/978-3-031-78937-3_42

making them suitable for dermatoscopic image analysis. Transfer learning technique utilizes knowledge from pre-trained models leading to enhanced accuracy and faster convergence of the model. It helps to initialize Swin Transformer models with features learned from vast image datasets, like ImageNet. This process leads to better convergence of model training and enhances the models ability to extract relevant patterns from dermatoscopic images.

The proposed approach integrates the capabilities of Swin Transformers and transfer learning to address the challenge of skin cancer diagnosis. A highly accurate model capable of classifying benign and malignant skin lesions can be created by fine-tuning Swin Transformers on a diverse dataset of skin lesions. In this paper, we outline the proposed methodology, which encompasses data collection, selection of model, transfer learning approach, and fine-tuning processes. Also, the proposed work contributes to dermatology image analysis utilizing deep learning architectures for early detection of skin lesions leading to better patient care.

2 Background Work

This section involves reviewing existing literature related to skin cancer prediction, deep learning architectures, transformer models, and transfer learning approach. Skin cancer represents a major health concern worldwide, with early prediction being difficult for proper treatment. Recently, deep learning architectures especially convolutional neural network (CNN) applications shown better results in automating the process of skin cancer prediction. Also, attention has turned to the usage of transformer-based architectures, such as Swin Transformer, considering their ability to capture complex patterns in dermatoscopic images. Utilization of deep learning for skin cancer diagnosis has gained attention due to its capability to learn complex features from high resolution images [1–6, 25–29]. CNNs achieved better accuracy in identifying malignant and benign lesions [14, 15]. The attention mechanism of transformer models offers an opportunity for improving diagnostic accuracy. The authors [20, 23] introduced Swin Transformer, to capture dependencies in images without relying on CNNs. Swin Transformer adopts a hierarchical approach to divide images into patches and capture contextual information efficiently [8–13]. This method can be used to perform classification and object detection. Recent works [16–24] have revealed the capability of transformers for skin disease detection utilizing the self-attention mechanisms to capture complex features and textures within images. Transfer learning, when combined with transformer architectures, has become a better approach particularly in vision based tasks to enhance model training. Vision Transformer (ViT) [17], demonstrated the utilization of transformers for image classification considering images as sequences of patches. This model achieved better results on ImageNet data. The authors [7] discussed that Transfer learning with Transformers provides better accuracy in performing computer vision tasks. Considering these aspects a Swin transformer based model with transfer learning approach has been proposed to predict skin cancer.

3 Methodology

This section, briefs about the steps of the proposed architecture for predicting skin lesions considering dematoscopic images utilizing Swin Transformers with transfer learning. The steps for the proposed architecture for predicting skin cancer have been specified in Fig. 1. Initial step is to collect dataset consisting of dermatoscopic images. Proposed work utilizes ISIC dataset with samples of skin lesions including multiple classes for processing. Next step is to perform data annotation ensuring the availability of labels indicating lesion type for each image to provide ground truth information. Then the input image samples will be resized, normalized and data augmentation such as random rotations, flips, and brightness adjustments, is applied to enhance model training process. Performing required preprocessing ensures that the data is compatible for Swin Transformer model construction. Also, Swin transformers can handle high resolution images such as dermatoscopic images and excel in feature extraction from images capturing local and global features. Considering these aspects Swin transformer model is selected since it plays an important role in performing computer vision tasks including skin lesion detection. Swin-B a mid-sized base model with minimum requirements of computational resources is used to process the dermatoscopic images. As specified in Fig. 1 swin transformer blocks plays vital role in capturing hierarchical features of input images and transforming them in a hierarchical manner.

Swin transformer model involves key processing elements such as patch partitioning, linear embedding, swin transformer blocks, and patch merging. During patch partitioning the swin transformer architecture splits high-resolution images into smaller nonoverlapping regions termed as patches and performs processing [20]. In Swin Transformer, linear embedding is used to flatten pixel values within each patch to convert them into token representations. This transformation helps to reduce the dimensionality capturing required information to build the model. Linear embedding is implemented utilizing a learnable linear projection layer involving matrix multiplication followed by applying normalization and activation function. The output will be the token embeddings that represent the content of the patch. These steps are required to work with the high resolution images.

Swin transformer blocks used to process the patches introducing the concept "vision tokens" to capture local features in patches. For token interaction shifted window mechanism is used to attend tokens in non-adjacent positions. Layered Normalization (LN) is utilized to normalize the activations including self-attention mechanism and multi-layer perceptron (MLP) with feed-forward network. Window-based Multi-Head Self-Attention (W-MSA) is a variant of self-attention, a mechanism used in transformers to weigh and combine information from various parts of a sequence used for efficient processing of high resolution images. Also, swin architecture consists of multiple layers, each containing W-MSA blocks stacked to form deep layered architectures to capture hierarchical features of images. MLP contains fully connected dense layers processing with nonlinear activation functions such as ReLU (Rectified Linear Unit). This can be used perform transformations and feature extraction on the Vision Tokens in order to capture complex features. Nonlinearity helps the model to learn complex relationships and patterns in image data. Transfer learning approach is used by Swin transformers to get the capability of using pre-trained weights from a large ImageNet dataset used for computer vision based tasks.Also, it helps in faster convergence, reduces overfitting of the model, provides the model valuable features and patterns. During this process fine tuning of the model is performed using different learning rates if required, fine tuning different layers of the model at different rates to prevent overfitting. Next step is to perform hyperparameter tuning using grid search to optimize learning rate, window decay and batch size. Then perform data augmentation to increase the size of data and enhance generalization of model using operations such as random rotations, flips, zoom, and adjusting brightness levels. Final step is to perform model evaluation using performance metrics as specified in Table 1. Also, validation of the model is performed to verify the efficiency of model to distinguish the benign and malignant skin lesions. Thus, Swin Transformers with transfer learning enhances the accuracy and efficiency of skin cancer detection.

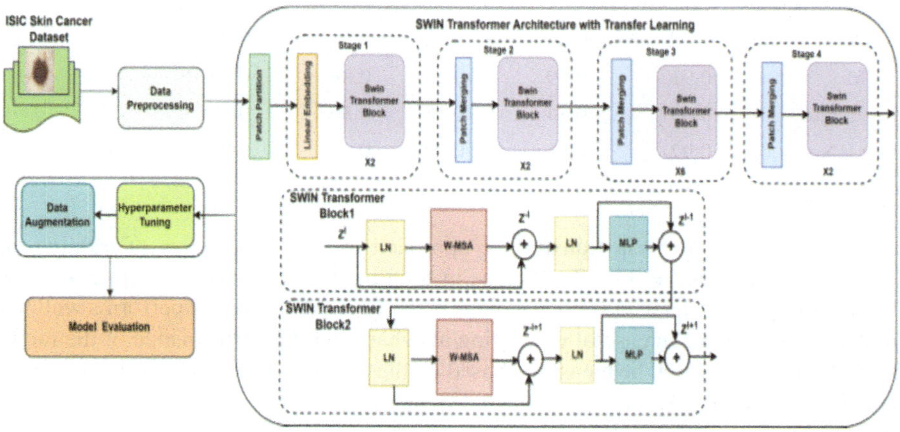

Fig. 1. Skin Cancer Prediction Architecture Using Swin Transformers with Transfer Learning

4 Results and Discussions

The dataset used in proposed approach is ISIC (International Skin Imaging Collaboration) dataset widely used collection of high-resolution dermoscopic images utilized for tasks related to skin cancer detection. The dataset can be used for training and evaluating deep learning architectures for skin-related medical image analysis. This dataset contains a diverse set of skin images, with both benign and malignant skin lesions. The dataset is annotated with important clinical information and labels, including information about lesion types, color, borders, and other related features. Deep learning architectures trained using ISIC dataset helps dermatologist with early skin cancer detection of the model.

Performance of swin transformer based approach is evaluated utilizing metrics like accuracy and more and the values are specified in Table 1. Accuracy of the proposed model is promising when compared to the methods that are in existence. The proposed Swin Transformer approach for skin cancer detection shows better performance, compared to existing methods based on evaluation metrics. The accuracy of 0.94 and ROC-AUC of 0.96 is remarkable. The proposed model excels in both accurately identifying skin cancer cases and differentiating between malignant and benign cases With a specificity value of 0.91, the proposed method proves to be a better tool for skin lesion diagnosis by significantly reducing the occurrence of misclassifications.

Table 1. Evaluation of Swin Transformer Model Using Performance Metrics

Method	Accuracy	Precision	Recall	F1-Score	AUC-ROC	Specificity
Proposed (Swin Transformer)	0.94	0.93	0.95	0.94	0.96	0.91
CNN	0.88	0.85	0.92	0.88	0.90	0.80
SVM	0.91	0.89	0.89	0.89	0.94	0.92
Ensemble	0.85	0.84	0.88	0.86	0.89	0.82
DenseNet	0.92	0.91	0.94	0.92	0.95	0.90

The Accuracy of Swin transformer approach during the process of training and validation is specified in Fig. 2. Increase in training accuracy based on epochs reveals the capability of model to classify the data correctly. Also, the model performs well with high accuracy values in case of overfitting also harming the performance of the model. Validation accuracy will be measured based on the performance of the model for unseen data. High validation accuracy reveals that swin transformer based model can generalize the learning from training process.

Fig. 2. Accuracy of Swin Transformer Model during Training and Validation

Figure 3 specifies the loss of Swin transformer model during training and validation. The model adjusts the hyperparameters such as weights, bias during learning process to decrease the loss. Also, decrease in validation loss specifies that the model performs better with new data. Training and validation accuracy and loss are re-quired parameters to measure the model's performance and provide information about the learning efficiency of the model on new data. Fine tuning hyperparameters such as learning rate, batch size, patch size, transformer, convolution layers, drop-out rates, and attention heads plays a major role in model performance improvement. Grid search mechanism is used to select the best possible value for the hyperparameter to improve the accuracy of the model. The results reveal significance and impact of proposed Swin Transformer model in predicting skin cancer accurately and can be utilized in dermatological practice.

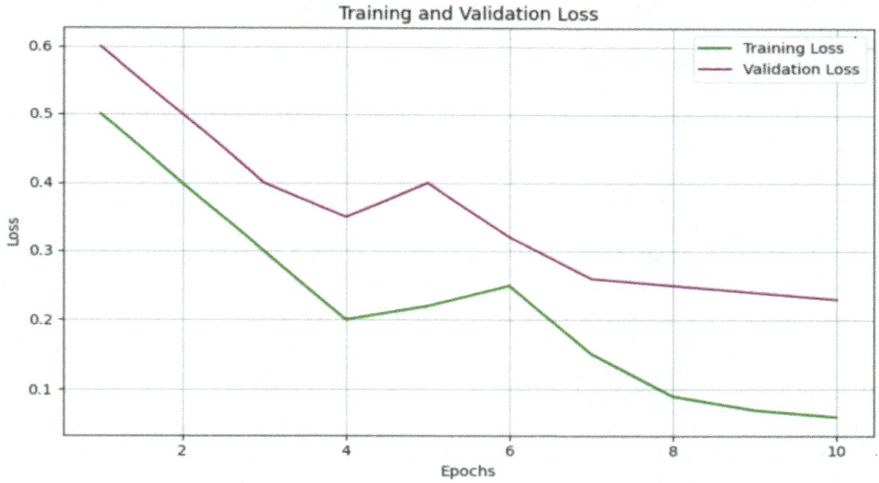

Fig. 3. Loss of the Proposed Model during Training and Validation

5 Conclusion

The proposed swin transformer based approach with transfer learning, represents a significant advancement in early prediction of skin cancer identifying the benign and malignant skin conditions properly. The proposed model performs better than the existing approaches with the accuracy of 0.94. Also, the proposed method minimizes the false positives and false negatives through balanced specificity. The proposed Swin Transformer approach not only holds potential to revolutionize skin cancer diagnosis but also embodies a promising step toward more accurate, reliable, and responsible healthcare practices. Thus the proposed swin transformer based approach can be used by medical practitioners for accurate skin cancer prediction. Further, future research directions include performing tasks like lesion segmentation, disease subtype classification, and multi-modal data integration. Future work is to extend the model for large-scale clinical validation and enhance the interpretability assisting dermatologists in their decision-making process.

References

1. Nahata, H., Singh, S.P.: Deep learning solutions for skin cancer detection and diagnosis. Mach. Learn. Health Care Perspect. Mach. Learn. Healthc.159–182 2020
2. Hosny, K.M., Kassem, M.A., Foaud, M.M.: Skin cancer classification using deep learning and transfer learning. In: 2018 9th Cairo international biomedical engineering conference (CIBEC), pp. 90–93. IEEE (2018)
3. Mazhar, T., et al.: The role of machine learning and deep learning approaches for the detection of skin cancer. Healthcare, **11**(3), 415 (2023)
4. Tembhurne, J.V., Hebbar, N., Patil, H.Y., Diwan, T.: Skin cancer detection using ensemble of machine learning and deep learning techniques. Multimed. Tools Appl. 1–24 (2023)
5. Hossin, M.A., Rupom, F.F., Mahi, H.R., Sarker, A., Ahsan, F., Warech, S.: Melanoma skin cancer detection using deep learning and advanced regularizer. In: 2020 International Conference on Advanced Computer Science and Information Systems (ICACSIS), pp. 89–94. IEEE (2020)
6. Esteva, A., et al.: Dermatologist-level classification of skin cancer with deep neural networks. Nature, **542**(7639), 115–118 (2017)
7. Rashid, J., et al.: Skin cancer disease detection using transfer learning technique. Appl. Sci. **12**(11), 5714 (2022)
8. Demir, A., Yilmaz, F., Kose, O.: Early detection of skin cancer using deep learning architectures: resnet-101 and inception-v3. In: 2019 Medical Technologies Congress (TIPTEKNO), pp. 1–4. IEEE 2019
9. Emara, T., Afify, H.M., Ismail, F.H., Hassanien, A.E.: A modified inception-v4 for imbalanced skin cancer classification dataset. In: 2019 14th International Conference on Computer Engineering and Systems (ICCES), pp. 28–33. IEEE 2019
10. Atta, A., Khan, M.A., Asif, M., Issa, G.F., Said, R.A., Faiz, T.: Classification of skin cancer empowered with convolutional neural network. In: 2022 International Conference on Cyber Resilience (ICCR), pp. 01–06. IEEE (2022)
11. Hasan, M.R., Fatemi, M.I., Monirujjaman Khan, M., Kaur, M., Zaguia, A.: Comparative analysis of skin cancer (benign vs. malignant) detection using convolutional neural networks. J. Healthc. Eng. **2021** (2021)
12. Periannasamy, S.M., et al.: Analysis of artificial intelligence enabled intelligent sixth generation (6G) wireless communication networks. In: 2022 IEEE International Conference on Data Science and Information System (ICDSIS), Hassan, India, pp. 1–8 (2022). https://doi.org/10.1109/ICDSIS55133.2022.9915945
13. Gothane, S.: Content summarization: journey toward word clouds. In: Kumar, A., Ghinea, G., Merugu, S., Hashimoto, T. (eds.) Proceedings of the International Conference on Cognitive and Intelligent Computing. Cognitive Science and Technology. Springer, Singapore (2022). https://doi.org/10.1007/978-981-19-2350-0_46
14. Simi Margarat, G., et al.: Early diagnosis of tuberculosis using deep learning approach for IoT based healthcare applications. Comput. Intell. Neurosci. **2022** (2022)
15. Monga, C., et al.: Secure techniques for channel encryption in wireless body area network without the certificate. Wirel. Commun. Mob. Comput. **2022** (2022)
16. Patel, P., Sivaiah, B., Patel, R.: Approaches for finding optimal number of clusters using K-means and agglomerative hierarchical clustering techniques. In: 2022 International Conference on Intelligent Controller and Computing for Smart Power (ICICCSP), Hyderabad, India, pp. 1–6 (2022). https://doi.org/10.1109/ICICCSP53532.2022.9862439
17. Madala, S., Suvarna Vani, K., Jalapally, P., Srujan Raju, K., Somasekhar, G.: Framework for diabetic retinopathy classification. In: Satapathy, S.C., Bhateja, V., Favorskaya, M.N.,

Adilakshmi, T. (eds.) Smart Intelligent Computing and Applications, Volume 2. Smart Innovation, Systems and Technologies, vol. 283. Springer, Singapore (2022). https://doi.org/10.1007/978-981-16-9705-0_58

18. Kolli, C.S., Raghunath, M.P., Meenakshi, S., Maheswari, K., Britto, C.F., Kushwaha, S.: Efficient development of supervised learning algorithm for kidney stone prediction. In: 2022 International Conference on Inventive Computation Technologies (ICICT), Nepal, pp. 1373–1379 (2022). https://doi.org/10.1109/ICICT54344.2022.9850573
19. Sathishkumar, A., Majji, S., Radhika Patnala, T., Rao Karanam, S., Kumar, A., Malyadri, M.: Experimentation methodology of orthogonal frequency division multiplexing signals process using radio over fiber (RoF) system. In: 2022 International Virtual Conference on Power Engineering Computing and Control: Developments in Electric Vehicles and Energy Sector for Sustainable Future (PECCON), Chennai, pp. 01–05 (2022). https://doi.org/10.1109/PECCON55017.2022.9850996
20. Avanija, J., Madhavi, K.R., Sunitha, G., Sangapu, S.C., Raju, S.: Facial expression recognition using convolutional neural network. In: 2022 First International Conference on Artificial Intelligence Trends and Pattern Recognition (ICAITPR), Hyderabad, pp. 1–7 (2022). https://doi.org/10.1109/ICAITPR51569.2022.9844221
21. Tiwari, L., Awasthi, V., Patra, R.K., Miri, R., Raja, H., Bhaskar, N.: Lung cancer detection using deep convolutional neural networks. In: Bhateja, V., Khin Wee, L., Lin, J.CW., Satapathy, S.C., Rajesh, T.M. (eds.) Data Engineering and Intelligent Computing. Lecture Notes in Networks and Systems, vol. 446. Springer, Singapore (2022). https://doi.org/10.1007/978-981-19-1559-8_37
22. Peng, L., et al.: Analysis of CT scan images for COVID-19 pneumonia based on a deep ensemble framework with DenseNet, swin transformer, and RegNet. Front. Microbiol. **13**, 995323 (2022)
23. Liu, B., Liu, Y., Zhang, W., Tian, Y., Kong, W.: Spectral swin transformer network for hyperspectral image classification. Remote Sens. **15**(15), 3721 (2023)
24. Liu, Y., Zhao, J., Luo, Q., Shen, C., Wang, R., Ding, X.: Automated classification of cervical lymph-node-level from ultrasound using depthwise separable convolutional swin transformer. Comput. Biol. Med. **148**, 105821 (2022)
25. Sunitha, G., Rani, B.S., Bhukya, S.N., Mohammad, H., Vittal, R.H.S.: Political optimizer-based automated machine learning for skin lesion data. In: Proceedings of Third International Conference on Advances in Computer Engineering and Communication Systems: ICACECS 2022, pp. 487–495. Springer, Singapore (2023)
26. Sowmya, T.S., Narasimhulu, T., Sunitha, G., Manikanta, T., Venkatesh, T.: Vision transformer based ResNet model for pneumonia prediction. In: 2023 4th International Conference on Electronics and Sustainable Communication Systems (ICESC), pp. 316–321. IEEE (2023)
27. Swathi, B., Kannan, K.S., Chakravarthi, S.S., Ruthvik, G., Avanija, J., Reddy, C.C.M.: Skin cancer detection using VGG16, INCEPTIONV3 and ResUNet. In: 2023 4th International Conference on Electronics and Sustainable Communication Systems (ICESC), pp. 812–818. IEEE (2023)
28. Avanija, J., Reddy, C.C.M., Reddy, C.S.C., Reddy, D.H., Narasimhulu, T., Hardhik, N.V.: Skin cancer detection using ensemble learning. In: 2023 International Conference on Sustainable Computing and Smart Systems (ICSCSS), pp. 184–189. IEEE (2023)
29. Kora, P., et al.: Automatic segmentation of prostate cancer using cascaded fully convolutional network. In: E3S Web of Conferences, vol. 309, p. 01068. EDP Sciences (2021)

An Assessment on the Detection and Diagnosis of Brain Cancers with Support Vector Machines

Mahesh Kotha[1](✉), Ratnababu Jyothi[1], S. Kirubakaran[2], P. Sravanthi[2], Sayyad Rasheeduddin[3], and V. Sravani Kumari[3]

[1] CMR Technical Campus, Hyderabad, Telangana, India
kotha.mahesh528@gmail.com, cmrtc.paper@gmail.com
[2] CMR College of Engineering and Technology, Hyderabad, Telangana, India
{drskirubakaran,p.sravanthi}@cmrcet.ac.in
[3] CMR Engineering College, Hyderabad, Telangana, India
{srasheeduddin,sravanivenna}@cmrec.ac.in

Abstract. The diagnosis of brain tumors is a crucial undertaking in the field of medical imaging, necessitating the utilization of precise and dependable methodologies to facilitate early identification and intervention. In recent times, the utilization of Support Vector Machines (SVMs) has been recognized as a viable methodology for the automated detection of brain tumors. Support Vector Machines (SVMs) represent a robust machine learning method that exhibits notable efficacy in classification tasks. In comparison to alternative machine learning algorithms, including neural networks and decision trees, SVMs provide distinct advantages. This manuscript provides a comprehensive overview of recent scholarly investigations pertaining to the application of Support Vector Machines (SVMs) in the context of brain tumor diagnosis. The present study encompasses an examination of the benefits associated with the utilization of Support Vector Machines (SVMs) in the context of brain tumor identification. Additionally, it addresses the difficulties that arise in the process of selecting appropriate SVM parameters, as well as the rates of accuracy reported in various researches. The findings indicate that Support Vector Machines (SVMs) have demonstrated notable levels of accuracy, ranging from 90% to 95%, as reported in several research publications. Nevertheless, it is imperative to exercise caution while selecting parameters in order to attain the most favorable performance outcomes. Additional investigation is required in order to substantiate the efficacy of brain tumor detection techniques based on Support Vector Machines (SVM) when applied to more extensive datasets. Furthermore, it is imperative to tackle the obstacles related to the selection of SVM parameters. In summary, the utilization of Support Vector Machine (SVM) for brain tumor identification exhibits considerable potential in enhancing the precision and efficacy of brain tumor diagnosis.

Keywords: SVM · brain cancer · MRI · CNN

D. Kriksciuniene et al. (Eds.): IBICA 2023, LNNS 1228, pp. 396–406, 2025.
https://doi.org/10.1007/978-3-031-78937-3_43

1 Introduction

A brain tumor is a medically significant disorder characterized by the presence of an anomalous proliferation of cells within the brain. This condition poses a significant risk to an individual's life and necessitates prompt identification and appropriate intervention. Brain tumors can be categorized into two primary classifications: malignant and benign. Malignant brain tumors provide a greater risk due to their tendency to exhibit accelerated growth and the potential for metastasis to both intracranial and extra cranial sites. In contrast, benign tumors often have a modest rate of growth and infrequently metastasize to distant anatomical sites. In the domain of brain tumor detection, Support Vector Machines (SVMs) can be employed as a classification technique to categorize Magnetic Resonance Imaging (MRI) pictures into two distinct classes, namely normal and abnormal. The Support Vector Machine (SVM) algorithm acquires information from a collection of labeled Magnetic Resonance Imaging (MRI) images and employs this acquired knowledge to categorize novel, unseen images as either normal or abnormal. Support Vector Machines (SVMs) possess several advantages as compared to alternative machine learning algorithms, including neural networks and decision trees. Convolutional neural networks (CNNs) exhibit computational efficiency, possess a reduced likelihood of overfitting, and offer ease of interpretation. In summary, the utilization of Support Vector Machines (SVMs) for brain tumor identification exhibits considerable potential in enhancing the precision and efficacy of brain tumor diagnosis. Support Vector Machines (SVMs) possess numerous advantages when compared to other machine learning methods, as evidenced by several studies that have demonstrated their ability to attain notable levels of accuracy. However, additional research is required to tackle the difficulties linked to SVM parameter selection and to verify the effectiveness of brain tumor detection algorithms based on SVM on more extensive datasets.

2 Relevant Review

The present section discusses the existing literature and research that is relevant to the topic at hand. Numerous researchers conducted multiple studies pertaining to the classification of Alzheimer's disease (AD) and the corresponding methodologies employed. This chapter presents a complete measurement for some research.

The study conducted by the authors [3] introduced a highly efficient model for the categorization of Alzheimer's disease (AD). This model integrates a novel approach for feature selection and ranks the selected features using the t-test. The proposed model consists of five stages. Firstly, a morphometry model based on voxel is employed to assess the global and local variations in gray matter.

The categorization approach presented in reference [4] aims to leverage the complementary nature of multi-modal data for improved analysis. The main approach used to characterize the features modality includes the utilization of couple-wise similarity.

. In addition, the utilization of the ITL technique to transfer source domain samples has the potential to augment the training samples in the target domain, hence contributing to a substantial enhancement in classification accuracy [5].

identified tumor has been successfully accomplished through the utilization of the Gabor filter. The aforementioned features are utilized in the training and classification process of brain tumors through the implementation of an Artificial Neural Network classifier. According to reference [10, 11], the system demonstrates a notable enhancement in the precision of brain tumor identification.

3 Methodology

Methodology employed in this study refers to the systematic approach used to collect and analyze data. The brain tumor detection method under consideration can be divided into three primary stages. During the initial phase, a filtering technique known as the Anisotropic Filter (AF) is employed to effectively eliminate noise present in the brain MRI picture. Subsequently, a segmentation procedure based on adjustment is employed, wherein a structuring element is utilized to segment the filtered tumor image [12]. In the third stage, morphological operations are employed to ascertain the precise location of the tumor. The depiction of the model's flow can be observed in Fig. 1.

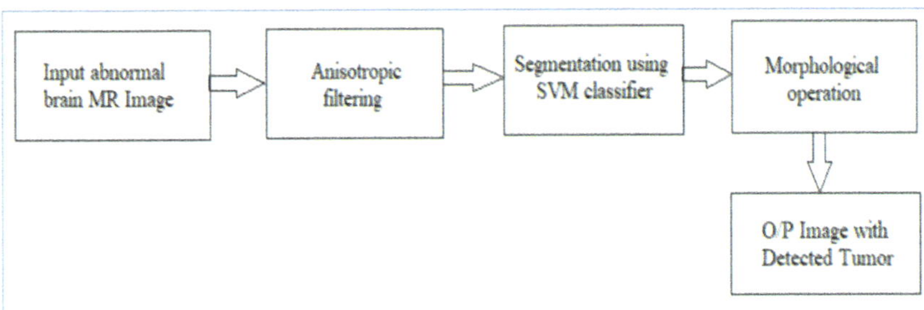

Fig. 1. Proposed approach flow chart.

A. Dataset

One of the atypical visual representations Magnetic resonance imaging (MRI) has been regarded as a potential modality for tumor recognition, as depicted in Fig. 2. The input image is composed of pixels measuring 256 by 256, with each pixel representing a shade of gray using 8 bits of information.

B. Anisotropic-Filtering

The principal objective of image filtering is to effectively remove noise from digital photographs. The presence of noise can have a substantial effect on the overall quality of an image, leading to suboptimal performance of many image processing methods. Consequently, image filtering is utilized as a preliminary processing technique. Anisotropic Filtering is a technique employed to reduce noise in photographs, and its application for denoising has been demonstrated in Fig. 3 of the referenced thesis. The utilization of the general formula of anisotropic diffusion is being employed in order to determine the process of image diffusion.

$$\frac{dx}{dt} = \mathrm{div}(c(m, n, t)\nabla I) = \nabla c.\nabla x + c(m, n, t)\nabla x \tag{1}$$

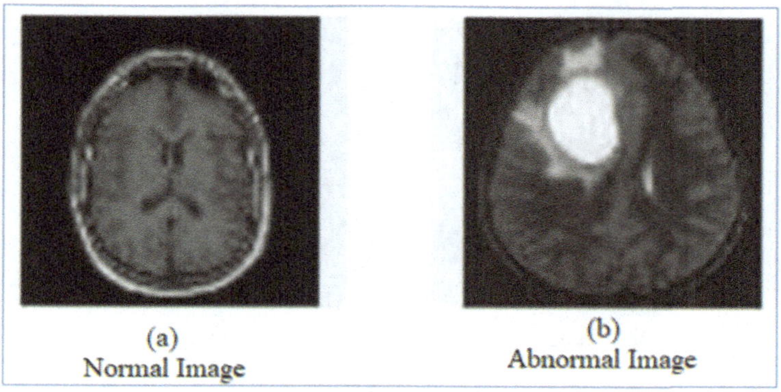

Fig. 2. MRI images of brain

The symbol ∇x is typically used in mathematics to represent the gradient operator with respect to the variable x. The term "signifies" refers to the gradient of a picture, whereas the notation & c(m, n, t) represents the coefficient of diffusion. The following diagram illustrates the estimation of discretized variances using both backward and forward approximations.

$$I_{i,j}^{t+1} = I_{i,j}^{t} + dt \sum (k,1) \in N4g(I_{k,1}^{t} - I_{i,j}^{t}).(I_{k,1}^{t} - I_{i,j}^{t}) \quad (2)$$

$$h(I_{k,1}^{t} - I_{i,j}^{t}) = \frac{C_{k,l}^{t} + C_{i,j}^{t}}{2} \quad (3)$$

The set N4, which represents the central pixel neighborhood, is defined as {(i-1, j), (i + 1, j), (i, j-1), (i, j + 1)}. It is evident where $I^{t}_{i,j}$ is that the noise pixel exhibits a strong diffusion effect, while the signal pixel demonstrates a relatively mild diffusion effect. Therefore, the elimination of noise and the placement of signals would be achieved. Various methods have been proposed for implementing diffusion algorithms that involve using a fixed step size for each iteration or processing the full image iteratively. The projection of the phase of iteration is denoted as Eq. (4).

In this context, the fraction ¼ has been employed to ensure the convergence of Eq. (2). Furthermore, the final result of the image has been achieved by the process of iteration. The technique has been iterated using the iteration error (IE) to control the number of iterations and determine their formula:

$$IE = \frac{//I^{n} - I^{n-1}//}{//I^{n}//} \leq T_{ie} \quad (4)$$

The process of iteration may be halted or temporarily suspended when the error estimate (I_E) reaches a value that is equal to or smaller than the predetermined tolerance threshold (T_{ie}).

C. Segmentation of image using SVM

Image segmentation is the process by which an image is divided into many segments or sections, with the purpose of identifying and extracting objects and

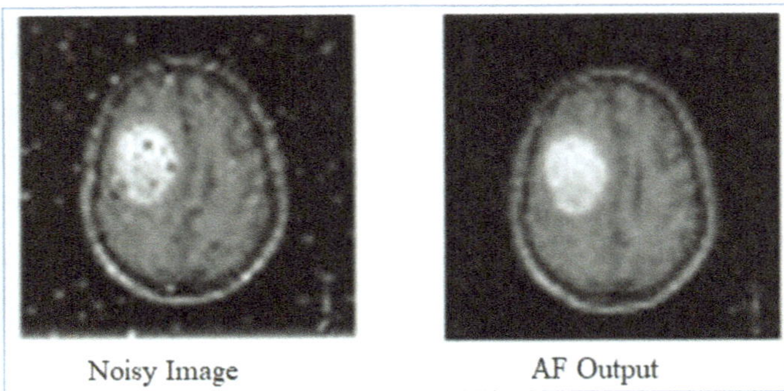

Fig. 3. I/O for AF

related information within the image. Support Vector Machines (SVM) is a widely utilized method employed for the purpose of segmentation. In this methodology, a solution is derived for a specific issue by determining an ideal hyperplane, as depicted in Fig. 4. This is exemplified by the subsequent illustration of a basic problem:

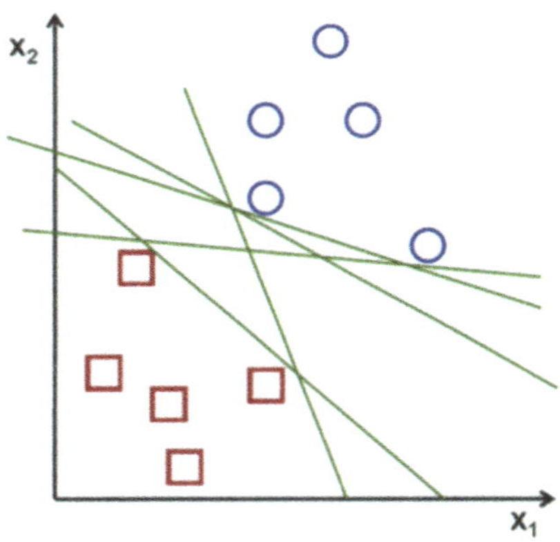

Fig. 4. Hyperplane in SVM

The equation presented below shows a mathematical expression used to determine a hyperplane.

$$p(x) = \alpha \cap + \alpha^T z \tag{5}$$

$$\backslash \alpha \cap + \alpha^T z \backslash = 1 \tag{6}$$

In this context, the variable z represents the occurrences of training examples in close proximity to a hyperplane. Traditionally, support vectors have been regarded as the closest training instances to hyperplanes. The utilization of the geometry result allows for the determination of the variance between a hyperplane (∝, ∝ 0) and a point z

$$D = \frac{\backslash\alpha \cap +\alpha^T z\backslash}{//\alpha//} \tag{7}$$

The numerator has exhibited similarities to the canonical plane, while the variance for support vectors has demonstrated a comparable pattern.

$$D_{\text{support vectors}} = \frac{\backslash\alpha \cap +\alpha^T z\backslash}{//\alpha//} = \frac{1}{//\alpha//} \tag{8}$$

The equation presented above represents the margin denoted as M, which is twice the variance with respect to the nearest examples.

$$M = \frac{2}{//\alpha//} \tag{9}$$

The issue of M has been one of maximizing, analogous to the issue of minimal for the R(∝) function under different constraints. The utilization of the limitations method on the hyperplane is being employed to accurately categorize instances of training z.

$$min\alpha, \alpha_0{}^{R(\alpha)} = \frac{1}{2}//\alpha//^2 subject\ to\ y_i(\alpha^T Z_i + \alpha_0) \geq I\forall_i \tag{10}$$

The variable yi represents the labels of each training instance. The problem at hand pertains to Lagrangian optimization, which can be effectively addressed by employing Lagrange multipliers to obtain the weight-vector (∝) and bias hyperplane (∝0).

4 Experimental Study

We have made diligent efforts to reliably identify brain tumors in aberrant brain MRI pictures. In order to achieve the desired objective, the application of anisotropic filtering has been employed for noise removal, while morphological procedures and support vector machines (SVM) have been utilized for segmentation [13].This study aims to conduct a comprehensive performance analysis of AF, focusing on its various aspects and evaluating its effectiveness in achieving desired outcomes. Three types of noise, namely speckle, salt and pepper, and Gaussian, were introduced to the input image. Subsequently, the Peak Signal-to-Noise Ratio (PSNR) and Mean Squared Error (MSE) values were determined using the following methodology:

$$\text{MSE} = \frac{1}{pq}\sum\nolimits_{i=0}^{p-1}\sum\nolimits_{i=0}^{q-1} //h(i,j) - g(i,j)//^2 \tag{11}$$

$$\text{PSNR} = 20\log_{10}\left(\frac{MAX_f}{\sqrt{MSE}}\right) \tag{12}$$

The variable "h" represents the original image matrix data.
The letter "g" represents a matrix of degraded image data.

The variable p denotes the intensity values of the rows in the image, with i being the index of each row.

The variable q is used to represent the column number of intensity values, whereas the variable j is used to represent the column index.

The term “MaxF” is commonly used to refer to the highest value of a signal that is found inside an image.

Similarly, the variables ‘i’ and ‘j’ are employed to indicate the associated indices of these values. The maximum signal value in the original image is represented by the symbol ‘MAXf’. The findings of the assessment are presented in Table 1 below.

Table 1. Measurement of MSE over several filters

Vlaues of MSE			
Method	Gaussian Noise	Speckle Noise	Salt pepper
Average Filter	0.0042	0.00445	0.00321
Median filter	0.003	0.00256	0.001
Mean	0.00561	0.00508	0.00421
Wiener	0.0051	0.00423	0.00512
HPF filter	0.00728	0.00713	0.00658
AF filtering	0.00358	0.00244	0.00168

Performance of AF has been exhibited in the below figures.

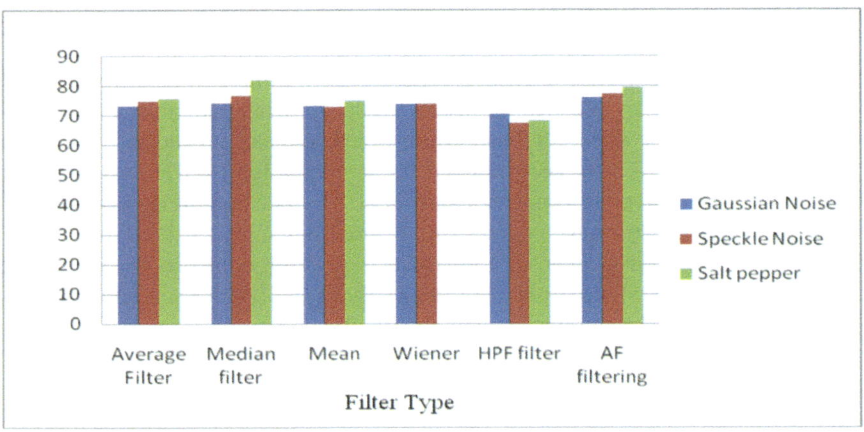

Fig. 5. Performance of various filters

Figure 5 illustrates a comparison of peak signal-to-noise ratio (PSNR) values for various types of diverse noises, including salt and pepper, Gaussian, and speckle.

Figure 5 presents a comparison of peak signal-to-noise ratio (PSNR) values across various filters for different types of noise. The filters that have been taken into account are presented in the table. We have taken into account the presence of Gaussian, salt and pepper, and speckle noises. Table 2 displays the PSNR values, with the x-axis representing various filter types (Fig. 7).

Table 2. Measurement of PSNR over several filters

Values of PSNR (dB)			
Method	Gaussian Noise	Speckle Noise	Salt pepper
Average Filter	73.24	74.82	75.52
Median filter	74.21	76.72	81.9
Mean	73.1	72.95	75.1
Wiener	73.94	73.85	73..08
HPF filter	70.32	67.22	68.12
AF filtering	75.89	77.21	79.5

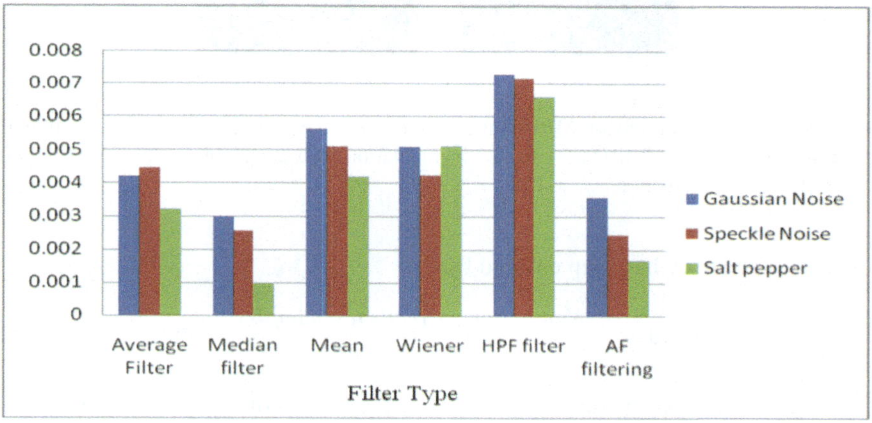

Fig. 6. Comparison of MSE value for pepper & salt, Gaussian as well as speckle noise among numerous filters.

Figure 6 above illustrates a comparative analysis of mean squared error (MSE) values across various filters for different types of noise. The filters that have been taken into account are presented in the table. We have taken into account the presence of Gaussian, salt and pepper, and speckle noises. The horizontal axis depicts different filter types, while the vertical axis represents mean squared error (MSE) values. Following the elimination of noise, the image classifier SVM effectively segments the image, after which morphological procedures are conducted. Furthermore, the assessment of correctness has served as a criterion for determining the effectiveness of classification. The

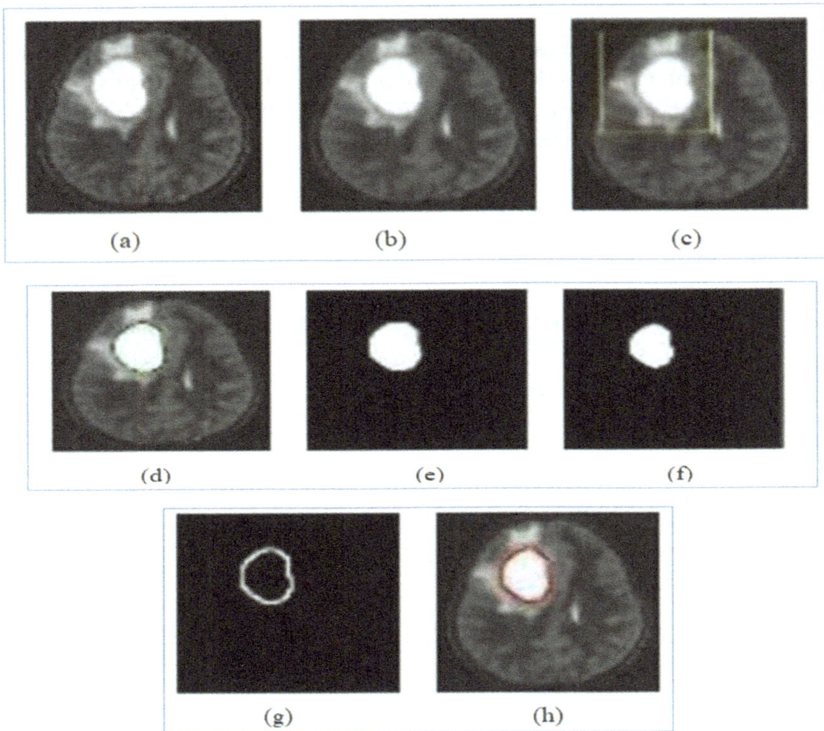

Fig. 7. Numerous operations on MRI image of brain (1) image of input (2) image of filtered © Bounding Box locating (d) SVM (e) Tumor segmentation (f) image, which is eroded (g) Outline of tumor (h) Tumor detection

provision of accuracy has been ensured by:

$$\text{Accuracy} = \frac{\textit{Number of correctly test samples}}{\textit{Total samples}} * 100\% \tag{13}$$

The objective of segmentation entails the examination of 380 pixels, of which 320 have been correctly identified.Therefore, the accuracy may be determined by calculating the ratio of 320 to 369 and multiplying it by 100. This yields an accuracy of 86.7%.The numerical representation of 7 demonstrates the final result associated with each level of reaction.

5 Conclusion

To achieve precise identification of brain tumors in MRI scans, it is imperative to eliminate any extraneous distortions present in the input image. Anisotropic filtering was employed for this specific objective. Next, the segmentation process utilized Support Vector Machine (SVM) to classify pixels into two distinct classes. Given that the system was intended to be compatible with any MRI brain imaging, an unsupervised learning kernel was selected for the support vector machine (SVM). Morphological operations were employed to excise the tumor from the segmented area. Ultimately, the employed methodology demonstrated proficiency in accurately identifying the presence of the tumor.

References

1. Altaf, T., Anwar, S.M., Gul, N., Majeed, M.N., Majid, M.: Multi-class Alzheimer's disease classification using image and clinical features. Biomed. Signal Process. Control **43**, 64–74 (2018)
2. Kanthi, M., Rao, K.V., Sekhar Reddy, L.C., Sarma, T.H., Bhaskar, N., Vasundhara, N.: "3D-CAN: a 3D convolution attention network for feature extraction and classification of hyperspectral images. In: 2023 International Conference on Network, Multimedia and Information Technology (NMITCON), Bengaluru, pp. 1–6 (2023). https://doi.org/10.1109/NMITCON58196.2023.10275985
3. Bharath Kumar, K., Swathi Priya, R., Renu Yamini, K., Venkata Bhargavi, K., Dheeraj Rao, N.: Implementation of smart bore-well child rescue system. In: Kumar, A., Ghinea, G., Merugu, S. (eds.) Proceedings of the 2nd International Conference on Cognitive and Intelligent Computing. ICCIC 2022. Cognitive Science and Technology. Springer, Singapore (2023). https://doi.org/10.1007/978-981-99-2746-3_51
4. Tamilselvan, R., Prabhu, A., Rajagopal, R.: An enhanced K-means algorithm for large data clustering in social media networks. Artif. Intell. Sustain. Appl. 147–162 (2023)
5. Chandramowleeswaran, G., Choubey, C., Pendem, S., Pal, S., Verma, A.: Implementation of human robot interaction with motion planning and control parameters with autonomous systems in Industry 4.0. In: 2023 Second International Conference on Augmented Intelligence and Sustainable Systems (ICAISS), Trichy, pp. 1805–1810 (2023). https://doi.org/10.1109/ICAISS58487.2023.10250484
6. Murthy, N., Spandan, G., Aswani, I., S. Nayak Bhukya, I., AC, R.: Deep learning and MRI improve carotid arterial tree reconstruction. In: 2023 International Conference on Data Science and Network Security (ICDSNS), Tiptur, India, pp. 1–7 (2023). https://doi.org/10.1109/ICDSNS58469.2023.10245397
7. Rajani, A., Kora, P., Madhavi, R., Jangaraj, A.: Quality improvement of retinal optical coherence tomography, pp. 1–5 (2021). https://doi.org/10.1109/INCET51464.2021.9456151
8. Reddy Madhavi, K., et al.: "COVID-19 detection using deep learning". In: 20th International Conference on Hybrid Intelligent Systems-HIS 2020, at Machine Intelligence Research (MIR) labs, USA, Springer AISC, 1375, pp 1–7 (2020)
9. Prabhakar, T., Srujan Raju, K., Reddy Madhavi, K.: Support vector machine classification of remote sensing images with the wavelet-based statistical features. In: Fifth International Conference on Smart Computing and Informatics (SCI 2021), Smart Intelligent Computing and Applications, Volume 2. Smart Innovation, Systems and Technologies, vol. 283. Springer, Singapore (2022)
10. Kora, P., et al.: EEG-based brain-electric activity detection during meditation using spectral estimation techniques, pp. 687–693 (2021). https://doi.org/10.1007/978-981-16-1941-0_68
11. Hemanand, D., Reddy, G.V., Babu, S.S., Balmuri, K.R., Chitra, T., Gopalakrishnan, S.: An intelligent intrusion detection and classification system using CSGO-LSVM model for wireless sensor networks (WSNs). Int. J. Intell. Syst. Appl. Eng. **10**(3), 285–293 (2022)

12. Bathula, A., Merugu, S., Skandha, S.S.: Academic projects on certification management using blockchain-a review. In: 2022 International Conference on Recent Trends in Microelectronics, Automation, Computing and Communications Systems (ICMACC). IEEE (2022)
13. Prashanthi, M., Chandra Mohan, M.: Hybrid optimization-based neural network classifier for software defect prediction. Int. J. Image Graph. **24**(04), 2450045 (2024). https://doi.org/10.1142/S0219467824500451

Analysis on Evaluating the Efficacy of Machine Learning Algorithms in Predicting Cardiac Disease

G. Vinoda Reddy[1(✉)], Bagam Laxmaiah[2], V. Venkataiah[3], Ch. Raja Kishore Babu[3], P. Sandhya Reddy[4], and M. S. S. Lakshmi Lavanya[5]

[1] Department of CSE (AI&ML), CMR Technical Campus, Hyderabad, Telangana, India
cmrtc.paper@gmail.com

[2] Department of CSE, CMR Technical Campus, Hyderabad, Telangana, India

[3] Department of CSE, CMR College of Engineering and Technology, Hyderabad, Telangana, India
{vvenkataiah,chrajakishorebabu}@cmrcet.ac.in

[4] Department of CSE (AI&ML), CMR Engineering College, Hyderabad, Telangana, India
patana.sandhyareddy@cmrec.ac.in

[5] Department of CSE (DS), CMR Engineering College, Hyderabad, Telangana, India
lavanya5814@cmrec.ac.in

Abstract. In the field of medical science, the utilization of historical data and patient prescriptions is employed to facilitate preventative healthcare measures and enhance diagnostic capabilities, ultimately contributing to the overall well-being of patients. The accessibility of a patient's historical medical records is a important obstacle in the field of medical sciences. Manual preventive measures are necessary for this purpose, but they are prone to errors, laborious, and consume a significant amount of time. Regression analysis is a statistical technique commonly employed in the field of machine learning to ascertain the association between a solitary dependent variable and several independent factors. Its primary objective is to generate predictions by considering various combinations of the independent variables. Within the field of Machine Learning, numerous models are employed for the purpose of regression analysis. This study focus on the utilization and assessment of the k-NN, random forest, and logistic regression models in order to model and predict occurrences of heart attacks in patients. The test runs for all regression models utilized a dataset consisting of the past cardiac illness history of 300 patients who were under treatment. This dataset was acquired by the UCI Machine Learning Repository. The logistic regression model had a peak accuracy of 92.50% in predicting heart illness.

Keywords: k-Nearest Neighbor · Random Forest · Logistic Regression · Preventive Health Care and Feature Selection

D. Kriksciuniene et al. (Eds.): IBICA 2023, LNNS 1228, pp. 407–415, 2025.
https://doi.org/10.1007/978-3-031-78937-3_44

1 Introduction

Preventive healthcare refers to the application of healthcare strategies aimed at averting the occurrence of diseases [1]. Primal, primary, secondary, and tertiary preventive strategies constitute the fundamental pillars of disease prevention [2].

Regression analysis is a collection of statistical techniques employed in statistical modeling for the purpose of ascertaining the associations among a needy variable and one or more sovereign variables, also known as "predictors," "covariates," "explanatory variables," or "features" [3]. Linear regression is a commonly employed method of regression analysis wherein the objective is to determine a line, or maybe a more intricate linear combination, that best aligns with the data based on a specified mathematical criterion. There are two predominant uses of regression analysis that can be logically distinguished [4, 5]. There exists a notable convergence in the utilization of regression analysis and machine learning in the domains of prediction and forecasting. Furthermore, in certain cases, regression analysis can be employed to deduce causal relationships among independent and dependent variables. It is significant to highlight that regressions, in isolation, solely reveal associations among a dependent variable and a set of independent variables within a certain dataset.

2 Relavant Study

The present section present a comprehensive review of the existing literature on the chosen topic.In order to classify cardiovascular disease, a study by [6, 7] conducted empirical research utilizing various machine learning algorithms for example logistic regression, KNN Classifier, RF, SVM, Decision Tree, and Gauss Nave Bayesian. The study employed data sourced from UCI. The logistic regression model achieve an accuracy rate of 85% in its predictions.Based on an analysis of the 4238 records contained within the Framingham dataset, [8] proposed the utilization of a logistic regression model as a way of categorizing heart disease. The logistic regression model had an accuracy rate of 86.66%.This study focuses on feature selection strategies and algorithms, utilizing several datasets related to heart disease for experimentation analysis and accuracy improvement demonstration. Feature selection techniques such as Decision Tree, Logistic Regression, Logistic Regression SVM, Naïve Bayes, and Random Forest are employed with the Rapid Miner tool to improve accuracy and show improvement in the findings [9, 10]. Saba recommended the utilization of logistic regression, decision trees, and support vector machines, among other techniques. The user's text does not contain any information to rewrite in an academic manner [11, 12]. The research employed the UCI dataset to evaluate the performance of logistic regression and logistic regression support vector machine models. The accuracy rates achieved by these models were 82% and 84%, respectively.

3 Proposed Method

The present study will employ a rigorous methodology to investigate the research question at hand. This methodology will involve a systematic approach to data collection. The investigation of the UCI dataset on heart illness is conducted within the proposed system, employing suitable methods for data collection and preparation. This involves cleaning the data and afterwards selecting the characteristics that exhibit a strong link with the target function. The k-Nearest Neighbors (k-NN), Random Forest, and Logistic Regression models are subsequently trained and evaluated to ascertain their ability to detect cardiac illness.

The methodology employed in the development of k-NN, random forest, and logistic regression models for the classification and forecast of cardiac disorders is demonstrated in form 1 (Fig. 1).

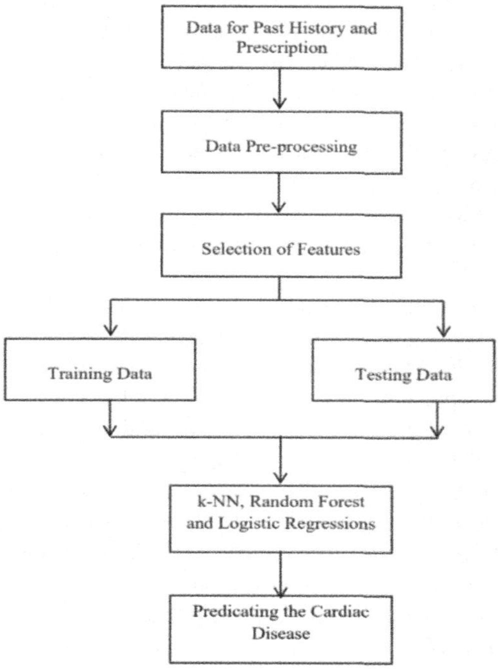

Fig. 1. Regression Analysis for Predicating the Cardiac Disease

Procedure for Predicating the Cardiac Disease

The data that has undergone a process of feature selection.

The optimal approach for classification and prediction.

For each value of i from 1 to k.

For each case, di, in the dataset worn for training and testing purposes,

Establish the desired goal value for the k-Nearest Neighbors (k-NN), Random Forest, and Logistic Regression models.

Set the initial weight of the instance.

Conclude the data by incorporating class values.

The categorization of label decision.

Please assign the class label as 1. If the conditional probability of P(yes|d) is greater than 0.5, then the class label is 2.

Features Selected for Predicating the Cardiac Disease

The features that have been chosen for analysis include the patient's age, which is deliberated in years.

Gender: (male/female)

The Classification of Chest Pain Types

a. normal angina refers to chest pain that take place appropriate to a reduction in blood flow to the heart. b. A normal angina episode involves chest pain that is not associated with any heart-related issues. c. Non-anginal pain commonly manifests as esophageal spasms, which are unrelated to the heart.

Asymptomatic refers to the absence of observable symptoms, particularly in the context of chest pain that does not exhibit indications of underlying disease.

The variable "Trest_bps" represents the resting blood pressure, measured in millimeters of mercury (mm Hg), upon admission to the hospital. It is generally considered worrisome if the recorded value exceeds the range of 130–140 mm Hg.

The variable "Chol" represents the serum cholesterol levels measured in milligrams per deciliter (mg/dL).

The serum can be calculated by summing the low-density lipoprotein (LDL), high-density lipoprotein (HDL), and 0.2 times the triglycerides.

A value exceeding 200 should be regarded as a matter of concern.

The variable "fbs" represents the fasting blood sugar level, with a value of 1 indicating a fasting blood sugar level greater than 120 mg/dl, and a assessment of 0 indicating a fasting blood sugar level under or equal to 120 mg/dl.

A blood glucose level exceeding 126 mg/dL is indicative of diabetes.

The resting electrocardiographic data, often known as restecg, are being discussed.

a. No significant observations were made. b. ST-T wave abnormalities can manifest with a spectrum of clinical presentations, ranging from minor symptoms to severe manifestations, indicating an irregular cardiac rhythm.

Possible or confirmed left ventricular hypertrophy (LVH) can be observed.

The primary ventricular chamber of an enlarged heart.

The variable "Thalach" represents the maximal heart rate obtained.

Exang refers to exercise-induced angina, with a value of 1 indicating its presence and 0 indicating its absence.

The Oldpeak phenomenon refers to the occurrence of ST depression that is elicited by physical work out in comparison to a state of rest.

The examination of the impact of exercise on heart stress reveals that a sick heart tends to experience a greater degree of stress.

The slope refers to the inclination of the peak exercise ST segment.

a. The upsloping pattern is characterized by an improved heart rate during activity, which is considered to be uncommon. b. The flatsloping pattern, on the other hand, is associated with minor changes in heart rate, which is typically observed in individuals with a healthy heart.

c. Downsloping: Indicators of Cardiac Dysfunction.

The variable "Ca" represents the count of main vessels (ranging from 0 to 3) that are colored by fluoroscopy.

a. The term "colored vessel" refers to a vessel that allows the doctor to visually observe the flow of blood. b. Optimal blood flow is characterized by increased mobility without the presence of any blood clots.

The thalium stress result is indicative of the outcome of a thallium stress test.

a. In the first instance (1,3), the observed condition is within the normal range. b. The previously identified flaw has been rectified and is now functioning properly. c. The defect observed is reversible and is characterized by inadequate blood flow during exercise. Hypothesis: Determine the presence or absence of an illness based on a binary classification, where 1 represents the existence of the disease and 0 represents its absence.

Regression Analysis for Predicating the Cardiac Disease

The data pertaining to heart diseases that has been collected from the UCI serves as a valuable resource for examining past medical records and prescribing appropriate treatments. The dataset comprises 13 distinct variables and encompasses the medical records of 300 individuals currently undergoing therapy.

The term "data" refers to a collection of facts, statistics, or information that is The initial step in data analysis is pre-processing, which involves cleaning and transforming the raw data The initial step involves loading a dataset related to cardiovascular health. Subsequently, all the records within the dataset undergo processes such as data cleansing and searching for missing values. The dataset contains all the requisite data. The attributes of the dataset exhibit a multiclass variable nature and are categorized in a twofold manner.

The process of selecting features: The patient record is distinguished by two specific dataset attributes, namely sex and age, which are extracted from a pool of 13 dataset variables and associated with unique person identifiers. All of the subsequent elements pertain to the field of medicine. Medical data have a crucial role in the prognostication of cardiovascular disease.

Classification and Predicating the Cardiac Disease

The k-Nearest Neighbors Regression is a non-parametric, supervised learning method worn for regression analysis. It utilizes nearness to categorize or forecast the alignment of an individual data point, specifically in the context of predicting cardiac illness. The method employed in this study involves estimating the relationship among independent factors and a continuous outcome by calculating the average of observations within a certain neighborhood.

The Random Forest Regression is a supervised ensemble learning method utilized for regression analysis. The utilization of ensemble learning methods involves the amalgamation of predictions generated by numerous machine learning algorithms, with the aim of enhancing the accuracy of heart illness forecasts.

Logistic regression models: Additionally, it is employed for the purpose of estimate the correlation among a dependent variable and one or more independent variables. Furthermore, it is utilized to produce prognostications about a categorical variable in comparison to a continuous variable in the context of cardiac disease.

Experimental Setup

The system under consideration has been constructed utilizing Python machine learning libraries such as scikit-learn, pandas, and matplotlib within a Python 3.7 programming environment. To execute the code, utilize Google Colaboratory and validate it on the data-bed dataset, which comprises cardiac disease data from a total of 300 patients obtained from the UCI data repository (https://archive.ics.uci.edu/ml/datasets/Heart+Disease). The dataset is utilized to evaluate the efficacy of testing carried out on k-NN, random forest, and logistic regressions.

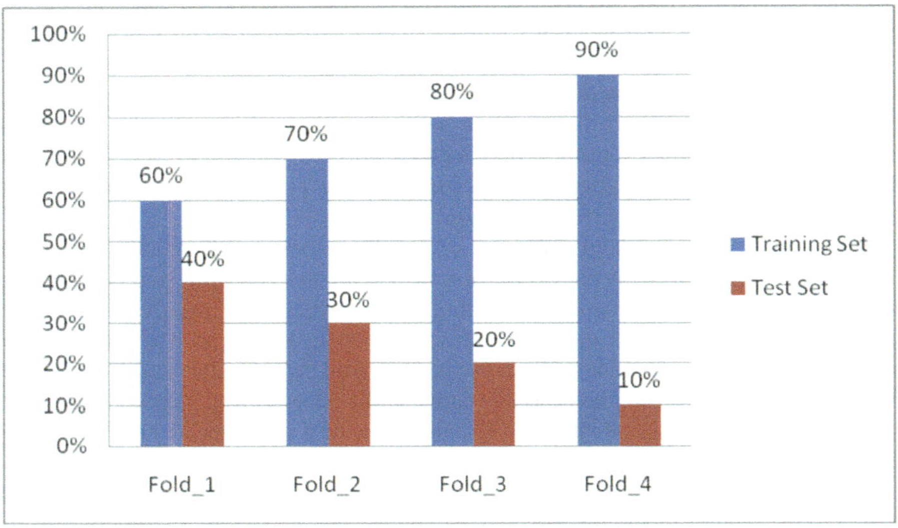

Fig. 2. Split Percentage of Training and Test Set.

So as to assess the effectiveness of the projected heart illness prediction system, the training and testing data were partitioned into four folds, each consisting of 300 records, as depicted in Fig. 2. The subsequent measures have been documented and utilized in the estimation of conventional automated metrics.

The proper prediction of heart disease (TP).

The false negative rate, which represents the number of cardiac diseases that were not correctly predicted, is of interest.

The false positive (FP) rate refers to the amount of healthy people who are incorrectly identified as having a heart condition.

The true negative (TN) represents the accurate identification and elimination of healthy patients.

Classical Metrics Estimated

Exactness of Cardiac Disease Prediction: The accuracy rate refers to the proportion of correctly predicted cases of heart illness. The estimation is conducted by employing

Eq. 1.

$$\text{Exactness of Disambiguation} = \left(\frac{TP}{TP + FP}\right) \times 100 \tag{1}$$

Sensitivity of Cardiac Disease Prediction: The metric denotes the proportion of accurately anticipated instances of heart illness that are deemed relevant. The estimation is conducted by employing Eq. 2.

$$\text{Sensitivity of Disambiguation} = \left(\frac{TP}{TP + FN}\right) \times 100 \tag{2}$$

Specificity of Cardiac Disease Prediction: The metric denotes the accuracy with which heart illness records that are not relevant are identified and accurately classified as such. The estimation is conducted by employing Eq. 3.

$$\text{Specificity of Disambiguation} = \left(\frac{TN}{TN + FP}\right) \times 100 \tag{3}$$

Accuracy of Cardiac Disease Prediction: The exactness of cardiac disease prediction refers to the proportion of correctly identified cases of cardiac disease, while the specificity pertains to the proportion of non-relevant cardiac disease records that are accurately excluded. The ability to anticipate cardiac illness is represented by the variable being calculated using Eqs. 4.

$$\text{Disambiguation Accuracy} = \left(\frac{TP + TN}{\textit{Total Results}}\right) \times 100 \tag{4}$$

The k-Nearest Neighbors (k-NN), Random Forest, and Logistic Regression algorithms were evaluated using a 4-fold cross-validation technique on the UCI dataset. The resulting accuracies of these algorithms are presented in Table 1. The logistic regression model achieves the best exactness of 0.93 when the training and testing data are split in a ratio of 90:10. In contrast, the k-NN model has the lowest accuracy of 0.81, while the random forest model does moderately well with an accuracy of 0.84. The augmentation of training data leads to enhanced precision in all models.

Table 1. Performance evaluation using classical metrics

Regression Technique	Exactness in %	Sensitivity in %	Specificity in %	Accuracy in %
Random Forest	92.86	96.3	84.62	92.5
Logistic Regression	83.93	92.16	68.97	83.75
KNN	81.82	90	66.67	81.25

Figure 3 illustrates the performance assessment of k-NN, random forest, and logistic regression models based on metrics such as exactness, specificity, and sensitivity, and

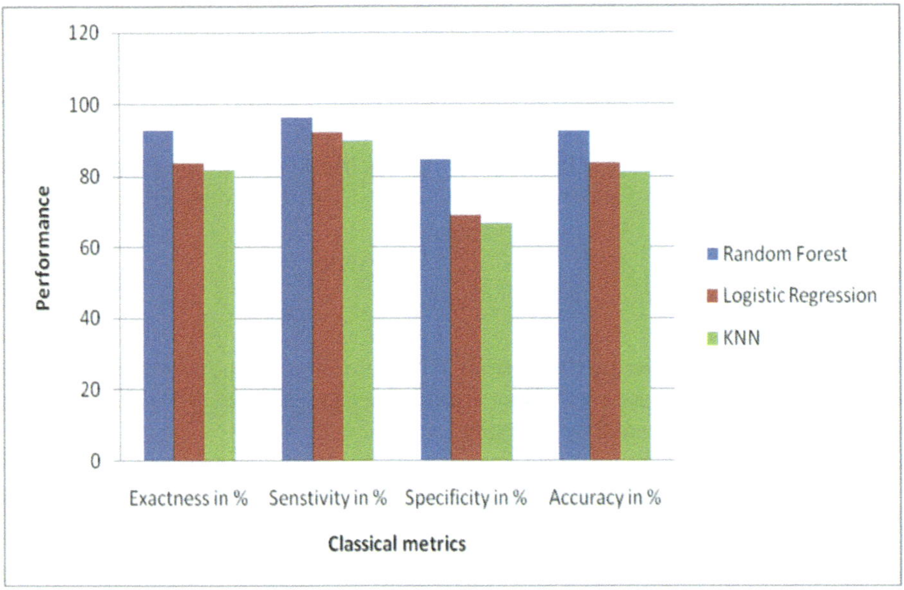

Fig. 3. Performance Evaluation using Classical Metrics

accuracy. Based on the empirical evaluation, it is evident that logistic regression outperforms k-NN and random forest in terms of exactness, sensitivity, specificity, and accuracy. This superiority can be attributed to the parametric nature of logistic regression, wherein pre-assumed parameters are utilized for variable testing or feature selection, enabling strong support for linear solutions. In contrast, k-NN is also parametric but exhibits a lazy execution during prediction, resulting in choosing of required features. Consequently, k-NN demonstrates a lower level of specificity and, as a result, lowers accuracy. Similarly, random forest, being a non-parametric model, exhibits moderate levels of exactness, sensibility, particularity, and precision.

4 Conclusion

One of the primary focuses within the medical field involves the utilization of a patient's preexisting data to forecast the occurrence of cardiovascular disease. Cardiovascular disease can be accurately anticipated in terms of its presence or absence. This study employs the k-NN, random forest, and logistic regression algorithms to construct a machine learning model that precisely predicts the presence or absence of cardiac illness by analyzing relevant clinical factors. The empirical findings and corroborating data consistently demonstrate that the utilization of regression analysis enables the prediction of cardiac risk, hence potentially leading to life-saving interventions. It is conceivable to further enhance efficiency and precision by augmenting the training dataset, as well as fine-tuning the hyper parameters by randomized search and grid search methods in the regression modeling procedures (Patil S., 2023).

References

1. Hugh, R., Leavell, E.: The science and art of preventing disease, prolonging life, and promoting physical and mental health and efficiency. Robert E. Krieger Publishing Company (1979)
2. Kanthi, M., Rao, K.V., Sekhar Reddy, L.C., Sarma, T.H., Bhaskar, N., Vasundhara, N.: 3D-CAN: a 3D convolution attention network for feature extraction and classification of hyper-spectral images. In: 2023 International Conference on Network, Multimedia and Information Technology (NMITCON), Bengaluru, India, pp. 1–6 (2023). https://doi.org/10.1109/NMITCON58196.2023.10275985
3. Tamilselvan, R., Prabhu, A., Rajagopal, R.: An enhanced K-means algorithm for large data clustering in social media networks. In: Artificial Intelligence for Sustainable Applications, pp. 147–162 (2023)
4. Chandramowleeswaran, G., Choubey, C., Pendem, S., Pal, S., Verma, A.: Implementation of human robot interaction with motion planning and control parameters with autonomous systems in Industry 4.0. In: 2023 Second International Conference on Augmented Intelligence and Sustainable Systems (ICAISS), Trichy, India, pp. 1805–1810 (2023). https://doi.org/10.1109/ICAISS58487.2023.10250484
5. Murthy, N., Spandan, G., Aswani, I., Bhukya, S.N., AC, R.: Deep learning and MRI improve carotid arterial tree reconstruction. In: 2023 International Conference on Data Science and Network Security (ICDSNS), Tiptur, India, pp. 1–7 (2023). https://doi.org/10.1109/ICDSNS58469.2023.10245397
6. Reddy Madhavi, K., Kora, P., Venkateswara Reddy, L., Avanija, J., Soujanya KLS., Prabhakar, T.: Cardiac arrhythmia detection using dual-tree wavelet transform and convolutional neural network. Soft Comput. Impact factor **3**(64) (2021)
7. Abbagalla, S., Rupa Devi, B., Anjaiah, P., Reddy Madhavi, K.: Analysis of COVID-19-impacted zone using machine learning algorithms. In: Springer series – Lecture Notes on Data Engineering and Communication Technology, vol. 63, pp. 621–627 (2021)
8. Reddy Madhavi, K., Vijaya Sambhavi, Y., Sudhakara, M., Raju, S.K.: COVID-19 isolation monitoring system. In: Springer series – Lecture Notes on Data Engineering and Communication Technology, vol. 63, pp. 601–609 (2021)
9. Bashir, S., Khan, Z.S., Khan, H., Anjum, A., Bashir, K.: Improving heart disease prediction using feature selection approaches. In: Proceedings of the 16th International Bhurban3 Conference on Applied Sciences and Technology (IBCAST) (2019)
10. Aelgani, V., Vadlakonda, D., Lendale, V.: Performance analysis of predictive models on class balanced datasets using oversampling techniques. In: Soft Computing and Signal Processing: Proceedings of 3rd ICSCSP 2020, vol. 1. Springer, Singapore (2021)
11. Kumar, A., et al.: Artificial intelligence bias in medical system designs: a systematic review. Multimedia Tools Appl., 1–53 (2023)
12. Yalawar, M.S., Vijaya Babu, K., Mahender, B., Singh, H.: A brain-inspired cognitive control framework for artificial intelligence dynamic system. In: Kumar, A., Mozar, S., Haase, J. (eds.) Advances in Cognitive Science and Communications. ICCCE 2023. Cognitive Science and Technology. Springer, Singapore (2023). https://doi.org/10.1007/978-981-19-8086-2_70

Assessment on Deep Learning Framework for Forecasting Patient Expenditure in Healthcare Sector

K. Srinivas[1(✉)], Mudimela Madhusudhan[2], G. Mallikarjun[3], M. Shiva Kumar[4], Chandrugonda Malleshwar Rao[5], and Botcha Kishore Kumar[6]

[1] Departmentof CSE (DS), CMR Technical Campus, Hyderabad, Telangana, India
cmrtc.paper@gmail.com
[2] Department of CSE, CMR Technical Campus, Hyderabad, Telangana, India
[3] JNTUH, Department of CSE, Research Center, CMR College of Engineering and Technology, Hyderabad, Telangana, India
g.mallikarjun@cmrcet.ac.in
[4] Department of CSE, CMR College of Engineering and Technology, Hyderabad, Telangana, India
mshivakumar@cmrcet.ac.in
[5] Department of CSE (AI&ML), CMR Engineering College, Hyderabad, Telangana, India
228rapr6611@cmrec.ac.in
[6] Departmentof CSE (DS), CMR Engineering College, Hyderabad, Telangana, India

Abstract. Some of the activities encompassed in this category include the practice of business profiling, the implementation of accountability care systems, and the adjustment of capitated scientifically pricing models using patient spending estimates. The present systems in use have significant challenges related to their predictive capabilities and data requirements. These approaches heavily depend on manually constructed models based on point and linear regression techniques. This article presents a multi-view deep learning system that use past claims data to forecast individual-level healthcare expenses. Our multi-view technique is capable of successfully representing several types of data, including patient demographics, clinical codes, drug execution, and facility utilization. Budget prediction study was conducted with data obtained from an authentic paediatric database. The strategy we offer demonstrates superior performance compared to all baseline methods in accurately estimating research expenditures, as supported by empirical evidence from the obtained data. The results of this research enhance the advancement of healthcare by enhancing approaches for both preventative and curative measures.

Keywords: Deep learning · patients · paediatric database

1 Introduction

One of the primary obstacles faced by healthcare practitioners and care organizations pertains to the escalating healthcare expenditures. The management of healthcare expenditure fluctuations, both in terms of upward and downward trends, holds significant

D. Kriksciuniene et al. (Eds.): IBICA 2023, LNNS 1228, pp. 416–422, 2025.
https://doi.org/10.1007/978-3-031-78937-3_45

importance. Billing purposes frequently make use of claims data, which is a distinct form of electronic health records (EHR). It encompasses comprehensive and continuous records pertaining to the health status of individuals, including their demographic information, diagnoses, medical procedures, prescribed drugs, and details on the healthcare facilities involved. Based on the data provided by the National Health Expend Data, it can be inferred that information serves as a highly complete resource for assessing the physical fitness levels of patients. The extensive reservoir of claims data introduces a fresh and hopeful approach to tackle the rising healthcare costs dilemma. By leveraging prior claims, it is possible to enhance the efficacy of data-driven methodologies in uncovering significant observations pertaining to expenditure trends. In order to identify patients who are at a heightened scientific risk and deliver a more elevated level of treatment, the utilization of an accurate scientific price prediction model at the character level can be advantageous. The existing approaches for forecasting patient expenditure generally depend on the utilization of manually constructed data points and models derived from linear regression [1–3]. An example of a model that relies solely on diagnostic classifications is the DCG (Diagnostic Classification Group) model [4]. Healthcare cost projections are formulated using linear regression models, which were manually constructed in collaboration with specialists from the respective country. Author developed a classification and regression tree (CART) model by incorporating a combination of scientific programming and user-generated fee elements. These tendencies contribute to the facilitation of healthcare expense forecasting. Nevertheless, these constraints present significant challenges to overcome.

The process of categorizing high-dimensional clinical codes into meaningful groupings heavily relies on the utilization of local knowledge.

They are not harnessing the valuable insights within the data assertions, such as facility usage, time-related information, and the correlation of scientific codes.

The utilization of linear regression in the model imposes constraints on its predictive capacity and results in suboptimal model performance.

This article provides an overview of the endeavors made to eliminate the obstacles hindering the process of reading. Our endeavors are motivated by a diverse array of observations. One notable aspect pertains to the diverse range of information that may be derived from insurance claims statistics.

1.1 Literature Review

The assessment of patient expenditure in the healthcare sector is a crucial undertaking with diverse possible uses include professional characterization, responsible healthcare administration, and alteration of fixed medical reimbursements. The existing techniques in use at now depend on the utilization of manually constructed characteristics and models utilizing linear regression. However, these approaches necessitate a substantial level of skill in the medical field and exhibit limited accuracy in their predictions. The present study presents a novel approach in the field of healthcare cost prediction by developing a multi-view deep learning system. This system utilizes historical claims data to anticipate future healthcare expenses at the individual level [1].

Various categories of pharmaceuticals that had been prescribed over a span of six months for the purpose of treating and controlling persistent medical conditions were employed in the process of enhancing and confirming the accuracy of a chronic disease assessment tool. The chronic disease score was developed using prescription data, along with age and sex as additional factors. Each of the three results, which include overall expenses, outpatient expenses, and primary care visits, were given weights that were empirically determined [5]. An empirical examination was undertaken to gauge the prognostic effectiveness of the updated chronic sickness rating concerning healthcare utilization, costs, hospital stays, as well as death. This assessment encompassed a juxtaposition with the initial rendition of the prolonged disease score, which was formulated via clinical evaluations of disease severity. Furthermore, the precision of the chronic illness score in forecasting healthcare outcomes was also scrutinized with regard to ambulatory care groups, which employ diagnosis for outpatients to delineate distinct diagnostic classifications [6].

The study utilized population-based automated pharmacy data to examine the utilization patterns of specific prescription prescriptions over a span of one year. These patterns were found through a consensus judgement procedure. The researchers then developed a metric called the Chronic Disease Score, which served as an indicator of an individual's chronic disease condition. The stability of this score throughout time and its correlation to various health status variables were assessed [7].

This current investigation seeks to create and assess the RxRisk framework, a pioneering risk appraisal tool that employs automated ambulatory pharmacy data to detect chronic diseases and anticipate forthcoming healthcare costs. The effectiveness of the RxRisk model in projecting expenses is appraised in contrast to a demographics-centric model, the Ambulatory Clinical Groups (ACG), and the Hierarchical Coexisting Conditions (HCC) ICD-9-CM diagnosis-oriented risk evaluation tools. An examination of every model's capacity to foretell healthcare resource usage is also carried out [8].

The goal of this study is to explore and evaluate the efficacy of models that utilize inpatient encounter data and outpatient pharmacy claims data in forecasting future healthcare expenditures. The pharmacy-based and inpatient-diagnosis models exhibit similar overall performance, with the former demonstrating greater efficacy in identifying individuals with low-cost profiles, while the latter excels in identifying a tiny subset of individuals with exceptionally high future costs. The combined model, which incorporates both types of data, had superior performance compared to each individual model. This was evidenced by an R2 value of 11.8 percent [9].

This ongoing research introduces an innovative multi-perspective deep learning structure with the aim of predicting future personal medical care costs, utilizing past claims data as the primary input. The multi-view strategy employed in our study efficiently captures and models the diverse and disparate information sources, this involved data elements like patient demographic information, medical identifiers, medication usage, and healthcare facility use. Expenditure prediction activities were carried out using a real-world dataset focused on pediatric patients, encompassing over 450,000 individuals. The empirical findings demonstrate that the method we have suggested exhibits superior performance compared to the baseline models in the prediction of medical spending. The aforementioned findings contribute to the advancement of preventative care and accountable care within the healthcare sector [10, 11].

In this investigative study, a versatile and resilient healthcare apparatus is suggested for monitoring patient well-being and managing illnesses in a home setting [12–15]. The device uses equipment often found in medical devices, such as dialysis machines, ventilators, and patient monitors. For patients who live at home and are susceptible to power disruptions, this gadget is both reasonably priced and appropriate.

2 Methodology

Multi-view learning is a well-known framework in which heterogeneous input can be expressed by multiple modalities, and each modality is decoded through a distinct learning perspective or module. The heart of a multi-perspective learning structure encompasses the analysis of patient representations that is suitable for terrain conditions. This is achieved by including three various viewpoints, namely demographics, scientific canons, and installation applications, within The structure. The layout of our model is depicted in Fig. 1, wherein each distinct color represents an individual learning component. The initial viewpoint is the demographic encoder, as illustrated in Fig. 1. This method entails the use of a feedforward neural network to depict demographic details in the form of a vector illustration. The second perspective relates to the utilization encoder, which utilizes an attention-driven bidirectional recurrent neural network to infer the application vector based on the application sequence. The construction of the scientific law portrayal involves the utilization of a hierarchical interest neural network, which operates within the 0.33 perspective of the scientific law encoder. The last stage in the process of projecting research funding involves the amalgamation of vectors generated by all three perspectives. The model that has been developed is comprehensive in all aspects. The dataset comprises a collection of values. The dataset was obtained from the Kaggle website.

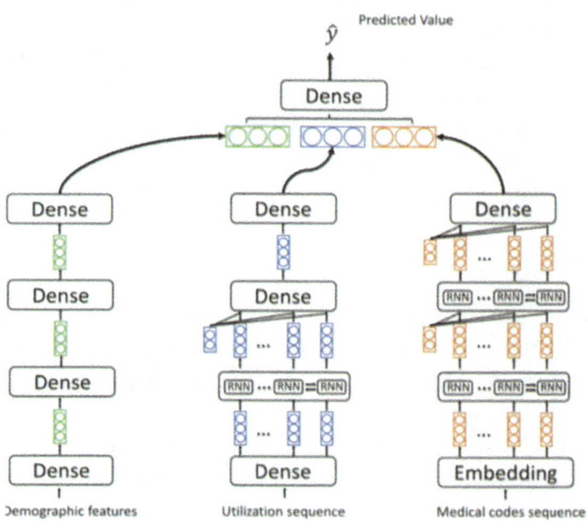

Fig. 1. Multi view framework [11]

3 Implementation

3.1 Data Collection

The work initial stage and a very vital module is data collection. The process often entails the compilation of an appropriate dataset. The dataset utilized for market forecasting must satisfy certain criteria. The process of data collection additionally facilitates the enhancement of datasets through the integration of external information.

The primary source of our data is derived from the medical Premium. The initial step involves doing an analysis of the Kaggle dataset, followed by the utilization of a model that aligns with the data's correctness. This approach will enable us to accurately examine the forecasts.

3.2 Pre-processing

Preparation is a pivotal stage in data analysis and machine learning undertakings. It encompasses converting unprocessed.

The pre-processing of unstructured data constitutes an integral component of the data mining process. The raw data may contain inaccuracies, inconsistencies, or a combination of both.

The process of data pre-processing involves several steps, including identifying missing values, handling categorical variables, partitioning the dataset into training and test sets, and performing feature scaling to normalize the variables and enable meaningful comparisons. The preprocessing technique employed in this study involves the utilization of the Min Max Scalar method.

3.3 The Process of Training the Machine

Just as the process of training a machine involves providing it with data, the act of feeding data to an algorithm enhances its performance when tested.

The training sets are employed for the purpose of modifying and fitting the models.

Due to the inherent inaccuracy of assessing a model's performance using hypothetical data, the test sets remain unaltered. The model training process incorporates a technique known as cross-validation, which allows for the estimation of the model's performance by using the training data.

3.4 Results and Analysis

The aim of this study is to forecast medical expenditure.

In order to construct prediction models, a selection of algorithms including NN (Neural Network), LSTM (Long Short-Term Memory), Random Forest, Linear Regression, Voting, and Decision Tree are employed.

Once the training of all the algorithms is completed, it becomes necessary to evaluate their precision. The most precise technique for predicting costs will be utilized, as illustrated in Fig. 2.

After identifying the optimal method, the user's data is inputted, enabling the algorithm to utilize this information in order determine the most suitable medical expense for the individual using it.

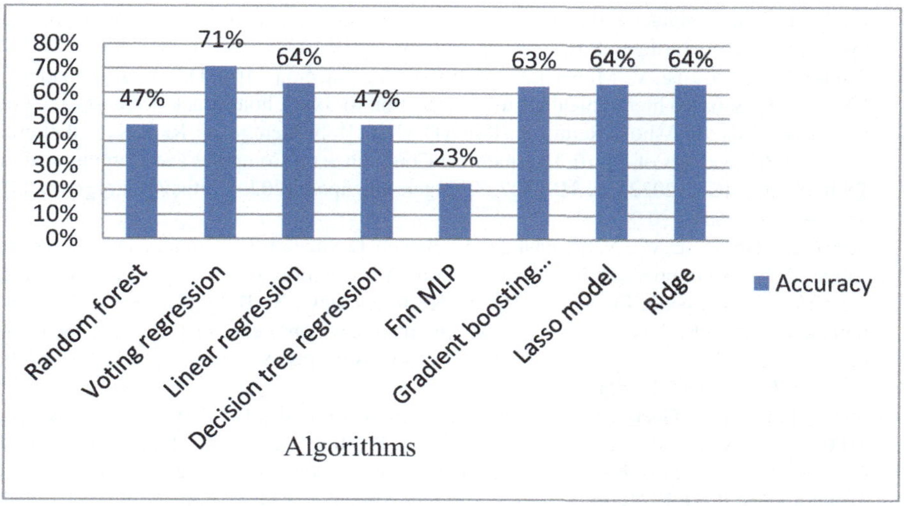

Fig. 2 Various prediction models

3.5 Conclusion

The method of data pre-processing in data mining can enhance the comprehensibility of unstructured data. Regrettably, the raw data frequently contains inaccuracies and inconsistencies.

Data preprocessing involves several steps, including the identification and handling of missing data, the identification and categorization of specific data categories, the partitioning of Partitioning the data into distinct training and testing datasets, and the application of feature scaling techniques to normalize the data for improved comparability. The preprocessing phase involves the utilization of the Minmax Scalar tool.

References

1. Morid, M.A., Sheng, O.R.L., Kawamoto, K., Ault, T., Dorius, J., Abdelrahman, S.: Healthcare cost prediction: leveraging fine-grain temporal patterns. J. Biomed. Inf. **91**, 103113 (2019)
2. Ash, A.S., et al.: Using diagnoses to describe populations and predict costs. Health Care Finan. Rev. **21**, 7–28 (2000)
3. Cowen, M.E., Dusseau, D.J., Toth, B.G., Guisinger, C., Zodet, M.W., Shyr, Y.: Casemix adjustment of managed care claims data using the clinical classification for health policy research method. Med. Care **36**, 1108–1113 (2021)
4. Reddy Madhavi, K., Jyothi Jarugula, G., Karuna, S.K., Srujan Raju, K., Sunitha, G.: Bird species recognition using deep transfer learning. In: Ashoka Reddy, K., Rama Devi, B., Boby George, K., Raju, S., Sellathurai, M. (eds.) Proceedings of Fourth International Conference on Computer and Communication Technologies: IC3T 2022, pp. 473–479. Springer, Singapore (2023). https://doi.org/10.1007/978-981-19-8563-8_46
5. Narasimharao, J., Laxmaiah, B., Arumalla, R., Unnisa, R., Fatima, T., Nazare, S.S. (2023). Restoration and Deblurring the Images by Using Blind Convolution Method. In: Reddy, K.A., Devi, B.R., George, B., Raju, K.S., Sellathurai, M. (eds) Proceedings of Fourth International

Conference on Computer and Communication Technologies. Lecture Notes in Networks and Systems, vol 606. Springer, Singapore. https://doi.org/10.1007/978-981-19-8563-8_32
6. Kumar, V.N., Srisuma, V., Mubeen, S., Mahwish, A., Najeema Afrin, D.B.V., Jagannadham, J.N.: Anomaly-based hierarchical intrusion detection for black hole attack detection and prevention in WSN. In: Ashoka Reddy, K., Rama Devi, B., Boby George, K., Raju, S., Sellathurai, M. (eds.) Proceedings of Fourth International Conference on Computer and Communication Technologies: IC3T 2022, pp. 319–327. Springer, Singapore (2023). https://doi.org/10.1007/978-981-19-8563-8_30
7. Fernandes Dimlo, U.M., Narasimharao, J., Bagam Laxmaiah, E., Srinath, D.S., Rani, S., Kumar, V.N.: An improved blind deconvolution for restoration of blurred images using ringing removal processing. In: Ashoka Reddy, K., Rama Devi, B., Boby George, K., Raju, S., Sellathurai, M. (eds.) Proceedings of Fourth International Conference on Computer and Communication Technologies: IC3T 2022, pp. 357–366. Springer, Singapore (2023). https://doi.org/10.1007/978-981-19-8563-8_34
8. Shaik Munawar, A., Geetha, K.S.: Arrhythmia classification using BiLSTM with DTCWT and MFCC features. In: Ashoka Reddy, K., Rama Devi, B., Boby George, K., Raju, S., Sellathurai, M. (eds.) Proceedings of Fourth International Conference on Computer and Communication Technologies: IC3T 2022, pp. 311–318. Springer, Singapore (2023). https://doi.org/10.1007/978-981-19-8563-8_29
9. Zhao, Y., et al.: Measuring population health risks using inpatient diagnoses and outpatient pharmacy data. Health Serv. Res. **36**, 180–193 (2001)
10. Subburaj, T., et al.: Discover crypto-jacker from blockchain using AFS method. In: Ashoka Reddy, K., Rama Devi, B., Boby George, K., Raju, S., Sellathurai, M. (eds.) Proceedings of Fourth International Conference on Computer and Communication Technologies: IC3T 2022, pp. 145–156. Springer, Singapore (2023). https://doi.org/10.1007/978-981-19-8563-8_15
11. Jagadeesh, S., Shrimali, M., Prasad, P.S., Maheswari, K., Narendrakumar, A., Reedy, G.K.: Automated arrhythmia classification using harris hawks optimization with deep learning model on IoT environment. In: 2023 Third International Conference on Artificial Intelligence and Smart Energy (ICAIS), Coimbatore, India, pp. 30–35 (2023). https://doi.org/10.1109/ICAIS56108.2023.10073736
12. Thaseen, A., Unnisa, R., Sultana, N., Madhavi, K.R., NagaJyothi, G., Kirubakaran, S.: Breast cancer detection using deep learning model. In: Brahmananda Reddy, A., Nagini, S., Balas, Vlentina E., Srujan Raju, K. (eds.) Proceedings of Third International Conference on Advances in Computer Engineering and Communication Systems: ICACECS 2022, pp. 669–677. Springer, Singapore (2023). https://doi.org/10.1007/978-981-19-9228-5_57
13. Reddy Madhavi, K., Sandya Vooradi, P., Mounika, S.Y., Tangudu, N.: Emotion analysis using convolutional neural network. In: Brahmananda Reddy, A., Nagini, S., Balas, V.E., Srujan Raju, K. (eds.) Proceedings of Third International Conference on Advances in Computer Engineering and Communication Systems: ICACECS 2022, pp. 523–531. Springer, Singapore (2023). https://doi.org/10.1007/978-981-19-9228-5_44
14. Grande Nagajyothi, G., Kumar, Pavan, Kumar, B.S., Deepak Kumar, B.P., Damodaram, A.K.: High-speed low area 2D FIR filter using vedic multiplier. In: Brahmananda Reddy, A., Nagini, S., Balas, V.E., Srujan Raju, K. (eds.) Proceedings of Third International Conference on Advances in Computer Engineering and Communication Systems: ICACECS 2022, pp. 569–578. Springer, Singapore (2023). https://doi.org/10.1007/978-981-19-9228-5_48
15. Kumara Swamy, M., Krishna Reddy, P.: A model of concept hierarchy-based diverse patterns with applications to recommender system. Int. J. Data Sci. Anal. **10**, 177–191 (2020)

A Dynamic Effective Class Balanced Approach for Semantic Segmentation of Imbalanced Medical Image Data

Sreedevi Chikkudu[1], Konkala Divya[2], A. Jyothi Babu[3], Ranjith Kumar Painam[4], K. Reddy Madhavi[5](✉), and V. Mahalakshmi[6]

[1] Department of CSE, B V Raju Institute of Technology, Narsapur, Telangana, India
[2] Department of Computer Science and Engineering (AI & ML), Vardhaman College of Engineering (Autonomous), Hyderabad, Telangana, India
[3] Department of Computer ApplicationsSchool of Computing, Mohan Babu University, Tirupati, India
[4] Department of ECE, Kallam Haranadhareddy Institute of Technology (Autonomous), Guntur, Andhra Pradesh, India
[5] School of Computing, Mohan Babu University, Tirupati, India
kreddymadhavi@gmail.com
[6] Department of Computer Science, College of Computer Science and Information Technology, Jazan University, Jazan, Saudi Arabia

Abstract. The current research challenge is to build algorithms that are output-focused and adhere to the trend of quality assurance. Inter- and intra-behavioral changes in the system have an impact on the output quality. This is a result of the class imbalance problem (CIP) that exists in the data sets and is caused by the reading of unknown values. When data sets are in such a state, predictive models that rely on them become severely unbalanced. As selection of samples and candidates for the algorithms is the first step in the CIP on uncertain data, the chosen random samples predict a high degree of class variation, a very small class, or perhaps a class that is much larger than anticipated. The semantic segmentation of images can be accomplished using a dynamic weighting method that is based on the effective sample idea. The results show that this weighting approach can improve the minimal-class segmentation accuracy while guaranteeing that the segmentation performance overall in multi-class segmentation tasks is verified in two separate semantic segmentation tasks.

Keywords: imbalanced data · deep neural network · convolutional neural network

1 Introduction

Machine learning ushered into Deep learning, classification faces challenges with real-time data, which consists of various classes of data. The uncertain presence of certain attributes in real-time data sets can cause imbalances in decision making, as instances

D. Kriksciuniene et al. (Eds.): IBICA 2023, LNNS 1228, pp. 423–432, 2025.
https://doi.org/10.1007/978-3-031-78937-3_46

are considered differently from another class. This problem is related to dealing with imbalanced datasets in deep learning. A positive class, or positive instances, is a target class with less high intensity instances. In deep learning algorithms, the prediction of positive instances to the positive class becomes less important, and additional exercises are needed to maximize the accuracy of predicting positive classes.

To handle class imbalances, methods such as sampling techniques are proposed to transform imbalanced datasets into well-balanced distributed datasets suitable for deep learning classification using effective classifiers. One simple approach is to repeat the duplication process of positive instances in the positive class until the number of positive instances and negative instances are equal in quantity. The classifier must be cautious of overfitting and synthesize positive instances with other positive instances in the periphery.

In a typical medical classification the severity of class imbalance problem is overcome with the images that appear as similar but different in minor number of features. This kind of medical image classification is often handled by two methods, sampling of small sizes and labeling due to cost of labeling issues, identification of prevalent rare classes or common classes, which could only be expected to be handled by the medical experts. The members of rare class appear in small percentages whereas the common class appears in high percentages. So, a large sample with broad features can ignore the rarity and commonality of the features in the positive classes.

Classification in Deep Learning faces challenges with real-time data. Classification in deep learning is intended to characterize the data sets towards a problem of decision making. Real-time data consists of various classes of data. The real-time data sets are composed with features often representing attributes that generally will not come into consideration of computations in deep learning. The uncertain presence of some attributes causes imbalances in decision making, as instances of datasets are considered into one target class differently from another class [2, 3]. Therefore, the problem of class imbalance is related with dealing of imbalanced datasets in deep learning. Further in the study related to the problem of class imbalance is the target class with positive presence, i.e., positive class which may contain less number of high intensity instances. The instances are lesser but contain high intensities in the positive class; they are also called positive instances or minor instances.

In deep learning most of the classification algorithms focus on maximization of the performance parameters related to accuracy [4, 5]. On iterative performance of the algorithms in deep learning for classification, the instances appear as of less importance instances and the classes as the less significant classes. In such situations the prediction of positive instances to the positive class becomes less importance of the real objective of the problem of classification in deep learning [4–6]. In order to consider such instances for boosting the accuracy of prediction additional exercises have to be considered for maximization of the accuracy of predicting positive classes.

The objective of handling class imbalances is to include uncertain positive instances into the positive class and improve the prediction rates. Many methods are proposed in the literatures for effective and practical prediction rate convergence. Major methods of handling class imbalances work with sampling techniques, to transform the imbalanced datasets into well-balanced distributed datasets suitable for classification in deep learning using effective classifiers [5–7]. The sampling methods [7–9] are mostly favorable for their portability in various applications, where a researcher is not restricted to a specific category of classifier. One of the simple approaches for handling class imbalances using sampling methods is to repeat the duplication process of positive instances in the positive class, until the number of positive instances and negative instances are equal in quantity. The classifier is forced to learn every specific instance with all the general features which are uncertain causing imbalance, and shall be beware of overfitting. The synthesizing of positive instances with other positive instances in the periphery, by inheriting the features that fit to designate the instance as positive is generally applied [10].

2 Related Work

The problem of class imbalances are handled by machine learning and deep learning using three methods, viz., active learning, cost-sensitive learning and sampling methods. Active learning is widely popular in deep learning, which uses small training data sets. With minimum effort and domain knowledge for labeling by using a strategic querying methods unlabeled datasets are extracted and synthesized as labeled datasets. Active learning is again a cognitive and iterative solution [11, 12]. At each iteration, the classifier is enriched with new feature information in order to collect iteratively as many datasets as more possible into the class as members. Different kinds of cost matrices can be used in cost-sensitive learning method, in order to describe cost of unclassified samples of different classes [13]. Sampling based methods, effortlessly aim to build the datasets considering the proportions of classes by examples from the real-time training samples, therefore preparing a balanced datasets. Sampling based methods or cost-sensitive learning method are not encouraged for solving the class imbalance problem on large image datasets.

Image segmentation is the fundamental process in the data augmentation of image processing with machine learning and deep learning methods. There conventional techniques and artificial intelligence based techniques. The conventional image segmentation techniques include methods based on threshold, edges, regions and graph-theoretic approaches [8]. The majority of threshold-based segmentation techniques first define thresholds based on the outcomes of band operations or common feature indices in the image [11]. Following that, they classify the pixels into the proper groups. As a result of discontinuities in the local features of the image, edge-based segmentation algorithms frequently identify edges between various parts of the image by detecting changes in the image's grayscale values and feature indexes [12]. A simile of watershed transformation is methodically applied as a technique employed popularly for the detection and segmentation of edges within the context of mathematical morphology theory [13]. This program simulates flooding to establish region borders by using two-dimensional photographs as the elevation data. In order to discriminate between various regions, region-based segmentation algorithms emphasize similarity within regions. The region-growing technique [14, 16] begins with the selection of seed pixels, links nearby pixels based on a similarity criterion, and repeats this process until the full region is generated. The region segmentation and merging algorithm divides the image into several subregions before merging them together in accordance with their individual characteristics

[12, 15, 17]. The image is mapped as a weighted, undirected graph using the graph-based segmentation technique. The segmentation of the image is accomplished by cutting and removing particular edges, and in the graph, the weights on each edge reflect the differences between pixels. To maximize similarity within subgraphs and reduce similarity between subgraphs is the segmentation principle [12, 13, 18]. It is also possible to combine the aforementioned techniques, for example, by merging related segments using a region-based algorithm and extracting the initial segments using an edge algorithm [12, 19], which accomplishes the goal of taking into account both internal spatial information and regional border information.

Deep learning semantic segmentation method is suitable for image segmentation tasks due to large data volume and complex backgrounds. However, images have larger image sizes and smaller target proportions, leading to foreground-background imbalance and inter-category imbalance. Deep neural networks are more effective at segmenting target categories with more pixels, weakening segmentation ability for categories with few pixels, resulting in information interpretation failure in images.

Research in image segmentation addresses class imbalance problem, proposes combined sampling methods, Deeplab-v3 + model, Tanimoto variant, tree species classification, and synthetic minority oversampling technique-based rotation forest algorithm.

Natural images face class imbalance in sample data, affecting the frameworks involved with detection of targets, classification, annotation and instance segmentation in images. Deep learning methods include class-rebalancing, information enhancement, and module improvement. Re-weighting, balanced meta-softmax, label distribution disentangling, and Meta-Wheel-Net optimize model classification performance by learning optimal sample distribution parameters and updating weighting functions.

The Dual Focal Loss function modifies the Focal Loss' loss scaling mechanism to increase classification accuracy in unbalanced images. This paper proposes a Dynamic Effective Class Balance (DECB) weighting method for semantic segmentation of images, demonstrating its effectiveness in highlighting minimal cases of classes as juxtaposed to compromising overall effect in image segmentation.

3 Proposed Work

A dermoscopic data set is used in the proposed work. Dermoscopic images from various demographics were obtained for the lesion analyses for the detection of melonama and mastocytosis as a data set [1], which included 10,015 dermoscopic images of patients.Actinic keratoses and intraepithelial carcinoma/Bowen's disease (akiec), basal cell carcinoma (bcc), benign keratosis-like lesions (bkl), dermatofibromas (df), melanomas (mel), melanocytic nevi (nv), and vascular lesions (vasc) are all included in this representative group of pigmented lesions.

The ratio during data assignments methods with the unlabeled and labeled data is 2:8. Labeled data is crucial for extracting class imbalance factors and combining it with unlabeled data for training. The total number of clients labeled is set to 0 and the parameter control of clients for the unlabeled data assigned between 0 and 8. This allows for flexible training patterns, allowing for pre-trained models and comparing differences between federated and ordinary semi-supervised learning modes. The optimal number of unlabeled clients can be determined through experiments for optimal model performance (Fig. 1).

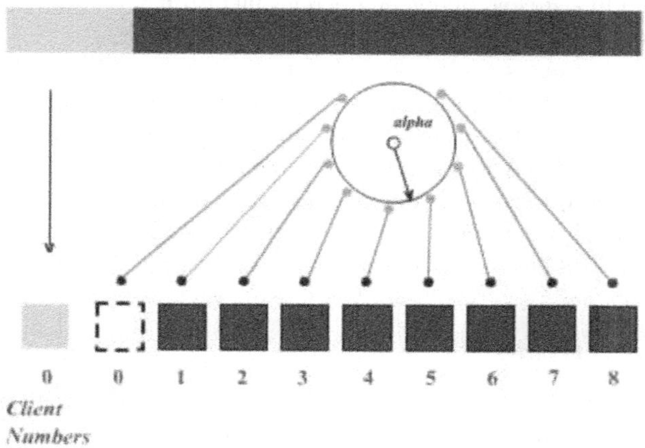

Fig. 1. Controlled label-free data distribution method

Handling the Class Imbalance Problem is one of the important challenges faced in deep learning while data preprocessing and augmentation [4, 9, 12, 13, 20]. The challenge of 'dealing with rarity' starts from identifying the features from the benchmark datasets. The outcome of the solution that is arrived from the class imbalance problem is to reduce the majority classifier criterion or to enhance the minority classifier criterion by the considering the change of distribution in samples. General method of handling class imbalances are undersampling and oversampling.

Most of the reasons for class imbalances is the skewness in the datasets. However, the class are equally sized or numbered with images, the number of characteristic features that are represented by the images matters important for an ideal class. Random sampling generated headed data or tailed data which is not suitable for study of image data sets for classification in deep learning. Naturally based on the criterion of the classifiers the classes generated are into two types, minority classes and majority classes. Minority class focus on less number of characteristic features, majority class focus on more number of characteristic features. In order to balance the representability of the characteristic features, the number of positive instance in the classes are added. While adding the number of positive instances in order to increase the size of class with more instances of characteristic features, the possibility of the error lies in generating the headed data, which leads to positive skewness of the data. While stabilizing the majority class, when some of the positive instances are removed and shared with other classes, the majority class is added with least characteristic features, where the possibility of the error lies in generating tailed data, with a negative skewness of the data.

The technique of undersampling underlies the principle of selecting a particular set of images randomly that belongs to a class with a number that is more than actual. The greater number of images will deal with the class with the same count of images with other class or classes in the image datasets. This technique is more properly used to balance the uneven datasets of classes by retaining the data in the minority class and further by decreasing the sizes of majority classes. However, the complications that arise due to undersampling are loss of potentially relevant information from the majority class as they are reduced.

The technique of oversampling underlies the principle of duplication of randomly selected images from the minority classes. Synthesizing would help to enhance more images with the like features of the existing minority classes, therefore enhancing the sizes of all classes and making the sizes all similar. The complications that arise due to oversampling is the 'overfitting of the model', due to duplicates in the minority classes and making the model learn only specific patterns in the large population of samples, where the model would never learn generalized patterns.

3.1 Effective Sample Space

Boundary effects appear during the expansion of the unbalanced datasets, which indicates that the classification performance of the model for a certain category won't improve when the number of samples in that category exceeds a certain number range.

$$E_S = \frac{(1 - H^S)}{(1 - H)}$$

where S is the number of samples and the hyperparameter H, is the probability of overlaps between the new sample and the existing sample, which is denoted as follows:

$$H = \frac{N - 1}{N'}$$

where N is the sample subspace and as the ideal sample space is infinite, H cannot be calculated and only hyperparameters are taken. To compute the hyperparameter H, the effective samples space size E_N can be obtained by combining the above two equations as follows:

$$E_N = \frac{(1 - H^N)}{(1 - H)} = H^0 + H^1 + H^2 + \cdots + H^{m-1}$$

Considering the first m terms of the equation, and if the sum of first m terms is 10p times the sum of the next m terms, the second n-m terms can be ignored and the value range of p is considered to be as more than or equal to 3. The following inequality is derived.

$$0 < H \leq \sqrt[n]{\frac{1}{10^p + 1'}}$$

Therefore the size of the effective sample subspace varies between the values of m and p. There is a positive correlation between the *m* and the negative correlation with *p* with respect to the E_m (Fig. 2).

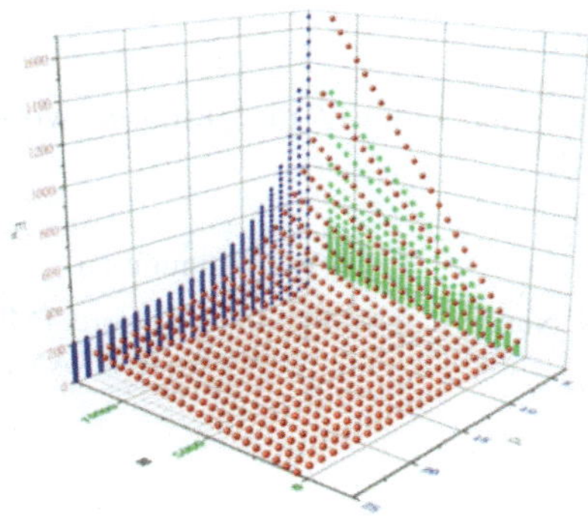

Fig. 2. The fitted surface demonstrated in scatterplot.

The formula calculates probability and effective sample space size for each sample class based on known subspace number, p, and maximum effective sample subspace, allowing deep learning methods to calculate these subspaces. The paper proposes an effective sample space algorithm that is closer to the ideal state and more suitable for semantic segmentation tasks with large numbers of samples, despite a constant H reaching an upper limit.

3.2 Dynamic Class Balance Weighting Method

The dynamic class balance weighting method (DCB) is a widely used technique in semantic segmentation tasks to balance the contribution of each classification in an unbalanced dataset. Weights of ith instance are computed as:

$$W_i = 1 - \frac{b_i}{b_{batch}}$$

where b_{batch}, is the sum of sample sizes within the batch and b_i is the sample size of class i, in the same batch. This paper proposes a minimal class in the neural network training process to increase the weight of categories with small sample sizes. The DECB method is determined by the proportion of each category in the sample subspace, with the weight of each category negatively correlated with its proportion.

Batch Size	b_{batch}	H	E_{bbatch}	n_{bi}	DCB Wts	E_{nbi}	DECB Wts
4	1046876	0.9999936	1046883	104600	0.099916	20937.52	0.80067
8	2098242	0.9999951	2098252	209800	0.099988	41964.84	0.800092
12	3172634	0.9999965	3172645	317200	0.09998	63452.68	0.80016
16	4178624	0.9999981	4178632	417800	0.099985	83572.48	0.800119
18	5097432	0.9999995	5097435	509700	0.099991	101948.6	0.800068

4 Results

The experimental evaluation and setup of the proposed work includes the following configurations:

Coding Tool	Libraries	H/W Configuration
Python 3.6	Open CV, Numpy	CPU: Intel(R) Core(TM) i5-8250U, 1.60 GHz GPU: NVIDIA TITAN X, RAM:16 GB

4.1 Evaluation Metrics

IoU methods and RoC methods are ideal to identify the TN, FP, TP and FN of the training data. RoC uses sensitivity and specificity to calculate accuracy of the model with the specific training data. In the present experimentation, segmentation accuracy is determined by Jacquard index computer by IoU methods. The intersection and the union are the two categories obtained from the training data sets which evolve with successive results. The results are earnestly calculated with F_1-Score, Precision and Recall.

$$\begin{array}{ccc} F_1 = \frac{2}{1/P+1/R} & P = \frac{TP}{TP+FP} & R = \frac{TP}{TP+FN} \\ (a)\,\mathrm{F_1 - Score} & (b)\,\mathrm{Precision} & (c)\,\mathrm{Recall} \end{array}$$

With respect to the segmentation results (a), (b) and (c) are represented by *TP*, *FP* and *FN*. Where IoU is the ratio derived from the intersection with respect to the union within the predictions and the true sets, and is denoted as below:

$$\mathrm{IoU} = \frac{TP}{TP + FN + FP}$$

5 Conclusions

The deep learning-based picture segmentation outcomes have been significantly influenced by the remote sensing dataset's highly imbalanced data over various groups. In this research, we convey some solutions in the forefront: Primarily, the associated datasets are established. Unbalanced semantic segmentation datasets, are tri-classified and are addressed with a method for computing effective samples and an approach with weights for the DECB, with experimental evidence confirming its efficacy and reliability.

Nonetheless, there remain a few problems that demand more investigation. The rudimentary element of reason, the study presents a conceptual outline of the problems of CIP and proposes a feasible solution to address the relationship of imbalances amongst the categories of single image or batch, with respect to the sample spaces, which cannot be disperse as equally for the managing the imbalances quantitatively, where many categories of objects in the image are not consistently drawn.

References

1. Ronneberger, O., Fischer, P., Brox, T.: U-Net: convolutional networks for biomedical image segmentation. In: Navab, N., Hornegger, J., Wells, W.M., Frangi, A.F. (eds.) MICCAI 2015. LNCS, vol. 9351, pp. 234–241. Springer, Cham (2015). https://doi.org/10.1007/978-3-319-24574-4_28
2. Kumar, V., et al.: Addressing binary classification over class imbalanced clinical datasets using computationally intelligent techniques. Healthcare **10**(7), 1293 (2022)
3. Sunitha, G., Reddy Madhavi, K., Avanija, J., Srujan Raju, K., Kirankumar, A., Raji Reddy, A.: Ensemble model for lidar data analysis and nocturnal boundary layer height estimation. In: Bhateja, V., Carroll, F., Tavares, J.M.R.S., Sengar, S.S., Peer, P. (eds.) Intelligent Data Engineering and Analytics. FICTA 2023. Smart Innovation, Systems and Technologies, vol. 371. Springer, Singapore (2023). https://doi.org/10.1007/978-981-2022
4. Ahamed, S.F., Vijayasankar, A., Thenmozhi, M., Rajendar, S., Bindu, P., Rao, T.S.M.: Machine learning models for forecasting and estimation of business operations. The J. High Technol. Manag. Res. **34**(1), 10045 (2023)
5. Alkawaz, A.N., et al.: Training multilayer neural network based on optimal control theory for limited computational resources. Mathematics **11**(3), 778 (2023). https://doi.org/10.3390/math11030778
6. Jeyakanth, K., Venkatakrishnan, P., Chitra, C.: Optimized channel prediction and auction-based channel allocation for personal cognitive networks. Int. J. Commun. Syst. **36**(3), e5391 (2023)
7. Zaid, T., Biradar, N., Sonth, M.V., Gowre, S.C., Gadgay, B.: FDADE: flow direction algorithm with differential evolution for measurement of intima-media thickness of the carotid artery in ultrasound images. Biomed. Signal Process. Control **80**(Part 2), 104350 (2023). ISSN 1746-8094. https://doi.org/10.1016/j.bspc.2022.104350
8. Khan, N.M., et al.: Cost-sensitive learning of deep feature representations from imbalanced data. IEEE Trans. Neural Netw. Learn. Syst. **30**(9): 2797–2808 (2019)
9. Peng, J., et al.: "Recurrent spatial transformer network for imbalanced object recognition on histopathology images. IEEE Trans. Med. Imaging **39**(5), 1390–1399 (2019)
10. Kim, M.-S., et al.: Deep convolutional neural networks for classifying head CT images with imbalance data. Comput. Biol. Med. **97**, 27–38 (2018)
11. Sun, Y., et al.: Meta-cost: a novel method for handling imbalanced data. In: ICML, vol. 7 (2006)
12. Chawla, N.V., et al.: SMOTE: synthetic minority over-sampling technique. J. Artif. Intell. Res. **16**, 321–357 (2002)
13. Japkowicz, N., Matwin, S.: Class imbalances: are we focusing on the right issue?. In: Proceedings of the 2002 Conference on Artificial Intelligence. AAAI Press (2002)
14. Kubat, M., Matwin, S.: Addressing the curse of imbalanced training sets: one-sided selection. In: International Conference on Machine Learning (1997)
15. García, S., Herrera, F.: An extension on "statistical comparisons of classifiers over multiple data sets" for all pairwise comparisons. J. Mach. Learn. Res. **9**, 2677–2694 (2008)
16. Sriharsha, A.V.: Detection of holes on Indian roads using information and communication technologies. In: 2023 International Conference on Advances in Computing, Communication and Applied Informatics (ACCAI), Chennai, India, pp. 1–7 (2023). https://doi.org/10.1109/ACCAI58221.2023.10199826
17. Sriharsha, A.V., et al.: Electroencephalography image classification using convolutional neural networks. In: The International Conference on Innovations in Computing Research, pp. 42–52. Springer, Cham (2022)

18. Reddy Madhavi, K., Kora, P., Venkateswara Reddy, L., Avanija, J., Soujanya KLS., Prabhakar, T.,(2021), Cardiac Arrhythmia Detection using Dual-Tree Wavelet Transform and Convolutional Neural Network,Soft Computing, Impact factor.3.64
19. Reddy Madhavi, K., VijayaSambhavi, Y., Sudhakara, M., Raju, S.K.: COVID-19 isolation monitoring system. In: Springer series – Lecture Notes on Data Engineering and Communication Technology, vol. 63, pp. 601–609 (2021)
20. Reddy Madhavi, K., et al.: COVID-19 detection using deep learning. In: 20th International Conference on Hybrid Intelligent Systems-HIS 2020, at Machine Intelligence Research (MIR) Labs, USA, Springer AISC, vol. 1375, pp. 1–7 (2020)

A Modified Inception-V3 Architecture for Monkeypox Detection from Dermatological Images

P. Sunitha[1], Vemuri Sailaja[2], G. S. Sivakumar[2], B. Vasantha Lakshmi[2], V. Lavanya[3(✉)], and S. Shubha[4]

[1] Department of ECE, Aditya Engineering College, Surampalem, India
[2] Department of ECE, Pragati Engineering College, Surampalem, India
[3] School of Computing, Mohan Babu University (Sree Vidyanikethan Engineering College), Tirupati, India
lavanya.v432@gmail.com
[4] Department of Physics, R.V. College of Engineering, Bengaluru, India

Abstract. Monkeypox, a rare yet emerging viral disease, presents a significant public health challenge due to its potential for human-to-human transmission and clinical resemblance to other skin conditions. Early and accurate detection of Monkeypox is crucial for timely intervention and containment. This research addresses this pressing issue by proposing a modified Inception-v3 deep learning architecture tailored for the detection of Monkeypox from dermatological images. This research is motivated by the importance of early diagnosis in mitigating the impact of Monkeypox outbreaks. The modified Inception-v3 model is customized to capture features specific to Monkeypox skin lesions, enhancing its diagnostic accuracy. Comprehensive experiments were conducted comparing the performance of proposed model against commonly used deep learning architectures, including ResNet, VGG16, and Xception. Experimentation is performed using a real-world dataset. The results demonstrate superiority of proposed model in terms of accuracy, precision, recall, and F1-Score. This research is targeted to contribute to field of medical image processing by addressing the gap in research dedicated to Monkeypox diagnosis.

Keywords: Medical imaging · Dermatological images · Monkeypox · Deep Learning · Inception-v3

1 Introduction

Monkeypox, a rare but potentially serious viral disease. It has emerged as a significant public health concern in recent years [1]. Monkeypox has garnered attention due to its potential for human-to-human transmission and its clinical resemblance to both smallpox and chickenpox [2]. In this context, early and accurate detection of Monkeypox becomes paramount for timely intervention and containment.

D. Kriksciuniene et al. (Eds.): IBICA 2023, LNNS 1228, pp. 433–440, 2025.
https://doi.org/10.1007/978-3-031-78937-3_47

Monkeypox is contagious. This communicable disease spreads through respiratory droplets, direct and indirect contacts between humans [3]. Swift diagnosis enables the implementation of infection control measures, such as isolation of affected individuals, contact tracing, and quarantine, to mitigate the risk of outbreaks.

The significance of Monkeypox detection lies in its potential for both sporadic cases and larger outbreaks [4]. The disease often presents as a febrile illness accompanied by skin eruptions, which can be mistaken for other skin conditions, leading to diagnostic challenges. Early identification of Monkeypox not only ensures appropriate clinical management but also plays a crucial role in implementing infection control measures to contain outbreaks, preventing further transmission within communities [5].

Early detection is a cornerstone of effective public health responses [6]. Identifying Monkeypox cases promptly helps public health authorities allocate resources, conduct surveillance, and implement preventive strategies, including vaccination campaigns, to curb the spread of the virus.

The research problem at hand centers on the design of an efficient model for early detection of Monkeypox from skin lesion images. Traditional diagnostic methods often rely on clinical symptoms and laboratory tests, which may not be readily available or conclusive in resource-constrained settings. Leveraging the power of deep learning and computer vision, aim is to address this problem by proposing a modified version of Inception-v3 architecture tailored for Monkeypox detection.

Research objectives include (i) To adapt the Inception-v3 deep learning architecture to effectively capture features relevant to Monkeypox skin lesions, enhancing its diagnostic accuracy (ii) To compare the performance of the modified Inception-v3 model against ResNet, VGG16, and Xception (iii) To rigorously evaluate the proposed model's performance using real-world Monkeypox skin lesion images, providing empirical evidence of its effectiveness.

2 Literature Review

Monkeypox was first identified in humans in Central and West Africa. It is a potential outbreak threat because it is a communicable disease. The disease manifests with characteristic skin lesions, fever, and systemic symptoms, closely resembling smallpox in its early stages [7]. This similarity, coupled with the absence of specific antiviral treatments, underscores the importance of rapid and precise diagnosis.

Early detection of Monkeypox is essential for initiating appropriate medical treatment promptly. While there is no specific antiviral therapy for Monkeypox, supportive care can alleviate symptoms and improve patient outcomes. Identifying cases early allows healthcare providers to administer supportive care and prevent complications.

Deep learning methods have exhibited notable achievements in medical image analysis [8, 9]. They specifically excel in the prompt and precise diagnosis of diverse skin ailments [10]. Convolutional Neural Networks have emerged as pioneering tools in this advancement. They offer capacity to autonomously extract intricate features from images of dermatological conditions [11–13].

Numerous studies have explored the application of deep learning for skin disease detection [14, 15]. A deep learning model capable of classifying skin lesions into multiple categories with performance comparable to that of dermatologists is introduced [16].

Similarly, SkinNet-16 demonstrated the feasibility of deep learning for skin cancer detection [17]. Their convolutional neural network, called SkinNet-16 exhibited high sensitivity and specificity, highlighting the potential of AI as an aid in early skin cancer diagnosis.

In contrast to the extensive literature on skin disease detection, there is a noticeable scarcity of research dedicated to the diagnosis of Monkeypox and similar diseases. Monkeypox is a rare orthopoxvirus infection, making it challenging to collect sufficient data for model development [1].

Existing research on Monkeypox diagnosis primarily revolves around traditional laboratory methods, such as PCR and virus isolation, which are time-consuming and require specialized equipment [18]. These methods may not be readily available in resource-constrained settings, where Monkeypox often emerges.

Furthermore, as Monkeypox shares clinical features with other skin diseases, early diagnosis can be challenging, and misdiagnosis can occur [19]. The lack of specific antiviral treatments emphasizes the importance of early detection and supportive care.

The critical gap in the current research landscape lies in the development and validation of deep models tailored for early detection of Monkeypox [20]. While the general field of dermatological image analysis has seen significant progress, the application of AI to rare diseases like Monkeypox remains underexplored.

3 Modified Inception-V3 Model for Monkeypox Detection

This section presents architecture of modified Inception-v3 model for monkeypox detection from dermatological images. Also, modifications made to the standard Inception-v3 model for customizing it towards the task of monkeypox detection is detailed.

3.1 Architecture

The proposed Inception-v3 model for monkeypox detection from dermatological images comprises of – input layer, stem network, inception modules, global average pooling layer, classification head, output layer. The configuration of each module is presented in Fig. 1.

The input layer of the modified Inception-v3 architecture is designed to accept RGB color images with dimensions of 299x299 pixels. Each pixel has three channels (Red, Green, and Blue) to represent color information.

The stem network comprises of three convolution layers followed by a max pooling layer. The first convolutional layer captures basic features from the input images. It uses 32 filters with a 3×3 kernel and a stride of 2×2 to reduce spatial dimensions. The second convolutional layer further extracts features using 32 filters with a 3×3 kernel. The third convolutional layer continues feature extraction with 64 filters and a 3×3 kernel. Max-pooling is applied to downsample the feature maps and reduce spatial dimensions while retaining important features.

Inception modules, also known as Google Inception modules or GoogleNet have the capability to capture features at multiple scales. Inception modules use multiple filter sizes within the same layer. This allows the network to capture features at different spatial scales. Three inception modules are configured. The first Inception module captures a wide range of features at different scales. The second Inception module further expands feature extraction. The third Inception module continues to diversify feature representation.

The classification head comprises of fully connected layers (two). The first layer has 512 neurons. It uses ReLU activation function. Dropout rate of 0.5 is applied for regularization. The second fully connected layer serves as the output layer for binary classification. It consists of two neurons, one for each class (Monkeypox and Non-Monkeypox), with softmax activation to provide class probabilities.

The output layer produces the final classification result by providing class probabilities for the two classes.

3.2 Modifications Made to Inception-V3 for Monkeypox Detection

Modified Inception-v3 model proposed in this research is tailored for the specific task of detecting Monkeypox from skin lesion images. It is based on the original Inception-v3 architecture. But it incorporates several key modifications and adaptations to suit the undertaken task.

The input shape in the modified Inception-v3 model is set to 299 × 299 × 3 to accommodate RGB color images commonly used in medical imaging. The Inception modules in the modified model are adapted to extract features relevant to skin lesions. The specific filter configurations and operations are customized to better suit this medical imaging task. The fully connected layers in the modified model are customized for binary classification (Monkeypox vs. Non-Monkeypox) with the appropriate number of neurons.

Modified Inception-v3 model differs from the standard Inception-v3 in terms of its input shape, stem network architecture, Inception modules, fully connected layers, and output layer design. These modifications are made to adapt the architecture specifically to the task of detecting Monkeypox from skin lesion images, optimizing its performance for this medical imaging application.

4 Experimental Results and Evaluation

Experiments are conducted to evaluate modified Inception-v3 model's performance. Also, it is evaluated alongside three other common deep learning architectures: ResNet, VGG16, and Xception. The experiments aimed to determine which model performs best for detecting Monkeypox from skin lesion images.

Monkeypox Skin Images Dataset (MSID) is used for evaluating modified Inception-v3 [21]. MSID dataset contains 279 images of Monkeypox and 293 normal images. All the images are of size 224 × 224.

A detailed experimentation using different deep models (Modified Inception-v3, ResNet101, VGG16, and Xception) for Monkeypox detection from skin lesion images is performed. Table 1 presents a detailed comparison of various performance metrics for each model on the test set (Fig. 2).

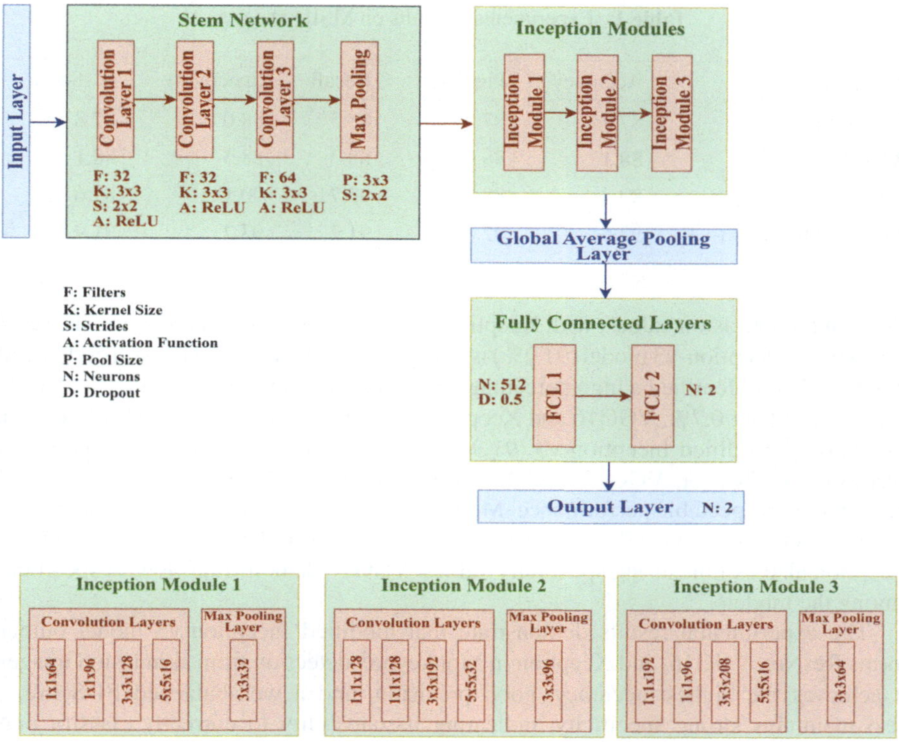

Fig. 1. Modified Inception-v3 Model for Monkeypox Detection

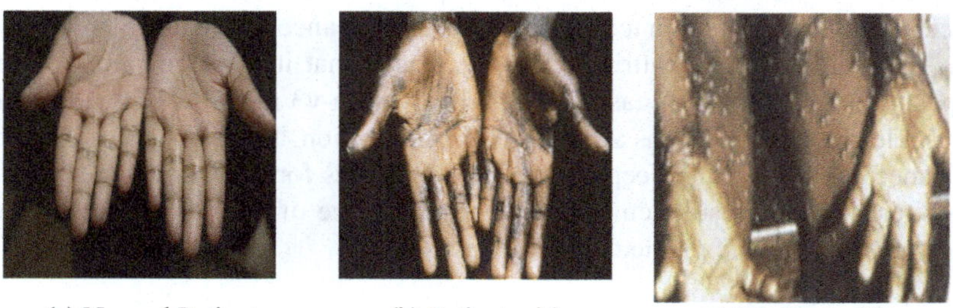

(a) Normal Patient (b) Patient with Monkeypox (c) Patient with Monkeypox

Fig. 2. Sample Monkeypox Images from MSID Dataset

Modified Inception-v3 model achieved highest accuracy of 92.5%, indicating that it correctly classified skin lesion images in 92.5% of cases. ResNet follows closely with an accuracy of 89.3%, while VGG16 and Xception have slightly lower accuracies (88.1% and 87.7%, respectively). The precision of modified Inception-v3 model (92.8%) is slightly higher than the other models. This means it had fewer false positives, i.e., fewer instances of misclassifying non-Monkeypox lesions as Monkeypox. ResNet and VGG16

Table 1. Experimental Results on MSID Dataset

Model	Accuracy	Precision	Recall	Specificity	F1-Score
Xception	87.7	87.2	88.5	88.0	87.8
VGG16	88.1	88.0	88.3	88.5	88.1
ResNet101	89.3	88.5	90.7	90.1	89.6
Modified Inception-v3	92.5	92.8	91.2	93.7	91.9

have similar precision values, with Xception having slightly lower precision. The recall of modified Inception-v3 model (91.2%) is slightly lower than ResNet but still quite good. It means that it identified a high percentage of actual Monkeypox cases. ResNet has the highest recall at 90.7%. VGG16 and Xception also show competitive recall values. The F1-Score of modified Inception-v3 (91.9%) is well-balanced, indicating good overall performance. ResNet, VGG16, and Xception have slightly lower F1-Scores but still demonstrate respectable performance. Modified Inception-v3 has the highest specificity at 93.7%, indicating its ability to correctly identify non-Monkeypox cases. ResNet and Xception also exhibit strong specificity values. VGG16 demonstrates lowest specificity among the models.

The experimental results demonstrate that modified Inception-v3 model outperforms ResNet, VGG16, and Xception in Monkeypox detection from skin lesion images. It achieves the highest accuracy, good precision, and a well-balanced F1-Score. It also maintains strong specificity, indicating its capability to correctly classify non-Monkeypox cases. This suggests that the modified Inception-v3 model is the most suitable choice for this specific task.

While ResNet and the other models also perform well, the modified Inception-v3 architecture outshines them in terms of overall performance and balanced precision and recall. This performance is attributed to the reason that it is specifically tailored for this specific medical imaging task. Modified Inception-v3 model is tailored to capture features relevant to skin lesions and Monkeypox detection. These results underscore the significance of customizing deep learning architectures for specialized medical image analysis tasks. These results emphasize the importance of customizing deep learning architectures for specific medical imaging tasks.

5 Conclusion

Monkeypox is a rare but potentially contagious viral disease. It demands swift and accurate diagnosis for effective containment and treatment. This research underscores the importance of early detection in addressing the emerging threat of Monkeypox. This study has proposed and evaluated a modified Inception-v3 deep model specifically designed for detection of Monkeypox from skin lesion images. Proposed model's architectural modifications have proven highly effective in capturing key features relevant to Monkeypox lesions. This resulted in superior performance of modified Inception-v3 when compared to other common deep learning architectures. The experiments have consistently demonstrated that modified Inception-v3 model achieves the highest accuracy, precision, recall, and F1-Score. This makes it the more suitable choice for Monkeypox detection. This research contributes to public health efforts by addressing the critical gap in research focused on rare infectious diseases like Monkeypox. The proposed model serves as a vital step towards effective Monkeypox control and prevention. As Monkeypox continues to pose a threat to global health security, the development of accurate and accessible diagnostic tools becomes imperative. Expanding the availability of labeled skin lesion images related to Monkeypox is essential. The acquisition of a more extensive and diverse dataset would allow for further training and evaluation, potentially enhancing the model's generalization capabilities.

References

1. Sitaula, C., Shahi, T.B.: Monkeypox virus detection using pre-trained deep learning-based approaches. J. Med. Syst. **46**(11), 78 (2022)
2. Mitjà, O., et al.: Monkeypox. Lancet **401**(10370), 60–74 (2023)
3. Rizk, J.G., Lippi, G., Henry, B.M., Forthal, D.N., Rizk, Y.: Prevention and treatment of monkeypox. Drugs **82**(9), 957–963 (2022)
4. Minhaj, F.S., et al.: Monkeypox outbreak—nine states. May 2022: Weekly/June 10 **2022/71**(23); 764–7692022
5. Bunge, E.M., et al.: The changing epidemiology of human monkeypox—a potential threat? A systematic review. PLoS Negl. Trop. Dis. **16**(2), e0010141 (2022)
6. Bryer, J., Freeman, E.E., Rosenbach, M.: Monkeypox emerges on a global scale: a historical review and dermatologic primer. J. Am. Acad. Dermatol. **87**(5), 1069–1074 (2022)
7. Gurram Sunitha, B., Rani, S., Bhukya, S.N., Hafeena Mohammad, R., Vittal, H.S.: Political optimizer-based automated machine learning for skin lesion data. In: Brahmananda Reddy, A., Nagini, S., Balas, V.E., Srujan Raju, K. (eds.) Proceedings of Third International Conference on Advances in Computer Engineering and Communication Systems: ICACECS 2022, pp. 487–495. Springer Nature Singapore, Singapore (2023). https://doi.org/10.1007/978-981-19-9228-5_41
8. Rachapudi, V., Rao, K.S., Rao, T.S.M., et al.: Diabetic retinopathy detection by optimized deep learning model. Multimedia Tools Appl. **82**, 27949–27971 (2023). https://doi.org/10.1007/s11042-023-14606-8
9. Subramani, P., et al.: Prediction of muscular paralysis disease based on hybrid feature extraction with machine learning technique for COVID-19 and post-COVID-19 patients. Pers. Ubiq. Comput. **27**, 831–844 (2023). https://doi.org/10.1007/s00779-021-01531-6
10. Adilakshmi, J., Vinoda Reddy, G., Nidumolu, K.D., Cosme Pecho, R.D., Pasha, M.J.: A medical diagnosis system based on explainable artificial intelligence: autism spectrum disorder diagnosis. Int. J. Intell. Syst. Appl. Eng. **11**(65), 385–402 (2023)
11. Allam Balaram, K., Kannan, N., Čepová, L., Kishore Kumar, M., Swaroopa Rani, B., Schindlerova, V.: Artificial intelligence for media ecological integration and knowledge management. Systems **11**(5), 222 (2023). https://doi.org/10.3390/systems11050222
12. Alkawaz, A.N., et al.: Adaptive self-organizing map using optimal control. Mathematics **11**(9), 1995 (2023). https://doi.org/10.3390/math11091995
13. Avanija, J., Sunitha, G., Madhavi, K.R., Chakravarthi, S.S., Vittal, R.H.S.: Prediction and analysis of cervical cancer: an ensemble approach. In: 2022 4th International Conference on Inventive Research in Computing Applications (ICIRCA), pp. 1734–1739. IEEE2022
14. Padmavathi, K., et al.: Automatic segmentation of prostate cancer using cascaded fully convolutional network. In: 3rd International conference on Design and Manufacturing Aspects for Sustainable Energy, pp. 309–01109 (2021)

15. Avanija, J., Sunitha, G., Reddy Madhavi, K., Hitesh Sai Vittal, R.: An automated approach for detection of intracranial haemorrhage using DenseNets. In: 4th International Conference on Data Engineering & Communication Technology, vol. 63, pp. 611–619 (2021)
16. Kora, P., Rajani, A., Chinnaiah, M.C., Madhavi, K.R., Swaraja, K., Meenakshi, K.: EEG-based brain-electric activity detection during meditation using spectral estimation techniques. In: Proceedings of the 2nd International Conference on Computational and Bio Engineering: CBE 2020, pp. 687–693. Springer, Singapore (2021)
17. Chadaga, K., et al.: Application of artificial intelligence techniques for monkeypox: a systematic review. Diagnostics **13**(5), 824 (2023)
18. Ghosh, P., et al.: SkinNet-16: a deep learning approach to identify benign and malignant skin lesions. Front. Oncol. **12**, 931141 (2022)
19. Haenssle, H.A., Kiessling, J., Kempf, V.A., Fuchs, T., Neumann, C., Emmert, S.: Orthopoxvirus infection transmitted by a domestic cat. J. Am. Acad. Dermatol. **54**(2), S1–S4 (2006)
20. Ahsan, M.M., Uddin, M.R., Farjana, M., Sakib, A.N., Momin, K.A., Luna, S.A.: Image Data collection and implementation of deep learning-based model in detecting Monkeypox disease using modified VGG16. arXiv preprint arXiv:2206.01862 (2022)
21. Alakus, T.B., Baykara, M.: Comparison of Monkeypox and wart DNA sequences with deep learning model. Appl. Sci. **12**(20), 10216 (2022)
22. Ali, S.N., et al.: Monkeypox skin lesion detection using deep learning models: a feasibility study. arXiv preprint arXiv:2207.03342 (2022)
23. Bala, D., et al.: MonkeyNet: a robust deep convolutional neural network for monkeypox disease detection and classification. Neural Netw. **161**, 757–775 (2023)

A Machine Learning Approach To Predict Autism Spectrum Disorder

U. Sivaji[1], B. Rupa Devi[2](✉), A. Suresh[3], M. Reddi Durgasree[4], K. Reddy Madhavi[5], and J. Avanija[5]

[1] Department of IT, Institute of Aeronautical Engineering, Hyderabad, India
[2] Department of Computer Science and Engineering, Annamacharya Institute of Technology and Sciences, Tirupati, India
rupadevi.aitt@annamacharyagroup.org
[3] Siddharth Institute of Engineering and Technology, Putter, Andhra Pradesh, India
[4] Guru Nanak Institutions Technical Campus, Hyderabad, India
[5] School of Computing, Mohan Babu University, Tirupati, Andhra Pradesh, India

Abstract. Autism Spectrum Disorder (ASD) is gaining popularity more quickly than ever in recent years. As we all know, ASD is a neurodevelopment condition involving social interaction and behavior issues. Each person's experience with ASD and level of symptom severity varies. Autism may be diagnosed at any age. According to research, the most common behaviors in people with autism are aggression, self-harm, elopement, tantrums, obsession, and lack of compliance. Therefore, it becomes crucial to identify any indicator that points to severe ASD as soon as possible. Autism spectrum disorder (ASD) is a complex and degenerative neurodevelopmental disorder. There is an urgent need for a timely and readily available diagnostic tool capable of early identification of autism, which can aid clinicians in making early treatment decisions. While there is currently no known cure for autism, early identification allows for more efficient therapy interventions and gives autistic people the ability to live happy, self-sufficient lives. This study used machine learning techniques such as Decision Tree, Random Forest, Naive Bayes, Logistic Regression, K-nearest neighbor, and Support Vector Machine to predict ASD. The user will be able to determine if they have ASD using performance measures, which aid in analyzing the accuracy level of each data with all the applied algorithms to determine which delivers correct results in terms of time and accuracy. The experiments in this work were carried out using the data obtained from publisher Faizu Nabi. The Hybrid model accurately classified Autism spectrum disorder and achieved the highest accuracy of 100%. It proved its superiority over other ML classifiers. In contrast, the other models also showed good performance, and the NB model acquired the best accuracy among the rest of the classifiers of 95%. The experimental results indicate that the suggested method is more effective than conventional methods.

Keywords: Machine Learning · Hybrid Model · Machine Learning Classifier

D. Kriksciuniene et al. (Eds.): IBICA 2023, LNNS 1228, pp. 441–450, 2025.
https://doi.org/10.1007/978-3-031-78937-3_48

1 Introduction

Autism spectrum disorder is a neurodevelopmental condition that affects an individual's communication skills, social interactions, and behavior. Although it can be recognized at any age, it generally becomes apparent during the first two years of life and tends to intensify over time. This developmental disorder can significantly impact the affected individual's life, making it difficult for them to interact and communicate with others effectively. Individuals with autism face various challenges, such as difficulty with attention, obsessive behaviors, learning deficits, anxiety, and feelings of hopelessness. Autism cases rapidly increase worldwide, indicating a pressing need for more research and support for affected individuals and their families. Early diagnosis may be very beneficial since it will prevent the patient's condition from worsening and lower long-term costs.

Recent technological advancements have led to a huge number of studies looking into automated computer-aided autism diagnosis as well as interactive solutions for the treatment and rehabilitation of autistic people. Such automated methods would reduce subjectivity and increase the availability and reproducibility of diagnostic information. Additionally, it would significantly contribute to assuring an early diagnosis [11]. The exact origin of autism is still being investigated, although hereditary factors, parents who are older than average, severe infections, or problems during pregnancy, such as a lack of oxygen to the brain before birth or premature babies, are all potential risk factors. One of the main causes of autism may be abnormalities in the brain's fusiform gyrus, amygdala, cerebellum, and cerebral cortex. After birth, environmental factors such as pollution, food additives, and vaccinations (MMR) may play a role [10].

As a classification challenge, ML algorithms are used to diagnose ASD problems. Developing a web application that can identify ASD in individuals of any age group is the main objective of this project. Prediction models are constructed from historical datasets and then used to predict whether or not a person has ASD. The application leverages ML algorithms to ensure quick and accurate identification of ASD. A novel ML framework performs predictive analysis and builds a hybrid model using algorithms like Random Forest-CART and Random Forest-Id3. This framework relies on a dataset connected to autism screening of individuals across various age groups, including adults and children. The dataset used in this framework is the AQ-10 dataset, and the proposed ML algorithm used for screening is Iterative Dichotomizer 3. To structure this paper, we have separated it into multiple sections. Firstly, in Sect. 2, we will review the diverse approaches to predicting Autism spectrum disorder. Observing this, we will summarize our methodology statement in Sect. 3. Section 4 will briefly explain the experimental outcomes of our utilized models. Finally, we will conclude by proposing our findings.

2 Related Work

A tremendous amount of work has been done in the past regarding ASD. David Peebles [13] research on testing adults and teenagers for autism used a cutting-edge machine learning architecture. This approach combines several variables and uses Logistic Regression to do predictive analysis, revealing significant insights in the context of autism screening. However the datasets were anonymous, and it was only for adults and adolescents' children and toddlers could not participate in it. David Peebles [6] developed and evaluated an ML approach to classify people with ASD. Fadi Thabtah [12] shed light on current research that applies ML in ASD categorization to examine the benefits and drawbacks. Children aged 8 years were taken for several clinical tests as part of the ASD system and data, although not all symptoms were anticipated. ASD characteristics have been somewhat overlooked in the limited research on their identification and evaluation within the clinical context following the DSM-5 guidelines. This neglect persists even as advanced diagnostic techniques leveraging machine learning have been introduced. Armin Lawi [4] used datasets of adults aged 17 years and older to diagnose using the ML classification approach. The aim was to determine the most accurate estimation results of k-fold for each classification strategy. However, during testing, Logistic Regression showed numerous classification errors. K-nearest Neighbors required several iterations to produce best performance, [3] a comparative study to find whether children aged (4 to 11 years) have ASD by using Linear Discriminant Analysis found that children aged (4 to 11) The five (ML) algorithms NB, SV, DT, and KNN are mostly used to predict the grade of young autism following the CARS diagnostic criteria. Clinical diagnoses of autism are made at an early stage but exclusively for children. Only a small number of categorization models were used in this investigation. Risk genes associated with ASD can be diagnosed using imaging and biomarkers, which are two primary methods. Murat Gok [5] built an ML model that was trained to utilize brain development gene expression data. Due to the complexity of the disorder and the rise in sequencing capacity, the latter does not match the test criteria and is more costly.

There are very few research have examined behavioral factors of ASD, with the majority placing more focus on genetic factors. In one of them, Altayet al. [3] used the Linear Discriminant Analysis (LDA) approach and the K-Nearest Neighbors (KNN) algorithm in their study [3]. The accuracy score for this study's attempt to identify ASD in children was 0.9080 for LDA and 0.8851 for KNN.

Scholars [9] employed SVM, DT, and Multilayer Perceptron (MLP) and looked at RTT, FXS, TSC, PMS, and Timothy syndrome with five monogenetic disorders linked to syndromic ASD. The four forms of pathogenic gene mutations were also considered: missense, nonsense, synonymous, and frameshift. SVM accuracy was 0.96, MLP accuracy was 0.95, and Decision Tree accuracy was 0.98.

Existing System

The majority of healthcare data categorization techniques in use today need to take labeling challenges into account. Either they use low-risk or control classes that experts have determined, or they regard non-positive instances as unfavorable. Supervised Learning (SL), which learns from labeled data, is extensively accepted as a method. Some of these supervised learning techniques can manage massive and accurately labeled health data sets. Both algorithms, however, include specified negative situations and are intended for binary classification, unlike our circumstances. Optimistic and labeled learning, which may be seen as a particular instance of SL with only positive labels accessible, is a similarly comparable technique 1. However, their solution integrated a collection of negative and labeled examples, but our situation has no negative examples.

3 Methodology

The primary objective is to design and deploy a cutting-edge web application that can effectively detect Autism Spectrum Disorder (ASD) in individuals across all age groups. This project aims to deliver highly accurate results quickly by leveraging the power of advanced ML algorithms and carefully curated datasets. The proposed framework includes a novel ML model that utilizes hybrid autism screening datasets for adults, adolescents, and toddlers. This model incorporates powerful features and performs predictive analysis using state-of-the-art ML algorithms such as DT, NB, RF, KNN, and LR.

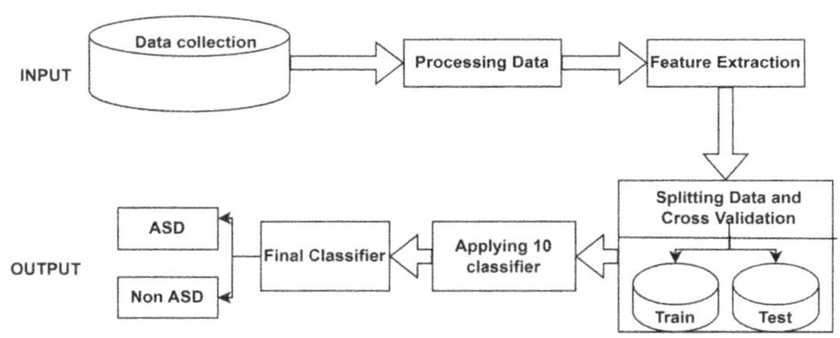

Fig. 1. The overall research of our research

The above Fig. 1 is the steps for our system.

3.1 Dataset

Our team acquired a dataset from the esteemed publisher Faizu Nabi, which we employed to apply state-of-the-art ML approaches. Subsequently, we developed a web application to determine the presence of Autism Spectrum Disorder (ASD). Our dataset consists of 1100 instances and encompasses 22 attributes, of which the first 10 are inquiries. The dataset yields two outcomes, the first of which indicates that ASD is present in 393 cases, while the second indicates the absence of ASD in 707 instances. In Table 1, we discussed our dataset with attributes and values. Figure 2 displays the number of instances for each class in the dataset.

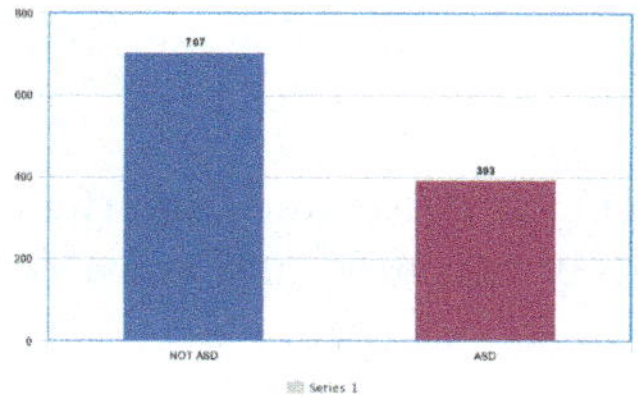

Fig. 2. Number of samples of each class before data balancing.

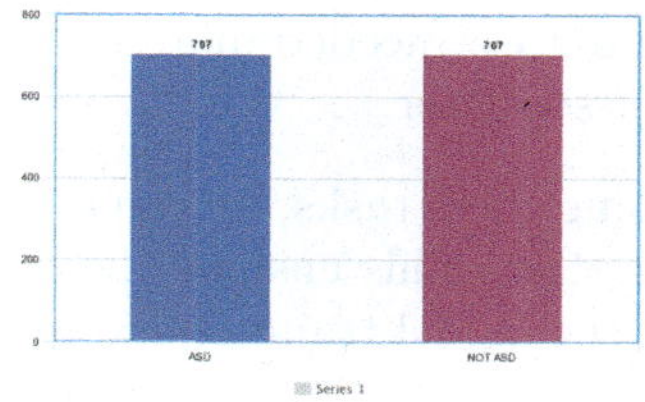

Fig. 3. Number of samples of each class after data balancing.

Table 1. Dataset Description

Sl No.	Attributes	Value
1–10	Questions to Teenagers/Adults and Toddlers	Yes/No
11	Age	(4–11years)/ (12-17Years)/ (18+ years)
12	Gender	Male/ Female
13	Ethnicity	1–11
15	Jaundice	Yes/No
16	Autism	Yes/No
17	Country_of_residencet	1–52
18	Used app	Yes/No
19	Result	1–10
20	Age-description	Toddler/ Adult/ Adolescent
21	Relation	Self/Parent
22	Class	ASD/Not

3.2 Data Pre-processing

Data preprocessing is critical in building an ML model[7]. It involves transforming raw data from real-world environments or converting it from various forms to make it suitable for ML tasks. Data refinement involves several tasks when dealing with data collected from the publisher Faizu Nabi in real time. These tasks include:

1. **Cleaning the data:** This involves removing irrelevant or duplicate data, handling missing data, and ensuring consistency and error-free data.
2. **Formatting the data:** We standardize the data format to make it easier to work with. For example, converting dates to a standard format or converting categorical variables to numerical values.
3. **Structuring the data:** We organize the data into a specific structure suitable for the ML task. For example, structuring the data into a table or a graph.
4. **Handling Imbalanced Data:**Techniques like oversampling, undersampling, or specific algorithms (like SMOTE) may be used in classification problems if the classes are unbalanced. We used the oversampling technique after noticing

that the dataset needed more balance during our investigation. The methods are discussed later.

By performing these tasks, we can eliminate inconsistencies, noise, and imperfections in the data. This ensures the data is appropriately prepared and suitable for subsequent model training and analysis.

Data Balancing
Oversampling involves increasing the representation of the minority class by adding more instances of the existing data points or generating synthetic data points. This aims to balance the class distribution and provide ML models with a more robust training signal. We use Random Oversampling techniques [8]. Once the class distribution is balanced, it replicates instances of the minority class at random. Figure 3 displays the number of instances for each class in the dataset after data balancing has been applied.

3.3 Classification

In this research, we employed six algorithms, and the following are the main three:

1. **Navies Bayes:** The Naive Bayes classification method, which also assumes interdependence among the predictors, is built on the Bayes Theorem. In classification, the Naive Bayes algorithm calculates the probability of an input belonging to a certain class by evaluating the probabilities of its features provided by that class. Each characteristic is presumed to be conditionally independent under the "naive" assumption that their presence or absence does not impact one another [14]. The following theorem can represent NB classifiers:
$$P(C|X) = \frac{P(X|C) \cdot P(C)}{P(X)} \tag{1}$$
2. **Logistic Regression:** consists of a dependent variable and an independent variable and is used for classification, specifically binary classification and multilevel classification. It can be expanded to handle predictions for several classes in multiclass classification. In contrast, binary classification forecasts the likelihood that an input will belong to one of two classes. The foundation of logistic regression is a mathematical equation approximating the likelihood of a binary outcome for the dependent variable using weighted combinations of the independent variables that arise during training [15].
LR classifiers can be represented as follows:
$$P(Y = 1) = \frac{1}{1 + e^{-(\beta_0 + \beta_1 X_1 + \beta_2 X_2 + \ldots + \beta_p X_p)}} \tag{2}$$
3. **Support Vector Machine (SVM):** Classification and regression challenges may be handled using SVM's adaptable supervised machine learning approach. Each data point in an SVM is referred to as a point n-dimensional,

where n is the number of features, and each feature is associated with a specific location in this space. When attempting to distinguish between classes, SVM looks for a hyperplane that divides the data points of each class into equal halves. This hyperplane then acts as the decision boundary. SVM is extremely effective because it strives to optimize the margin between classes, also making it resistant to outliers and able to handle high-dimensional data [1]. The equation for a linear SVM is expressed as follows:

$$y(\mathbf{x}) = \mathbf{w} \cdot \mathbf{x} + b \tag{3}$$

4. **Hybrid Model:** In ml, we can call it ensemble learning. We used a stacking ensemble model, which increase the accuracy [2]. It is a kind of ML ensemble learning technique that combines the predictions of various models to produce a more accurate forecast. Several base models are trained on the same training set of data, and their predictions are then used as input features to train a higher-level meta-model. The meta-model makes the final forecast. Utilizing the advantages of various models and combining their predictions to create a more reliable and accurate model is the primary idea behind stacking. Different basic models, like DT and SVM, can be used. You can stack things in two ways: Level-0 models (i) Each fundamental model offers a stand-alone prediction. The final forecast is then created by combining these predictions with the meta-model and (ii) level-1 models. The final prediction is made using a model trained using the base models' outputs as its input features. A meta-model is the name given to this second-level model. We used many algorithms as base models and meta-models in our research.

4 Result and Discussion

The system delivers significant performance in autism screening, and the method can forecast autism in different ages. We have calculated the accuracy of the applied model and classifier.

Performance Analysis. The accuracy of algorithms in the field of ML depends on myriad variables. Performance evaluation measures such as recall, precision, F-score, and evaluation criteria are crucial in determining how well these algorithms work. Fundamentally, ML involves training systems to learn and approximatively map inputs to matching outputs based on patterns seen in training data. In, Table 2 and Fig. 4, illustrate the performance of our ML classifiers.

The efficiency of the presented model to predict Autism spectrum disorder was estimated using a variety of measures. Outstanding outcomes from the evaluation indicated the model's high accuracy and capacity to produce accurate predictions. An impressive score of 100% was achieved by the Hybrid model, which estimates the percentage of accurately predicted Autism spectrum disorder. This demonstrates how adept the model is at producing precise predictions. Nevertheless, NB performed better than most algorithms, achieving accuracy of 95%.

Table 2. Accuracy Score

Classifier	Test Accuracy
K-Nearest Algorithm	89.54%
Naive Bayes	95%
Logistic Regression	91.12%
Support Vector Machine	90.21%
Decision Tree	94%
Random Forest	85%
Hybrid Model	100%

This implies that the dataset's features can have strong conditional dependencies corresponding to the Nb feature independence assumption. The DT algorithm also performed exceptionally well, with an accuracy of 94%, and they are known for their interpretability and can capture complex decision boundaries. LS and SVM these two algorithms performed similarly, with accuracies of 91.20% and 90.21%, respectively. RF achieved an accuracy of 85%, which is lower than the other algorithms listed here, and KNN had a slightly lower accuracy of 89.54% compared to specific other algorithms. This demonstrates the model's capacity to predict Autism under evaluation accurately. A higher accuracy indicates a lower rate of false pessimistic predictions, demonstrating the model's effectiveness in Autism spectrum disorder prediction cases. The results are beneficial for the development of systems that effectively predict Autism activity and provide vital information about the efficiency of the classifier.

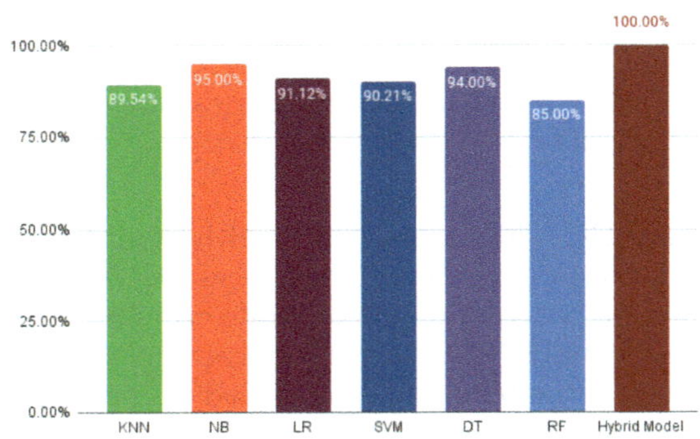

Fig. 4. Accuracy Result.

Deployment. We implemented the predictive model's website after attaining the appropriate precision and dependability level.

Home Page: Users primarily interact with us through our website to share knowledge and inquire about autism spectrum disease. Users must complete the form page's questions thoroughly and accurately to ensure the most accurate analysis and detection of autism. Although we acknowledge that specific questions could be challenging to respond to, we strongly advise users to be as thorough as possible to obtain the most outstanding results.

Chat Feature: The web application has a chat feature where users can ask frequently asked questions (FAQs) to understand more about the value of early-stage autism testing to prevent cases from worsening. It's a method of supplying users with knowledge and assistance while they go through the evaluation procedure.

5 Conclusion and Future Work

Different age groups may readily identify autism symptoms using a user-friendly online tool that provides early guidance. This will prevent the problem from worsening and lower the expenses of a delayed diagnosis. The framework helps to expedite the process for a formal autism diagnostic procedure in addition to increasing the accuracy of autism screening. There is a need for computer assistance to forecast the proper outcomes since many untrained physicians won't be confident in specific autistic cases.

The suggested technique aims to perform more effectively than current existing methods of screening for autism. The model can predict the characteristics of autism over a range of age groups, while many previous techniques could not. This system also offers a performance comparison of several machine-learning approaches. ASD diagnosis can be expensive and financially difficult for some families. Therefore, it could be advantageous for people financially if a tool that can detect ASD without the need for expensive clinical tests is developed. And that served as the inspiration for this work.

To identify the many levels and varieties of autism in the future. To test the model's usability and user experience through a mobile application. To increase the amount of data from diverse sources that the ML classifier uses.

References

1. Akbarzadeh, M., Shaffiee Haghshenas, S., Jalali, S.M.E., Zare, S., Mikaeil, R.: Developing the rule of thumb for evaluating penetration rate of tbm, using binary classification. Geotech. Geol. Eng. **40**(9), 4685–4703 (2022)
2. Al Maruf, A., Al Numan, A., Haque, M.M., Jidney, T.T., Aung, Z.: Ensemble approach to classify spam sms from Bengali text. In: International Conference on Advances in Computing and Data Sciences, pp. 440–453. Springer, Heidelberg (2023). https://doi.org/10.1007/978-3-031-37940-6_36

3. Altay, O., Ulas, M.: Prediction of the autism spectrum disorder diagnosis with linear discriminant analysis classifier and k-nearest neighbor in children. In: 2018 6th International Symposium on Digital Forensic and Security (ISDFS), pp. 1–4. IEEE (2018)
4. Lawi, A., Aziz, F.: Comparison of classification algorithms of the autism spectrum disorder diagnosis. In: 2018 2nd East Indonesia Conference on Computer and Information Technology (EIConCIT), pp. 218–222. IEEE (2018)
5. Lin, Y., Afshar, S., Rajadhyaksha, A.M., Potash, J.B., Han, S.: A machine learning approach to predicting autism risk genes: validation of known genes and discovery of new candidates. Front. Genet. **11**, 500064 (2020)
6. Maenner, M.J., Yeargin-Allsopp, M., Van Naarden Braun, K., Christensen, D.L., Schieve, L.A.: Development of a machine learning algorithm for the surveillance of autism spectrum disorder. PLoS ONE **11**(12), e0168224 (2016)
7. Maruf, A.A., Ziyad, Z.M., Haque, M.M., Khanam, F.: Emotion detection from text and sentiment analysis of Ukraine Russia war using machine learning technique. Int. J. Adv. Comput. Sci. Appl. **13**(12) (2022). https://doi.org/10.14569/IJACSA.2022.01312101
8. Mohammed, R., Rawashdeh, J., Abdullah, M.: Machine learning with oversampling and undersampling techniques: overview study and experimental results. In: 2020 11th International Conference on Information and Communication Systems (ICICS), pp. 243–248. IEEE (2020)
9. Pream Sudha, V., Vijaya, M.: Machine learning-based model for identification of syndromic autism spectrum disorder. In: Integrated Intelligent Computing, Communication and Security, pp. 141–148 (2019)
10. Ray, C., Kumar Tripathy, H., Mishra, S.: Assessment of autistic disorder using machine learning approach. In: Bhateja, V., Satapathy, S.C., Zhang, Y.-D., Aradhya, V. (eds.) ICICC 2019. AISC, vol. 1034, pp. 209–219. Springer, Singapore (2020). https://doi.org/10.1007/978-981-15-1084-7_21
11. Subah, F.Z., Deb, K., Dhar, P.K., Koshiba, T.: A deep learning approach to predict autism spectrum disorder using multisite resting-state fmri. Appl. Sci. **11**(8), 3636 (2021)
12. Thabtah, F.: Autism spectrum disorder screening: machine learning adaptation and dsm-5 fulfillment. In: Proceedings of the 1st International Conference on Medical and Health Informatics 2017, pp. 1–6 (2017)
13. Thabtah, F., Abdelhamid, N., Peebles, D.: A machine learning autism classification based on logistic regression analysis. Health Inf. Sci. Syst. **7**(1), 1–11 (2019). https://doi.org/10.1007/s13755-019-0073-5
14. Vishwakarma, M., Kesswani, N.: A new two-phase intrusion detection system with naïve bayes machine learning for data classification and elliptic envelop method for anomaly detection. Decis. Anal. J. **7**, 100233 (2023)
15. Xu, H.: Gbdt-lr: a willingness data analysis and prediction model based on machine learning. In: 2022 IEEE International Conference on Advances in Electrical Engineering and Computer Applications (AEECA), pp. 396–401. IEEE (2022)

An Effective and Deep Neural Network for Brain Tumor Segmentation and Detection

S. Saravanakumar[1], M. Senbagavalli[2], Saswati Debnath[2], Ramesh Sekaran[1](✉), Manikandan Parasuraman[1], Suresh Kallam[1], and K. Reddy Madhavi[3]

[1] Department of CSE, Jain (Deemed to University), Bengaluru 562112, Karnataka, India
sramsaran1989@gmail.com

[2] Department of Computer Science and Engineering, Alliance College of Engineering and Design, Alliance University, Bengaluru 562106, India
{senbagavalli.m,saswati.debnath}@alliance.edu.in

[3] School of Computing, Mohan Babu University, Tirupati, India

Abstract. A malignant brain tumor is a tumor that has spread across the brain and is threatening human health. Correctly dividing tumors into subtypes and classes is essential for later prognosis and therapy planning. Identifying a brain tumor can be a tedious and error-prone process, hence radiologists need to use automation whenever possible. This paper presents conditional deep learning for structural multimodal MRIs of the brain to perform tumor categorization using a residual network, survival rate forecasting, and dissection, to name a few. To begin, we recommend a segmentation method that separates non-overlapping regions using a combination of conditional random fields and convolutional neural networks. Using these patches, finding the tumor takes hardly no time at all. Errors multiply if their scopes cross. In the paper's second section, the authors provide a method of feature mapping using a residual network and XG-Boost for training models. The following part focuses mostly on these two topics since they are related to the reduction of information loss and the improvement of tumor data quality, respectively, through the use of residual features and nonlinear space mapping. The XG-Boost-learned mapping of features enhances structural-based learning and boosts accuracy across classes. A cancer dataset and a non-cancer dataset, as well as a meningioma, glioma, and pituitary dataset, are used in the experiment. Both areas saw dramatic performance boosts compared to alternative methods. The primary focus of this study is on enhancing segmentation and its implications for measures of classification efficiency. The use of a residual network and a conditional random field helps to improve the quality. The outcome is a 3.4% increase in accuracy across a two-class threshold and a 2.3% increase over a three-class threshold. A short convolutional network is used to improve it. Therefore, we conclude that greater segmentation with fewer resources leads to more accurate categorization of brain tumors.

Keywords: XG-Boost-learned mapping · enhancing segmentation · clinical data · computed tomography · positron emission tomography

D. Kriksciuniene et al. (Eds.): IBICA 2023, LNNS 1228, pp. 451–460, 2025.
https://doi.org/10.1007/978-3-031-78937-3_49

1 Introduction

Medical Our brain acts as our body's control centre. Through a complex system of neurons and synapses, it controls every vital bodily function. A brain tumor [1] is a serious medical issue. Brain function is compromised by the growth of abnormal cells, which can spread to the spinal cord and neurological system. It expands slowly, stays within clearly delineated borders, and rarely invades neighboring areas. It's believed that 700,000 Americans are afflicted with it. The vast majority of brain tumors are secondary tumors [2, 3]. It poses a serious hazard to human life. A secondary brain tumor [3] occurs when malignant cells from another organ, such as spread from the breast or lung to the brain. Brain tumors, especially malignant ones, can invade healthy tissue, develop at an alarming rate, and have irregular margins. Cancerous cells are present in it. According to the World Health Organisation, benign tumors are categorised as grades 1 and 2, while malignant tumors are classified as grades 3 and 4. Brain tumors have previously been classified using machine learning and deep learning methods. As far as machine learning goes, Support Vector Machine (SVM) [6, 7], the Fisher kernel function [8, 9], and deep learning methods like Even poorer results can be obtained by combining a convolutional neural network with a genetic algorithm [6, 7] (Fig. 1).

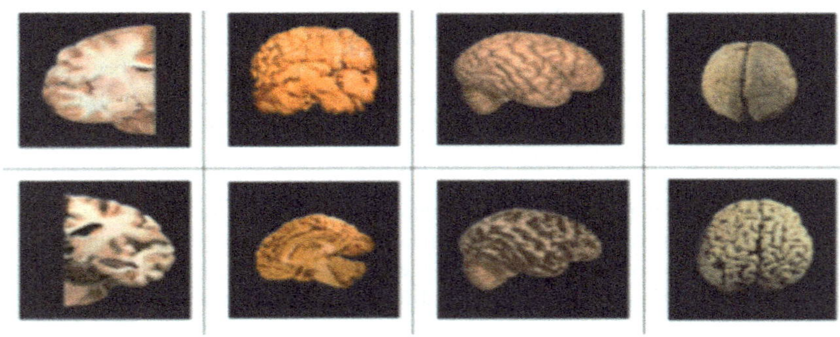

Fig. 1. The Brain tumors images

The purpose of this research is to determine how different types of GBM failure affect the outcomes of subsequent treatment. Clinical factors influencing OS and PRS after a second diagnosis of GBM were studied in a prospective cohort of patients. The recurrence pattern was considered as one of the clinical aspects. Kaplan-Meier survival curves were calculated, and the log-rank test was used to compare curves and determine whether or not there were statistically significant variations in survival times. Survival times were considerably longer for patients who had a local recurrence, (24.1 months vs. 18.2 months) than those who experienced a distant recurrence ($P = 0.015$, $HR = 1.856$ (1.130–3.010)). Benefits were greatest for patients who had a cancer recurrence in close proximity the prognosis for patients with recurrent GBM who are suitable for a second procedure varies according to the pattern of recurrence [6]. This paper proposes using conditional deep learning on structural multimodal MRI images of brain tumors to perform separation, classification using residual networks, and overall survival forecasting.

This article's structure continues below. Section 2 contains the supplementary materials needed to understand the classification strategy for brain tumors. In Sect. 3, we lay out the suggested methodology in great detail. The experimental outcomes and debates are elaborated on in Sect. 4, and conclusions are drawn in Sect. 5.

2 Related Work

In In order to find non-overlapping regions, conditional random fields were used, they devised a convolution network-based segmentation technique. These patches need to be put as soon as feasible because of the tumor. In the event of duplication, the blunders should be amplified. The last sections of the paper suggest a method for feature mapping employing residual networks and XG-PCA-Boost-based learning. Below, we'll cover the rest of this article. Second, we discussed several characteristics of maps (for references, see [10]).

Depending on the anatomy being imaged, MRI can provide pictures with a wide range of characteristics [8]. MRI scans are highly regarded for their high level of detail and accuracy in identifying tumors. Brain tumors need to be segmented in order to be classified, hence MRI alone isn't enough for detection or diagnosis [9, 10]. Segmentation of brain tumors is essential for medical image analysis. Diagnostic medical imaging using MRI [11]. There are many challenges in deciphering MRI scans of brain tumors [12]. Thanks to recent developments in CI and ML, brain tumors can now be detected at an earlier stage. There are a variety of ways to diagnose brain cancer from MR images. Some methods include Fuzzy C Means (FCM), Support Vector Machines (SVM), and expectation-maximization (EM) [7, 9]. These are effective algorithms for identifying regions in medical images and extracting relevant information [12]. Non-Negative Matrix Factorization, Gaussian Mixture Modelling, and Hidden Markov Random Fields have all been used to detect and categorise brain tumors [13, 14]. Due to the ineffectiveness of prior research and approaches, segmentation is a difficult issue in image processing. These methods increase overlaps and decrease feature efficiency.

3 Proposed Methodology

The following research contribution forms the basis of the proposed strategy. Since the tumor is often the smallest possible size, improving or minimising the amount of overlapping patches during the segmentation procedure is important. During the procedure, it must not intersect with any other objects or patches. The suggested study refines segmentation's edges with conditional input from a conditional random field and generates patches that are grounded in context and semantic relationship. Residual learning and feature mapping are two methods that can be used to improve feature mapping; the combined results are used to create a feature matrix. When the structure is optimised and the complexity is reduced, XG-Boost produces superior training outcomes.

Proposed Ensemble CNN with CFR-XHBOOST
Input: Dataset D = { (input number of features; total number of RAs; (1),...,x (n)) } ;
assign {Yes/No} as the prediction class; a number of training items, 'k';
Results: Accuracy of classification based on hypotheses ;
Begin
Step 1: Separating the real-time dataset into a training set and a testing set;
Step 2: Evaluate the quantity of RSS feeds; (L)
Step 3: Collective learning: ANN incorporating RSS;
Step 4: Data processing for input;
Step 5: Partition (training, testing) = (80:20);
Step 6: Instruction New = choose (Instruction set, D);
Step 7: Testing New = choose (Testing set, D);
Step 8: Combine ANN with RSS;
for $i = 1$ $to\ L$; $D\ i = (Trainingset, D\ )$; end for
Step 9: Testing:
for all in testing;
'x' $H(\ ) = argma\ x\ y \in Y \sum$; $Li = 1\ (\ h\ i\ (\ x\ ) = y\ )$
end for
Step 10: Compute ‘k’ nearest data points of random subspace with original subspace ‘ ’
Step 11: Put a label on each data point

End

Machine learning training networks in an adapted ANN model. Several studies have described this process in great detail. Typically, the model operation starts with the generation of a random seed. At this point, we know where the particles are that represent the ANN’s linked biases and masses. To choose the particles, a random procedure is usually employed. The Modified ANN model’s training network is trained using the particles’ initial locations, biases, and weights. This is followed by an evaluation of the combined skill system. The difference between the two sets of results, as measured either directly or using mathematical algorithms, can be used to do this. The calculated error is decreased with each repetition by moving the particle to a new location. By adjusting the smallest error acquired until that point, every particle, and each electron in the universe, in gigabytes, the velocity equation is fine-tuned. Particle locations can be prioritised based on the resulting rate. As you go from one step to the next, you can anticipate a decreasing error rate. In Algorithm 2, the procedure is iterated whenever one of the stopping conditions is satisfied. Data for the optimal ANN architecture were collected after the sixth neurone, as demonstrated in Fig. 2, and In the eleventh neurone, the RMSE value was determined to be the lowest, as shown in Fig. 3. The method’s effectiveness was assessed by compiling data from multiple GIS model runs. This investigation’s pertinent solutions are presented after an explanation of the generated record and ANN procedures. In terms of optimisation, the PSO-ANN method was identical to the ANN method.

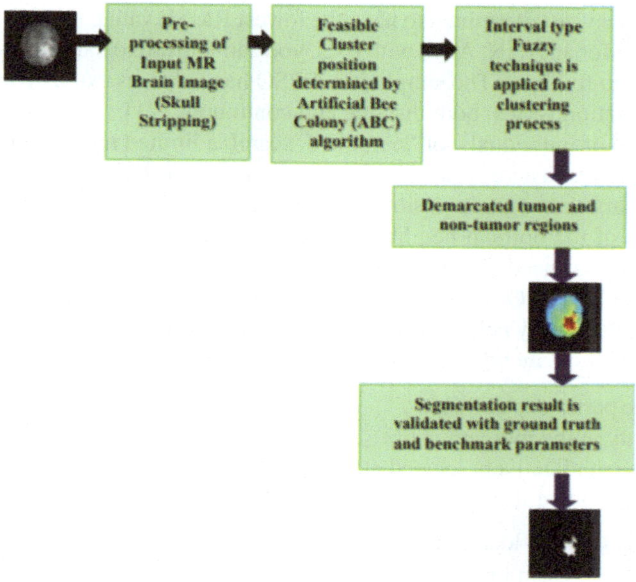

Fig. 2. The Brain tumors proposed diagram

Figure 3 Information was gathered from multiple GIS model simulations to ascertain the method's effectiveness. This research presents important solutions and then goes on to explain the produced record and ANN techniques. The optimisation strategy was identical in the PSO-ANN method and the ANN method. The PSO model [64] of stochastic optimisation uses a population-based method to mimic the social behaviour of flocks. It employs fitness functions to assess individuals and is population-based, similar to EA. Beyond the sixth neurone, further, mostly meaningless variance was being stored, the

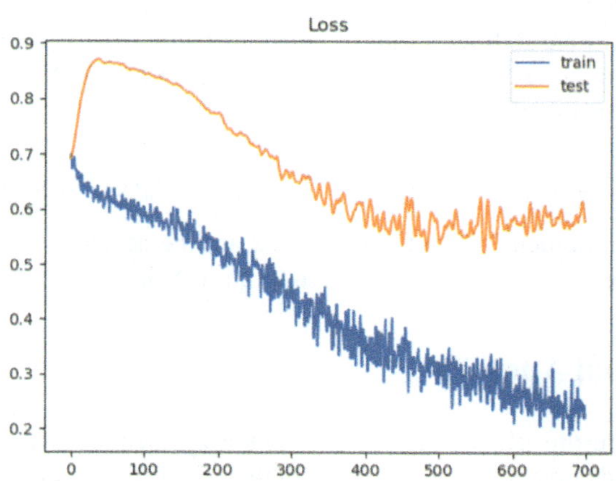

Fig. 3. Simulation of CNN-CRF-XGBOOST Accuracy

eleventh neurone was determined to have the lowest RMSE value, making it the optimal practical design for an ANN. Also, with EAs, you can't afford to make mistakes and then figure out how to fix them. The simplicity of PSO has led to its extensive application in solving optimisation issues, both discrete and continuous [21].

A big group of individuals, or "particles," search a limited region in particle swarm optimisation (PSO). There is an ideal solution to the problem represented by every particle. The particle's cosmic position is determined With relation to its own position as well as that of the optimal neighboring component, If you want to know how long a particle lives, you need to know where it is in the cluster. The chosen algorithm is referred to as G best PSO. In recent years, this approach has become famous for being the "optimal PSO for tiny neighbourhoods." The optimisation approach, fitness function, and ultimate solution are all affected by the particle distance.

Right now, the particle is at this ai
Particle velocity at the moment is bi.
Particles are most effectively located at: ci
It is given by (b_i) that the particle is optimally located.

Finding the optimal location for a particle is all about maximising its fitness value. F is a possible representation of the goal function [22]. Therefore, at time t, the ideal value of the A particle is (1):

$$RSO_i(o+2) = \begin{cases} RSO_i(s) if \ \ wg(a_i(s+1)) > dg(b_i(s)) \\ SO_i(s+1) if \ \ wg(a_i(s+1)) < dg(b_i(s)) \end{cases} \tag{1}$$

The swarm as a whole decided which particle was best for the gbest model in the course of determining the best place to begin. Assuming b is a global representation of the ideal particle's location, we get (2):

$$\widetilde{zb}\{zb_0, b_1 \ldots . zb_k\} = \min\{g\big(zb_1(s)\big), \ldots, (zb_k(\mathrm{s}))\} \tag{2}$$

We can say that the swarm is of size k if that is the size of the hive.

Velocity updates are defined in all dimensions from 1 to Nd. Bi, j stands for that value since the velocity vector of the ith particle includes a zjth component. The Eq. (3) can be used to determine the particle's speed:

$$zb_{i,j}(s+1) = qzb_{i,j}(s) + c_1 m_{1,j}(s) b_{i,j}(s)) - zb_{i,j}(s)) + c_2 m_{2,j}(s)(\widetilde{zb(s)} - zb_{i,j}(s)) \tag{3}$$

While w is the moment of inertia, c1 and c2 are the acceleration constants, and In this context, the reference frames are c1_2 m_(1,zj) (s) and m_(3,zj)U (0, 2).

4 Experimental Analysis

The experimental results of the suggested system are presented here. As was previously noted, the proposed approach is evaluated with the use of two separate data sets. Experimental outcomes are evaluated using supplementary tests such the F-score, recall, precision, and accuracy. For classification and feature mapping, the proposed system utilises the Resnet-XG-Boost framework, while segmentation is handled by a CNN-CRF based network. It matched the original dataset against the larger one. The experimental results of the proposed system are compared to those of existing brain tumor detection methods, and vice versa. Tables 2 and 3 show the confusion matrices for the classification results achieved by the proposed BTC system for the two datasets.

Tumor diagnosis via random display of examples from Table 3's three classes. Experiments with CNN-CRF-RESNET on both the original and enhanced versions of two datasets are shown. Table 4 and Fig. 4 show the results of experiments conducted on the datasets using various classes, with 10 iterations of cross-validation used to fine-tune the validation. Accuracy, precision, recall, and F-score are used to rank the results of the suggested approach. It compared the base dataset to the expanded one. When comparing the original and enhanced datasets, Three-classes increases in accuracy by around 1.2%, precision by about 1.5%, recall by about 1.3%, and F-score by about 1.5%. Improvements in accuracy, precision, recall, and F-score are seen when the number of instances is increased for two-class datasets. It compared the base dataset to the expanded one. The F-score, recall, and accuracy on two classes all improve by 0.3 percentage points when comparing the original dataset to the supplemented one Table 1.

Table 1. .

Classification	F-Measure	Sensitivity	Precision	Accuracy
ANN-ICA	65.12	66.12	69.45	71.23
KNN-LDA	66.17	55.67	64.04	69.45
CNN-SVM	71.45	66.78	90.13	87.05
EW-PCA	72.67	73.89	85.78	89.18
FCM-RF	76.89	71.67	87.56	91.56
CNN-CRF-XGBOOST	98.1	98.45	97.12	98.89

Table 1 compares the proposed model to some other popular classifiers so that we may get a feel for how well it performs. Table 1 compares the precision, responsiveness, and selectivity of a number of algorithms (Fig. 5).

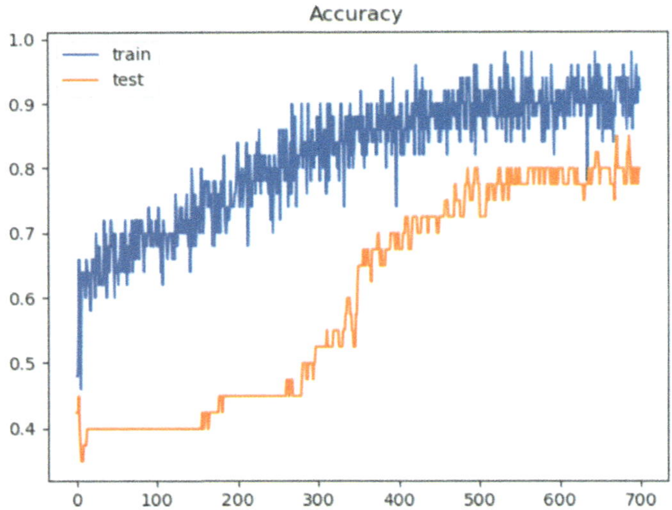

Fig. 4. Simulation of CNN-CRF-XGBOOST F-measure value

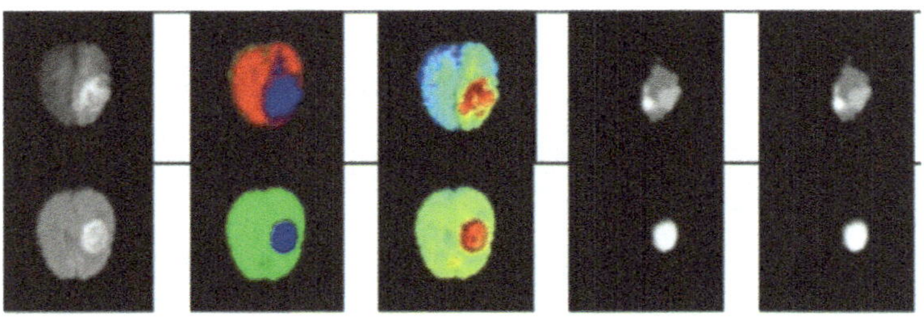

Fig. 5. Output images of Brain tumor detection by the CNN-CRF-XGBOOST

5 Conclusion

In this research, we describe a fully automated approach to classifying brain tumors that requires minimal human intervention. In the proposed approach, brain MRI images are processed by Deep Residual Learning to extract characteristics. The suggested method uses a CNN-CRF based methodology to minimise overlapping edges, allowing for more effective segmentation. In this method, CRF gives conditional judgements for boundary segments. Overlapping tiny tumors are diminished by the addition of patches with distinct borders. As a result, the ability of a residual network to map features and the XG-Boost base ensemble classifier to train are both improved in the absence of overlapping patches. Both cancer and non-cancer datasets, as well as meningioma, glioma, and pituitary datasets, are used in the investigations. We applied CNN-CRF-Resnet using the proposed approach to both datasets, performed 10-fold cross-validation, and analysed the results. We were able to attain meaningful measures for both classes, including an accuracy of 98.89%, precision of 97.12%, recall of 98.45%, and an F-Score of 98.1%.

The following investigations including three classes demonstrated an improvement in reliability (98.56%), validity (97.56%), usability (98.12%), and performance (98.45%). When more examples from all classes were included, the enhanced method's findings got better. There was a similar improvement in all three groups. The two-class and three-class methods provide a huge computational and resource improvement over the status quo. The suggested study has as its goal the enhancement of patch overlaps and its influence on the effectiveness of feature mapping by residual network. The proposed work additionally contributes a focusing approach and a learning mechanism that enhance learning and features.

References

1. Abdel-Maksoud, E., Al-Awadi, R.: Brain tumor segmentation based on a hybrid clustering technique, pp. 71–81 (2015)
2. Aswathy, S., Kumar, S.: Quick Detection of Brain Tumor Using a Combination of E. M. And Level Set Method (2015)
3. Despotović, I., Goossens, B., Philips, W.: MRI Segmentation of the Human Brain: Challenges, Methods, and Applications, pp. 1–23 (2014)
4. Chandra, G.R., Rao, K.R.H.: Tumor Detection in Brain Using Genetic Algorithm, pp.449–457 (2016)
5. Capelle, A.S., Colot, O., Fernandez-Maloigne, C.: Evidential segmentation scheme of multiecho M.R. images for the detection of brain tumors using neighborhood information. Inf. Fusion **5**, 103–216 (2004)
6. Sharma, M.M.: Brain tumour segmentation techniques: a survey. Brain **4** (2016)
7. Ravi, A., Sreejith, S.: A review on brain tumor detection using image segmentation. Int. J. Emerg. Technol. Adv. Eng. **5**, 60–64 (2015)
8. Shen, D., Wu, G., Suk, H.-I.: Deep learning in medical image analysis. Annu. Rev. Biomed. Eng. **19** (2017)
9. Qayyum, A., Anwar, S.M., Awais, M., Majid, M.: Medical Image Retrieval Using Deep Convolutional Neural Network, arXiv preprint arXiv:1703.08472 (2017)
10. Sun, Y., Wang, C.: A Computation efficient CNN system for high quality brain tumor segmentation. Biomed. Signal Process. Control **74**, 103475 (2022)
11. Zhou, Z., He, Z., Jia, Y.: AFPNet: a 3D fully convolutional neural network with atrous-convolution feature pyramid for brain tumor segmentation via MRI images. Neurocomputing **402**, 235–244 (2020)
12. Parra, C., Kozma, R.: Automated Brain Data Segmentation and Pattern Recognition Using ANN (2003)
13. Papageorgiou, E.I., Spyridonos, P.P.: Brain Tumor characterization using the soft computing technique of fuzzy cognitive maps. Appl. Soft Comput. **8**, 820–828 (2008)
14. Gao, J., Xie, M.: Skull-Stripping MR Brain Image Using Anisotropic Diffusion Filtering and Morphological Process, pp. 1–4 (2009)
15. El-Dahshan, E.S.A., Hosny, T., Salem, A.B.M.: Hybrid intelligent techniques for MRI brain image classification. Digital Signal Process. **20**, 433–441 (2010)
16. Othman, F., Basri, A.: Probabilistic Neural Network for Brain Tumor Classification, pp. 136–138 (2011)
17. Rathi, V.P., Palani, S.: Brain Tumor MRI Image Classification with Feature Selection and Extraction Using Linear Discriminant Analysis, pp. 1–17 (2012)
18. Islam, A., Iftekharuddin, K.M.: Multi-fractal Texture Estimation for Detection and Segmentation of Brain Tumors. IEEE (2013)

19. Agn, M., Law, I.: Brain Tumor Segmentation by a Generative Model with a Prior on Tumor Shape. IEEE (2014)
20. Hooda, H., Singhal, T.: Brain Tumor Segmentation: A Performance Analysis Using K-Means, Fuzzy C-Means and Region Growing Algorithm, pp. 1621–1626 (2015)
21. Dinesh, G., Venkatakrishnan, P., Jeyanthi, K.M.A.: Dispersed spectrum sensing and scheduling in cognitive radio network based on SSOA-RR. Int. J. Commun. Syst. **33**, e4588 (2020). https://doi.org/10.1002/dac.4588
22. Senthilkumar, R., Venkatakrishnan, P., Balaji, N.: Intelligent based novel embedded system based IoT enabled air pollution monitoring system. Microprocess. Microsyst. **77**, 103172 (2020). ISSN 0141-9331. https://doi.org/10.1016/j.micpro.2020.103172
23. Chilakala, S., Renuka, N., Mubeen, S.: Automation of traffic lights through IoT. In: 2020 7th International Conference on Smart Structures and Systems (ICSSS), Chennai, India, pp. 1–5 (2020). https://doi.org/10.1109/ICSSS49621.2020.9202275
24. Raju, S., Jagtap, V., Kulkarni, P., Ravikanth, M., Rafeeq, M.: Speech recognition to build context: a survey. In: 2020 International Conference on Computer Science, Engineering and Applications (ICCSEA), Gunupur, India, pp. 1–7 (2020). https://doi.org/10.1109/ICCSEA49143.2020.9132848
25. Nagarani, N., Venkatakrishnan, P., Balaji, N.: Unmanned Aerial vehicle's runway landing system with efficient target detection by using morphological fusion for military surveillance system. Comput. Commun. **151** (2020)

Evaluating Classification Algorithms for Breast Cancer Prediction: A Comparative Analysis Approach

Rajesh Tiwari[1], Radhe Shyam Panda[1], Abdul Subhani Shaik[2], M. Sirin Kumari[2], K. Srujan Raju[3](✉), and Ravi Regulagadda[3]

[1] CMR Engineering College, Hyderabad, Telangana, India
[2] CMR College of Engineering and Technology, Hyderabad, Telangana, India
[3] CMR Technical Campus, Hyderabad, Telangana, India
ksrujanraju@gmail.com

Abstract. Currently, breast cancer is a significant contributor to mortality among women, ranking second only to lung cancer in terms of cancer-related deaths. Breast cancer arises as a result of the fast proliferation of cells within the breast tissue. Breast cancer can manifest in any region of the breast and can be prevented with early initiation of treatment. Breast cancer is a malignant neoplasm characterized by the uncontrolled proliferation of cells originating from breast tissue. The management of breast cancer is contingent upon the specific subtype of cancer and its corresponding stage, ranging from stage 0 to stage IV. Treatment modalities commonly employed encompass surgical intervention, radiation treatment, and chemotherapy. The objective of this study is to employ various data mining algorithms and technologies to diagnose breast cancer. The identification and analysis of disease patterns in earlier cases have led to the discovery of a novel facet in the field of medical advancement. The dataset used in this study was obtained from the UCI machine learning repository. The research focused on evaluating the classification performance of various algorithms, including Binary Categorization, Pseudo random Forest, Regression Analysis, Multilayer Perceptron, and K-NN, for predicting the occurrence of Breast Cancer Disease. The analysis of the gathered data encompassed an examination of its dependability, true positive rates, F1-score, and Kappa statistics.

Keywords: Binary Categorization · Pseudo random Forest · Precision · Kappa statistics · KNN classifier · Recall · and ROC area

1 Introduction

Breast cancer is a prominent concern in contemporary society on a global scale. This issue is a significant health concern for women. The prevalence of breast cancer is second only to that of lung cancer in terms of comprehensiveness. The disease serves as the primary factor contributing to mortality from cancer in the female population. The observed phenomenon could potentially be attributed to either ancestral factors or genetic

D. Kriksciuniene et al. (Eds.): IBICA 2023, LNNS 1228, pp. 461–471, 2025.
https://doi.org/10.1007/978-3-031-78937-3_50

predispositions. Breast cancer has a hereditary component, occurring in approximately 6% to 10% of cases, specifically among individuals who possess mutations in the BRCA1 and BRCA2 genes. Both prolonged exposure to estrogen and the utilization of hormone replacement therapy (HRT) throughout the menopausal period have been linked to an increased risk.

Risk factors for many health conditions include obesity, alcohol consumption, infertility, and delayed initiation of pregnancy. Two clinical manifestations of breast cancer include nipple discharge and palpable tumor development. Breast cancer exhibits a higher propensity for metastasis to many organs, including the liver (56%), bones (71%), and lungs (61%).

2 Literature Survey

Sequential Minimum Optimisation (SMO) is an innovative approach utilized for the training of Support Vector Machines (SVMs). The Sequential Minimal Optimization (SMO) technique, developed by John Platt in 1998, is a highly effective and efficient method for training Support Vector Machines (SVMs). The essential principle is achieved by employing a dual quadratic iterative method that aims to maximize the smallest set that contains two components in each iteration. The individual and their team members utilized the IBK data mining technique, Best First Tree, and SMO algorithms in order to achieve classification accuracy [1]. Based on their investigation, it was concluded that the utilization of SMO would yield superior outcomes characterized by enhanced precision and a diminished incidence of errors.

According to the second source, Zarei et al. [2] employed various computational.

A cohort of 250 individuals diagnosed with breast cancer underwent examination at the Cancer Registry located in Adyar, Chennai, India. The examination involved the application of Ensemble learning, specifically the Classification and Regression Trees (CART) method, which utilizes various tree classification methods. The findings of the analysis have been made public. A data mining method was employed to generate a variety of measures for the investigation, including selectivity, sensitivity, and kappa statistics [3].

In this study, the Badger Tumor Dataset was utilized to employ Logistic Regression, Nearest Neighbours, and Support Vector Machines classifiers with the purpose of predicting breast cancer as either benign or malignant [4, 5]. The study demonstrated a correlation between the recall and precision metrics and the quantity of features present in the dataset.

Enhancement refers to the process of improving the quality of an image, resulting in a higher level of clarity and comprehensibility. The mammography enhancement approach encompasses several techniques, namely filtering, top hat operation, and discrete wavelet transform (DWT). The technique of contrast stretching is employed to enhance the contrast of the image [6]. The segmentation of mammography images has emerged as a crucial factor in enhancing the efficacy of breast cancer detection and diagnosis. Thresholding is the segmentation approach that is most frequently employed. The features are derived from the region of interest that has been delineated as the breast area. The subsequent phase involves the utilization of the Support Vector Machine (SVM) classifier to classify the regions.

In order to maximize the duration available for the implementation of safety measures, it is necessary to minimize the duration of seismic waveform elaboration. This research proposes an optimal data mining technique for accurately and efficiently estimating the level of hazard associated with an ongoing seismic event. A comprehensive analysis was conducted on a large dataset comprising numerous seismic recordings of earthquakes in Japan and Italy. Through the application of three data mining techniques, namely Artificial Neural Networks (ANN), Decision tree (C5), and Logical Regression, a model was derived. The model was then tested on the initial seismic recordings to expedite decision-making processes. The outcomes of the tests were highly favorable, indicating a significant reduction in decision time. This information is supported by reference [7, 8].

3 Proposed System

This process is closely associated with the field of taxonomy and involves the cognitive activities of perception, differentiation, and interpretation of concepts and entities as shown the Fig. 1. A rule-based classifier employs a collection of IF-THEN rules to perform the task of classification. The application of the IF predicate to a given condition may result in the deduction of a corresponding conclusion. The utilization of a rule-based technique in the diagnosis of breast cancer has exhibited its efficacy as a valuable tool in providing diagnostic help to healthcare professionals. The establishment of a system of this nature necessitates the formulation of classification criteria that are derived from historical diagnostic records. The ideal regulations should be characterized by a minimal quantity and effective functioning. The objective of this paper is to extract fundamental tenets using the Carcinoma data set. The original dataset was subjected to a series of runs in order to determine the optimal classifier between J48 decision tree and Random Tree.

Data Collection – The initial phase involves the collection of unprocessed data. The internet contains a vast array of datasets. The computer vision resource at UCLA, which is widely recognized, provides users with access to a diverse range of samples, including datasets related to pharmaceuticals and the iris. The dataset pertaining to cancer, available in the machine learning repository of California State University, Irvine, includes attribute 10.

User Interface – The subsequent procedure involves the utilization of a software tool known as WEKA.WEKA is a widely recognized machine learning and data mining program that was initially created at the University of Waikato in New Zealand. The WEKA software package encompasses a range of supervised and unsupervised learning methods, including but not limited to regression, clustering, classification, and association rule mining. The WEKA software package provides support for various file formats, including.arff,.csv, and.json. Weka offers a diverse range of applications for the aim of implementation, including Explorer, Experimenter, and Knowledge Flow, among others.

Data Pre-processing – The subsequent stage involves the process of data preparation. Once the dataset has been chosen, preprocessing techniques are employed to transform the unprocessed data into a structured format. Data cleaning and data transformation

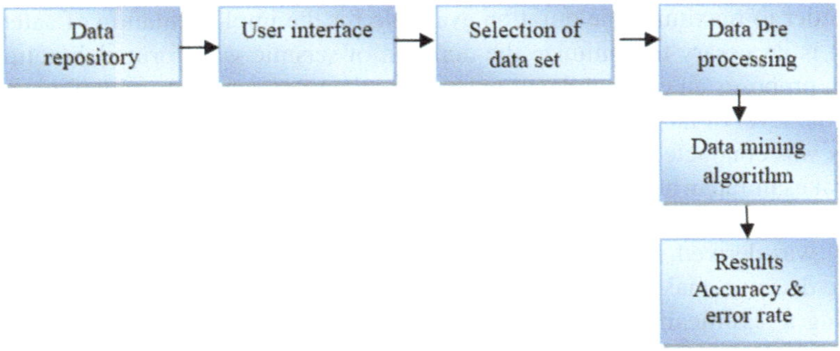

Fig. 1. Design flow of projected representation

are two important processes in the process of data preprocessing. The WEKA software provides support for data pre-processing filters that may be applied to both supervised and unsupervised learning tasks.

Data Mining Algorithm – Once the dataset has been cleansed by eliminating undesirable data and addressing missing values, the subsequent step involves the application of data mining tools. There exists a range of categorization methods, such as Decision tree, Random Forest, and Support Vector Machine, among others.

Performance Estimation – The assessment of effectiveness is conducted using metrics such as True Positive, False Positive, Precision, and Recall. The instance that is classified correctly is regarded as the measure of reliability.

Theoretical Description of the Classifiers

The research project encompassed five classifying methods. The classifiers encompassed in this set are the Multilayer Perceptron, Naive Bayes, Random Forest, and K-Nearest Neighbour.

3.1 Naïve Bayes

The fundamental principle underlying Naive Bayes classifiers is the independence assumption, which posits that the value of a parameter for a certain class is not influenced by the magnitudes of other variables. This assumption is commonly referred to as "fitted model independence." The purpose of this approach is to identify factors that are relevant to classification when the available training data is limited. The Native Bayes algorithm is derived from the integration of the Bayesian framework with a decision rule as shown the Fig. 2.

In the field of Natural Language Processing (NLP), the work of trend analysis involves the identification of the sensitive or emotional subtext within a given text corpus. To do this, a commonly employed approach is the utilization of prediction models.

The Bayesian approach is employed to ascertain the likelihood of a hypothesis given the presence of pertinent evidence.

The calculations were performed utilizing conditioned probabilities.

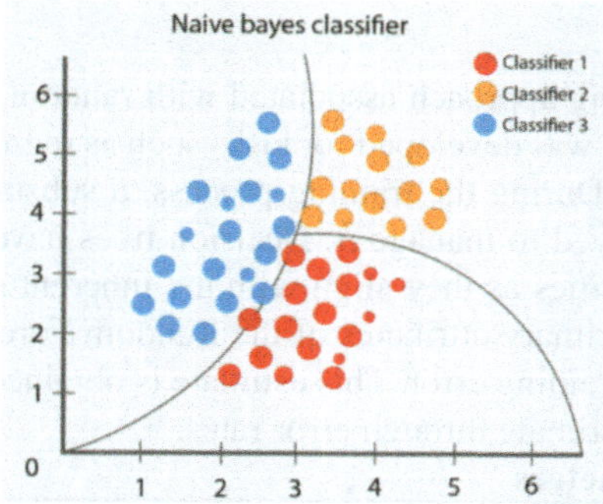

Fig. 2. Developing a Naive Bayes-based Bayes Classifier from the ground up for the purpose of Sentiment Analysis

3.2 Logistic Regression

Logistic regression is a frequently employed method in the context of binary classification. The value of the coefficients of logistic regression offers a concise and cumulative representation of the impact of a variable on the observed likelihood of an event possessing a specific characteristic. In this particular case, the outcome is contingent upon one or maybe more independent variables. The available options consist of the binary values 1 (representing true) or 0 (representing false). The data is elucidated through the description of the association among a single dependent binary parameter and one or maybe several independent variables.

3.3 Multilayer Perceptron

The algorithm known as k-nearest-neighbor is a classification tool. The process of machine learning can be characterized as rather straightforward. The k-nearest neighbor approach is frequently utilized in classification and regression tasks. The first stage of input was derived from the votes of the neighbors. The neighbor in closest proximity is identified by the point with the smallest distance. This particular approach can be employed for the purpose of categorizing data samples.

3.4 K-nearest Neighbour's Classifier

The method known as k-nearest-neighbor is a classification technique. The process of machine learning is characterized by its high degree of simplicity and ease of understanding. The k-nearest neighbor technique is commonly employed in both classification and regression tasks. The first input was derived by the voting of adjacent entities. The point with the smallest distance serves as an indicator of the nearest neighbor. The categorization of sample data can be accomplished by the utilization of these strategies.

3.5 Random Forest

The constructivist teaching approach associated with random forests, often referred to as random choice forests, was developed for tasks such as regression, classification, and other related objectives. During the training process, a substantial number of creative decision trees are employed to manage it. Decision trees have advantages from incorporating random possibilities as they align with the inherent structure of decision tree algorithms. One of the primary attributes of the Random Forest algorithm is its ability to provide an estimated training error. This estimate is obtained by cross-validation and is presented as a range estimate through error rates.

Performance Parameters

Kappa Statistics

Statistical analysis is employed for comparing and rank two distinct types of data. Greater numerical values indicate a broader range of values spanning from zero to 1, as well as a more robust inter-rate treaty. This projection exhibits a higher degree of accuracy compared to typical estimations. The acronym refers to the unambiguous monetary worth of the contractual opportunity. The user is able to express their thoughts or ideas effectively.

$$K = \frac{P(A) - P(E)}{1 - P(E)}$$

Nearer,

K = Statistics Kappa
P (A) = Accordance percentage
P (E) = Likelihood of agreement

TP Rate

The real positive rate refers to the ratio of patients having a currently existing ailment whose diagnostic test results were positive. The calculation involves the use of the variable Val, where FN represents the proportion of false negatives and TP represents the count of true positives. The sum of FN and TP is the overall count of positive instances. The calculation of a genuine positive rate, also known as sensitivity or true positive rate (TPR), is performed using the following formula.

$$\text{TPR} = \frac{TP}{TP - FN}$$

$$\text{Val} = \frac{FN}{FN + TP}$$

FP Rate

Whenever a diagnostic test produces a good outcome in the absence of a genuine illness, it tends to elicit heightened concerns. The calculation of the false positive rate involves determining the proportion of erroneous convictions, or mistakes made in correctly identifying cases, in relation to the sum of accurately classified frequencies and true negatives (FP + TN). This implies the potential occurrence of a distressing event, when a favorable conclusion is communicated despite the presence of a quantifiable value that aligns with the forecast.

Precision

Precision is a characteristic that pertains to the level of information provided by numerical values, specifically in relation to the amount of digits they contain. It serves as an indicator of the proximity between many measurements. The issue of accuracy is not a determining factor in this context.

Precision refers to the ratio of correctly identified positive instances to the total number of instances classified as positive (true positives).

Recall

The metric in question is a quantitative measure that assesses the proportion of accurate positive predictions relative to the total number of potential positive predictions.

Recall refers to the proportion of accurately identified positive instances in relation to the total number of positive instances as determined by correct positive algorithms (true positives).

Measure Lining

When evaluating the performance of a method utilizing the F-measure, factors such as memory and accuracy can be taken into account. The expression can be formulated with the letter F,

$$F = [\text{Precision} \times \text{Recall}/\text{Precision} + \text{Recall}] \times 2$$

The F-Measure of an unoptimized classification model. The utilization of this approach is prevalent due to the exclusive reliance on accuracy rate as the sole metric for evaluating the efficacy of the model. The F-Measure tool is designed to integrate precision and recall into a single metric, thereby encompassing both properties.

MCC

The Matthew's Correlation Coefficient (MCC) is a viable approach for selecting and representing a singular value confusion matrix. Typically, this method is employed whenever two replicas that are not easily compared give rise to confusion. While there is no definitive method for representing the matrix of confusion of true and false negatives and positives with a single value, the Matthews Correlation Coefficient (MCC) is commonly regarded as the most effective measure. The measure in question is essentially a correlation coefficient that quantifies the relationship among a predicted series and the corresponding actual series. The values range from -1 to $+1$, resulting in a profit. The coefficient of -1 signifies the complete disparity between the observed value and the predicted value, while a coefficient of $+1$ indicates a precise forecast. A coefficient of 0 implies that the prediction is nothing more than a random guess. The ensuing formula can be taken into consideration,

$$\text{MCC} = \frac{\text{TP} * \text{TN} - \text{FP} * \text{FN}}{\surd[(\text{TP} + \text{FP})(\text{TN} + \text{FN})(\text{TP} + \text{TN})(\text{FP} + \text{FN})]/1} \tag{1}$$

Receiver Operating Characteristics Area or ROC Area

The territory around the ROC curve is referred to as ROC area or Receiver Operating Characteristics area. This binary classifier is visually represented by a graphical plot. The central element in this illustration is formed through the representation of the substantial true positive rate alongside varying levels of false positives across several thresholds. Every point of information within the Receiver Operating Characteristic (ROC) curve symbolizes a specific combination of specificity and sensitivity, corresponding to a particular decision threshold. In this context, the true positive rate is referred to as the responsibility, while the false positive rate is categorized as exceptional (0.90–1), decent (0.80–0.90), acceptable (0.70–0.80), bad (0.60–0.70), or fail (0.50–0.60).

4 Experimental Results

The melanoma dataset utilized in the present study was sourced from the machine learning repositories at the University of California, Irvine (UCI) in the United States. The currently accessible data, consisting of a dataset containing 286 cases with 10 different characteristics, can be utilized for the purpose of analyzing and predicting breast cancer. The variables incorporated in our analysis encompass age, degree of malignancy, menopausal status, involvement of lymph nodes, extent of tumor destruction, tumor size, clustering, chest characteristics, classification, and age.

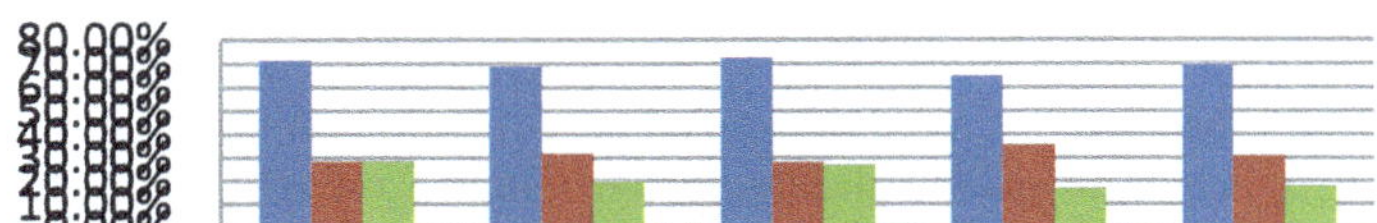

Classifiers	Naïve Bayes	Logistic Regression	K-nearestneighbour's classifier	Multilayer Perceptron	Random Forest
Correctly classified instances	71.82%	69.21%	72.89%	65.10%	70.01%
Incorrectly classified instances	28.62%	31.92%	28.31%	35.95%	31.28%
Kappa statistic	0.29	0.2	0.27	0.17	0.18

Fig. 3. KNN classifiers

In this particular case, the findings presented in Fig. 3 indicate that the KNN classifier exhibits a higher proportion of accurately identified cases (72.88%), surpassing the performance of the other classifiers mentioned. Notably, the KNN classifier's accuracy is only slightly comparable to that of the naive Bayes model (71.678%). Conversely, the KNN classifier misclassifies approximately 26.957% of cases (79 instances), which is relatively lower compared to the misclassification rates of the other classifiers. The Multilayer understanding classifier, in particular, illustrates the highest rate of erroneously categorized instances (35.314%), corresponding to 101 cases. It is worth noting that the entire number of instances in this analysis is 286.

Categorization of correctly and incorrectly-Following is another table that highlights how Random Forest displays the higher MAE number (0.3727). Additionally, the MAE's value using KNN's and Naive Bayes is smaller or equal to (0.3272 and 0.3257). The biggest relative margin of error (89.085%) is shown in the Random Forest.

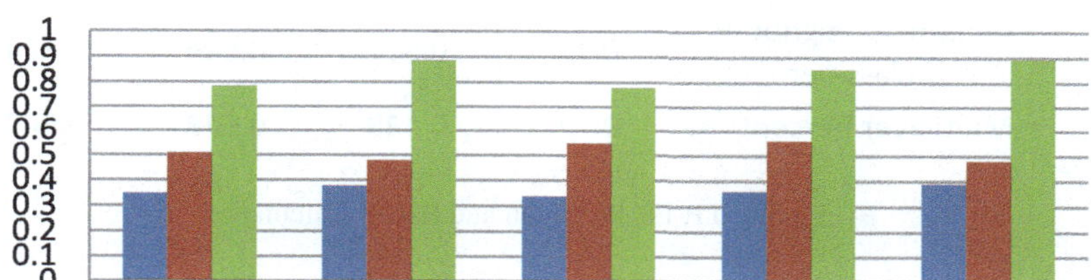

Classifiers	Naïve Bayes	Logistic Regression	K-nearestneighbour's classifier	Multilayer Perceptron	Random Forest
MSE	0.35	0.38	0.34	0.36	0.39
RMSE	0.51	0.48	0.55	0.56	0.48
RAE	78.64%	88.91%	78.21%	85.33%	89.51%

Fig. 4. RMSE, MSE, RAE, Calculation

As shown the Fig. 4 following demonstrate that the KNN has the highest actual positive rate (0.724), whilst Multilayer Perceptrons has the weakest real positive rate (0.647). In this scenario, Rf has a high False Alarm Rate although KNN and Naive Bayesian have decent rates of accuracy (0.699 and 0.704, correspondingly) (0.543).

Lastly the Fig. 5 demonstrates that Naïve Bayes exhibits the highest MCC and in contrary the multilayered perceptron is possessing low MCC value. The biggest percentage of the ROC area is encompassed by Naïve Bayes (70.1%).

The image presented below offers a comprehensive depiction of the facts and information pertaining to Benign and Malignant tumors. It highlights the contrasting characteristics and tissue composition of both tumors, while also illustrating the existence of Anaplasia inside the tissue mass.

As shown the Fig. 6 displays the differences of the normal and cancerous cell division and the structure of it as we offered data on the malignant as well as benign class that we utilized to predict the tumour.

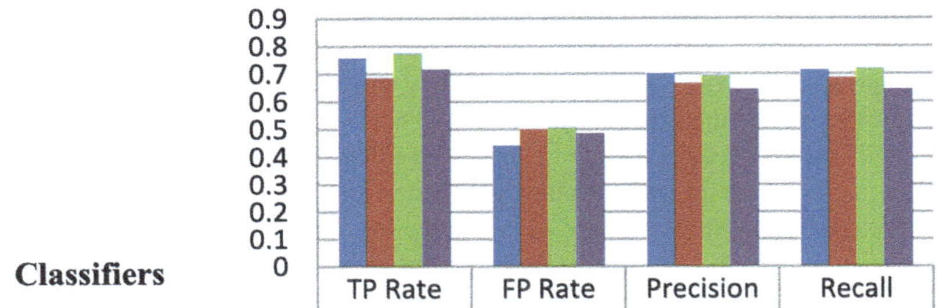

Classifiers	TP Rate	FP Rate	Precision	Recall
Naïve Bayes	0.76	0.446	0.704	0.717
Logistic Regression	0.69	0.505	0.668	0.689
K-nearest neighbour's classifier	0.78	0.511	0.699	0.724
Multilayer Perceptron	0.72	0.489	0.648	0.647

Fig. 5. TP, FP, Precision and Recall Calculation

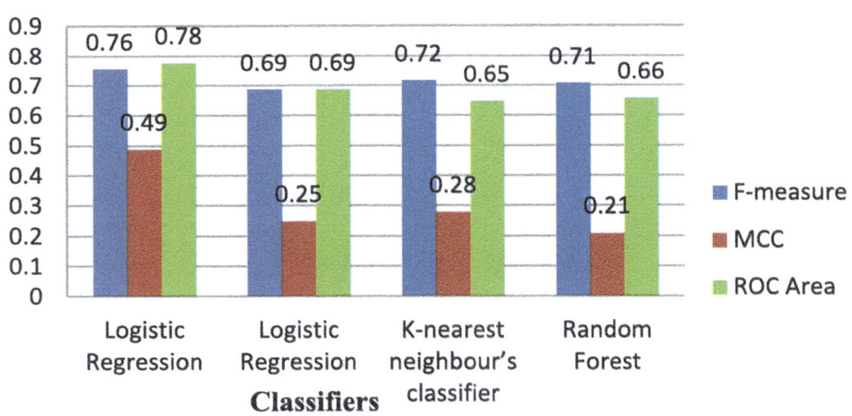

Fig. 6. MCC, F-measurement and ROC area Calculation

5 Conclusion

To predict and take into account breast cancer, a multitude of predictors, including Naive Bayes, Random Forest, Logistic Regression, Multilayer Perceptron, and K-nearest Neighbour's, might be utilized. The WEKA gathering tool was utilized for contrasting the demonstrations of various learning approaches. It is noted that the K-nearest Neighbour's categorization approach is excellent. Its overall classification rating is 97.9021%, indicating the highest. The subsequent classifier with the highest level of accuracy is the multilayer perceptron, which properly classified 96.5035% of the examples. The investigation's main purpose is to determine and assess a dataset of 286 breast cancer patients along with 10 attributes. It evaluates the performance of numerous classification algorithms while contrasted with distribution plots. In this pursuit, it was additionally able to observe. These include the Kappa statistic, the Mean Absolute Error, the F-measure, the MCC, the ROC Area, the Comparative Absolute Error, the FP rate, the TP rate, the root-average-squared Error, and the coefficient of precision Recall. The K-nearest Neighbour algorithm exhibited the highest ROC Area percentage (99.9%), followed by the Multilayer Perceptron algorithm which consistently achieved the second-highest percentage (98.0%).

References

1. Chaurasia, V., Pal, S.: A novel technique for diagnosing breast cancer using data mining methods. J. Emerg. Dev. Comput. Commun. Eng. 2464 (2014)
2. Zarei, S., Aminghafari, M., Zalie, H.: Application & comparison of different linear algorithms for breast cancer diagnosis. Particularly In the fields of Academic, Medicinal and Microbiology, vol. 4, no. 2, pp. 123–128
3. Padmapriya, B., Velmurugan, T.: Classification technique approach says of breast cancer data. J. Data Mining Tech. Appl. **5**(1), 43–49 (2016)
4. Tiwari, R., et al.: An artificial intelligence-based reactive health care system for emotion detections. Comput. Intell. Neurosci. **2022**, Article ID 8787023. https://doi.org/10.1155/2022/8787023
5. Sharma, A., Kulshrestha, S., Daniel, S.: Machine learning approaches for breast cancer diagnosis and prognosis. In: World Congress on Smooth Cryptography and its Engineering Disciplines (icSoftComp), held in Changa, India, 1–2 December 2017 (2017)
6. Rejani, A., Thamarai Selvi, S.: Early breast cancer identification using SVM classifier algorithm, Ireaneus, Y. J. Contemp. Comput. Eng. Sci. **1**(3), 127–130 (2009)
7. Lakshmi, K.R., Veera Krishna, M., Prem Kumar, S.: Performance comparison of data mining techniques for estimation & identification of breast cancer disease survivability. Asian J. Inf. Technol. Comput. Sci. **3**, 81–87 (2013)
8. Narendar, C., Prakash, D.B.: A real time image inverse and training process for future multimedia application with RFO machine learning technique. In: AIP Conference Proceedings, vol. 2477, no. 1. AIP Publishing (2023)

Analysis on Deep Convolution Neural Network Evaluation of Microaneurysms Identification on Retinal

A. Srinivasula Reddy[1], X. S. Asha Shiny[1], Y. Aruna Suhasini Devi[2], G. Karthik Reddy[2], G. Srikanth[3], and B. P. Deepak Kumar[3](✉)

[1] CMR Engineering College, Hyderabad, Telangana, India
dr.ashashinyxs@cmrec.ac.in
[2] CMR College of Engineering and Technology, Hyderabad, Telangana, India
yarunasuhasini@cmrcet.ac.in, gandhari.karthik@cmrcet.org
[3] CMR Technical Campus, Hyderabad, Telangana, India
Cmrtc.paper@gmail.com

Abstract. The amount of people with diabetes is rising quickly in India. Numerous illnesses, including neuropathy, renal failure, heart attacks, eyesight loss, and skin conditions are becoming more common among diabetic people. We are currently concentrating on visual loss in this research. Prompt sickness identification and management can help identify visual loss. Vision loss is caused by diabetic retinopathy, and one of the disease's early side effects is microaneurysms (MAs). Early on in the development of diabetic retinopathy, the fundus evaluation is highly beneficial. However, because MAs frequently appear as tiny faint patches, it might be difficult for medical professionals to recognize MAs on retinal images. As a result, a lot of research has been done on robotized MA recognition. The double-ring filter, the index of shape on the Hessian structure, and the Gabor filter are all techniques that are currently in use, and this study suggests a single MA identifier by combining them. However, this study uses deep convolutional neural networks (DCNN) to do automated MA identification because DCNN has demonstrated superior performance in photo acknowledgment tests. We structured the suggested approach using a three-layer insight with 48 elements for false positives (FPs) reduce features and a two-step DCNN. The major DCNN in the two-step DCNN is used to begin MA discovery, while the second DCNN is used to minimize FPs. Utilizing the DIARETDB1 data base, the suggested approach demonstrates flawless implementation. In order to guarantee the AI model's accuracy and efficacy in identifying and diagnosing diabetic retinopathy, it can be continuously enhanced over time. Deep learning algorithms can be used to train the AI model to recognize the MAs.

Keywords: Diabetic retinopathy · Micro aneurysms (MAs) · Deep convolutional neural networks (DCNN) · Image recognition · Fundus images

D. Kriksciuniene et al. (Eds.): IBICA 2023, LNNS 1228, pp. 472–479, 2025.
https://doi.org/10.1007/978-3-031-78937-3_51

1 Introduction

Age-related disorders like diabetes have become more prevalent in Sweden and other areas of the world in recent years. As per a recent poll [1], 4% of the nation's population has been diagnosed with diabetes alone, which is recognized as one of the primary causes of blindness in the nation if left untreated. With a focus on routine medical checkups and the use of specialized facilities for the detection and monitoring of the aforementioned condition, early identification and diagnosis have been recognized as one strategy to reduce the percentage of visual impairment caused by diabetes [1]. It is unnecessary to overstate how this has affected medical staff; to name a few, it has increased workloads for both staff and facilities and increased diabetes screening efforts. Many strategies have been proposed and found to help lessen the stress associated with these ongoing screening and check-up activities. One such tactic is the use of retinal pictures and medical digital image signal processing to diagnose diabetes-related disorders such diabetic retinopathy.

One metabolic disorder is diabetes. The body produces glucose during meal digestion, which gives it the energy it needs. The hormone insulin, which aids in the bloodstream entry of digested food, is produced by the pancreas, an organ situated close to the stomach. The pancreas normally produces the appropriate amount of insulin during food intake to aid in the blood's transportation of glucose into our cells. In individuals with diabetes, the pancreas either produces too little or no insulin, or when it does, the cells do not react properly.

Blood glucose accumulates, overflows into the urine, and eventually leaves the body. Consequently, even though the blood contains a significant amount of glucose, the body loses its primary fuel source. There are essentially three forms of diabetes. Type 1 diabetes is brought on by an autoimmune disorder. The body's immune system attacks the beta cells in the pancreas that produce insulin, causing the organ to produce either none at all or very little of the necessary amount. Type 2 diabetes is brought on by abnormalities in the beta cell itself. Examples of this malfunction include insulin resistance and decreased insulin production. Insulin resistance prevents muscles, fat, and other cells from responding to insulin that is produced. Type 3 diabetes, often known as gestational diabetes, only occurs during pregnancy. During this stage, the body fights against the consequences of the insulin that is produced.

2 Related Works

Digital image processing techniques have been more popular for DR screening since they were proposed as one of the screening approaches at the DR conference in Liverpool, UK in 2005 [1]. In an attempt to further improve the method's sensitivity and specificity, efforts to improve some of the currently used screening procedures have intensified, and new techniques have also been proposed.

Takeda Bozorg Other people [2] Image preparation techniques include thresholding and gradient weighting. When it comes to error filtering, this step is crucial. Shape, intensity, and local converge filters are the parameters that are taken out of the picture. Habib and others [3] The ensemble classifier technique is used in this study to identify the MA in retinal pictures. Candidate detection and post-detection categorization are carried out using Gaussian filters. This approach is useful for identifying cases that are false positives. Singular Spectrum Analysis is used in the approach for microaneurysm identification that Su Wang et al. [4] have given. In order to accomplish their goal, they used preprocessing techniques such as segmentation, thresholding, and morphological processes in conjunction with the identification of microaneurysms to remove exudate and optic nerve. [5] Antal et al. A method for microaneurysm detection and classification has been devised in this work. Using the Messidor database, 89% AUC was attained during the process's final round of classification.

It evaluate the outcomes of five distinct approaches, developed by five distinct research teams using the same dataset. Using an algorithm described in this study, the evaluation was carried out consistently [6].

J. Figueiredo and others [7] Binary classifier is used to derive contextual and numerical features. It employed the common elements of variational segmentation and Hessian multiscale analysis across many wavelet bands. With this approach, the sensitivity was 94% and the specificity was 89.3%. J. Larsent and others [8] Prior to using the various image processing techniques, the author employed pretreatment procedures. Through the use of image processing, hemorrhage was identified. The obtained results show a sensitivity of 98% and a specificity of around 72%.

Ram S. and others [9] The author eliminated the clutters in the early stages of the clutter rejection approach, which is employed for the early diagnosis of retinopathy. a similarity score determined by the actual MA similarity. The ROC (Retinopathy Online Challenge) and DIARETDB data sets are used. 88.56% specificity and sensitivity were attained using DIARETDB1 [10]. The intensity of each pixel varies, causing variations between MA and non-MA sections. The author also applied a morphological strategy. The intensity gradient characteristics are retrieved using the aforementioned approaches.

J. Lazar [11] Here, the author took a different strategy using a rotation-based cross sectional, and after using a peak-finding technique, local maximum pixels were chosen for the details. [12] Zhang Zang In this case, a multiscale gaussian filter was employed to distinguish between the MA and Non MA sections according to pixel intensity. Following filtering, dictionary-based learning and sparse representation classes were used for classification. Following categorization, image processing techniques are used to retrieve the vessel.

3 Proposed System

The suggested approach was organized around the use of 48 different types of features for FP reduction, DCNN for false positive (FP) reduction, and MAs identification. Using retinal pictures, multiple research groups have been developing automated approaches for detecting MA as shown the Fig. 1. The Retinopathy Online Challenge (ROC) database has been used to compare these methods. We also present a method that combines the three MA detectors: the double-ring filter, the Gabor filter, and the shape index based on the Hessian matrix. These older methods required the establishment of several parameters. However, since MAs were found using DCN, this work suggests an MA identification technique based on DCN. A revolutionary method for pattern recognition and object classification is the deep convolutional neural network (DCNN). As the DCNN, it was used. It uses an image patch, which in our study had a size of 21 by 21 pixels, to conduct MA prediction for each pixel as it scans the complete image. This DCNN was trained using 246 MAs. These comprised 83 and 163 MAs from the ROC and DIARETDB1 training image datasets, respectively [6]. We used parallel movement, smoothing, brightness adjustment, horizontal and vertical inversion, and smoothing to amplify from 246 to 2460 MA patches. Consequently, we trained our DCNN using a total of 2460 normal patches—none of which contained MAs.

Working

The individual neurons in a convolutional neural network (CNN), a type of feed-forward artificial neural network, are tiled such that they react to overlapping regions in the visual field. They are invariants of Multilayer Perceptrons (MLP), which are intended to require as little preprocessing as possible and are inspired by biology. The recognition of images and videos makes extensive use of these models. When CNNs are employed for image identification, they examine receptive fields—small regions of the input image—using multiple layers of microscopic neuron collections that are part of the model. For each layer that emerges from this collection, we tile the findings so that they overlap and offer a more accurate representation of the original image.

This explains why they can if there is any translation of the incoming image. Neuron cluster outputs can be aggregated by local or global pooling layers included in convolutional networks. Inspired by biological processes, convolutional networks are made up of various configurations of fully linked and convolutional layers, with point-wise nonlinearity added at the end or after each layer. Convolution is applied to small regions to prevent the circumstance where billions of parameters would exist if all the layers were properly coupled. Convolutional networks employ shared weights in their convolutional layers, which is useful since it lowers the amount of memory needed and improves speed by using the same filter (weights bank) for every pixel in the layer. CNNs learn the filters on their own, which are typically manually-engineered in other algorithms, because they require comparatively less pre-processing than other image classification techniques. The lack of reliance on past information and the challenging nature of hand-engineered features give CNNs a significant advantage over other models.

Measurement of Performance Parameters To obtain the assessment, many performance matrices are computed. There are several performance matrices in the red lesion detection system, including a) sensitivity (represented by the symbol SN), b) specificity (represented by the symbol SP), and c) accuracy (represented by the symbol ACC). Operating Characteristics of the Receiver Sensitivity testing is typically done on diseased or malfunctioning eyes, and specificity testing is done on healthy eyes. If there is an illness, this will show up as negative on tests (Fig. 2).

Larger values of specificity and sensitivity are preferable for diagnosis.

$$S.N = \frac{TP}{TP + FN} \tag{1}$$

$$S.P = \frac{TN}{TN + FP} \tag{2}$$

$$ACC = \frac{TN + TP}{TN + TP + FN + FP} \tag{3}$$

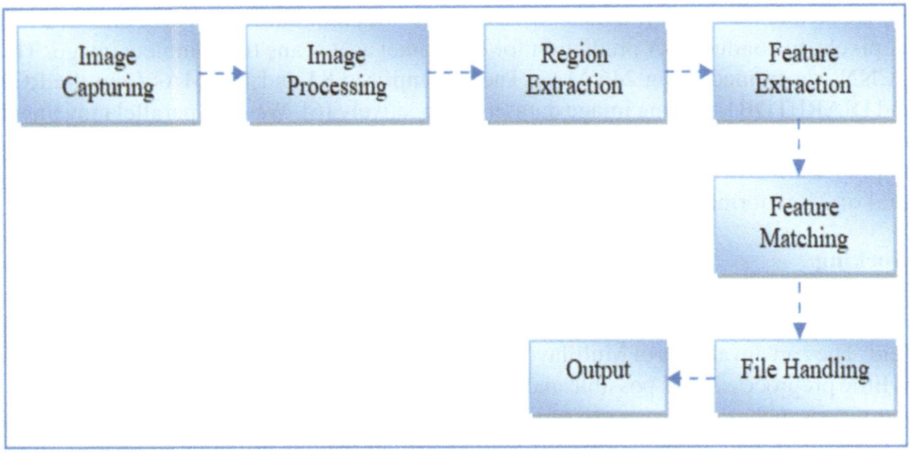

Fig. 1. Data Flow of the Proposed Method

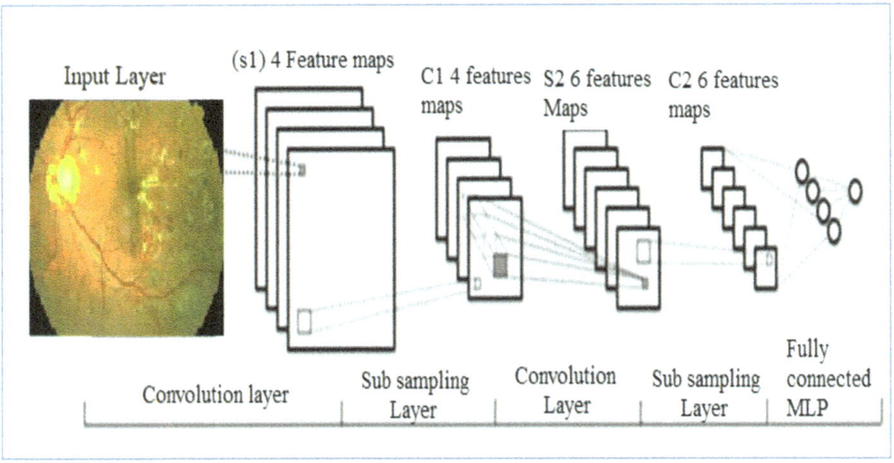

Fig. 2. Deep Convolution Neural Network (DCNN)

This is the TP (True Positive) representation that was previously used: If the fraction of anomalous FI found abnormal solely, then defective was detected as faulty. TN (true negative): Positive is detected as positive if Normal FI is detected as normal. False positives or positive cases mistaken for negative ones. False Negatives (FN): when a positive instance is found in a negative case Sensitivity is regarded as a legitimate positive and specificity as a valid negative.

4 Results and Discussion

At the outset of this chapter are the findings from the diagnostic of the twenty-five fundus images used for diagnosis and detection. For every piece of data, the Receiver Operator Characteristics (ROC) curve is also displayed. A brief analysis of the data is then provided, along with a summary of some of the thresholds that were used to obtain the ROC. As the input image for Fig. 3 is displayed.

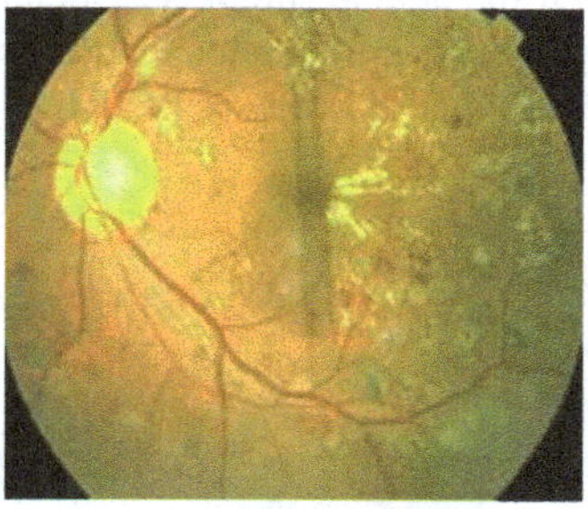

Fig. 3. Input Image

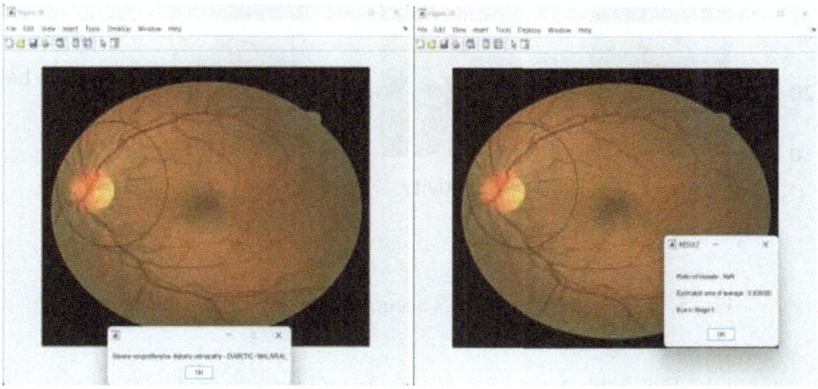

(a) (b)

Fig. 4. Output Images

Retinal Image Capture to obtain retinal images, a specialized camera known as a fundus camera is employed. Fundus cameras come in two varieties: mydriatic and non-mydriatic as shown the Fig. 4. Although it necessitates pupil dilation, the mydriatic camera produces fundus images of higher quality. On the other hand, the non-mydriatic camera requires no pupil dilation, is easier to use, and is smaller. Color fundus imaging (FI), red-free pictures, fluorescein angiograms (FA), and optical coherence tomography (OCT) are utilized in diagnostic procedures. Screening of the eyes is the sole method to identify the advancement of DR. In order to photograph the retinal blood vessels, sodium fluorescent dye is introduced into the bloodstream during FA. This emphasizes the blood vessels. More abnormalities are seen in FA than in FI, such as the formation of new blood vessels, leaks or blockages, hemorrhages, and MAs. However, there are some negative effects linked to FA, such as the pupil staying dilated for nearly twelve hours following testing, urine turning orange and darker for a few days, nausea, vomiting, fainting, swollen hives, etc. FI is hence favored these days.

Supervised learning, or SVM, is mostly used to solve data classification problems. SVM uses the maximum margin hyperplane to categorize the samples into distinct groups. Use kernel functions if the data problem is associated with linearly inseparable than SV. The sigmoid, linear and polynomial data, and the radical basis function (RBF) comprise the primary kernel function. SVM uses the RBF kernel.

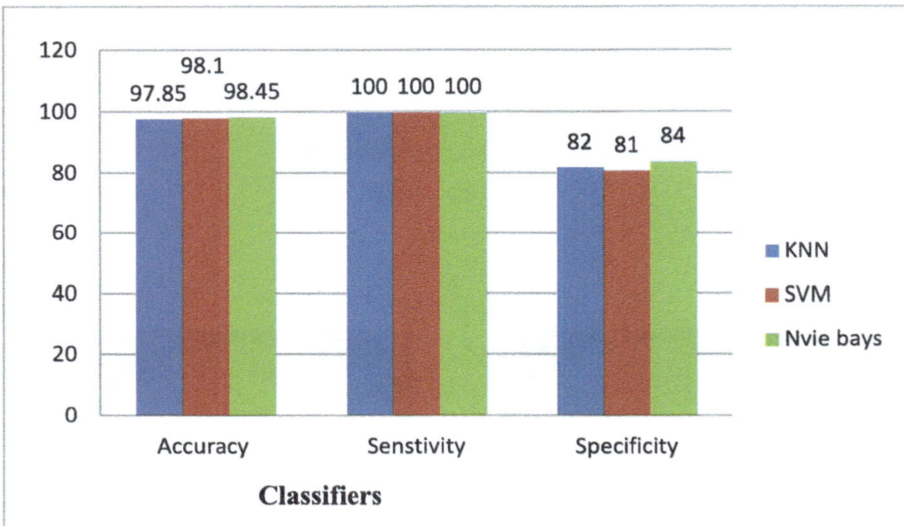

Fig. 5. The suggested system's accuracy, sensitivity, and specificity

As shown the Fig. 5 the Accuracy, specificity and Sensitivity bar chart b) Accuracy, specificity and Sensitivity pi chart. The experiment carried out in this work is evaluated using three metrics: accuracy, sensitivity, and specificity, which are used to calculate the accuracy of MA detection. For the proposed work, three classifiers—KNN, SVM, and Naive Bayesian—have their accuracy, sensitivity, and specificity calculated.

5 Conclusion

The research worked best when there were fewer than three FPs. However, our earlier approach performed somewhat better when there were more than three FPs. In this study, the suggested method was implemented using the same parameters as the prior FPs reduction method. We would therefore need to enhance that procedure. According to the suggested method's performance findings, the sensitivity was 84% of 8.0 FPs per picture. This outcome was identical to the last methods. An issue with the prior approach was its excessive number of rules. Test results were used to experimentally determine a number of parameters, including the shape index, Gabor filter, and double-ring filter threshold values, as well as the parameters for merging MA detectors. These parameters might be automatically learned in the proposed study. By adding a post-processing technique and refining the network architecture, the method could be further enhanced.

References

1. Cheung, N., Mitchell, P., Wong, T.Y.: Diabetic retinopathy. Lancet **376**(9735), 124–136 (2010)
2. Varalaxmi, G., Baddam, S.R., Yalamarthi, E.S., Swaraja, K., Madhavi, K.R., Sujatha, C.: Diagnosis of gastrointestinal diseases using modern CNN techniques. In: 2023 IEEE 8th International Conference for Convergence in Technology (I2CT), pp. 1–6. IEEE (2023)
3. Aparna, Y., Somasekhar, G., Bhaskar, N., Raju, K.S., Divya, G., Madhavi, K.R.: Analytical approach for soil and land classification using image processing with deep learning. In: 2023 2nd International Conference for Innovation in Technology (INOCON), pp. 1–6. IEEE (2023)
4. Pala, V.C.R., Kamatagi, S., Jangiti, S., Swaraja, K., Madhavi, K.R., Kumar, G.N.: Yoga pose recognition with real time correction using deep learning. In: 2023 International Conference on Sustainable Computing and Data Communication Systems (ICSCDS), pp. 387–393. IEEE (2023)
5. Madhavi, K.R., Vooradi, S., Mounika, P., Yedlla, S., Tangudu, N.: Emotion analysis using convolutional neural network. In: Reddy, A.B., Nagini, S., Balas, V.E., Raju, K.S. (eds.) Proceedings of Third International Conference on Advances in Computer Engineering and Communication Systems. Lecture Notes in Networks and Systems, vol. 612. Springer, Singapore (2023). https://doi.org/10.1007/978-981-19-9228-5_44
6. Niemeijer, M., Ginneken, B.V., Cree, M.J., et al.: Retinopathy online challenge: automatic detection of microaneurysms in digital color fundus photographs. IEEE Trans. Med Imag. **29**, 185–195 (2010)
7. Figueiredo, I.N., Kumar, S., Oliveira, C.M., Ramos, J.D., Engquist, B.: Automated lesion detectors in retinal fundus images. Comput. Biol. Med. **66**, 47–65 (2015)
8. Larsen, N., Lund-Andersen, H., Sjølie, A.K., Gardh, E., Owens, D.R.: Automated detection of fundus photographic red lesions in diabetic retinopathy. Invest. Ophthalmol. Vis. Sci. **44**(2), 761–766 (2003)
9. Seth, S., Agarwal, B.: A hybrid deep learning model for detecting diabetic retinopathy. J. Statist. Manage. Syst. **21**(4), 569–574 (2018)
10. Rajesh Tiwari, M., et al.: Enhanced power quality and forecasting for PV-wind microgrid using proactive shunt power filter and neural networkbased time series forecasting. Electr. Power Compon. Syst. (2023). https://doi.org/10.1080/15325008.2023.2249894
11. Jagtap, V., Srujan Raju, K., Rathnamma, M.V., Sasi Kiran, J.: Efficient lossy audio compression using vector quantization (ELAC-VQ). In: Satapathy, S., Bhateja, V., Janakiramaiah, B., Chen, Y.W. (eds.) Intelligent System Design. Advances in Intelligent Systems and Computing, vol. 1171. Springer, Singapore (2021). https://doi.org/10.1007/978-981-15-5400-1_42
12. Ashok Kumar, P.M., Subha Mastan Rao, T., Arun Raj, L., Pugazhendi, E.: An efficient text-based image retrieval using natural language processing (NLP) techniques. In: Satapathy, S., Bhateja, V., Janakiramaiah, B., Chen, Y.W. (eds.) Intelligent System Design. Advances in Intelligent Systems and Computing, vol. 1171. Springer, Singapore (2021). https://doi.org/10.1007/978-981-15-5400-1_52

Development and Implementation of a Scalable Architecture for Smart Healthcare Systems Employing Internet of Things

A. Srinivasula Reddy[1], Sunil Kumar Singh[1], P. Raveendra Babu[2], R. Bhargav Ram[2], Kanthi Murali[3], and S. V. S. V. Prasad Sanaboina[3](✉)

[1] CMR Engineering College, Hyderabad, Telangana, India
kumarsunil@cmrec.ac.in
[2] CMR College of Engineering and Technology, Hyderabad, Telangana, India
{praveendrababu,dr.r.bhargavram}@cmrcet.ac.in
[3] CMR Technical Campus, Hyderabad, Telangana, India
Cmrtc.paper@gmail.com

Abstract. Indeed, the traditional healthcare systems had a subservient nature. The patient was required to schedule numerous appointments with the doctor, regardless of minor concerns. The Internet of Things (IoT) has successfully materialized the vision of a more intelligent society and encompasses a diverse array of applications, one of which is healthcare. The resolution of this issue is being achieved through the advancement of intelligent healthcare systems, resulting in a significant reduction in mortality rates on a global scale. The implementation of Healthcare IoT has the potential to enhance patient well-being and elevate the quality of care provided by healthcare professionals, including doctors, nurses, clinicians, pharmaceutical companies, and the wider population. This article examines the planning and execution of a secured architecture for a smart healthcare system utilizing the Internet of Things (IoT). The efficacy of this intelligent healthcare model is assessed within the MATLAB environment in order to address challenges such as packet loss and other related issues. This study examines the Cloud Request View Ratio (CRVR), packet loss, and success rate of the Internet of Things (IoT) within the healthcare system. Subsequently, proponents of safety and confidentiality solutions engage in a discourse regarding the challenges associated with the implementation of safety mechanisms.

Keywords: IoT · Security · Smart health systems · privacy

1 Introduction

The rapid progress of digital processing technology and robust networking has facilitated the emergence of a diverse array of online services. To guarantee the sustainability of human social interaction, these services utilize the Internet of Things (IoT) as a medium connecting individuals, things, and operations [1]. The emergence of web-based service-based technologies has facilitated a transition in research practices from centralized to

D. Kriksciuniene et al. (Eds.): IBICA 2023, LNNS 1228, pp. 480–487, 2025.
https://doi.org/10.1007/978-3-031-78937-3_52

decentralized environments. The Internet of Things (IoT) has experienced significant advancements in the digital age due to its vast array of potential applications, including but not limited to smart agriculture, smart healthcare, and smart traffic management [2]. The integration of the Internet of Things (IoT) has resulted in enhanced productivity, decreased expenses, and generated valuable data, thereby contributing to the enhancement of quality of life across several domains. The utilization of Internet of Things (IoT) technology in the healthcare sector is being employed with the aim of enhancing patient monitoring, reducing costs, and promoting advancements in patient care.

Advancements in remote monitoring, autonomous assistive solutions, pharmaceutical administration, early warning and proactive treatment programs, asset management, and equipment maintenance have facilitated the implementation of these strategies. The Internet of Things (IoT) refers to a network of physical items and sensors that are networked, enabling the collection of data, processing, analysis, and decision-making processes. The utilization of data interchange to establish connections and facilitate interactions among Internet of Things (IoT) devices, objects, and the surrounding environment is crucial for both business and various socially significant operations [3]. It is noteworthy that a considerable number of contemporary IoT solutions exhibit a lack of emphasis and dedication towards ensuring robust security and privacy frameworks. The presence of these significant security vulnerabilities gives rise to concerns over the privacy of sensitive personal data. Smart healthcare is considered to be one of the primary application domains of the Internet of Things (IoT). Scholars and researchers have shown a growing interest in this field in recent years, aiming to explore the potential of the Internet of Things (IoT) within the healthcare sector [4]. Numerous services, applications, and prototypes have been developed within the field, with due consideration given to various practical considerations. The investigation of IoT-based Smart Healthcare encompasses various areas such as network topologies and their corresponding platforms, interoperability, heterogeneity, security, as well as the exploration of novel services and applications. Guidelines and rules have been created to facilitate the advancement of Internet of Things (IoT) technology within the healthcare sector. Notwithstanding all of these endeavors, the implementation of IoT in the healthcare sector remains nascent. The composition is structured in the following manner: The work concludes with Sect. 5, subsequent to Sect. 2's elucidation of the literature review and Sect. 3's depiction of the approach.

2 Literature Review

Research conducted by Satija, Ramkumar, Manikandan, and colleagues [5] shown a significant impact of wearable sensors and smartphones in the field of health tracking. Specifically, their research highlighted the substantial role of these technologies in emergency alerting and detection, leading to potentially life-saving advantages in remote areas with restricted access to high-speed internet connectivity. Furthermore, the authors discuss the necessity of prioritizing overseas patients within healthcare systems, since the limited capacity to provide prompt medical attention to every patient necessitates such a measure. Individuals originating from different regions find it exceedingly distressing that medical professionals with specialized training allocate precedence to individuals who physically present themselves at the emergency room. The strategic approach of prioritizing patients using telemedicine prior to their transfer to hospitals is considered to be a pragmatic method. Botezatu, Serban, Herghelegiu, Zal, and colleagues [6] delineated the procedures employed in the development and construction of a prototype for a cost-effective monitoring system. The approach focuses on the remote monitoring of patients in hospital wards subsequent to their release from the intensive care unit (ICU). The device offers mobile assistance and facilitates expedited and efficient medical care during emergency scenarios through the utilization of low-power sensor arrays specifically designed for temperature monitoring and mobility assessment.

In their study, Islam, Kwak, Kabir, and Kwag [7] discussed the utilization of advanced technologies such as big data, intelligence systems, and wearables within the healthcare domain. Furthermore, an individual also discussed the diverse aspects of Internet of Things (IoT) and electronic health (eHealth) as well as the legal frameworks governing them across different countries. This analysis explores the advancements in Internet of Things (IoT)-based healthcare technology, while also delving into the unique considerations around privacy and security. Specifically, it examines the security requirements, threat models, and assault taxonomy that are relevant under the context of healthcare.

3 Proposed Model

The current research focuses on the architectural planning and execution of the safe structure for a smart healthcare system utilizing the Internet of Things (IoT) Fig. 1.

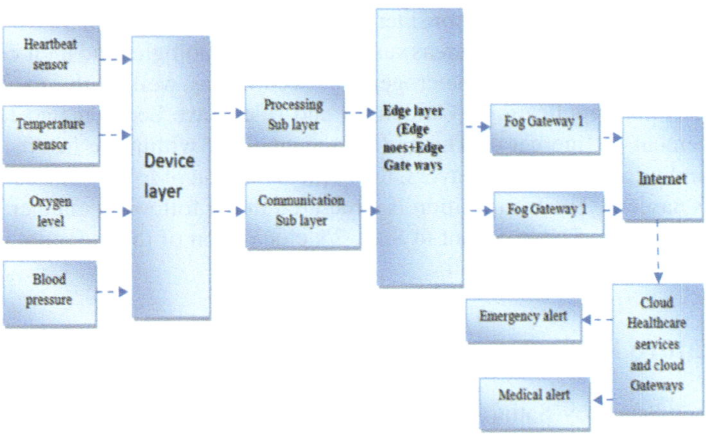

Fig. 1. System Architecture for projected model

The Internet of Things (IoT) has significantly enhanced the quality of healthcare products, services, and personalized therapies focused on the needs of individual patients. The acquisition of patients' raw data for later gathering and analysis methods is facilitated by the smart health environment. Most patient-centered technology is outfitted by sensors or microchips. The human body, or the patient's body, is interconnected with a variety of health sensor devices. The sensors have the capability to quantify physiological parameters such as blood pressure, body temperature, oxygen saturation, and heart rate.

To facilitate the transmission of healthcare data to gateways using Bluetooth, Zigbee, and WiFi, sensors and smart medical devices integrate cloud-based services with edge or node computing. The data generated by wide-area communication technologies such as 4G LTE, LoRaWAN, and NB-IoT is subsequently transmitted to the data center through the processing layer. To obtain contextualized personal medical data for individual patients, several physical devices including portable, implantable, and other tracking devices are connected to a miniature wireless module within the device layer [8, 9].

The data center undertakes the processing and analysis of a substantial volume of data prior to its dissemination to individual patients in a comprehensive manner. The gateway plays a crucial role in facilitating communication across several platforms by providing a route for communications. Additionally, it enables a range of services such as data processing, publishing, and subscribing to data provided by diverse sensors. The primary purpose of this layer is computational processing. Nevertheless, a significant portion of the data can be subjected to preprocessing on mobile devices through the utilization of the existing capabilities. The system is comprised of a network of intelligent gates that are interconnected by a fog layer.

The inclusion of this layer in the IoT network facilitates the provision of a localized storage system for interconnected sub networks and the dissemination of advanced computational capabilities to edge nodes. The collection of data from a sub network is facilitated by intelligent gateways, which additionally facilitate protocol conversion and offer enhanced services. Smart health gateways play a crucial role in securely and efficiently handling the authentication and authorization of remote end-users for medical sensors. During the initialization process, it is necessary to establish mutual authentication between the medical device and the smart gateway. The last layer of the cloud infrastructure has many components, including data warehouses, servers for big data processing, and a remote healthcare server that synchronizes periodically with the hospital database. The data that has undergone processing in the preceding layer is retained at this particular layer for the purpose of conducting analytics, facilitating decision-making processes, and supporting various other applications.

The collected data serves the purpose of acquiring insights and discerning patterns that can enhance the quality of patient treatment and well-being. Furthermore, it is utilized to acquire pertinent information regarding a patient's medical background and present health condition. Once the user or patient has been authenticated, the Information Alert Server transmits notifications to the Smart Gateway. Upon the successful completion of the authentication process, individuals in need are sent with health notifications that are encoded [10–12]. Location-based data and the Data Management Server are utilized to transmit emergency and medical notifications to ambulances and the relevant medical staff. Healthcare professionals are required to obtain proper certification and authorization in order to provide specific services. Hence, it can be asserted that the smart health service facilitated by the Internet of Things (IoT) is inherently secure. The privacy of the health indicators of the IoT client is ensured through the requirement of user identity.

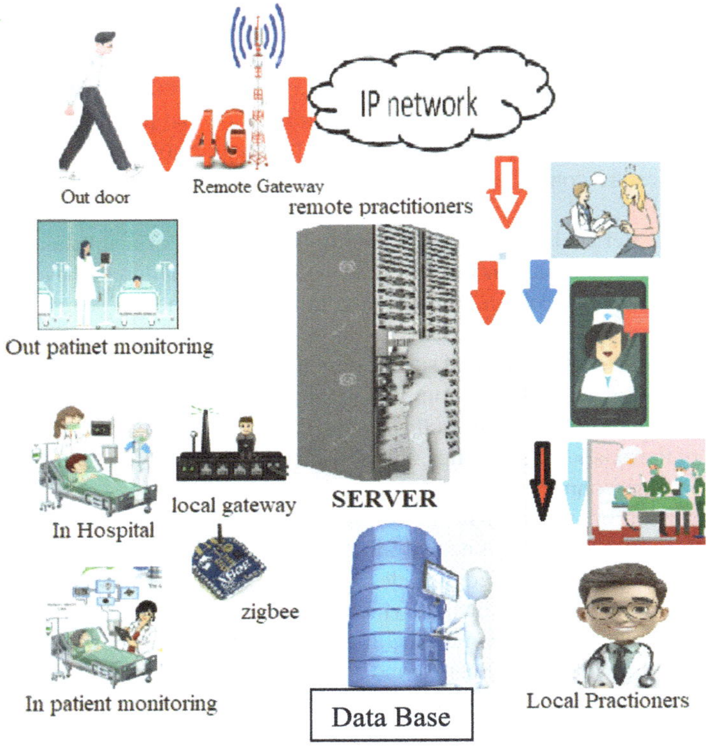

Fig. 2. Architecture of healthcare using IoT

The designing component primarily emphasizes the consumer-centric and architectural prerequisites for IoT in relation to intelligent as shown the Fig. 2, interconnected, and tailored healthcare services. The requirements are subsequently transformed into a functional architecture, followed by the allocation of its components onto physical infrastructure. The advancements in information, telecommunications, and network technologies have significantly impacted health care systems and have made valuable contributions to the creation of medical information systems [7]. Figure 3 depicts the architectural framework of healthcare systems employing Internet of Things (IoT) technology. In recent years, the phenomenon of population aging has led to a transformation in the healthcare landscape, with a heightened focus on the domains of home care and e-health. The primary objective is to offer diverse healthcare services to patients residing at their residences, as opposed to hospitals, with the intention of enhancing their quality of life through the provision of treatment within familiar surroundings. The healthcare industry is currently experiencing a significant transition due to the utilization of the Internet of Things and advancements in mobile and wearable technologies. The newly developed approach is specifically designed to enhance the holistic well-being of the patient. This is achieved by promoting and facilitating active engagement of the patient, utilizing mobile devices and multi-channel technology for effective implementation.

4 Results and Analysis

The implementation of this intelligent healthcare model utilized the MATLAB R2016a software, operating on a machine equipped with an Intel Core I5 processor running at a clock speed of 2.27 GHz, 8 GB of RAM, and the Windows 8 operating system. The evaluation of the outputs of this model is conducted in relation to widely utilized methodologies that are specifically tailored for healthcare applications. Figure 3 illustrates the performance analysis in relation to the conventional healthcare system. The performance analysis considered a sample size of 1000 service requests or service hits. The observed packet drop rate of 0.2% is significantly lower than the success rate of 99.6%. The conventional healthcare system demonstrates a success rate of 94.4% and a packet drop rate of 5.6%.

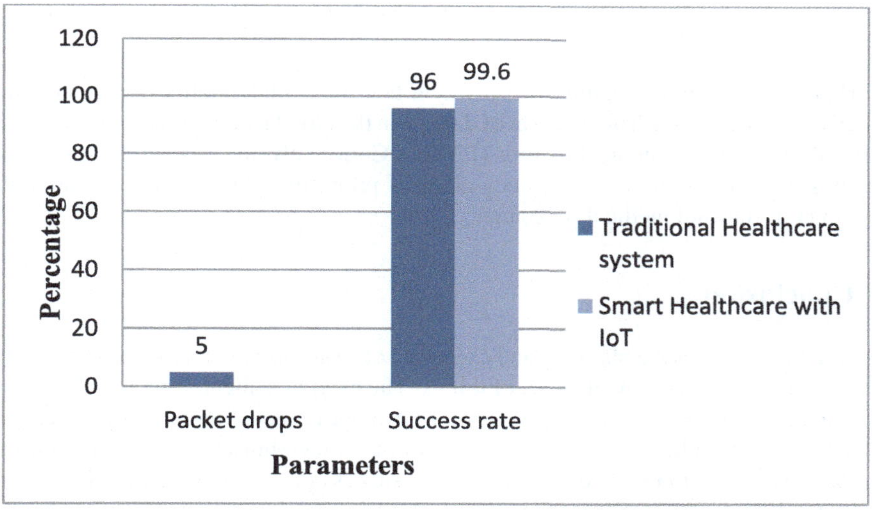

Fig. 3. Comparative Performance

The analysis of both conventional healthcare systems and smart healthcare about IoT involves the examination of cloud request view ratio (CRVR) values as shown the Fig. 4. It has been demonstrated that traditional healthcare systems possess a comprehensive understanding of superfluous requests, although potentially exhibiting lower CRVR values in comparison to the smart healthcare model. The significant discovery underscores the utilization of cloud consensus. The current iteration of this model exhibits a reduced CRVR and well manages the aspect of providing an accessible perspective of the EMR (Electronic Medical Record). The decreased collision resistance vulnerability rate (CRVR) is believed to have advantages in terms of privacy considerations.

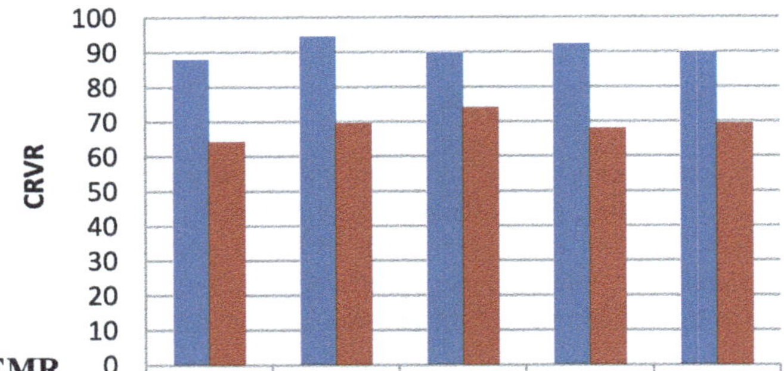

Number of EMR	2	4	6	8	10
Traditional Healthcare system	88.2	94.8	90.1	92.5	90.2
Smart Healthcare with IoT	64.6	70	74.4	68.3	69.8

Fig. 4. CRVR Computation

Hence, based on the findings, it is evident that the confidentiality of patient data is effectively safeguarded. The retrieval of data from the cloud results in a rapid and precise retrieval of electronic medical records (EMRs). Conversely, in cases where sensor data deviates from the norm, an emergency alarm is promptly transmitted to the attending physician or relevant medical personnel.

5 Conclusion

This article addresses the planning and execution of a safe architecture for a smart healthcare system that utilizes the internet of things. The need of safeguarding healthcare data necessitates an examination of strategies for developing secure and intelligent applications. The smart healthcare model underwent evaluation within a MATLAB environment to assess its ability to manage packet loss and success rate. A more favorable outcome is observed, with a success rate of 99.8% and a minimal occurrence of packet drops at 0.2%. The decreased collision resistance vulnerability rate (CRVR) is believed to confer advantages in terms of privacy considerations. If there is an anomaly in the sensor data, a medical professional or a relevant individual will be notified through an emergency alert. Hence, it is apparent based on the findings that patient data is safeguarded with a significant emphasis on privacy.

References

1. Zholdas, N., Postolache, O., Mansurova, M.: Health monitoring system using internet of things. In: 2021 IEEE International Conference on Smart Information Systems and Technologies (SIST) (2021)

2. Al-Mahmud, O., Khan, K., Roy, R., Alamgir, F.M.: Internet of things (IoT) based smart health care medical box for elderly people. In: 2020 International Conference for Emerging Technology (INCET) (2020)
3. Gupta, A.K., Johari, R.: IOT based electrical device surveillance and control system. In: 2019 4th International Conference on Internet of Things: Smart Innovation and Usages (IoT-SIU) (2019)
4. Akshat, Gaurav, Zahid, Bhupendra, Aditi, Sachin Kumar, Maneesha, Praveen Pandey: A smart healthcare monitoring system using smart phone interface. In: 2018 4th International Conference on Devices, Circuits and Systems (ICDCS) (2018)
5. Satija, U., Ramkumar, B., Manikandan, M.S.: Real-time signal quality-aware ecg telemetry system for iot-based health care monitoring. IEEE Internet Things J. **4**(3), 815–823 (2017)
6. Archip, A., Botezatu, N., Şerban, E., Herghelegiu, P. C., Zală, A.: An IoT based system for remote patient monitoring. In: 17th International Carpathian Control Conference, pp.1–6 (2016)
7. Reddy Madhavi, K., Vijaya Sambhavi, Y., Sudhakara, M., Srujan Raju, K.: COVID-19 Isolation Monitoring System. Lecture Notes on Data Engineering and Communication Technology, vol. 63, pp. 601–609 (2021)
8. Gyrard, A., Da Costa, R.P.F., Boudaoud, K.: Applying Internet of Things for Personalized Healthcare in Smart Homes (2015)
9. Tejeswara Kumar, M., Vikram G, N.V.R., Patel, P.: Analysis of an IoT-based SDN smart health monitoring system. In: Microelectronics, Circuits and Systems: Select Proceedings of Micro2021, pp. 325–334. Springer, Singapore (2023)
10. Jagtap, V., Srujan Raju, K., Rathnamma, M.V., Sasi Kiran, J.: Efficient lossy audio compression using vector quantization (ELAC-VQ). In: Satapathy, S., Bhateja, V., Janakiramaiah, B., Chen, Y.W. (eds.) Intelligent System Design. Advances in Intelligent Systems and Computing, vol. 1171. Springer, Singapore (2021). https://doi.org/10.1007/978-981-15-5400-1_42
11. Ashok Kumar, P.M., Subha Mastan Rao, T., Arun Raj, L., Pugazhendi, E.: An efficient text-based image retrieval using natural language processing (NLP) techniques. In: Satapathy, S., Bhateja, V., Janakiramaiah, B., Chen, Y.W. (eds.) Intelligent System Design. Advances in Intelligent Systems and Computing, vol. 1171. Springer, Singapore (2021). https://doi.org/10.1007/978-981-15-5400-1_52
12. Returi, K.D., Radhika, Y., Mohan, V.M., Srujan Raju, K.: A method of speech signal analysis using multi-level wavelet transform. In: Satapathy, S., Bhateja, V., Janakiramaiah, B., Chen, Y.W. (eds.) Intelligent System Design. Advances in Intelligent Systems and Computing, vol. 1171. Springer, Singapore (2021). https://doi.org/10.1007/978-981-15-5400-1_67

Improving Hybrid Shaft Clustering for Brain Tumor Classification and Area Evaluation

Rajesh Tiwari[1], Sheo Kumar[1], E. N. V. Purna Chandra Rao[2], A. Pradeep Kumar[2], Badam Prashanth[3], and Najeema Afrin[3](✉)

[1] CMR Engineering College, Hyderabad, Telangana, India
[2] CMR College of Engineering and Technology, Hyderabad, Telangana, India
drapradeepkumar@cmrcet.ac.in
[3] CMR Technical Campus, Hyderabad, Telangana, India
Cmrtc.paper@gmail.com

Abstract. A key component of quantitative brain image interpretation is the precise classification of brain tumors in magnetic resonance (MR) images, which has attracted a lot of scientific interest. Conventional techniques for segmenting MR brain images, including K-means and fuzzy C-Means (FCM) algorithms for clustering, handle every pixel independently and do not integrate spatial information between nearby pixels. Therefore, the noise and degree of homogeneity in brain magnetic resonance imaging poses challenges to the precision of these classification methods. We present a novel way to classification that tackles this problem by utilizing the hybrid shaft clustered technique, that blends Fuzzy Kernel C-Means (FKCM) and adaptive K-means clustering. Furthermore, we determine the area by figuring out how many cells the tumor occupied and how segmented its area is. The results of our simulation show that, in comparison to traditional methods, our suggested method delivers greater segmentation and area estimating reliability.

Keywords: MRI · K-means · brain tumor segmentation · area estimation · quantitative analysis

1 Introduction

One of the more well-known and accessible point brain disorders is the tumor. According to estimations with the World Health Organization (WHO), assessment and treatment are essential to the over five lakh people who worldwide suffer from tumors each year. Advances in therapeutic imaging systems enable their application in some medical fields, such as workstation-assisted pathology diagnosis, surgical planning and guidance, and longitudinal dissection. Magnetic Resonance Imaging (MRI) and Computed Tomography (CT) are two of the last imaging modalities that can support the mossycup oak imaging tactics that are occasionally employed in conjunction with neurosurgery and neurology. Segmenting objects, especially anatomic structures and additional pathologies, based on MRI images may be a critical task, as the results occasionally became the basis for new requests. Segmentation systems are entirely dependent on those particular

D. Kriksciuniene et al. (Eds.): IBICA 2023, LNNS 1228, pp. 488–499, 2025.
https://doi.org/10.1007/978-3-031-78937-3_53

requirements and picture modalities. Moreover, it will prove difficult to segment medical images because they generally include a large amount of data. In addition, there may occasionally be some artifacts from patients' restricted mobility through fragile tissue boundaries, which are often poorly defined. The management of brain tumors presents unique challenges that complicate their categorization. Tumor types can come in an infinite number and in a variety of shapes and sizes. It could appear at any range and with varying picture intensities. Some of them distort the surrounding structures or may be associated with edema, which alters the picture intensity surrounding the tumors. Furthermore, the majority of the brain's data is divergent because to the presences from claiming a few MRI procurement norms. The stage that comes after segmentation and preprocessing is known as "characteristic extraction," and it can be done once Tom has finished reading through "character determination." When those necessary offers are selected, the segmentation transformation may apply. The problems with selecting appropriate chK-meansels for denoising, segmentation algorithms, feature extraction, and prediction, as well as the computation of those orders for brain MRI images, are still genuine.

2 Literature Review

The Grab cut method is utilized in the work suggested [1] to accurately segment actual lesion indicators, and the Transfer learning model visually geometry group (VGG-19) is fine-tuned after collecting the features. The hand-crafted (impact and texture) features are subsequently combined with the features by means of a serial depending technique. Entropy is utilized to maximize these characteristics for rapid and precise classification, and classifiers are given a fused vector.

Although it might seem sense that more characteristics might be more beneficial than less ones, this isn't actually the case in real-world applications for the three main reasons that are shown in the preceding diagram [2]. The classifier's complexity and computing cost increase significantly right away. Furthermore, it has been shown that adding more features to the classifier preparation process will unavoidably increase the amount of misclassified data, even while the amount of misclassified data may decrease [3]. Thirdly, it is more likely that characteristics containing little to no separating power would cause a commotion, undermining the classifier's conjecture to disguise information due to a preset amount of available data and a huge amount of accessible features [4]. Therefore, feature extraction is a crucial step toward identifying the more informative characteristics and fine-tuning the classifier's capacity to consistently categorize ambiguous data. Various K-means are used [5] for image order breakdown. It also explains why the classical neural networks do not have a faster union rate. This highlights the need for optimized neural networks having the highest meeting rate possible for picture order applications. Four distinct tumor kinds have been arranged by the authors in [6] using the LDA approach. However, the segmentation precision that the work reveals is at request of 80%, which is often poor. This also suggests the many motivations for missegmentations. Authors in [7, 8] carried out FCM-based characterisation of various MR glioma image levels. This approach is certain to be better than any structure that employs principles, although the accuracy of the FCM-based segmentation is poor. Another drawback of this system is that just glioma images were segmented using FCM, which means there is no capability for summing up.

To segment images, authors in [8, 9] must use those Kohonen neural networks. Here, a percentage modification of the acknowledged Kohonen brain structure is also implemented, and it has the potential to perform far more effectively than those acknowledged neural networks. Authors in [10–12] propose a modification to the standard FCM, such as any rate as square FCM (LS-FCM), for brain tumor differentiation. Both multiclass order and bi-level segmentation require assistance that hints at the superior method of the proposed system compared to the conventional classifiers. This also provided a crucial point: as the amounts regarding classes increase, the differences between the various computations get more pronounced. In light of this, this method recommended multi-class segmentation procedures as opposed to bi-level segmentation systems. Likewise, writers in [13] effectively materialize and advocate adaption for LS-FCM. A comprehensive comparative analysis will be conducted between those FCM, neural classifier, and factual classifier. The preferences from claiming FCM in terms of order exactness were proposed by 63 effects. Only bi-level order is carried out, which is inadequate for determining the characteristics of the robotized framework.

The authors use the modified PKFCM for tumor image order in [15–18]. Abnormal pictures, such as those in which meningioma, glioma, and metastases are distinguished using PKFCM based on any rate of square characteristic conversion. Additionally, an analogous assessment might be carried out for FCM. According to this methodology, in terms of segmentation accuracy, the convert-based PKFCM might perform better than the FCM.

3 Proposed Model

The crucial step in the classification and assessment of malignancies is the separation of brain images. The suggested tumor segmentation technique utilizing the hybrid shaft clustering technique is shown more clearly in Fig. 1. It combines the techniques of fuzzy kernel C-Means (FKCM) and adaptable K-means clustering.

By using the adaptive cluster indexing localization, the challenges seen in traditional K-means clustering are resolved and improved. This successfully provides the mean feature of cluster centers and proximity matching. The novel adaptive K-means clustering will be introduced through the use of this adaptive clustering technique. For efficient brain tumor segmentation, each cluster will be then subjected to the FKCM clustering procedure. Additionally, area estimation was performed by determining the number of cells occupied by each segmented tumor and the area of the tumor in relation to the distinct characteristics of the primary and secondary tumors.

Alphabet 1: K-means clustering with adaptation.

The adaptable K-means clustering technique consists of two stages. The adaptable K-means clustering technique is employed in the first step to choose the initial centroids. When initial centroids are fixed in this manner, they stay fixed throughout the process of execution. It results in less iterations and consistency when grouping related things. Finally, the K-means method is able to produce a local optimum due to the distinct initial centroids selection. This ultimately results in the K-means clustering algorithm's global

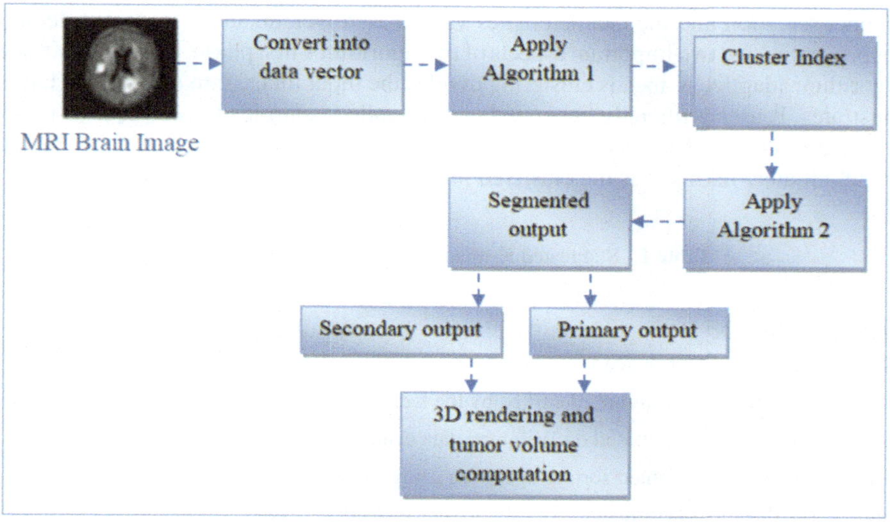

Fig. 1. Block schematic of the projected model

best solution. The second stage involves grouping comparable objects using a weighted adaptive K-means clustering algorithm based on Euclidean distance. Weights related to each range of attribute values in the data set are taken into account for processing in both phases. The weighted ranking strategy is the first stage in the suggested improved weighted adaptive K-means clustering algorithm. The distance between the origin and every weighted attribute of a data point in a data set is added up and used in the weighted ranking algorithm. Equation is used for calculating weighted data points. 1.

$$U_{j=1}^{n} = \sum\nolimits_{i=1}^{n} W_i X_i \tag{1}$$

while the weight of the th attribute is represented by *Wi*. Equation provides the framework for calculating n numbers of U values for n data points 2.

$$U_{j=1}^{n} = \begin{pmatrix} U_1 \\ U_2 \\ U_3 \end{pmatrix} \tag{2}$$

Next, the sorted distances are applied to the weighted data points (df_weighted). Following sorting, the data points are divided into k equal sets, whereby k is the total amount of clusters that have been predetermined. The beginning centroids in every set are the middle points or the mean value of the middle points. This algorithm's initially chosen centroids selection results in cluster members that are consistent. As the K-means method uses distance measurements to determine the final members of the group, the equation for distance also serves as the foundation for the suggested weighted ranking method dependent initial centroids selection. Therefore, the improvement of the starting center choosing is positively related to the principle of the proposed weighted ranking. The consistency among group members as a result of the initial centroids' distinct selection and the characteristics value-based weights taken into consideration for processing are two important aspects of the algorithm's first phase. Using a partitioned algorithm, adaptive K-means clustering divides the input image into clusters, as Table 1 illustrates. It uses artificially generated entities called centroids to represent its cluster and re-estimates it. The cluster seed, which is the initial object in a cluster, is compared to the initiator. Taking S as the clustered image and I as the input picture.

Table 1. Suggested K-means clustering with adaptation

Input: I; Output: S
Step1: convert the input image I into data vectors
Step 2: The initial centroids are identified by the k data points known as cluster centers
Step 3: Label the clusters based on Weighted data centroids as presented using Eq. 1
Step 4: perform the weighted sorting operation using Eq. 2
Step 5: The distance is computed from 'U' to the centroid
Step 6: The closest centroid is assigned as segmented cluster
Step 7: repeat the process step 3 to step 6 until all clusters covered
Step 8: combine all the segmented clusters and concatenate output S

The suggested clustering algorithm is depicted in Fig. 2, and we used the fuzzy algorithm to maximize the k-means clustering.

Algorithm 2: FKCM

Projected Kernel Function

Data is mapped to a larger feature space using non-linear projections created with the kernel function. One of the extra techniques that work well in pairing with linear models to solve non-linear issues that generalize is the kernel trick. Assume the initial dataset in $Rnisx1, x2, .., xn \in Rn$. Next, data are mapped by function ϕ to an entirely novel feature space with a higher F:

$$\phi = Rn \rightarrow F \tag{3}$$

Thus, this is how the kernel function is defined:

$$K(x, y) = \langle \phi(x), (x) \rangle \tag{4}$$

Thus the definition of the measurement function is as the following:

$$\begin{aligned} \mathrm{d}^2(x, y) &= ||\phi x - \phi y||2 = \langle \phi(x), \phi(x) \rangle - 2\langle \phi(x), \phi(y) \rangle + \langle \phi(y), \phi(y) \rangle \\ &= K(x, x) - 2K(x, y) + (y, y) \end{aligned} \tag{5}$$

Numerous varieties of kernels are frequently employed, such as: a kernel that is linear:

$$K(x, y) = x \cdot y \tag{6}$$

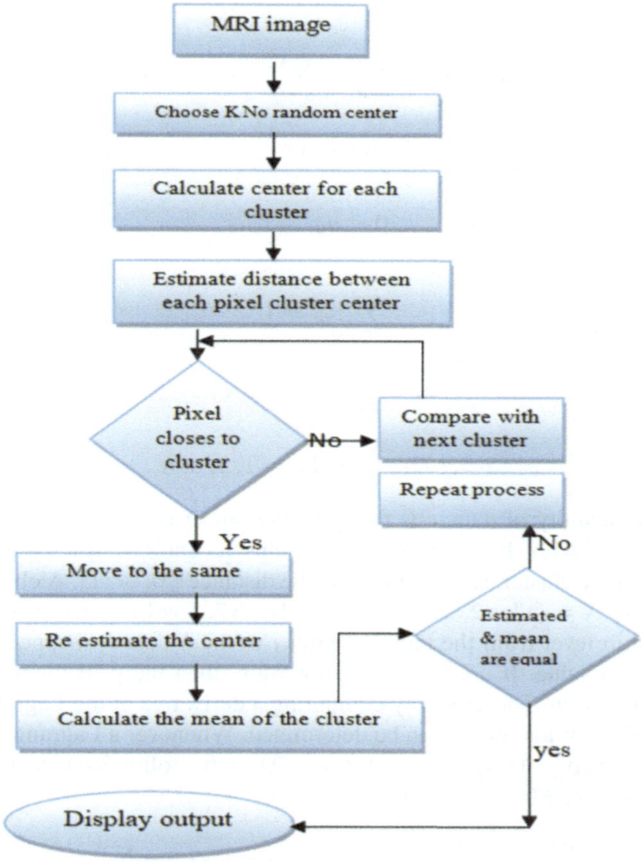

Fig. 2. Flow chart of K-means clustering

RBF kernel function:

$$K(x, y) = \exp\left(-\frac{\|x - y\|^2}{\sigma^2}\right) \tag{7}$$

Polynomial Kernel Function:

$$K(x, y) = (x \cdot y + 1)^{\mathrm{d}} \tag{8}$$

The Fuzzy C-Means technique is modified by the Fuzzy Kernel C-Means method. Employing kernel functions, this technique groups imperfect data employing non-linear functions over a considerably higher dimensional space. The kernel function has demonstrated remarkable power in retrieving data stored in the database by translating real data from low-dimensional spaces into higher-dimensional ones. This function is the transition between linear and nonlinear. This is a mathematical representation of fuzzy

C-Means:

$$J(V, U, X, c, m) = \sum_{i=1}^{c} \sum_{k=1}^{N} (u_{ik})^m \|x_k - v_i\|^2 \tag{9}$$

Subject to

$$0 \leq u_{ik} \leq 1 \tag{10}$$

$$\sum_{i=1}^{c} u_{ik} = 1 \tag{11}$$

$$0 \leq \sum_{i=1}^{c} u_{ik} \leq N \tag{12}$$

$$i = 1, 2, .., c \; and \; k = 1, 2, .., N$$

where d is the quantity of data, c is the total amount of clusters (c > 2), and d is the degree of fuzziness with ❑ > 1. Data xk and the cluster center vi, which is the dataset that is going to be in the cluster, are separated by the distance $\|xk - vi\|$. A cluster center set is represented by $X = \{x1, x2,.., xn\}$, and $V = \{v1, v2,.., vc\}$ is a membership function. The membership level from the cluster center i to the k data is mapped by $U = [ui]$ as a membership value. If uik and vi are optimum, then the goal function will reach its optimum condition. Therefore, by computing a derivative of the Lagrange function, the optimal values of uik and vi can be determined. Whenever a Lagrange multiplier λ, according to the theory of Lagrange multipliers. Then, the following is how the Lagrange function for Fuzzy C-Means might be created:

$$L = \sum_{i=1}^{c} \sum_{k=1}^{N} (u_{ik})^m \|x_k - v_i\|^2 + \lambda\left(1 - \sum_{i=1}^{c} u_{ik}\right) \tag{13}$$

in order every parameter's Lagrange function derivative to equal zero.

$$\frac{dL}{du_{ik}} = 0 \tag{14}$$

in order to achieve the ideal circumstances for the cluster center and membership value:

$$u_{ik} = \frac{1}{\sum_{j=1}^{c} \frac{(\|x_k - v_i\|^2)\frac{1}{m-1}}{\|x_k - v_j\|^2}} \tag{15}$$

$$v_i = \frac{\sum_{k=1}^{N} ((u_{ik})^m x_k}{\sum_{k=1}^{N} (u_{ik})^m} \tag{16}$$

The category of data determines how accurate the FCM approach is. The process of convergence can appear sluggish and imprecise if the data cannot be separated linearly. To solve issues like these, a different approach is therefore required as a connection. The kernel method is one approach that can be applied. In order to finish the categorization of thalassemia information employing Fuzzy Kernel C-Means (FKCM), this study will use the kernel approach to the FCM. Let $x_1, x_2, \ldots x_n \in R^p$ be the collection of data in R^p. Then, there is an additional dimension to the function ϕ, which maps data from R^p to H as an entirely novel aspect space.

$$\phi = R^n \rightarrow F \tag{17}$$

Consequently, the following are the fuzzy kernel C-Means goal functions:

$$J(V, X, c, m) = \sum_{i=1}^{c} \sum_{k=1}^{N} (u_{ik})^m \|\phi(x_k) - \phi(v_i)\|^2$$

subject to 0

$$0 \leq u_{ik} \leq 1 \tag{19}$$

$$\sum_{i=1}^{c} u_{ik} = 1 \tag{20}$$

$$0 \leq \sum_{i=1}^{c} u_{ik} \leq N \tag{21}$$

While c is the amount of data, m is the degree distinct fuzziness with m > 1, and C is the amount of clusters (C ≥ 2). The kernel mapping distance among the cluster center vi and the xk data is denoted by $\|\emptyset(xk) - \emptyset(vi)\|$. The set of data to be clustered is represented by $X = x_1, x_2, \ldots x_n\}$, the cluster center set is represented by $V = [uik]$ and the membership function is denoted by $U = [u_{ik}]$. Similar to Fuzzy C-Means, we will provide a Lagrange multiplier λ and use it to identify the optimal uik and vi. The Lagrange function is going to be utilized to get the ideal answer for Fuzzy Kernel C-Means:

$$L = \sum_{i=1}^{c} \sum_{k=1}^{N} (u_{ik})^m \|\phi(x_k) - \phi(v_i)\|^2 + \lambda\left(1 - \sum_{i=1}^{c} u_{ik}\right) \tag{22}$$

It might be expressed in the following way, considering every parameter's Lagrange function derivative equal to zero:

$$\frac{dL}{du_{ik}} = 0 \tag{23}$$

Hence, $K(xi, xi) = K(vi, vi) = 1$ determined by the RBF kernel function structure to be utilized. In order for each member and cluster center to be in the best possible state:

$$u_{ik} = \frac{1}{\sum_{j=1}^{c} \left(\frac{1-k(x_k v_i)}{1-k(x_k v_i)}\right) \frac{1}{m-1}} \tag{24}$$

$$v_i = \frac{\sum_{k=1}^{N}((u_{ik})^m k(x_k v_i) x_k}{\sum_{k=1}^{N}(u_{ik})^m k(x_k v_i)} \tag{25}$$

$$m = mi + \frac{t}{T}(mf - mi) \tag{26}$$

Every of the beginning and final values of m are represented by mi and mf.It is anticipated that m will drop if the mf value is minimal and mi is sufficiently large.

Area Calculation
The segmentation output image $f_{x,y}$ can be expressed as the total amount of black and white pixels added together.

$$M = \sum_{x=1}^{L}\sum_{y=1}^{L} f_{x,y}(0) + f_{x,y}(1) \tag{27}$$

where L = 1,2,3,……,256

$$fx, y(0) = black\ pixel\ having\ the\ value\ of\ zero,$$

$$fx, y(1) = write\ pixels\ having\ the\ value\ of\ one$$

$$P = \sum_{i=1}^{L}\sum\nolimits_{J=1}^{L} f_{x,y}(1)$$

$$Where\ P = number\ of\ write\ pixels$$

Presently, utilizing the aforementioned formula and the typographic and digital imaging units, where one pixel is equivalent to 0.264583 mm, we can determine the area of the segmented tumor. For example, 1 pixel = 0.264583 mm.

The tumor's area can therefore be stated to be the following:

$$ATumor = \left(\sqrt{P}\right) * 0.2646mm^2 \tag{29}$$

4 Results and Analysis

The following section provides a thorough simulation examination of the suggested approach utilizing the MATLAB software tool on a variety of test datasets of brain pictures. Evaluation is done in a variety of ways as shown the Fig. 3 and comparisons between the different methodologies have been examined.

K-means clustering, the suggested hybrid clustering techniques, and the segmentation outputs of individual tumors from MR brain images are displayed in Fig. 3, 4, and 5. We can see from the data that the suggested hybrid clustering approach has identified the tumor more successfully and accurately. Even though our suggested technique will take a little longer to execute than k-means clustering, the segmentation output's reliability will be higher—that is, the tumor area would be approximated accurately to aid in subsequent diagnosis.

The assessment of the efficacy of the suggested hybrid clustering algorithm is shown in Fig. 6 and Fig. 7 above, as compared to the traditional clustering algorithms that have been published in the academic literature. Performance time in seconds and tumor area in millimeter-squared were computed. Comparing the suggested technique to FCM and K-means clustering techniques, Fig. 6 shows that it takes very little time for the segmented process, and Fig. 6 shows that it offers the highest reliability in the segmented region.

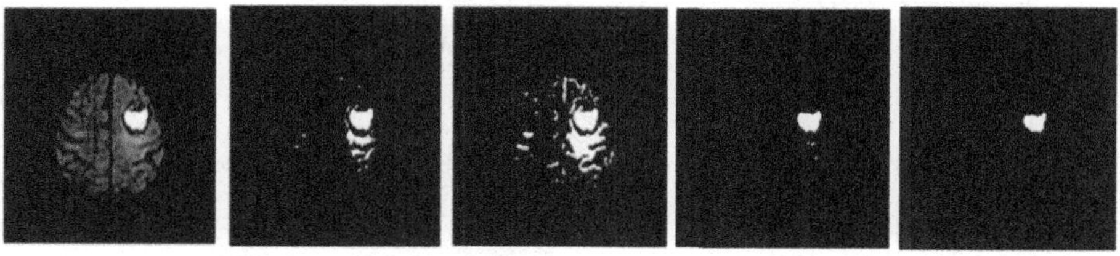

(a) (b) (c) (d) (e)

Fig. 3. Single tumor case (a) Original Image (b) manual segmentation (c) FCM (d) K-means (e) proposed clustering

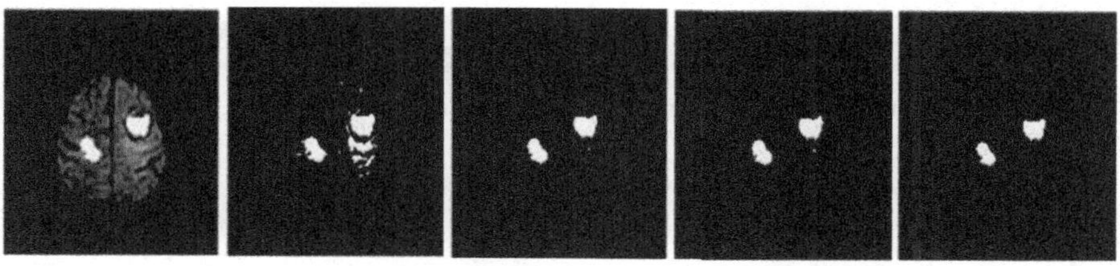

(a) (b) (c) (d) (e)

Fig. 4. Two tumor case (a) Original Image (b) manual segmentation (c) FCM (d) K-means (e) proposed clustering

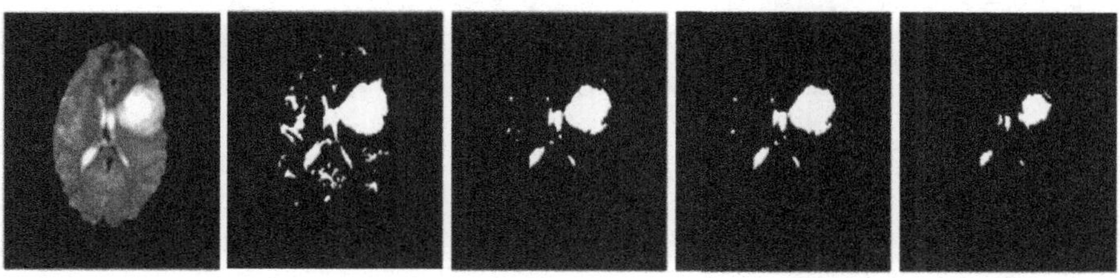

(a) (b) (c) (d) (e)

Fig. 5. Multiple tumor s (a) Original Image (b) manual segmentation (c) FCM (d) K-means (e) proposed clustering

5 Conclusion

This article suggested a productive method for utilizing the hybrid shaft clustering technique to segment and identifies malignancies from brain medical images. Furthermore, this method uses fuzzy selection clustering in conjunction with the K-mean shifting procedure to calculate the weight for every picture element. The results of the simulation demonstrated that, despite requiring less time for segmentation, the suggested segmentation technique outperformed the current segmentation algorithms with respect of each ocular and quantitative analysis. By evaluating the suggested improved weighted hybrid shaft clustering technique against state-of-the-art K-means and FCM with regard to a

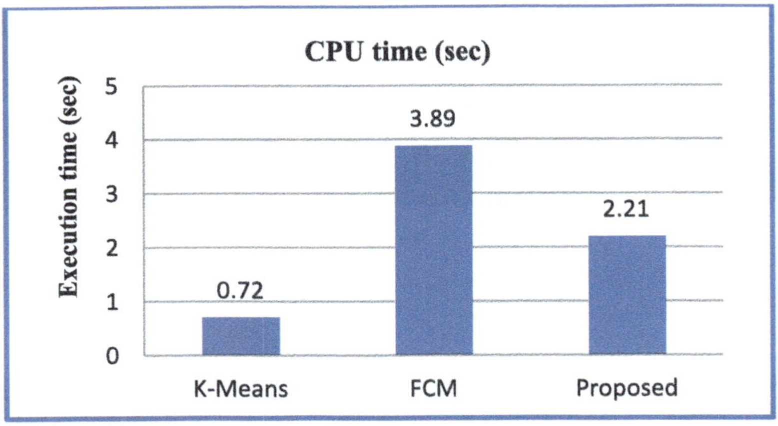

Fig. 6. Evaluation of efficiency for identifying of multiple tumors using CPU running time

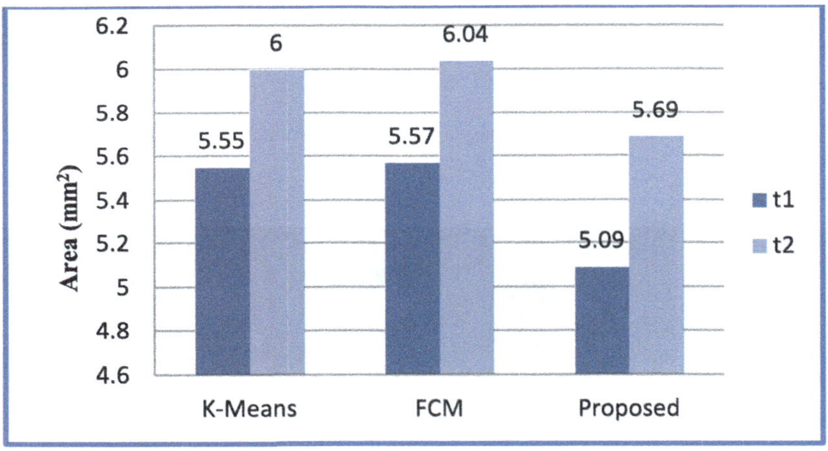

Fig. 7. Predictable area of segmented tumors T1 = tumor 1 and T2 = tumor 2

reliable area estimate, the robustness of the algorithm is demonstrated. The work might be expanded to create a segmentation strategy, focusing on increasing segmentation accuracy and creating a classifier that is consistent at all stages for certain application requirements. Moreover, with more precise and efficient clustering methods, this can be expanded to 3D multimodal medical image segmentation.

References

1. Saba, T., et al.: Brain tumor detection using fusion of hand crafted and deep learning features. Cogn. Syst. Res. **59**, 221–230 (2020)
2. Sharif, M., et al.: A unified patch based method for brain tumor detection using features fusion. Cogn. Syst. Res. **59**, 273–286 (2020)

3. Toğaçar, M., Ergen, B., Cömert, Z.: BrainMRNet: brain tumor detection using magnetic resonance images with a novel convolutional neural network model. Med. Hypotheses **134**, 109531 (2020)
4. Kumar, P.M.K.: Satellite image denoising using local spayed and optimized center pixel weights. Int. J. Electr. Comput. Eng. **4**(5), 751 (2014)
5. Wu, M.-N., Lin, C.-C., Chang, C.-C.: Brain tumor detection using color-based k-means clustering segmentation. In: Third International Conference on Intelligent Information Hiding and Multimedia Signal Processing (IIH-MSP 2007), vol. 2. IEEE (2007)
6. Kumar, P.M.K.: A new encryption and decryption for 3D MRT images. Asian J. Electr. Sci. **6**(1), 1–6 (2017)
7. Kalyanapu Srinivas, P.M.K., Jagadeeswara Rao, A.: A new video watermarking using redundant discrete wavelet in singular value decomposition domain. J. Mech. Continua Math. Sci. **14**(5), 476–486 (2019)
8. Tiwari, R., et al.: An artificial intelligence-based reactive health care system for emotion detections. Comput. Intell. Neurosci. **2022**, Article ID 8787023. https://doi.org/10.1155/2022/8787023
9. Gumaste, P.P., Bairagi, V.K.: A hybrid method for brain tumor detection using advanced textural feature extraction. Biomed. Pharmacol. J. **13**(1), 145–157 (2020)
10. Chahal, P.K., Pandey, S., Goel, S.: A survey on brain tumor detection techniques for MR images. Multimedia Tools Appl. (2020)
11. Chandra, S.K., Bajpai, M.K.: Brain tumor detection and segmentation using mesh-free super-diffusive model. Multimedia Tools Appl. **79**(3), 2653–2670 (2020)
12. Rinesh, S., et al.: Investigations on brain tumor classification using hybrid machine learning algorithms. J. Healthc. Eng. **2022** (2022)
13. Sandya, V., Baligeri, V., Lal, B., Petli, V., Kumar, P.: Deep learning based brain tumor detection with internet of things. In: 2023 IEEE International Conference on Integrated Circuits and Communication Systems (ICICACS), pp. 1–6. IEEE (2023)

Author Index

D. Kriksciuniene et al. (Eds.): IBICA 2023, LNNS 1228, pp. 501–503, 2025.
https://doi.org/10.1007/978-3-031-78937-3

The manufacturer's authorised representative in the EU is Springer Nature Customer Service Centre GmbH, Europaplatz 3, 69115 Heidelberg, Germany. If you have any concerns regarding our products, please contact ProductSafety@springernature.com

Printed and bound by CPI Group (UK) Ltd, Croydon, CR0 4YY
30/09/2026
02225561-0015

Lecture Notes in Networks and Systems 1228

The series "Lecture Notes in Networks and Systems" publishes the latest developments in Networks and Systems—quickly, informally and with high quality. Original research reported in proceedings and post-proceedings represents the core of LNNS.

Volumes published in LNNS embrace all aspects and subfields of, as well as new challenges in, Networks and Systems.

The series contains proceedings and edited volumes in systems and networks, spanning the areas of Cyber-Physical Systems, Autonomous Systems, Sensor Networks, Control Systems, Energy Systems, Automotive Systems, Biological Systems, Vehicular Networking and Connected Vehicles, Aerospace Systems, Automation, Manufacturing, Smart Grids, Nonlinear Systems, Power Systems, Robotics, Social Systems, Economic Systems and other. Of particular value to both the contributors and the readership are the short publication timeframe and the world-wide distribution and exposure which enable both a wide and rapid dissemination of research output.

The series covers the theory, applications, and perspectives on the state of the art and future developments relevant to systems and networks, decision making, control, complex processes and related areas, as embedded in the fields of interdisciplinary and applied sciences, engineering, computer science, physics, economics, social, and life sciences, as well as the paradigms and methodologies behind them.

Indexed by SCOPUS, EI Compendex, INSPEC, WTI Frankfurt eG, zbMATH, SCImago.

All books published in the series are submitted for consideration in Web of Science.

For proposals from Asia please contact Aninda Bose (aninda.bose@springer.com).